Functional and Evolutionary Ecology of Fleas

Fleas are one of the most interesting and fascinating taxa of ectoparasites. All species in this relatively small order are obligatory haematophagous (blood-feeding) parasites of higher vertebrates. This book examines how functional, ecological and evolutionary patterns and processes of host–parasite relationships are realized in this particular system. As such it provides an in-depth case study of a host–parasite system, demonstrating how fleas can be used as a model taxon for testing ecological and evolutionary hypotheses. The book moves from basic descriptive aspects, to functional issues and finally to evolutionary explanations. It extracts several general principles that apply equally well to other host–parasite systems, so will appeal not only to flea biologists but also to mainstream parasitologists and ecologists.

BORIS R. KRASNOV is a senior research scientist in the Jacob Blaustein Institutes for Desert Research at Ben-Gurion University of the Negev. He has worked in the field of ecology for almost thirty years. He was awarded the Rector's Award for Outstanding Scientists by Ben-Gurion University in 2006.

Functional and Evolutionary Ecology of Fleas

A Model for Ecological Parasitology

BORIS R. KRASNOV
Marco and Louise Mitrani Department of
Desert Ecology
Jacob Blaustein Institutes for Desert Research
Ben-Gurion University of the Negev
and
Ramon Science Center

CAMBRIDGE UNIVERSITY PRESS
Cambridge, New York, Melbourne, Madrid, Cape Town, Singapore, São Paulo, Delhi

Cambridge University Press
The Edinburgh Building, Cambridge CB2 8RU, UK

Published in the United States of America by Cambridge University Press, New York

www.cambridge.org
Information on this title: www.cambridge.org/9780521882774

© B. R. Krasnov 2008

This publication is in copyright. Subject to statutory exception
and to the provisions of relevant collective licensing agreements,
no reproduction of any part may take place without
the written permission of Cambridge University Press.

First published 2008

Printed in the United Kingdom at the University Press, Cambridge

A catalogue record for this publication is available from the British Library

Library of Congress Cataloguing in Publication data

Krasnov, Boris R., 1955–
  Functional and evolutionary ecology of fleas : a model for ecological parasitology / Boris R. Krasnov.
    p. cm.
  Includes bibliographical references and index.
  ISBN 978-0-521-88277-4 (hardback)
  1. Fleas–Ecology.   2. Fleas–Evolution.   3. Host-parasite relationships.
I. Title.
QL599.5.K73   2008
595.77′517857 – dc22   2007053048

ISBN 978-0-521-88277-4 hardback

Cambridge University Press has no responsibility for the persistence or accuracy of URLs for external or third-party internet websites referred to in this publication, and does not guarantee that any content on such websites is, or will remain, accurate or appropriate.

Contents

Preface ix

Part I Brief descriptive ecology: what do fleas do?

1 Composition of the order   3
    1.1 Infraorders and families   3
    1.2 Temporal pattern of discovery of flea species   4

2 Hosts of Siphonaptera   9
    2.1 Avian and mammalian hosts   9
    2.2 'Realized' and available hosts   10
    2.3 Number of flea species among host orders   15
    2.4 Fleas, small mammals and biogeography   16
    2.5 Concluding remark   17

3 Geographical distribution of fleas   18
    3.1 General patterns of geographical distribution   18
    3.2 Fleas on islands   21
    3.3 Size of geographical range   22
    3.4 Relationship between flea and host(s) geographical ranges   25
    3.5 Concluding remark   28

4 Origin and evolution of fleas   29
    4.1 Ancestral and sister taxa   29
    4.2 Origin of flea parasitism   32
    4.3 Phylogenetic relationships within Siphonaptera   33
    4.4 Cophylogeny of fleas and their hosts   34

4.5 Flea diversification: intrahost speciation, host-switching or climate? 41
4.6 Concluding remarks 44

5 Life cycles 45
5.1 Mating and oviposition 45
5.2 Larvae 48
5.3 Pupae 51
5.4 Imago 52
5.5 Seasonality 54
5.6 Concluding remark 67

6 Fleas and humanity 68
6.1 Medical aspects 68
6.2 Veterinary aspects 72
6.3 Fleas in human habitats 74
6.4 Concluding remarks 76

Part II Functional ecology: how do fleas do what they do?

7 Ecology of sexual dimorphism, gender differences and sex ratio 79
7.1 Sexual dimorphism 79
7.2 Physiological gender differences 82
7.3 Gender differences in behaviour 90
7.4 Gender differences in responses to environmental factors 91
7.5 Sex ratio as an ecological consequence of gender differences 93
7.6 Concluding remarks 102

8 Ecology of flea locomotion 103
8.1 On-host locomotion 103
8.2 Off-host locomotion 104
8.3 Concluding remark 114

9 Ecology of host selection 115
9.1 Evolutionary scale: principal and auxiliary hosts 116
9.2 Ecological scale: host selection 122
9.3 Fleas and the ideal free distribution 130
9.4 Distribution of fleas on the body of a host 141
9.5 Time spent on- and off-host 147
9.6 Concluding remarks 153

10 Ecology of haematophagy   154
    10.1 Mouthparts and host skin   154
    10.2 Measures of feeding success   156
    10.3 Host-related effects   162
    10.4 Flea-related effects   173
    10.5 Environment-related effects   178
    10.6 Concluding remarks   181

11 Ecology of reproduction and pre-imaginal development   182
    11.1 Measures of reproductive success   182
    11.2 Host-related effects   185
    11.3 Flea-related effects   199
    11.4 Environment-related effects   206
    11.5 Concluding remarks   216

12 Ecology of flea virulence   217
    12.1 Host metabolic rate   218
    12.2 Host body mass and growth rate   223
    12.3 Host haematological parameters   227
    12.4 Host features related to sexual selection   228
    12.5 Host behaviour   229
    12.6 Host survival   232
    12.7 Host fitness   235
    12.8 Concluding remarks   237

13 Ecology of host defence   239
    13.1 First line of defence: avoidance   240
    13.2 Second line of defence: repelling or killing fleas   243
    13.3 Third line of defence: immune response against fleas   255
    13.4 Concluding remarks   278

Part III Evolutionary ecology: why do fleas do what they do?

14 Ecology and evolution of host specificity   283
    14.1 Measures of host specificity   283
    14.2 Variation in host specificity among flea species   285
    14.3 Host specificity and evolutionary success   295
    14.4 Host specificity and host features   303
    14.5 Evolution of host specificity: direction, reversibility and conservatism   308
    14.6 Applicative aspects of host specificity studies   314
    14.7 Concluding remarks   320

15 Ecology of flea populations 321

15.1 Measuring abundance and distribution 322
15.2 Is abundance a flea species character? 324
15.3 Aggregation of fleas among host individuals 328
15.4 Biases in flea infestation 338
15.5 Relationship between flea abundance and prevalence 352
15.6 Factors affecting flea abundance and distribution 356
15.7 Concluding remarks 374

16 Ecology of flea communities 375

16.1 Are flea communities structured? 377
16.2 Local versus regional processes governing flea communities 386
16.3 Inferring patterns of interspecific interactions 391
16.4 Negative interspecific interactions 401
16.5 Similarity in flea communities: geographical distance or similarity in host composition? 405
16.6 Concluding remarks 408

17 Patterns of flea diversity 410

17.1 Flea diversity and host body 411
17.2 Flea diversity and host gender 415
17.3 Flea diversity and host population 415
17.4 Flea diversity and host community 417
17.5 Flea diversity and host geographical range 424
17.6 Flea diversity and the off-host environment 425
17.7 Flea diversity and parasites of other taxa 431
17.8 Concluding remarks 435

18 Fleas, hosts, habitats 436

18.1 The Middle East 437
18.2 Central Europe 443
18.3 Other examples 451
18.4 Concluding remarks 454

19 What further efforts are needed? 455

19.1 Where we are now and what do we have? 456
19.2 What do we lack? 458
19.3 Not only a pure science . . . 463

References 466
Index 583

Preface

I was privileged to be introduced to the study of zoology in the Department of Zoology and Comparative Anatomy of Terrestrial Vertebrates at the Moscow State University in Russia. I began my scientific career studying behavioural mechanisms that influence the spatial structure of rodent populations in different landscapes, from the tundra and the Arctic shore of the Chukchi Peninsula to the rainforests of southern Vietnam. At the time, academic staff members and students of the department under the leadership of Professor Nikolai Naumov were working intensively on rodent ecology, aiming to understand their role in infectious zoonoses, mainly the plague. Consequently, every student who studied rodent ecology was introduced to fleas, as they are the principal vectors of the plague.

In the beginning of the 1990s, I started to work at Ben-Gurion University of the Negev and continued to study rodents and other desert-dwelling animals (tenebrionid beetles and lizards) in the Negev Desert. These studies resulted in a book, *Spatial Ecology of Desert Rodent Communities*, written together with my colleagues Georgy Shenbrot and Konstantin Rogovin, and published by Springer-Verlag in 1999 (Shenbrot *et al.*, 1999a). However, I also subliminally continued to collect fleas from every captured rodent, not being sure at that time why exactly I was doing this. In the mid 1990s, I read several papers by Robert Poulin, Serge Morand and Jean-François Guégan, which opened my eyes to an enthralling new world of parasites. I was so fascinated with the ideas and findings of ecological and evolutionary parasitology that, in the middle of my scientific career, I abruptly switched from studying behaviour and spatial ecology to studying the ecology of host–parasite relationships. Naturally, fleas and rodents were a familiar and very convenient model association that allowed me to combine the ecology of free-living organisms and parasitology, two parallel worlds, wherein scientists too often are not aware of each others' achievements.

Parasites are becoming increasingly important in studies of ecology and evolution. This is mainly due to the numerous advantages of using parasites to examine patterns and processes in animal communities because of, for example, the relative ease of obtaining replicated samples (e.g. host individuals or host species) and the fact that parasites of the same taxon share a trophic level. Another advantage of studying parasite communities is that most hosts are usually parasitized by several closely and/or distantly related parasite species that use the same resource. Thus, the study of the community organization of parasites allows a better understanding of the processes of competition and facilitation in biological communities. Ecological and evolutionary studies of parasites, in turn, are powerful tools for understanding the spread of dangerous zoonotic diseases and provide a theoretical basis for their control and prevention. All these issues have led to a sharp increase in empirical, comparative and theoretical studies of host–parasite relationships. Patterns and processes in host–parasite systems have been documented and studied at a variety of levels, across various habitats, in different biogeographical regions and for various parasite taxa. The goal of this book is to examine how functional, ecological and evolutionary patterns and processes of host–parasite relationships are realized in one particular host–parasite system. I attempt to demonstrate how Siphonaptera can be used as a model for testing ecological and evolutionary hypotheses.

My hope is that, on the one hand, this book will be of specific interest for biologists studying fleas, providing them with an up-to-date review of the biology of their study animals. On the other hand, I hope that the book will serve a much greater audience and be relevant to both parasitologists and ecologists. The book provides an in-depth case study of a model host–parasite system, looking at it from many angles, and extracting from it several general principles that apply equally well to other host–parasite systems. Often, a book with detailed information on one taxon inspires research on other taxa, and this book could become a guideline for further research into both parasitism and animal population and community organization.

Fleas represent one of the most fascinating taxa of ectoparasites. All species in this relatively small monophyletic order are obligatory haematophagous parasites of mammals and birds. From the ecological and evolutionary perspectives, fleas represent an interesting model. In particular, this is related to the characteristic *modus vivendi* of these insects. On the one hand, in contrast to endoparasites and permanent ectoparasites such as lice, they spend much time off their hosts and are therefore affected, not only by factors linked to the host per se, but also by the off-host abiotic environment. On the other hand, in contrast to temporary ectoparasites such as mosquitoes and ticks, they spend more time on their hosts than is required merely to obtain a blood meal. This creates a causal chain of flea–host–environment interactions, which in itself is an important and interesting subject for investigation. Another advantage of using fleas as a model taxon is the opportunity to manipulate flea infestation on living hosts both in the field and in the laboratory and to monitor changes in an individual host over time. Indeed, fleas, in contrast to many other parasites, can be counted on a live animal that itself than can be marked, released, recaptured and examined again.

Fleas serve as the vectors of many diseases dangerous to humans. Apart from this, the veterinary aspect of flea parasitism is also very important, with flea-bite allergies and hypersensitivity being serious problems for both livestock and pets. However, in spite of the importance of fleas and their convenience as models for ecological and evolutionary studies, there is a lack of literature dealing with flea bionomics from modern ecological and evolutionary perspectives. Although there have been several brilliant reviews dealing with flea life history (e.g. Marshall, 1981a; Traub, 1985; Vashchenok, 1988), most ecological and evolutionary approaches that have been developed during the last two decades have not been applied to these animals. This book is aimed at filling the gap between the descriptive biology of fleas and current ecological and evolutionary theory.

An additional issue of note is that fleas have been, and are being, extensively studied in countries of the former USSR; thus much flea literature is in Russian. Moreover, these papers were published in exotic journals, periodicals and collective volumes, making them difficult to obtain and to understand for the Western scientific community. Two reviews of Russian flea literature (Bibikova, 1977; Bibikova & Zhovty, 1980) were published in English, but both are outdated. Given that Western flea-related sources, at least, up to the late 1970s, were carefully reviewed by Adrian Marshall (1981a) and Robert Traub (1980, 1985), I tried to include as many examples as possible from studies done in the former USSR and post-USSR countries as well as in Eastern Europe. Many studies of fleas were done in China. I regret that the Chinese literature has not been as thoroughly reviewed as it should have been. Nevertheless, I did my best to use Chinese sources as well. In this endeavour I obtained help from one of my colleagues and collaborators, Dr Liang Lu, from the Chinese Centre for Disease Control and Prevention in Beijing.

I intentionally avoided the purely applied aspects such as, for example, the control of fleas on domestic animals, as this book is not meant to be either a medical or a veterinary text. Instead, the book moves from basic descriptive aspects, to functional issues and finally to evolutionary explanations. Part I provides a brief description of flea taxonomy, life cycles, and flea–host associations, addressing the question: *what* do fleas do? Part II addresses the functional ecology of fleas. It deals with proximate causes of flea responses to their hosts and environment and, thus, addresses the question: *how* do fleas do what they do? Finally, Part III deals with the evolutionary ecology of fleas and the ultimate explanations of observed patterns, addressing the question: *why* do fleas do what they do? In addition, (a) in contrast to many earlier texts on parasitology and (b) following Claude Combes's (2001) idea of a parasite and a host being involved in a durable and intimate interaction, I consider both fleas and their hosts together as two partners in the same game.

During my studies on fleas, and while writing this book, I was helped by many people. Robert Poulin, Allan Degen and Berry Pinshow (in chronological order) were the very first persons with whom I shared the idea of writing this book. It would not have been written without their encouragement. I would like to thank my collaborators and co-authors in publications (in alphabetical order): Zvika Abramsky, Allan Degen, Laura Fielden, Kevin Gaston, Michael Hastriter, Michael Kam, Irina Khokhlova, Tatiana Knyazeva, Natalia Korallo, Carmi Korine, Liang Lu, Sergei Medvedev, Dana Miklisova, Serge Morand, Ladislav Mošanský, David Mouillot, Berry Pinshow, Robert Poulin, David Saltz, Georgy Shenbrot, Marina Spinu, Michal Stanko, Valentin Vashchenok, Diego Vázquez and Maxim Vinarski. These people represent countries from Russia to New Zealand and from China to Canada. Members of my team in the Mitrani Department of Desert Ecology (Jacob Blaustein Institutes for Desert Research, Ben-Gurion University of the Negev), Sergei Burdelov and Nadezhda Burdelova, have worked with me during the past 12 years, in the field and in the laboratory, and I am very grateful for their help. I thank my research students and postdoctorate fellows (in alphabetical order): Marine Arakelyan, Dikla Bashary, Tatiana Demidova, Lusine Ghazaryan, Joëlle Goüy de Bellocq, Hadas Hawlena, Ana Hovhanyan, Mariela Leiderman, Maria Lizurume, Natella Mirzoyan, Isik Oguzoglu, Luis Rios, Michal Sarfati, Pirchia Sinai and Kelly Still. They represent not only Israel, but also (in alphabetical order) Argentina, Armenia, Guatemala, France, Russia, Turkey and the USA. I hope they learned not only to study fleas, but also to view them as interesting and charming animals rather than repulsive and aggravating pests. The ideas in this book were discussed over the years with colleagues who helped with their suggestions. They are (in alphabetical order): Vladimir Ageyev, Michael Begon, Frank Clark, Claude Combes, Natalia Darskaya, Katharina Dittmar de la Cruz, Lance Durden, Kenneth Gage, Terry Galloway, Heikki Henttonen, Matthias Kiefer, Michael Kosoy, Marcela Lareschi, Kim Larsen, Herwig Leirs, Douglas Morris, Kosta Mumcuoglu, Robert Pilgrim, Yigal Rechav, Michael Rosenzweig, Lajos Rózsa, Uriel Safriel, Arkady Savinetsky, Svetlana Shilova, Albert Survillo, Viktor Suntsov, Andrey Tchabovsky, David Ward and Michael Whiting. Omar Amin, Daniel Frynta, Ryszard Haitlinger, Liang Lu, Elena Naumova, Michal

Stanko and Elena Zharikova provided me with some rare literature. My colleagues (in alphabetical order): Allan Degen, Kirill Eskov, Megan Griffiths, Michael Hastriter, Irina Khokhlova, Burt Kotler, Serge Morand, Berry Pinshow, Robert Poulin and David Ward read earlier versions of chapters of this book and made helpful comments. Zoe Grabinar and Marcia Chertok improved the English prose. I thank Cambridge University Press and, in particular, Jacqueline Garget for the opportunity to publish this book with a leading scientific publisher. Finally, I thank Irina Khokhlova, who is not only my collaborator of many years, but also my spouse, and our children Helena and Alexander for their continuous support and patience.

For taxonomy and names, I followed Medvedev *et al.* (2005) for fleas, Clements (2007) for birds and Wilson & Reeder (2005) for mammals. Consequently, some species names differ from those in the original sources.

PART I   BRIEF DESCRIPTIVE ECOLOGY: WHAT DO FLEAS DO?

1

Composition of the order

Siphonaptera is a relatively small order of secondarily wingless holometabolous insects. According to a recent taxonomic scrutiny by Medvedev *et al.* (2005), the order includes 2005 species and 828 subspecies belonging to 242 genera and 97 subgenera. In addition, 409 specific, 147 subspecific, 65 generic, and seven subgeneric names are considered to be synonyms. However, some recently discovered species are still absent from Medvedev *et al.*'s (2005) database (e.g. Barriere *et al.*, 2002; Durden & Beaucournu, 2002; Pampiglione *et al.*, 2003; Beaucournu & Wells, 2004; Hastriter & Haas, 2005). Nonetheless, at present, this database is the most complete one available. In this chapter, I briefly outline the composition of the order and provide basic information on the higher-level flea taxonomy.

1.1 Infraorders and families

Cladistic analysis of fleas using 50 morphological features of the head, thorax and abdomen (Medvedev, 1994) resulted in the above-generic taxonomic scheme including four infraorders and 18 families as follows.

Order Siphonaptera
  Infraorder Pulicomorpha
    Family Pulicidae
    Family Tungidae
    Family Malacopsyllidae
    Family Rhopalopsyllidae
    Family Vermipsyllidae
    Family Coptopsyllidae
    Family Ancistropsyllidae

4   Composition of the order

> Infraorder Pygiopsyllomorpha
>   Family Lycopsyllidae
>   Family Pygiopsyllidae
>   Family Stivaliidae
> Infraorder Hystrichopsyllomorpha
>   Family Hystrichopsyllidae
>   Family Chimaeropsyllidae
>   Family Macropsyllidae
>   Family Stephanocircidae
> Infraorder Ceratophyllomorpha
>   Family Ceratophyllidae
>   Family Leptopsyllidae
>   Family Ischnopsyllidae
>   Family Xiphiopsyllidae

Species composition of different flea families ranges from 594 in Hystrichopsyllidae to two in Malacopsyllidae (Medvedev *et al.*, 2005) (Fig. 1.1).

Large siphonapteran families (Pulicidae, Tungidae, Rhopalopsyllidae, Pygiopsyllidae, Stivaliidae, Hystrichopsyllidae, Stephanocircidae, Ceratophyllidae and Leptopsyllidae) contain species that exploit different host orders and even host classes (e.g. pulicids, ceratophyllids and leptopsyllids parasitize both mammals and birds). Other families are more host-specific. For instance, fleas of the family Ischnopsyllidae are associated exclusively with bats (Chiroptera), Chimaeropsyllidae with elephant shrews (Macroscelidea) and Malacopsyllidae with armadillos (Cingulata). Soricomorph hosts are parasitized mainly by fleas from the hystrichopsyllid tribe Doratopsyllini, whereas lagomorphs by fleas from the pulicid tribe Spilopsyllini.

1.2   Temporal pattern of discovery of flea species

The earliest description of a flea species dates back to 1758, when Linnaeus described the house flea *Pulex irritans* and the sand (chigoe) flea *Tunga penetrans*. The next descriptions of flea species were made 40 years later by Bosc (*Nosopsyllus fasciatus*) and Schrank (*Ceratophyllus gallinae* and *Monopsyllus sciurorum*). Discoveries and descriptions of new flea species are continuing at present (e.g. Hastriter, 2000a, b, 2001a, b, 2004; Lewis & Haas, 2001; Lewis & Stone, 2001; Barriere *et al.*, 2002; Durden & Beaucournu, 2002; Hastriter & Whiting, 2002; Hastriter *et al.*, 2002; Hastriter & Eckerlin, 2003; Beaucournu & Wells, 2004; Lewis & Eckerlin, 2004; Acosta & Morrone, 2005; Hastriter & Haas, 2005; Beaucournu *et al.*, 2006).

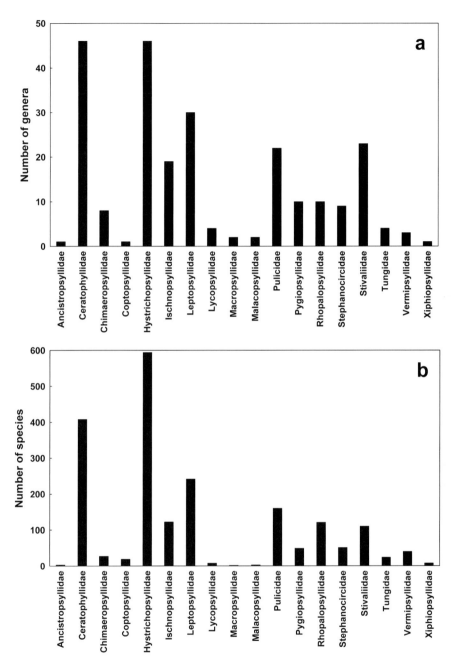

Figure 1.1 Number of valid (a) genera and (b) species in siphonapteran families. Data from Medvedev *et al.* (2005).

6 Composition of the order

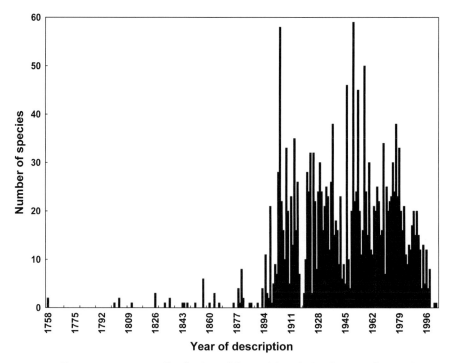

Figure 1.2 Frequency distribution of dates of description for 2005 flea species. Data from Medvedev *et al.* (2005).

The frequency distribution of dates of flea descriptions demonstrates that the majority of flea species was discovered in the first half of the last century (Fig. 1.2). However, periods when many flea species were described alternated with periods with few descriptions (even reaching zero per year). The three highest peaks in the twentieth century (Fig. 1.2) coincide with the descriptions of many new flea species by highly productive taxonomists (e.g. Jordan & Rothschild, 1915a, b; Rothschild, 1915a, b; Wagner, 1929; Ioff *et al.*, 1946). For example, K. Jordan and N. N. Rothschild, separately and together, described more than 600 flea species, F. Smit and J. Wagner each described about 140 species, I. Ioff and R. Traub each described about 100 species and K.-C. Li, R. Lewis, J.-C. Beaucournu, G. Holland and C. Baker each described about 50 species. Interestingly, the number of researchers involved in description of flea species grew steadily until the beginning of the 1990s and then dropped sharply (Fig. 1.3a). However, 'description effort' per researcher does not follow this pattern. Number of flea species described per researcher per year peaked between the 1920s and 1960s (Fig. 1.3b).

The rate of discovery of new flea species is not caused only by the above-mentioned 'subjective' reason. Biological parameters of flea species as well as

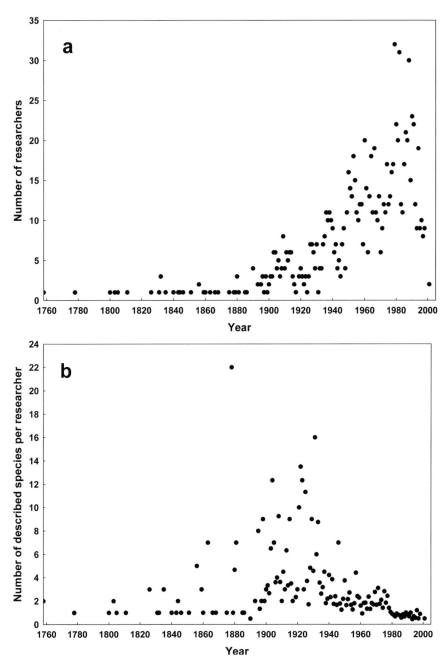

Figure 1.3 Distribution of (a) number of researchers who provided taxonomic description of new flea species (including synonyms) and (b) 'description effort' (number of descriptions per researcher) from 1758 to 2001. Data from Medvedev et al. (2005).

those of their hosts are also involved (see Chapter 13). In addition, the latest discoveries of the new species of Siphonaptera were made mainly in remote geographic regions such as Tasmania, southeastern Asia, central Africa and South America. However, a new species, *Jordanopsylla becki*, was discovered in North America (Nevada) as late as in 2000 (Hastriter, 2000a), and a new genus (*Psittopsylla*) was described recently in Mexico (Lewis & Stone, 2001).

More information on flea biology and their host associations is still needed. About 600 flea species (about 30% of the total number of known species) are known from only a single host and a single record (Medvedev, 2002). The highest numbers of such species were found in the Madagascar, Eastern African, Papuan, Malayan and Caribbean biogeographical subregions. These poorly known species were collected mainly from rodents (murines and cricetines) and shrews, although some species were recorded from bats, lagomorphs, carnivores and marsupials.

2

Hosts of Siphonaptera

Fleas are characteristic parasites of birds and mammals. Occurrences of fleas on reptiles are accidental (e.g. Tillyard, 1926; Dunnet & Mardon, 1974), although they are able to digest blood of these hosts (Belokopytova *et al.*, 1983; Vashchenok, 1988). Accidentally, fleas are able to feed even on haemolymph of ticks (Bilyalov *et al.*, 1989). In this chapter, I address general patterns of distribution of fleas within the two classes of higher vertebrates.

2.1 Avian and mammalian hosts

The majority of fleas parasitize mammals (more than 94% of species: Vashchenok, 1988; Beaucournu *et al.*, 2005), whereas their association with birds is much less frequent. Analysing host associations of 1951 flea species, Medvedev (1997a, b) found that fleas were recorded on 16 mammalian and 21 avian orders. Furthermore, 1835 of the 1951 species analysed were harboured by 1606 mammalian hosts, whereas only 214 species were recorded on 543 bird hosts. Among the latter, only 60 species (about 3% of the total number of flea species) can be considered as specific bird parasites (Medvedev, 1997a, b).

The number of flea–mammal associations compared with flea–bird associations suggests that fleas are mainly parasites of mammals. Parasitism on birds is, thus, secondary, and fleas parasitic on birds are commonly thought to have originated from fleas parasitic on mammals, with this switch occurring at least 16 times during the siphonapteran evolution (Hopkins, 1957; Holland, 1964). In particular, species characteristically parasitic on birds are found in six of 18 siphonapteran families (Holland, 1964), suggesting that the switch from mammalian to avian hosts was related to ecological rather than to phyletic reasons (Beaucournu *et al.*, 2005). Some bird fleas have likely originated from ancestors parasitic on arboreal mammals (Traub et al., 1983). For example, 18 specimens of five species of rodent fleas were collected from nests of woodpeckers (Haas & Wilson, 1985). Interestingly, the occurrence of mammalian flea species in bird nests increases in winter (Roman & Pichot, 1975). This supports the idea that the switch from mammals to birds is associated with ecological factors. Bird fleas could also switch from burrow-dwelling mammals to burrow-dwelling birds (see Holland (1964) for examples with *Actenopsylla* and *Ornithopsylla*) or could exploit birds of prey or owls switching from the mammalian victims of these predators (Scharf, 1998). Traub et al. (1983) suggested that fleas of the family Ceratophyllidae probably evolved from ancestors infesting squirrels (Sciuridae) in the early Eocene (40–45 million years ago). Furthermore, within the family, the adaptation to exploit avian instead of mammalian hosts probably occurred independently in different genera. However, the opposite was also suggested. For example, it is thought that formerly a bird flea *Ceratophyllus lunatus* reverted to mammals (it is the only species of the genus that does not exploit birds) (Hopkins, 1957). An ability to feed successfully on mammals has been reported for some other bird fleas (e.g. *Frontopsylla frontalis*: Shevchenko et al., 1976).

2.2 'Realized' and available hosts

Distribution of fleas among mammalian and avian orders in six different regions located on different continents is summarized in Table 2.1. Even from a superficial glance at this table, it becomes clear that (a) most flea species exploit mammals (in particular, rodents and, in Australia, marsupials), and (b) there are no data on fleas for many bird and mammal species. This is still true for well-studied regions such as central Europe or North America. The reason for the lack of reports of fleas from some host species or orders can be that either these species do not harbour fleas and/or they merely have not been examined for parasites. In some cases, the former appears to be true (for example, for cetaceans and proboscids), whereas it is impossible to distinguish between the two reasons in other cases. Nevertheless, if the former reason is true, this means that fleas under-use the available pool of host species. Some host species may not be used by fleas due to obvious reasons, such as aquatic life (e.g. pinniped carnivores, although some flea species are associated with penguins and many sea birds: Holland, 1964), absence of pelage (e.g. elephants) and lack of shelters necessary for successful development of pre-imaginal stages of many flea species. Even if a host species possesses a shelter, but its physical conditions (physical structure, bedding material, temperature and/or relative humidity) are unfavourable for pre-imaginal fleas, this host would probably be unsuitable for fleas, as is the case

Table 2.1 *Number of flea species recorded on birds and mammals belonging to different orders in six geographical regions*

Region	Class	Order	Number of host species infested with fleas	Number of flea species recorded	Number of flea species per host species	Proportion of host species infested with fleas	Number of orders with no flea records
Poland	Aves	Podicipediformes	1	1	1	0.2	11
		Anseriformes	1	3	3	0.02	
		Falconiformes	2	2	1	0.06	
		Galliformes	1	2	2	0.1	
		Charadriiformes	2	1	0.5	0.02	
		Columbiformes	1	2	2	0.2	
		Cuculiformes	1	1	1	0.5	
		Piciformes	1	2	2	0.1	
		Passeriformes	37	18	0.5	0.22	
	Mammalia	Erinaceomorpha	1	6	6	0.5	0
		Soricomorpha	8	19	2.4	0.9	
		Chiroptera	12	10	0.8	0.6	
		Carnivora	9	19	2.1	0.6	
		Lagomorpha	2	2	1	0.7	
		Rodentia	23	37	1.6	0.9	
		Artiodactyla	2	3	1.5	0.3	
China	Aves	Anseriformes	1	1	1	0.02	15
		Falconiformes	2	2	1	0.03	
		Charadriiformes	2	2	1	0.02	
		Columbiformes	2	4	2	0.07	
		Apodiformes	1	2	2	0.09	
		Piciformes	1	1	1	0.03	
		Passeriformes	29	25	0.9	0.04	

(cont.)

Table 2.1 (cont.)

Region	Class	Order	Number of host species infested with fleas	Number of flea species recorded	Number of flea species per host species	Proportion of host species infested with fleas	Number of orders with no flea records
	Mammalia	Erinaceomorpha	3	12	4	0.43	5
		Soricomorpha	16	45	2.8	0.37	
		Scandentia	1	7	7	1.00	
		Chiroptera	10	24	2.4	0.11	
		Carnivora	23	46	2	0.41	
		Lagomorpha	18	74	4.1	0.56	
		Rodentia	106	309	2.9	0.62	
		Artiodactyla	12	12	1	0.3	
South Africa	Aves	Sphenisciformes	1	1	1	0.2	18
		Falconiformes	1	1	1	0.01	
		Galliformes	1	1	1	0.06	
		Columbiformes	1	1	1	0.07	
		Coraciiformes	5	3	0.6	0.1	
		Passeriformes	9	2	0.2	0.03	
	Mammalia	Macroscelidea	4	9	2.2	0.6	5
		Erinaceomorpha	1	3	3	1	
		Soricomorpha	5	12	2.4	0.3	
		Chiroptera	20	12	0.6	0.3	
		Primates	1	2	2	0.2	
		Pholidota	1	2	2	1	
		Tubulidentata	1	1	1	1	
		Hyracoidea	2	6	3	0.7	
		Carnivora	29	25	0.9	0.8	

Region	Class	Order				
Australia		Lagomorpha	6	11	1.8	1
		Rodentia	49	78	1.6	0.6
		Artiodactyla	9	5	0.6	0.2
	Aves	Sphenisciformes	2	2	1	0.2	14
		Procellariiformes	8	6	0.7	0.1
		Charadriiformes	4	3	0.7	0.04
		Columbiformes	2	1	0.5	0.07
		Passeriformes	8	3	0.4	0.02
	Mammalia	Monotremata	2	8	4	0.5	3
		Dasyuromorphia	19	30	1.6	0.3
		Paramelemorphia	7	20	2.9	0.3
		Diprotodontia	33	34	1.03	0.2
		Chiroptera	11	5	0.4	0.2
		Rodentia	22	34	1.5	0.4
Canada, Alaska, Greenland	Aves	Podicipediformes	1	1	1	0.2	9
		Procellariiformes	1	1	1	0.04
		Pelicaniformes	4	4	1	0.2
		Anseriformes	6	3	0.5	0.1
		Falconiformes	6	4	0.7	0.2
		Galliformes	7	5	0.7	0.4
		Gruiformes	1	1	1	0.07
		Charadriiformes	8	7	0.9	0.05
		Strigiformes	7	12	1.7	0.4
		Piciformes	6	11	1.8	0.4
		Passeriformes	67	23	0.3	0.2
	Mammalia	Didelphimorphia	1	2	2	1	1
		Soricomorpha	19	40	2	1
		Chiroptera	9	6	0.7	0.5

(cont.)

Table 2.1 (cont.)

Region	Class	Order	Number of host species infested with fleas	Number of flea species recorded	Number of flea species per host species	Proportion of host species infested with fleas	Number of orders with no flea records
		Carnivora	22	73	3.3	0.9	
		Lagomorpha	10	38	3.8	1	
		Rodentia	87	116	1.3	1	
		Artiodactyla	2	1	0.5	0.2	
Argentina	Aves	Sphenisciformes	2	3	1.5	0.3	17
		Charadriiformes	1	2	2	0.01	
		Columbiformes	1	1	1	0.04	
		Psittaciformes	1	1	1	0.04	
		Strigiformes	1	2	2	0.05	
		Piciformes	1	1	1	0.03	
		Passeriformes	7	7	1	0.01	
	Mammalia	Didelphimorphia	7	18	2.6	0.4	3
		Cingulata	8	4	0.5	0.5	
		Chiroptera	6	8	1.3	0.1	
		Carnivora	9	7	0.8	0.4	
		Rodentia	51	78	1.5	0.4	
		Lagomorpha	2	3	1.5	0.7	
		Artiodactyla	3	2	0.7	0.2	
		Perissodactyla	1	1	1	1	

Sources: Data from Skuratowicz (1967) for Poland, Liu *et al.* (1986) for China, Segerman (1995) for South Africa, Dunnet & Mardon (1974) for Australia, Holland (1985) for Canada, Alaska and Greenland, and Autino & Lareschi (1998) for Argentina. Numbers of bird and mammal species inhabiting each region are taken from Banfield (1974), Strahan (1983), Skinner & Smithers (1990), Redford & Eisenberg (1992), Zhang *et al.* (1997), Mitchell-Jones *et al.* (1999) and Lepage (2006).

for many mammals in the lowland tropics (Wenzel & Tipton, 1966). On the other hand, a historical component cannot also be ruled out. Some host species may not harbour fleas merely because they have not been colonized by fleas either due to their recent invasion into a given region or because of unknown reasons or by chance. For example, among 189 bird species examined for ectoparasites in Azerbaijan, only 40 species harboured fleas (Gusev et al., 1962). Flea-free and flea-harbouring species were found not only within the same bird order, but also within the same family and even genus.

Unfortunately, the absence of parasites from specimens that were parasitologically examined is rarely reported, especially in regional monographs (but see Dunnet & Mardon, 1974). As a result, the proportion of host species for which data on flea species exist ('realized' as opposed to potential flea hosts) may be underestimated. Sometimes, this proportion seems to be extremely low. In most mammalian orders it varies from about 17% (Primates) to 70% (Lagomorpha). However, the percentage of rodent species reported to harbour fleas is much greater, ranging from 40% in Argentina to 100% in Canada. Nevertheless, in general, the percentage of mammal species with known flea fauna is higher than that of birds (56% versus 12%, respectively). On the other hand, both within and among regions, the number of flea species parasitic on mammals of a particular order is higher in many cases than the number of potential host species of the order. In contrast, the number of flea species parasitic on birds is always much lower than the number of available hosts. On a global scale, the ratios of flea and host species are 1.1 : 1 for mammals and 1 : 2.5 for birds (Medvedev, 2005).

2.3 Number of flea species among host orders

As already mentioned, the number of flea species recorded on mammals is much higher than that recorded on birds (27.5 ± 7.0 versus 4.0 ± 0.8, respectively, as averages across regions). Among mammals, the largest number of flea species were recorded on rodents, and the smallest on perissodactyls (e.g. tapirs) and an aardvark (108.7 versus 1.0, respectively, as averages across regions). Among birds, most fleas were found on passerines and the least on parrots, cuckoos and cranes (13.0 versus 1.0, respectively, as averages across regions). Regional data reflect global patterns. For example, analysis of the global list of flea–host associations (Medvedev & Lobanov, 1999; Medvedev, 2002, 2005; Medvedev et al., 2005) demonstrated that 70.3% of flea–mammal associations involved rodents, whereas 55.3% of flea–bird associations involved passerines (Medvedev, 2002). In addition, rodents composed 82% of all specific and/or principal hosts for fleas. Among other mammalian orders, 9.5% of associations involved fissiped carnivores, 6.5% involved erinaceo- and soricomorphs and 3.7% involved bats. Among non-passerine birds, main associations with fleas were characteristic for the tube-nosed seabirds (Procellariiformes: 7.7% of associations), diurnal birds of prey (Falconiformes: 6.5% of associations) and shorebirds (Charadriiformes: 7.7% of associations).

The number of flea species differs drastically among different host orders and, thus, comparison of flea fauna among host orders without taking into account this difference can mask interesting patterns. When the number of flea species is recalculated per one mammalian host species, among-order differences appear to be less evident (from 0.5 in Cingulata to 7.0 in Scandentia with Rodentia taking a median position of 1.75; see Table 2.1). Among avian orders, the mean number of flea species per host species appears to be similar and ranges from about 0.5 in Passeriformes to 2.0 in Strigiformes and Piciformes. Therefore, although the highest absolute number of flea species exploiting mammals and birds is characteristic for rodents and passerines, respectively, flea species are 'diluted' among hosts of these species-rich taxa.

2.4 Fleas, small mammals and biogeography

Primary association of fleas with small mammals is observed in all parts of the world. In the Palaearctic region, fleas exploit mainly voles (Arvicolinae), gerbils (Gerbillinae) and hamsters (Cricetinae), whereas in the Nearctic region, they exploit mainly voles, New World rats and mice (Sigmodontinae), pocket gophers (Geomyidae) and kangaroo rats and pocket mice (Heteromyidae). Mammals from other orders that characteristically harbour fleas in both parts of the Holarctic are pikas and hares (Lagomorpha) as well as hedgehogs (Erinaceomorpha) and shrews and moles (Soricomorpha). Flea hosts in the Neotropics are sigmodontine and caviomorph rodents (Caviidae, Chinchillidae, Capromyidae, Octodontidae etc.) as well as representatives of two marsupial orders, American opossums (Didelphimorphia) and shrew opossums (Paucituberculata). In the Afrotropics, fleas parasitize mainly rats and mice (Murinae), bamboo rats (Rhizomyidae) and African mole-rats (Bathyergidae). Other flea hosts in this region are hyraxes (Hyracoidea) and elephant shrews (Macroscelidea). Fleas from the Oriental region (including Wallacea and the Southern Pacific islands) infest mainly murines and squirrels (Sciuridae) as well as a number of marsupial orders (Dasyuromorphia, Paramelemorphia and some Diprotodontia). Finally, in the Australian region, fleas parasitize murine rodents and some marsupials (Paramelemorphia and Diprotodontia).

2.5 Concluding remark

The high number of flea species associated with small mammals and the high number of species within most small mammalian orders suggest that, from the evolutionary perspective, diversification of fleas was a response to diversification of their hosts. Further discussion of this issue will be addressed in Chapter 4.

3
Geographical distribution of fleas

Geographical range is an immanent feature of every recent or extinct species. Two main characteristics of a geographical range of a species are its position and its size. Many important ecological and evolutionary questions involve these two parameters. What determines the limits of species occurrences? What are the causes of variation in the size of geographical range among and within taxa? How are position and size of geographical ranges related? Is the size of geographical range heritable, i.e. are geographical ranges of sister species similar in size? Do fluctuations in the abundance of a species affect its geographical range? Is the degree of specialization associated with the position and size of geographical range? Basic knowledge on geographical distribution of a taxon of interest is necessary for answering these questions.

Fleas are distributed around the world, although they most probably were introduced by humans and their pets and livestock to some oceanic islands. Therefore, these insects would be a very convenient model for biogeographical studies. Some examples of such studies will be presented later in this book. In this chapter, I focus on taxonomically related patterns of distribution of fleas around the world, variation in the size of geographical range of fleas, and on the relationship between geographical range of a flea and that of its host(s).

3.1 General patterns of geographical distribution

Fleas are found on all continents and on most oceanic islands. They even inhabit Antarctica where the endemic *Glaciopsyllus antarcticus* occurs on a number of seabird species (e.g. Bell *et al.*, 1988; Steele *et al.*, 1997). Birds of the sub-Antarctic islands are also parasitized by fleas (e.g. *Notiopsylla kerguelensis* and *Parapsyllus heardi* on the Kerguelen Islands: Chastel & Beaucournu, 1992).

The geographical distribution of fleas is characterized by highly unequal numbers of flea species among different regions (Medvedev, 1996, 1998, 2000a, b). The flea fauna of the Palaearctic appears to be the most diverse and includes 892 species (approximately 38% of the total number of known species). The numbers of species in the Nearctic, Afrotropical and Neotropical realms are similar (299, 275 and 289 species, respectively), whereas in the Oriental and Australian realms, the numbers are considerably less (191 and 68 species, respectively). It should be noted, however, that such regions as the Philippine Islands, Wallacea and Papua–New Guinea and islands eastward are largely unexplored.

Medvedev (1998, 2000a, b, 2005) suggested a classification scheme for the geographical distribution of fleas and divided flea families according to their occurrence in the World. Medvedev (2005) distinguished among (a) families characteristic for the northern hemisphere; (b) families occurring mainly in the northern hemisphere but with a few representatives in the southern hemisphere; (c) one family characteristic of tropics and subtropics of both Old and New Worlds (Pulicidae); (d) families characteristic of tropics of the Old World; (e) families distributed mainly in the southern hemisphere but with a few representatives in the northern hemisphere; and (f) families characteristic for the southern hemisphere.

In general, the flea fauna of the southern hemisphere is characterized by small families and subfamilies such as Malacopsyllidae, Rhopalopsyllidae and Craneopsyllinae in South America, Xiphiopsyllidae and Chimaeropsyllidae in Africa, and Macropsyllidae, Lycopsyllidae and Stephanocircidae in Australia. Pygiopsyllidae are distributed in southern and southeastern Asia, Wallacea, Australia and South America. In contrast, the largest flea families (Hystrichopsyllidae, Ceratophyllidae, Leptopsyllidae and Ischnopsyllidae) inhabit mainly the northern hemisphere. In these flea families, the highest number of species occurs in the Palaearctic, although representatives of the families are also distributed in other parts of the world. For example, there are only two hystrichopsyllid species in Australia, whereas this family is represented by 332 species in the Palaearctic. The number of ceratophyllids in the Palaearctic is 1.2 times greater than the number of ceratophyllids in the rest of the world (Medvedev, 2005). Ceratophyllidae are absent from Australia and New Zealand, whereas the number of species of this family in the Afrotropical realm is the lowest among the regions (although all afrotropical ceratophyllids are endemic).

The degree of flea endemism (at least, at the generic level) varies less among realms. The percentage of endemic genera reaches 61% in the Afrotropical realm, and is slightly lower in the Neotropical and Australian realms (56% and 58%, respectively). Endemic genera comprise about 45% of all flea genera in the Palaearctic, 37% in the Nearctic and 42% in the Oriental realms. The main host

Table 3.1 *Main host taxa for flea species endemic of a biogeographical realm*

Realm	Main host taxa for endemic flea species
Oriental	Dasyuromorphia: Dasyuridae
	Diprotodontia: Petauridae, Pseudocheiridae
	Paramelemorphia: Paramelidae
	Rodentia: Sciuridae, Murinae
Palaearctic	Soricomorpha: Soricidae
	Rodentia: Arvicolinae, Gerbillinae, Cricetinae
	Lagomorpha: Ochotonidae
Nearctic	Soricomorpha: Soricidae
	Rodentia: Arvicolinae, Sigmodontinae, Heteromyidae, Geomyidae
Neotropics	Rodentia: Sigmodontinae, Caviidae
Afrotropics	Macroscelidea
	Hyracoidea
	Rodentia: Murinae, Rhizomyinae, Bathyergidae
Australian	Dasyuromorphia: Dasyuridae
	Diprotodontia: Burramyidae, Phalangeridae, Pseudocheiridae, Vombatidae
	Paramelemorphia: Paramelidae
	Rodentia: Murinae

Source: Data from Medvedev (2005) and Medvedev & Krasnov (2006).

groups for flea species endemic for a particular zoogeographical realm are presented in Table 3.1.

The geographical distribution of flea species is probably related to the plate tectonics and subsequent dispersal and redistribution of the host taxa (Traub, 1980). As a result, analysis of the geographical distribution of flea families, genera and species by Medvedev (1996) demonstrated apparent similarity between flea faunas of Eurasia and North America as well as between North and South Americas (Table 3.2). These similarities are probably related to the history of dispersals of some mammalian taxa (together with their fleas) via the land bridges between continents such as the Beringia Land Bridge and/or a land bridge across Greenland between North America and Europe. There is also evidence of the connection of the flea faunas of Australia and South America (e.g. distribution of stephanocircids) that seems to be related to the links between South America, Antarctica and Australia in the Upper Cretaceous. The distribution of Tungidae suggests links between the flea faunas of Africa and South America (African *Neotunga* and South American *Hectopsylla* and *Rhynchopsyllus*), although some taxonomists suspect that *Neotunga* is not necessarily related to *Tunga* (M. W. Hastriter, personal communication, 2006).

Table 3.2 *Jaccard similarity among flea faunas of seven biogeographical realms*

Realm	Oriental	Palaearctic	Nearctic	Neotropics	Afrotropics	Australian
Palaearctic	17.4					
Nearctic	3.4	22.3				
Neotropics	0.9	6.3	21.7			
Afrotropics	11.7	12.0	0.9	2.0		
Australian	13.0	1.6	0.0	1.2	6.1	
New Zealand	5.4	0.0	0.0	3.3	2.0	14.2

Source: Data from Medvedev (1996).

Some flea species from different families have cosmopolitan or, at least, very broad distribution. The most famous (and the most important from the medical and veterinary points of views) cosmopolitan fleas are several pulicids (*Pulex irritans*, *Xenopsylla cheopis*, *Ctenocephalides canis*, *Ctenocephalides felis*), ceratophyllids (*Nosopsyllus consimilis*, *Nosopsyllus fasciatus*) and one leptopsyllid (*Leptopsylla segnis*). Ubiquitous distribution of these species is related to dispersal via humans, their livestock, pets and commensals (mice and rats). For example, the house flea *P. irritans* is thought to originate in South America (Traub, 1980; Buckland & Sadler, 1989; Beaucournu *et al.*, 1993) and was introduced to the Old World by the Vikings (Rothschild, 1973; Buckland & Sadler, 1989) and/or through ancient cultural contacts between Japan and Ecuador (Traub, 1980). Nevertheless, the origin of any of these species is related to wild host species (e.g. see Beaucournu *et al.* (1997) for *L. segnis*). Moreover, despite cosmopolitanism, distribution of any of these fleas is not uniform. Instead, they are distributed in patches which are characterized by the host and environmental conditions that are favourable for each given species (see Beaucournu & Pascal (1998) for *N. fasciatus* and Beaucournu & Menier (1998) for *Ctenocephalides* species).

3.2 Fleas on islands

As mentioned above, fleas were likely introduced to many oceanic islands by humans and human-associated animals. However, flea faunas of continental islands are composed of both indigenous and introduced elements. The introduced fleas succeeded in switching to local hosts and in establishing themselves in the natural habitats on some islands (e.g. *P. irritans* on Santa Cruz and Santa Rosa Islands: Crooks *et al.*, 2001, 2004), but not on others (e.g. *C. felis* and *X. cheopis* on the Martinique and Guadeloupe Islands: Pascal *et al.*, 2004). Furthermore, introduction of a wildlife species to an island may lead to the establishment of new fleas. For example, the ranges of *Ceratophyllus ciliatus* and possibly *Hystrichopsylla dippiei* were probably extended to the Baranof Island by the introduction of red squirrels and martens, respectively (Haas *et al.*, 1980). *Monopsyllus anisus* has been unintentionally introduced to Honshu Island as a consequence of the intentional introduction of the red-bellied squirrel *Callosciurus erythraeus* (Shinozaki *et al.*, 2004). As recently as 1987, the rabbit flea *Spilopsyllus cuniculi* that arrived with the introduction of rabbits was established on one of the small islands of the sub-Antarctic Kerguelen Archipelago (Chekchak *et al.*, 2000).

In general, insular faunas are characterized by an impoverishment in species number (MacArthur & Wilson, 1967). Insular faunas of fleas are not excluded from this rule and, consequently, are species-poor. Depauperation of the insular flea faunas is manifested both within host species when flea assemblages of the island host populations are compared with those on the mainland (see Durden (1995) for fleas of the cotton mouse *Peromyscus gossipinus* and the eastern grey squirrel *Sciurus carolinensis* on St Catherine Island and Milazzo *et al.* (2003) for fleas of the black rat *Rattus rattus* and the house mouse *Mus musculus* on Sicily) and across host species when the entire or almost entire flea assemblages on an island are compared with those on the mainland (see Uchikawa *et al.* (1967) for the Oki Islands of Japan; Konkova & Timofeeva (1970) for the Kuril Islands; Bengston *et al.* (1986) for Iceland; Scharf (1991) for the islands of Lake Michigan; and Wilson & Durden (2003) for the Georgia Barrier Islands).

3.3 Size of geographical range

Classification scheme based on the size of the geographical range of fleas was proposed by Medvedev (1998, 2000a, b, 2005). Accordingly, the order contains (a) six families with narrow ranges (Coptopsyllidae, Xiphiopsyllidae, Chimaeropsyllidae, Malacopsyllidae, Macropsyllidae and Lycopsyllidae); (b) two families with moderate ranges (Ancistropsyllidae and Vermipsyllidae); (c) seven families with broad distribution (Leptopsyllidae, Pulicidae, Tungidae, Rhopalopsyllidae, Stephanocircidae, Stivaliidae and Pygiopsyllidae); and (d) three families with ubiquitous distribution (Hystrichopsyllidae, Ceratophyllidae and Ischnopsyllidae). This classification, however, was based on the number of biogeographical realms, regions and sub-regions in which flea species were recorded rather than on calculations of the size of geographical ranges.

A strong right skew of the frequency distribution of geographical range size is a common characteristic for almost all living taxa (see Gaston, 2003 for review). Fleas conform to this rule. Krasnov *et al.* (2005a) calculated the size

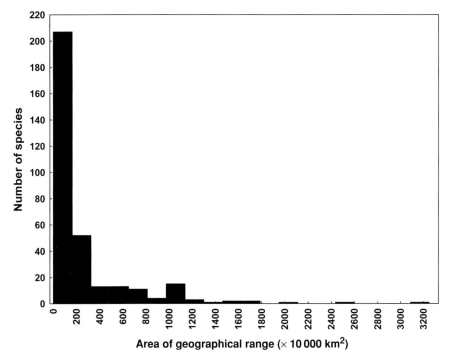

Figure 3.1 Frequency distribution of geographical range sizes of 326 flea species from seven geographical regions. Data from Krasnov et al. (2005a).

of the geographical range of 326 flea species parasitic on small mammals in seven large geographical regions. The frequency distribution of geographical ranges of fleas was highly right-skewed (Fig. 3.1) independent of whether the data were within a region or pooled across regions. At the within-region level, the most right-skewed distribution was shown by fleas from South Africa and the least right-skewed by fleas from Venezuela (Table 3.3). Distribution of geographical range sizes was also right-skewed within flea families (Table 3.4). In other words, there are more species with narrow geographical ranges than broadly distributed species both within a region and within a family.

Comparison of data from Table 3.4 with the classification of Medvedev (1998, 2000a, b, 2005) provides little support for the latter. For example, there is not much difference in the mean size of geographical range between coptopsyllid and hystrichopsyllid fleas which, according to Medvedev's (2005) classification, belong to different categories. On average, the largest geographical ranges are characteristic for ceratophyllids, whereas the smallest for pygiopsyllids. Furthermore, among-species variation in the size of geographical range is the highest in stephanocircids and hystrichopsyllids and the lowest in lycopsyllids

Table 3.3 *Skewness (γ_1) of the frequency distribution of geographical range sizes of fleas from seven geographical regions*

Region	Number of species	γ_1	Reference
Venezuela	25	1.23 ± 0.46	Tipton & Machado-Allison, 1972
Canada, Alaska and Greenland	103	2.73 ± 0.24	Holland, 1985
Australia	35	2.94 ± 0.40	Dunnet & Mardon, 1974
South Africa	63	6.82 ± 0.30	Segerman, 1995
Morocco	20	3.00 ± 0.51	Hastriter & Tipton, 1975
Mongolia	76	3.33 ± 0.27	Kiefer *et al.*, 1984
Asian Far East	19	1.65 ± 0.52	Yudin *et al.*, 1976
All species	326	3.18 ± 0.13	

Source: Data from Krasnov *et al.* (2005a).

Table 3.4 *Mean (± S.E.) size, coefficient of variation (CV) and skewness (γ_1) of the frequency distribution of geographical range sizes of fleas belonging to 12 families from seven geographical regions*

Family	Number of species	Mean size of geographical range ($\times$ 10 000 km^2)	CV (%)	γ_1
Ceratophyllidae	64	481.62 ± 70.11	116.45	2.52 ± 0.30
Chimaeropsyllidae	17	25.95 ± 7.02	111.45	1.62 ± 0.55
Coptopsyllidae	2	128.41 ± 115.00	126.65	—
Hystrichopsyllidae	95	199.29 ± 35.11	171.72	2.71 ± 0.25
Leptopsyllidae	53	224.55 ± 33.00	106.97	1.71 ± 0.33
Lycopsyllidae	4	41.89 ± 21.00	100.26	1.16 ± 1.01
Macropsyllidae	1	37.66	—	—
Pulicidae	48	233.82 ± 67.49	199.98	3.45 ± 0.34
Pygiopsyllidae	14	20.48 ± 5.95	108.99	1.38 ± 0.60
Rhopalopsyllidae	13	445.67 ± 125.46	101.50	0.55 ± 0.62
Stephanocircidae	12	36.45 ± 18.76	178.26	3.24 ± 0.64
Stivaliidae	3	29.36 ± 19.10	112.69	1.58 ± 1.22

Source: Data from Krasnov *et al.* (2005a).

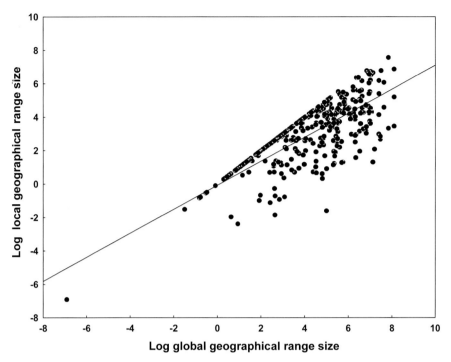

Figure 3.2 Relationship between sizes of the 'local' (within a region) and global geographical ranges of 326 flea species from seven geographical regions. Data from Krasnov *et al.* (2005a).

and rhopalopsyllids. Although pulicid fleas have, on average, relatively broad geographical ranges, the proportion of species with a narrow distribution in this family is higher than in the other flea families (as indicated by skewness values).

Furthermore, there was a strong positive relationship between geographical range size of a flea species within a given region (local geographical range) and the entire geographical range size of this species (global geographical range) (Fig. 3.2). In other words, only a few flea species were represented in the considered regions by margins of their geographical ranges.

3.4 Relationship between flea and host(s) geographical ranges

Geographical range limits of a species are commonly thought to be determined by hard barriers to dispersal and, in the absence of such barriers, by an interplay between physical factors (such as climate) and interspecific interactions (mainly competition) (Brown *et al.*, 1996; Case & Taper, 2000; Gaston,

2003). In other words, the geographical range of a species should reflect the abiotic environment, the ensemble of species with which it interacts, and its own capacity to respond via natural selection to those interactions. Interactions with competitors can lead to sharp deviations in range limits from those expected from individual climatic tolerances alone (Case et al., 2005).

The geographical distribution of a flea species should be a result of interaction between its responses to the geographical distribution of host(s) and to the off-host environment (see also Chapter 14). The latter includes both physical factors (because both imago and pre-imaginal fleas are affected by extrinsic physical environment; see Chapters 5 and 10–11) and presence of competitors (see Chapter 16).

Relative role of the off-host environment and potential competitors in determining geographical range may differ between host-specific and host-opportunistic fleas (see Chapter 14). Obviously, both host-specific and host-opportunistic fleas should be distributed within the geographical range of their hosts. However, if several closely related flea species are highly specific for the same host, they could display mutually exclusive (geographically vicariate) distribution as a result of competitive exclusion as they compete for the single resource (MacArthur & Levins, 1964; but see Chapter 16). In this case, the distribution of potential competitors will strongly affect geographical ranges of a target flea species (assuming that the highly host-specific parasite and its host coevolved to the same limits of tolerance to environmental gradients and have the same distributional limits). In contrast, competitive exclusion among host-opportunistic fleas is unlikely because they compete for several resources and should be able to coexist as a result of resource partitioning. In other words, the effect of the distribution of competitors on the size of geographical range should be stronger in host-specific than in host-opportunistic species. Furthermore, a host-opportunistic flea is unlikely to be adapted to the cumulative range of environmental tolerance of all its hosts. As a result, the geographical distribution of a host-opportunistic flea should be determined mainly by the limits of its own environmental tolerance within the geographical ranges of its hosts.

These predictions were tested by Shenbrot et al. (2007) on fleas of the genus *Amphipsylla* (in total, 32 species). Species of this genus are distributed mainly in the southeastern Palaearctic with some species spilling over into the western and northeastern Palaearctic and even into the Nearctic. They infest muroid rodents belonging mainly to subfamilies Arvicolinae, Calomyscinae, Cricetinae and Myospalacinae (in total, 51 species). Three types of topological relationships of geographical ranges of a flea and its host(s) can be envisaged among species of *Amphipsylla* as follows. The geographical range of a flea (a) is completely nested in the geographical range of a single host (host-specific species); (b) is nested in the geographical ranges of several geographically vicariate hosts (locally host-specific species); and (c) covers the geographical ranges of several non-vicariate hosts (host-opportunistic species). The results of Shenbrot et al.'s (2007) study showed that the range size of host-specific fleas correlated positively with the size of the host geographical range and negatively with the size of the geographical range of the fleas' potential competitors, whereas the range size of both locally host-specific and host-opportunistic fleas correlated positively with the size of the host geographical range only. This suggests that the main determinant of the size of the geographical range of a flea species is the size of the geographical range of its host(s) and, thus, supports the view of a geographical range of a species as a spatial reflection of its ecological niche (Brown & Lomolino, 1998; Pulliam, 2000). Nevertheless, the geographical range size of a flea appeared to be better predicted by the size of the hosts' geographical ranges rather than by the number of hosts (but see Chapter 14). This means that the spatial extent of a species' occurrence is affected by the spatial representation of available resources rather than by the breadth of its ecological niche. The role of potential competitors in determining the geographical range size was manifested in highly host-specific fleas only, suggesting generally low importance of interspecific competition in fleas (see Chapter 16).

Furthermore, host-specific fleas occupied 0.2–80.0% of the geographical range of their hosts, whereas these percentages were 0.9–83.7% in locally host-specific fleas and 16.6–63.7% in host-opportunistic fleas. There are several possible reasons for the absence of a flea from parts of the hosts' geographical range. First, our knowledge can be incomplete due to spatially irregular sampling. Second, a flea can be absent from some parts of the host's geographical range if there are hard barriers to a flea's dispersal such as disjunctions in the host's geographical range. For example, *Amphipsylla asiatica*, *Amphipsylla kuznetzovi* and *Amphipsylla primaris* are absent from the Arctic, eastern Siberian and central Chinese, respectively, parts of highly fragmented geographical range of the narrow-skulled vole *Microtus gregalis*. There are, however, examples when large disjunctions in the host's geographical range did not act as barriers for flea dispersal (trans-Beringian distribution of *Amphipsylla marikovskii*, trans-Atlantic distribution of *Amphipsylla sibirica*). Third, the limits of environmental tolerance of a flea may be narrower than that of its host(s). Finally, the absence of a host-specific flea from some parts of a host's geographical range may result from interactions with congeneric species resulting in competitive exclusion, although accounting for competitive interactions increased estimations of the part of a host geographical range occupied by a host-specific flea by 10% only.

3.5 Concluding remark

In general, the occurrence of fleas literally everywhere testifies to the high evolutionary success of this small insect order. The distribution of fleas all over the world and their occurrence in a variety of landscapes under most environmental conditions makes them a convenient model taxon for testing various macroecological hypotheses as will be shown in Part III of this book.

4

Origin and evolution of fleas

One of the main impediments in studies of the evolution and phylogeny of Siphonaptera is morphological specialization related to their ectoparasitic way of life (Whiting, 2002a). In particular, this specialization is reflected in the peculiar morphology of the head, thorax and genitalia. On the one hand, specialized characters are not especially informative for use in phylogenetic reconstructions. On the other hand, the sharing of these specialized characters by practically all flea species strongly suggests monophyly of the order. Indeed, the monophyletic origin of fleas is supported by both morphological (e.g. Medvedev, 2003a, b) and molecular evidence (Whiting, 2002a, b). Although many authors agree on the monophyly of Siphonaptera, there is still no consensus on other questions related to their phylogeny and the origin of their parasitism. What are the relationships of fleas with other insect taxa? Did specialized features such as winglessness, laterally compressed body and locomotory apparatus allowing jumping originate as adaptations to ectoparasitism or, alternatively, were they characteristic of flea ancestors and are thus pre-adaptations to parasitism (see Medvedev, 2005)? Did fleas coevolve with their hosts? What was the main driver of diversification of this order? In this chapter, I review these questions and summarize the relatively limited knowledge on the evolutionary history of fleas.

4.1 Ancestral and sister taxa

Although fossil fleas are extremely rare, there are several finds of fleas from the Baltic and Dominican amber. Fleas from the Baltic amber (*Palaeopsylla dissimilis*, *Palaeopsylla klebsiana*, *Palaeopsylla baltica*) are from the Upper Eocene (Dampf, 1911; Peus, 1968), whereas *Pulex larimerius* and *Rhopalopsyllus* sp. from the Dominican amber date back only to the Miocene (Lewis & Grimaldi, 1997).

Some evidence even suggests that fleas existed as early as in the Mesozoic, although their association with Mesozoic mammals is questionable. A number of Cretaceous or late Jurassic insects were sometimes considered as belonging to Siphonaptera or to their ancestral taxon (Riek, 1970; Ponomarenko, 1976; Rasnitsyn, 1992, 2002a; but see Willmann, 1981a, b). These considerations were based on the similarity in some morphological features between fleas and such fossil insects as the late Jurassic *Strashila incredibilis* (Rasnitsyn, 1992) from Siberia, the late Cretaceous *Saurophthirus longipes* from Siberia (Ponomarenko, 1976) and the late Cretaceous *Saurophthirodes mongolicus* from Mongolia (Ponomarenko, 1986), although the latter did not demonstrate clear morphological evidence of adaptations to jumping, and this insect is questioned as being parasitic and related to *Saurophthirus* (Rasnitsyn 2002a, b). However, the late Cretaceous *Tarwinia australis* (Jell & Duncan, 1986) from Australia (Koonwarra Fossil Bed), although probably different in habits from modern fleas, had the laterally compressed body and saltatory legs (Lukashevich & Mostovsky, 2003). *Niwratia elongata* which was also found in these deposits also had laterally compressed 'siphonapteran' body and pronotal comb (Jell & Duncan, 1986). Nevertheless, relationships between these Jurassic and Cretaceous insects and Siphonaptera are unclear. For example, the placement of *S. longipes* into Siphonaptera by Ponomarenko (1976) seems to be erroneous, and this insect is probably a panorpoid (Labandeira, 1997). Smit (1978) rejected the inclusion of the mentioned Cretaceous taxa into Siphonaptera (but this was prior to the description of the flea-like *Tarwinia*). Rasnitsyn (1992, 2002) also suggested being cautious about phylogenetic topology of *Strashila*, *Saurophthirus*, *Tarwinia* and fleas, although he noted the evidence indicating that the three former taxa may form a monophyletic group with fleas (Rasnitsyn, 1992). Furthermore, he termed these Mesozoic creatures as 'pre-fleas' and mentioned that they show many unusual features but are all conceivable as parasites of pterosaurs living permanently on their wing membrane (*Saurophthirus* and *Strashila*) or both on membrane and in the fur (*Tarwinia*) (Rasnitsyn, 2002; see also Ponomarenko, 1976). Rasnitsyn (2002) described the fossil 'pre-fleas' as being similar and supposedly synapomorph with true fleas in having a relatively short moniliform antenna and a triangular, hypognathous head equipped with a piercing beak (not seen clearly enough in *Tarwinia*). The similarity between *Saurophthirus* and *Strashila* is expressed in a weakly sclerotized, extensible abdomen and between *Saurophthirus* and true fleas in ctenidia (absent in *Strashila*, unknown in *Tarwinia*), whereas *Tarwinia* and true fleas are similar in having a laterally compressed body and sclerotized abdomen. Thus, due to the absence of ctenidia, *Strashila* may be hypothesized as a sister group of the ctenidiate forms, for which the long claws with small but distinct basal lobes (claw structure is unknown for *Tarwinia*) may be synapomorphic. *Saurophthirus* may have sister relationships with *Tarwinia* and true fleas combined, the two latter forming a group with putative synapomorphies in the sclerotized abdomen and somewhat compressed body.

The most common phylogenetic hypothesis favours the origin of fleas from the Mecoptera-like ancestors. This hypothesis was initially proposed by Tillyard (1935) and supported by Hinton (1958) on the basis of comparison of some larval characters. However, Boudreaux (1979) stated that Siphonaptera are more closely related to Diptera than to Mecoptera and argued that features that fleas seem to share with mecopterans can be instead either generally primitive insect characters or convergences (see also Byers, 1996). Nevertheless, close phylogenetic relationships between Siphonaptera and Mecoptera were further supported by analyses of various morphological characters (Kristensen, 1975, 1981; Rothschild, 1975; Beutel & Gorb, 2001) as well as by molecular data (Whiting, 2002a, b). Presumably synapomorphic characters that suggest sister relationships between Mecoptera and Siphonaptera include the absence of the outer group of microtubuli in the sperm flagellum (but see Jamieson (1987) about the presence of this feature in Bittacidae), the specific configuration of the acanthae in the proventriculus and the absence of extrinsic labral muscles (see Beutel & Pohl, 2006). However, the problem of sister relationships between Siphonaptera with either the entire Mecoptera or some of the within-Mecoptera taxa has not been resolved. The crucial point is the phylogeny of the Mecoptera *sensu lato*. Phylogenetic analyses of sequences of the four loci (18S and 28S ribosomal DNA, cytochrome oxidase II, and elongation factor-1α) of 69 taxa represented major flea and mecopteran lineages strongly supported a paraphyletic Mecoptera with two major lineages (Whiting, 2002b). According to this view, one lineage includes Nannochoristidae, Boreidae and Siphonaptera, whereas the other includes the remaining mecopteran taxa. Thus, among mecopteran families, Boreidae appeared to be the closest living relative to Siphonaptera. This, as well as paraphyly of Mecoptera, was further supported by the combined analysis of both morphological and molecular data (Whiting *et al.*, 2003), but has been challenged by comparison of mecopteran and siphonapteran sperm structure (Dallai *et al.*, 2003). The latter study supported the monophyly of Mecoptera and, thus, sister relationships between fleas and the entire mecopteran lineage as was suggested earlier (Kristensen, 1981). Nevertheless, the sister relationships between boreids and fleas are largely accepted (Hastriter & Whiting, 2003; Medvedev, 2005) and, in addition to molecular evidence (see Whiting, 2002a, b), are supported by several potential synapomorphies, such as similarities of the proventricular spines (Richards & Richards, 1969), wing reduction, loss of the arolium (Beutel & Gorb, 2001), multiple sex chromosomes (Bayreuther & Brauning, 1971), morphology of the ovaries (Kings & Teasly, 1980; Štys & Bilinski, 1990; Bilinski *et al.*, 1998; Simiczyjew & Margas, 2001) and specific process of resilin secretion (Rothschild & Schlein, 1975; Schlein, 1980).

4.2 Origin of flea parasitism

There are two main scenarios of the origin of flea parasitism. Did loss of wings, ability to jump and laterally compressed body evolve as an adaptation of parasitism or were they already present in free-living flea ancestors and, therefore, only facilitated the transition from free living to parasitism? Traditionally, the origin of fleas is associated with the appearance of a pelage and fossorial way of life of their hosts. For example, Smit (1972) suggested that flea ancestors were scavengers that lived in hosts' burrows and that they lost their wings during adaptation to parasitism. Hastriter & Whiting (2003) suggested the following scenario. When the boreid–flea ancestor shifted from free living in snowy, mossy habitats to living in burrows of its host, it probably lost its wings and acquired its jumping ability. Further adaptations to parasitism included lateral flattening and development of suctorial mouthparts and elaborate ctenidia and setae.

However, the morphology of the fossil fleas from amber suggests that in the late Eocene, when the modern mammalian orders started to appear, fleas already had their characteristic appearance possessing all the main features of morphological specialization to parasitism. This allowed Medvedev (2005) to challenge the above-mentioned 'adaptation-to-parasitism' scenarios (i.e. flea features resulted from the transition to parasitism). He argued that flea ancestors were wingless. Otherwise, if they made the transition to blood-sucking, it would result in another taxon of flying haematophages such as mosquitoes. Instead, Medvedev (2005) proposed the 'pre-adaptation-to-parasitism' scenario (i.e. some flea features were pre-adaptive and facilitated the transition to parasitism). He theorized that winglessness, jumping ability and the laterally compressed body of fleas are all associated with their life in spatially restricted conditions (e.g. host's burrow) rather than with parasitism *per se*, whereas other features (such as suctorial mouthparts, keel-like frontal part of the body, combs and bristles) are parasitism-related adaptations. Medvedev's (2005) scenario refuted the possibility that the pre-adaptations to spatially restricted conditions could evolve in boreid-like insects that live in mossy habitats.

Furthermore, Medvedev (2001, 2005) suggested that the origin of fleas was in the Lower Jurassic. This hypothesis is based on the geographical distribution of fleas with the basal taxa inhabiting South America, Africa, India and Australia, i.e. former parts of Gondwanaland (Medvedev, 2000a, b). However, the evolutionary heyday of fleas seems to start later, namely in the Eocene, because flea diversification probably evolved as a response to the diversification of mammals (see below). For example, Ischnopsyllidae are thought to originate at the Upper Eocene/Lower Oligocene boundary in southeastern Asia (Medvedev, 1990). From there, the tribes Chriopteropsyllini and Ischnopsyllini spread over into the Afrotropics and Holarctic, and Porribiini into Australia, whereas Sternopsyllini

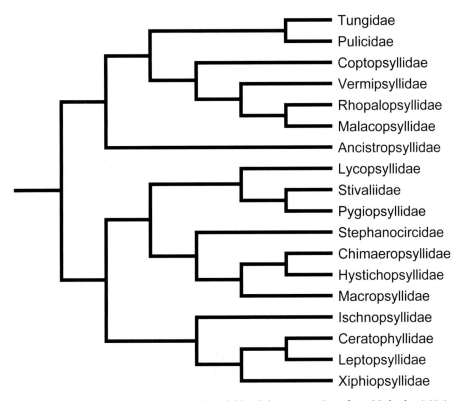

Figure 4.1 Cladistic relationships within Siphonaptera. Data from Medvedev (1994).

went into the Neotropics (via Australia and Antarctica). Nycteridopsyllini penetrated North America from Asia via the Beringia Land Bridge later (Medvedev, 1990). Whiting & Hastriter (M. W. Hastriter, personal communication, 2006) noted that the distribution of some families would suggest that ancestral fleas arose before the separation of Gondwanaland (between 140 and 200 mya).

4.3 Phylogenetic relationships within Siphonaptera

Among flea families, Hystrichopsyllidae and Pulicidae seem to be the oldest and Ceratophyllidae and Leptopsyllidae the youngest (Traub, 1980; Traub et al., 1983; Medvedev, 1996). Indeed, Hystrichopsyllidae and Pulicidae are thought to date back to the Lower Cretaceous or Upper Jurassic (Jellison, 1959; Traub et al., 1983), whereas Ceratophyllidae appeared no earlier than the Middle Eocene (Traub et al., 1983). Medvedev (1994, 1998, 2005) suggested that infraorders Pulicomorpha and Pygiopsyllomorpha are the most basal, whereas Ceratophyllomorpha is the most derived (Fig. 4.1). He also presented morphological evidence advocating a phylogenetically basal position of the family Ancistropsyllidae and its association with pulicomorphs, although earlier it was considered as a close relative of young Leptopsyllidae (Hopkins & Rothschild, 1971) or was even placed in the superfamily Ceratophylloidea together with Ceratophyllidae (Smit, 1982). Still, there is no agreement about phylogenetic relationships both among and within flea families (see Whiting *et al.*, 2002 versus Medvedev, 1994).

4.4 Cophylogeny of fleas and their hosts

The concept of a common evolutionary history between parasites and their hosts holds a central place in modern parasitology (Brooks, 1988; Barker, 1991; Klassen, 1992; Hoberg *et al.*, 1997; Paterson & Banks, 2001). The common history of host–parasite associations is often defined as 'coevolution', although there are some differences in macroevolutionary (e.g., Brooks, 1988) and microevolutionary contexts (e.g., Toft & Karter, 1990) of this term. If the only events during the process of reciprocal natural selection in the host and parasite lineage (i.e. coevolution) of a particular host lineage and a particular parasite lineage were those of cospeciation (contemporaneous speciation in the host and parasite lineages), then there would be full congruency of host and parasite phylogenies. However, most studies of cophylogeny of host–parasite associations have demonstrated that full congruence of phylogenies is not generally the case and that the common history of hosts and parasites is complicated with coevolutionary events other than cospeciation (Paterson *et al.*, 1993; Beveridge & Chilton, 2001; Roy, 2001). Potential events that lead to incongruence between parasite and host phylogenies have been discussed widely (e.g. Paterson & Gray, 1997; Page & Charleston, 1998; Paterson & Banks, 2001; Roy, 2001), although definitions differ slightly among authors. These events include: (a) host switching (when a parasite species colonizes a host taxon from a different lineage); (b) duplication (when a parasite speciates without host speciation and, as a result, closely related parasite species occur on the descendant host); (c) lineage sorting (when a parasite species is removed from a host species); and (d) inertia (when a parasite species does not change despite host's speciation and, as a result, the same parasite species occurs on multiple closely related hosts). For example, observations of Riddoch *et al.* (1984) on fleas of the sand martin *Riparia riparia* in Britain hinted at a duplication event. Northern populations of the sand martin were exploited mainly by *Ceratophyllus styx jordani*, whereas *Ceratophyllus styx styx* was characteristic of southern populations. The subspecies differed in morphology of the genitalia (lower length/breadth ratio of the clasper in males and a more pronounced indentation of the posterior edge of sternum VII in females in *C. s. jordani*) and variation at a polymorphic aminopeptidase locus (higher frequency of the $Ap\text{-}1^S$ allele in *C. s. styx*).

Fleas are commonly thought to coevolve with their hosts (Traub, 1980, 1985). This idea stems mainly from the observations that some morphological features of fleas are complementary with behaviour and skin and/or fur morphology of their hosts such that they are often exactly what we would expect to see if a flea adapts itself to its particular host species. For example, fleas possess various anatomical features that allow them to attach to the host's hairs and to resist the host's grooming (Traub, 1980; but see Smit, 1972). These features are represented by sclerotinized bristles as well as helmets, ctenidia, spines and setae. Furthermore, these structures, as well as head shape and modifications of shape and size of spines, correlate with particular characteristics of the host's fur (Traub, 1972a, 1985). For example, the Egyptian spiny mouse *Acomys cahirinus* is a specific host for *Parapulex chephrenis*. The coat of *A. cahirinus* is characterized by thick but very short hairs and widely spaced long, rigid keratin spines. This coat can be groomed efficiently and fleas can be easily detached. However, the entire body of *P. chephrenis* is covered with sclerotinized bristles which probably facilitate flea resistance to host grooming. Numerous examples of fleas' adaptive anatomy in relation to the ecology and behaviour of their hosts and of fleas themselves were reported by Traub (1980). For instance, fleas that exploit mammals with coarse fur (e.g. bandicoots) tend to have sharply pointed pronotal spines. Fleas of many ground-dwelling rodents that possess unspecialized fur and construct deep and complex burrows and nests usually lack combs and specialized bristles. In contrast, fleas parasitic on arboreal and gliding hosts (e.g. *Myoxopsylla* which exploits dormice) have well-developed combs with numerous spines. Furthermore, fleas that spend much time in the fur of their hosts tend to have better-developed or more numerous ctenidia, spines and bristles in comparison with fleas that visit a host only for a blood meal and spend most of their lives in the hosts' burrow or nest. Additional support for the hypothesis of the evolution of combs and spines as a tool to anchor a flea in the fur of a host is provided by the correlation of the distance between tips of the comb spines and the diameter of the host's hair (Humphries, 1966; Amin & Wagner, 1983; Medvedev, 2001b). For example, this relationships has been found in *Cediopsylla simplex*, *Ctenocephalides canis*, *Corrodopsylla curvata*, *Megabothris acerbus*, *Orchopeas howardi*, *Orchopeas leucopus*, *Oropsylla bruneri* and *Oropsylla arctomys* (Amin & Wagner, 1983).

Biogeography has also been used to illustrate flea–host coevolution. For example, Morrone & Gutiérrez (2004) applied panbiogeographical analysis (see Craw et al. (1999) for details on this approach) to the geographical distribution of 112 flea species belonging to 48 genera and eight families in the Mexican transition zone, and argued that a significant diversification of the flea taxa occurred in parallel with the diversification of their mammal hosts.

Despite numerous descriptions and evidence of flea–host coevolution from both flea morphology (Traub, 1980, 1985) and flea–host biogeographical patterns (Traub, 1980, 1985; Jameson, 1999), earlier attempts to present coevolutionary scenarios of the common history of fleas and their hosts were purely qualitative. Nevertheless, the complex traits such as combs, spines, helmets and bristles that suggest a purposive design are less likely to have evolved by chance and are more likely to be the adaptive products of selection.

Comparison of the evolutionary age of flea taxa and the taxa of their hosts does not add much for resolving the enigma of flea–host coevolution. For example, it was suggested that the association between *Hystrichopsylla* fleas and soricomorph hosts presents evidence of coevolution because *Hystrichopsylla* display primitive morphological characteristics among fleas as shrews and moles do among mammals (Lewis & Eckerlin, 2004). However, many fleas from basal families (e.g. Pulicidae) exploit evolutionarily young hosts, whereas evolutionarily young ceratophyllids exploit evolutionarily primitive hosts. A similar pattern may occur within taxonomic units of the lower order. For example, among Neotropical species belonging to subgenus *Alloctenus* of the genus *Ctenophthalmus*, the most basal species are parasites of sigmodontine rodents (Muridae), whereas the most derived species are exclusive parasites of the shrew genus *Cryptotis* (Morrone et al., 2000). In other words, Szidat's (1940) rule that the more primitive the host, the more primitive the parasites which it harbours does not seem to hold for fleas, although examples of the reciprocal can also be found across the order.

During the last two decades, several analytical methods for the reconstruction of coevolutionary history have been proposed (see review in Paterson & Banks, 2001). The rationale of all methods is the same, namely, to choose the most parsimonious evolutionary scenario of the association between a set of hosts and a set of parasites among a number of possible scenarios. However, approaches for this choice differ among methods. The Brooks parsimony analysis (BPA: Brooks, 1988) uses parasite associations with their hosts and parasite phylogeny as host's character states (see example in Fig. 4.2a). 'Parasite-based' host phylogeny is built from these characters using Wagner parsimony and is compared with the recognized host phylogeny. The tree reconciliation analysis (Page, 1990, 1993a, b, 1994a, b; Charleston, 1998, Page & Charleston, 1998) is based on comparison of topologies of host and parasite trees, creating a map between the two trees. The generalized parsimony method proposed by Ronquist (1995, 2001) assigns differential costs to four types of possible events (codivergence, duplication, sorting and host-switching). It transforms host phylogeny into a cost matrix and analyses it by searching for the reconstruction with minimal global cost. The ParaFit method (Legendre et al., 2002) implies the four-corner approach (Legendre et al.,

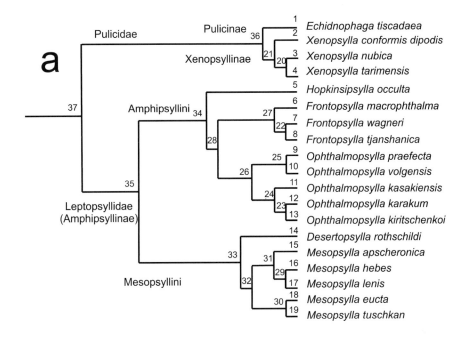

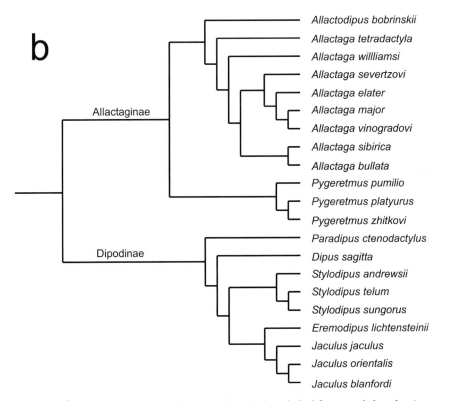

Figure 4.2 (a) Phylogeny of fleas parasitic on jerboas derived from morphology data (see Krasnov & Shenbrot, 2002 for details). Numbers above branches are additive binary coding used for the BPA. (b) Phylogeny of jerboas based on phallic morphology, the coronal structure of molars and bullar morphology (modified after Shenbrot et al., 1995). Redrawn after Krasnov & Shenbrot (2002) (reprinted with permission from Israel Science Journals).

1997) and tests the significance of the global null hypothesis that associations of host and parasite species are random and the evolution of the two groups was independent.

Availability of these new analytical methods has led to numerous studies of host–parasite cophylogeny on a variety of different host–parasite systems (Hafner & Page, 1995; Desdevises *et al.*, 2000, 2002a; Paterson *et al.*, 2000; Ronquist & Liljeblad, 2001). However, to the best of my knowledge, only two studies have applied these methods for fleas.

Krasnov & Shenbrot (2002) studied historical patterns of the association between fleas and jerboas (Dipodidae) using all the above-mentioned methods. In general, the conclusion was that host-switching was common in these associations. For example, a phylogenetic host tree reconstructed by the BPA (Fig. 4.3) was incongruent with the known jerboa phylogeny (Fig. 4.2b), but suggested that the common evolutionary history of fleas and jerboas was characterized mainly by inertia, host-switching and sorting events. An inertia event was exemplified by *Frontopsylla wagneri* parasitizing two closely related *Allactaga* species (*Allactaga sibirica* and *Allactaga bullata*, both belonging to the subgenus *Orientallactaga*). An example of a host-switching event was represented by *Xenopsylla* fleas occurring mainly on dipodine jerboas with one species switching to *Allactaga tetradactyla* belonging to another lineage (Allactaginae). Finally, it is likely that *Xenopsylla* was lost by a sorting event from *Jaculus blanfordi*. Additionally, there was a geographical pattern in parasite-based jerboa trees. Species with common geographical distributions and/or habitat preferences or common geographical origin tended to cluster together (Fig. 4.4). Species of sandy habitats of Turan (*Eremodipus lichtensteinii*, *Paradipus ctenodatylus*, *Dipus sagitta*) and those of non-sandy habitats of Turan (*Allactaga vinogradovi*, *Allactaga bobrinskii*, *J. blanfordi*) formed separate clades as did the North African jerboas (*Jaculus jaculus*, *Jaculus orientalis*, *A. tetradactyla*), Kazakhstanian and Minor Asian jerboas (*Pygeretmus pumilio*, *Pygeretmus platyurus*, *Pygeretmus zhitkovi*, *Stylodipus telum*, *Allactaga williamsi*, *Allactaga elater*), and jerboas of the Mongolian origin (*A. sibirica*, *A. bullata*).

The study suggested that the evolutionary history of the jerboa–flea associations involved association by colonization with frequent host-switching and linear sorting events, whereas widespread cospeciation was absent. The distribution of fleas on jerboas is, therefore, affected mainly by ecological and geographical factors. These factors can allow host-switching and override any tendency towards the strict cospeciation expected from the transmission mode of fleas.

For example, the distribution of pulicid fleas on jerboas is difficult to explain without invoking host-switching that occurred both among jerboas and between jerboas and other rodent taxa. An African origin of Pulicidae is commonly accepted (Traub, 1985; Medvedev, 1998), whereas both allactagine and dipodine

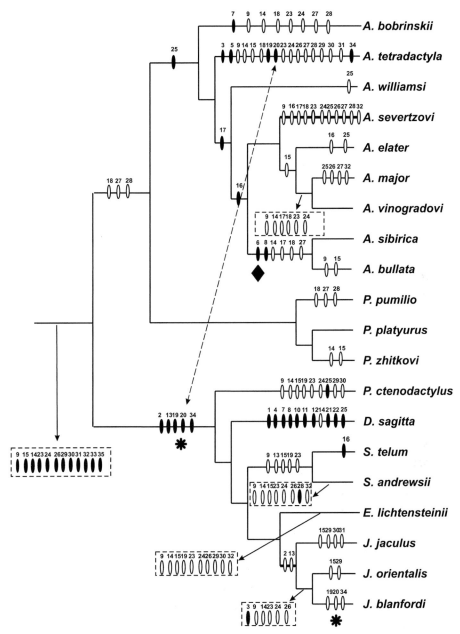

Figure 4.3 Additive binary codes of the flea phylogenetic tree (Fig. 4.2a) optimized onto the jerboa phylogenetic tree (Fig. 4.2b). Gains (filled ovals) and losses (open ovals) of 'characters' are shown. Asterisks, diamond, and dashed arrowed line illustrate examples of sorting event, inertia event, and host-switching event, respectively. Redrawn after Krasnov & Shenbrot (2002) (reprinted with permission from Israel Science Journals).

40 Origin and evolution of fleas

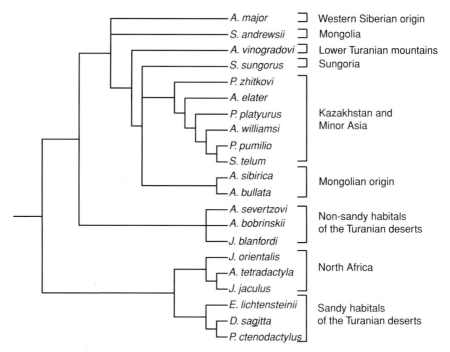

Figure 4.4 Strict consensus of three trees produced by the BPA for species of jerboas from additive binary coding of the distribution and phylogeny of their flea parasites. Redrawn after Krasnov & Shenbrot (2002) (reprinted with permission from Israel Science Journals).

jerboas originated in Central Asia (Shenbrot *et al.*, 1995). Two possible, not mutually exclusive, scenarios can be suggested. In the first, ancestors of the dipodine genus *Jaculus* dispersed to Africa (Black & Krishtalka, 1986) where they were presumably colonized by pulicids. Some species returned to Asia with these new parasites that then switched onto the sympatric dipodine *D. sagitta*. In the second scenario, pulicids colonized jerboas switching from Gerbillinae which originated in Africa and dispersed to Asia no later than in the Miocene (Wessels, 1998).

Another study compared the phylogeny of fleas from the genus *Geusibia* and their ochotonid hosts (pikas) (Lu & Wu, 2005). Reconciliation of the phylogenetic trees demonstrated that, as with jerboas, there was no strong evidence of cospeciation. Instead, host-switching seemed to be the main type of event during evolutionary history of this association.

Both these studies, together with evidence presented by Traub (1980, 1985), suggest that although fleas coevolved with their hosts they did not cospeciate with them.

4.5 Flea diversification: intrahost speciation, host-switching or climate?

Rodents harbour the highest diversity of fleas, which suggests that the most intensive flea diversification was associated with high diversification of Rodentia, i.e. from the Eocene. Hypothetically, this happened in a temperate zone (Medvedev, 1996). The geographical distribution of modern flea species supports this hypothesis as most flea species are distributed in regions of temperate and subtropical climate (see above) and predominate on mountain landscapes.

As was mentioned above, studies on flea–host cophylogeny (although not numerous) suggest that flea diversification did not result from cospeciation with their hosts. Consequently, flea diversification (increase in species diversity within a host lineage) over evolutionary time could result mainly from two different types of evolutionary event (Poulin, 2007a). First, a new flea species can be acquired via colonization from a different host lineage (host-switching) (e.g. Krasnov & Shenbrot, 2002). Second, the flea taxon can speciate on a host without an accompanying host speciation event and can, thus, produce multiple closely related flea lineages on the host's descendants (duplication) (e.g. Riddoch et al., 1984). Mouillot & Poulin (2004) proposed that the relative importance of these two processes in shaping the diversification of parasite assemblages can be indicated by the value of the exponent of the power relationship between the number of higher taxa (e.g. genera) and species richness. If this value across several comparable parasite assemblages is close to 1, this would indicate that host-switching was the main cause of diversification. Indeed, if each species in an assemblage is taxonomically independent of the other species, it must therefore have had a separate origin. In contrast, an exponent below 1 indicates that several species belong to the same genus or genera, suggesting that they have a common ancestor and that they may have radiated from this common ancestor within a host lineage. In other words, if the number of host switches in the past is approximately equal to the current number of species (the exponent is close to 1), diversification of parasite assemblages stems mainly from host-switching. If, however, there are currently much more species than there were host switches (the exponent is much less than 1), the diversification of parasite assemblages is likely the result of intrahost parasite speciation.

In woody plant communities, Enquist et al. (2002) found that these exponents were statistically invariant across and within biogeographical regions, types of plant physiognomy and geological time. However, this appeared not to be true for assemblages of intestinal helminth parasites, where the value of the exponents varies according to the identity of the vertebrate host taxa (Mouillot & Poulin, 2004).

42   Origin and evolution of fleas

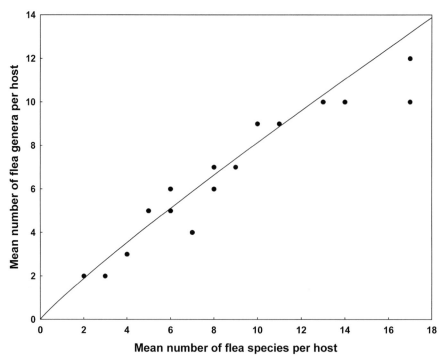

Figure 4.5 Relationship between the number of flea species and flea genera per host across 19 small mammalian host species in the Novosibirsk Region of Siberia. Data from Violovich (1969), analysis from Krasnov et al. (2005b).

What about fleas? The relationships between the number of flea species and the numbers of flea genera, tribes, subfamilies or families on a host species across 25 Holarctic regions, as well as in each region, were found to be well described by simple power functions (Krasnov et al., 2005b) (see Fig. 4.5 for an example from the Novosibirsk Region of Siberia). Furthermore, the relationship between the number of flea species and the number of flea genera per host was stronger (i.e. the exponent was higher) than that between the number of flea species and the number of flea tribes, subfamilies or families. The relationship between the number of flea species and the number of flea tribes per host was stronger than that between the number of flea species and the number of flea subfamilies, whereas the relationship between the number of flea species and the number of flea subfamilies per host was stronger than that between the number of flea species and the number of flea families. All this was true for both the whole data set and each region separately.

The exponents of species–genera relationships for flea assemblages ranged from a low of 0.74 to a high of 0.95. Moreover, values greater than 0.92 were found in only four of the 25 regions, and were lower than 0.88 in as many as 16 of the 25 regions. Thus, the values of the exponents were, in general, lower that those found by Enquist et al. (2002) for plant communities (0.94) and by Mouillot & Poulin (2004) for the communities of helminth parasites in fish and bird hosts (0.97 and 0.92, respectively), but were close to, albeit somewhat higher than, those for helminth parasites of mammals (0.83) (Mouillot & Poulin, 2004). This suggests that intrahost speciation seems to have played an important role in flea diversification. Although host-switching was suggested to have been the main event in the evolutionary history of flea–jerboa and flea–pika associations (see previous section), it seems that these two studies are not sufficient to make general conclusions about the main type of events during flea–host coevolution. Comparison of the values of exponents for flea–mammal and helminth–mammal associations also hints at similar mechanisms influencing the rate of intrahost speciation of ecto- and endoparasites in mammals, and these mechanisms can be related to still-unknown host features. However, the lack of invariance of the exponent value of the power function across different regions (0.74–0.95), in contrast to that found for plant communities (Enquist et al., 2002), suggests that some local conditions might strongly affect fundamental processes and mechanisms of diversification. These local conditions can be related, for instance, to climate.

The effect of climate on the evolution of parasite species diversity has been explained mainly by the assumption that higher energy input (e.g. measured as local solar radiation or temperature) determines evolutionary rates (Rohde, 1992, 1999). Presumably, a greater input of solar energy leads to faster evolution (Rohde, 1992). If this is so, we might expect that in relatively colder regions, the main way for a parasite assemblage to diversify is via host-switching, and this should lead to roughly only one species per genus on any given host species. In contrast, the number of species per genus can be expected to increase in relatively warmer regions, where warmer temperatures favour speciation (Rohde, 1992). Consequently, a negative relationship between local mean annual temperature and the value of the exponent of the power function between the number of species and the number of higher taxa per host species could be expected. This was the case for fleas (Fig. 4.6) (Krasnov et al., 2005b), suggesting that multiple congeneric species of fleas parasitic on the same host species occurred mainly in warmer regions.

The increase of evolutionary rates may be the outcome of an increase in mutation rate, the acceleration of physiological processes and/or shortened generation time (Rohde, 1992). All these can explain, at least partly, why flea assemblages in warmer rather than in colder regions diversified more via intrahost speciation. It cannot explain, however, why flea assemblages in the colder regions diversified mainly by host-switching, especially given that the dispersal abilities of

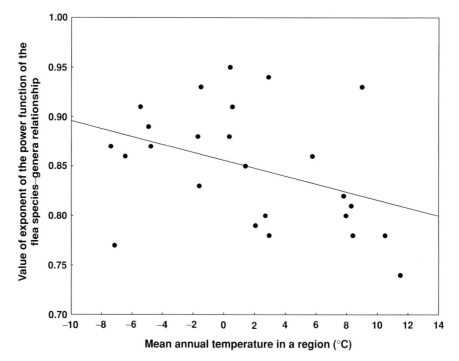

Figure 4.6 Relationship between mean annual temperature in a region and the value of the exponent of the power function between the number of flea species and the number of flea genera per host species, across 25 Holarctic regions. Redrawn after Krasnov *et al.* (2005b) (reprinted with permission from Blackwell Publishing).

parasite species are restricted at lower temperatures (Rohde, 1985, 1992, 1999). Nevertheless, flea transfers from host to host often occur when hosts visit each other's burrows (e.g. Ryckman, 1971). Rodent burrows in temperate and colder regions are deeper, more complicated and more frequently visited than those in warmer regions (Kucheruk, 1983). These processes can facilitate host-switching by fleas, independent of the effects of air temperature on the mobility of fleas.

4.6 Concluding remarks

Pathways of flea evolution and their transition to parasitism are still unclear. Scarcity of both fossil records and analytical studies of flea–host phylogeny does not allow us to determine explicitly (a) which features fleas inherited from their free-living ancestors; (b) which features evolved in fleas as adaptations to parasitic way of life; and (c) what were the main types of event during coevolution of fleas and their vertebrate hosts. Nevertheless, it is clear that (a) although fleas coevolved with their hosts, they did not cospeciate with them; and (b) local environmental conditions had a strong effect on flea evolution.

5
Life cycles

Fleas are typical holometabolous insects. The life cycle of any flea species consists of an egg that hatches into a larva, which generally undergoes three larval moults and an inactive pupal stage before emerging as an adult. Although Siphonaptera is a small order, fleas demonstrate high variability in some life-cycle details such as diversity of larval microenvironmental preferences, larval feeding mechanisms, and nutritional requirements. In this chapter, I present some information on flea life cycles.

5.1   Mating and oviposition

Newly emerged female fleas have underdeveloped ovaries blocked with a follicular plug (Kunitskaya, 1960, 1970; Vashchenok, 1966a), whereas newly emerged males of many species have a testicular plug which prevents the passage of sperm from the testes to the vas deferens (Akin, 1984; Dean & Meola, 1997). In general, the blood meal is a trigger for the development of ovaries in female fleas (Vashchenok, 1988; Liao & Lin, 1993) and for the dissolution of the testicular plug in males (Rothschild *et al.*, 1970; Kamala Bai & Prasad, 1979; Akin, 1984). Consequently, the majority of fleas mate after feeding. This was established, for example, for *Leptopsylla segnis* and *Leptopsylla taschenbergi* (Kosminsky, 1960) and *Citellophilus tesquorum* (Bryukhanova, 1966). Unfed *Nosopsyllus fasciatus* usually do not mate, but can be forced to copulate by increased air temperature (Iqbal & Humphries, 1970). Occasionally, only one sex has to be fed prior to copulation (e.g. *Echidnophaga gallinacea*: Suter, 1964; Marshall, 1981a). However, some species can mate immediately after emergence or, at least, prior to the first blood meal. In particular, copulation prior to feeding has been reported for *Ceratophyllus hirundinis* (Holland, 1955), *Ceratophyllus gallinae* (Humphries, 1967a), *Nosopsyllus mokrzeckyi* (Kosminsky, 1961), *Nosopsyllus consimilis* (Alekseev, 1961), *Ctenophthalmus dolichus* (Bgytova, 1963) and *Neopsylla setosa* (Bryukhanova, 1966). However, the reproductive output of such copulations is extremely low (if there is any at all). For example, unfed male *Ctenocephalides felis* are unable to inseminate females (Dean & Meola, 1997), whereas unfed females are unable to produce eggs (Zakson-Aiken *et al.*, 1996). Although females of several *Ceratophyllus* species sometimes copulate prior to the first blood meal, their eggs mature only after feeding (Holland, 1955; Darskaya, 1964; Humphries, 1967a). An exceptional example is represented by *Tunga monositus* in which a newly emerged male inseminates a female and dies without feeding (Barnes & Radovsky, 1969). However, males of other *Tunga* species require a blood meal before copulating (Witt *et al.*, 2004).

Mating may occur only on the host or both on-host and off-host. The former behaviour is characteristic, for example, for *L. segnis* and *L. taschenbergi* (Kosminsky, 1961), *Pectinoctenus pavlovskii* (Vasiliev, 1961), *Spilopsyllus cuniculi* (Rothschild & Ford, 1966) and *C. felis* (Hsu & Wu, 2000), whereas the latter behaviour is exemplified by *Xenopsylla cheopis* (Vashchenok, 1988). Copulation in some flea species is associated with courtship behaviour. Examples of the mating behaviour of fleas were reported for *C. gallinae* (Humphries, 1967a, b), *N. fasciatus* (Iqbal & Humphries, 1970, 1974, 1976) and *C. felis* (Hsu & Wu, 2001). Both sexes are active during copulation in most species (e.g., *C. gallinae*: Humphries, 1967a; and *C. felis*: Hsu & Wu, 2001), whereas only the male is active in species with sessile females (Marshall, 1981a). Mating pheromones are commonly thought to be involved in the mating behaviour of fleas (Iqbal, 1973; Yue *et al.*, 2002; Eisele *et al.*, 2003).

Multiple matings were recorded in some fleas such as *E. gallinacea* (Suter, 1964), *C. gallinae* (Humphries, 1967a), *N. fasciatus* (Iqbal & Humphries, 1976), *X. cheopis* (Tchumakova *et al.*, 1978), *C. tesquorum* (Tchumakova *et al.*, 1978) and *C. felis* (Hsu & Wu, 2000). For example, multiple-mated females of *C. felis* displayed higher fecundity and fertility than single-mated females, suggesting that multiple matings are advantageous (e.g. Hsu & Wu, 2000). Yue *et al.* (2002) recorded as many as 48 mating events during 8 h in a single male and as many as 27 mating events during 7 h in a single female *C. felis*. A positive effect of multiple matings on the fecundity of fleas can explain the increase in size of successive clutches reported for some flea species (e.g. *X. cheopis* and *C. tesquorum*: Tchumakova *et al.*, 1978; and *N. consimilis*: Vashchenok, 1988). It is unclear though, whether this increase was due to multiple matings or multiple blood meals as in, for example, *C. tesquorum* (Starozhitskaya, 1968).

After successful mating and fertilization, female fleas begin oviposition. Clutch size differs considerably among species, ranging from one or two in *Xenopsylla* species (Vashchenok, 1988; Krasnov *et al.*, 2004a) to several tens or hundreds in the chigoe fleas such as *Tunga penetrans* (Geigy & Herbig, 1949). For example, two flea species cohabitating in the Negev Desert but parasitic on different hosts differ in the size of their clutches. Female *Xenopsylla dipodilli* (parasite of Wagner's gerbil *Dipodillus dasyurus*) usually lays no more than two eggs, whereas female *Parapulex chephrenis* (parasite of the Egyptian spiny mouse *Acomys cahirinus*) oviposits when up to eight eggs have accumulated in the oviducts (Krasnov *et al.*, 2002a). One of the reasons for this is a difference in the size of the eggs (an egg of *X. dipodilli* is about three to four times larger than that of *P. chephrenis*). Indeed, species with very large eggs (e.g. *Sphinctopsylla ares* and species of *Hystrichopsylla*) never have more than two eggs within the oviduct at any one time. The chorions of these eggs are heavily chitinized, and their surfaces are characteristically structured (Linley *et al.*, 1994; M. W. Hastriter, personal communication, 2006).

The rate of oviposition varies both among and within species under influence of many host- and environment-related factors (see Chapter 11). Clutch size changes during the lifespan of a female flea (e.g. *C. felis*: Osbrink & Rust, 1984). Initially it increases sharply, but decreases gradually with age (Table 5.1; see Chapter 11).

Lifetime fecundity may vary among flea species from a low of 100 eggs (e.g. *X. cheopis*: Samarina *et al.*, 1968) to a high of many thousands (e.g. *T. penetrans*: Barnes & Radovsky, 1969). The highest lifetime fecundity is found among the sessile and stick-tight fleas. For example, lifetime egg output of *Dorcadia ioffi* attained 2479 (Grebenyuk, 1951), whereas that of *T. monositus* was about 1000 (Lavoipierre *et al.*, 1979). Darskaya *et al.* (1965) proposed a classification of flea species according to their daily egg output. They distinguished three groups, namely fleas with: (a) 0.6–3.6 eggs per female per day (*Stenoponia, Rhadinopsylla, N. setosa, Paradoxopsyllus repandus*); (b) 3.6–5.6 eggs per female per day (*Nosopsyllus laeviceps, Nosopsyllus iranus, N. mokrzeckyi, N. consimilis, X. cheopis, Xenopsylla hirtipes, Xenopsylla gerbilli*); and (c) 4–13 eggs per female per day (*L. segnis, L. taschenbergi, C. tesquorum, Nosopsyllus tersus, Xenopsylla conformis*). However, the overlap of the egg production values between the three groups makes this classification not especially useful. Nevertheless, it is interesting that group (a) is mainly composed of species that spend most of their time in the host nest/burrow rather than on the host body ('nest' fleas; see Chapter 9), whereas fleas of group (c) spend most of their life in the host's fur ('body' fleas; see Chapter 9). Egg production may vary also within species (e.g. clutch size varies from one to nine in *N. consimilis*: Vashchenok, 1967a) depending on such factors as season (Bibikova *et al.* (1971) for *Xenopsylla skrjabini* and *X. gerbilli*; Table 5.1), host species and environmental conditions. The most famous cases of the effect of a host on flea reproduction are represented by the rabbit fleas *S. cuniculi* and *Cediopsylla simplex*. In these fleas, reproduction is related to the reproductive and, consequently, hormonal cycle of their hosts, the European rabbit *Oryctolagus cuniculus* and the eastern cottontail

Table 5.1 *Fecundity (mean number of eggs per female per day) in five flea species in relation to female age*

Species	Age (days)	Fecundity	Reference
Leptopsylla segnis	2–5	14.0	Vashchenok, 2001
	41–45	3.7	
Xenopsylla gerbilli	5–6	6.2	Bibikova et al., 1971
(spring generation)	8–10	3.0	
	15–17	1.5	
Xenopsylla gerbilli	10	6.9	Bibikova et al., 1971
(summer generation)	21	5.7	
	25	4.5	
Xenopsylla skrjabini	9–10	1.3	Bibikova et al., 1971
(spring generation)	20	0.0	
Xenopsylla skrjabini	9–10	2.0	Bibikova et al., 1971
(summer generation)	20	0.0	
Xenopsylla skrjabini	6–7	8.8–9.8	Bibikova et al., 1971
(autumn generation)	18	4.0	
Xenopsylla skrjabini	1–5	5.5	Korneeva & Sadovenko, 1990
(laboratory colony)	6–20	8.1	
	21–30	2.4	
	>31	0.2	
Xenopsylla conformis	3–5	1.0	Grazhdanov et al., 2002
	7–8	5.5	
	19–21	2.0	
Nosopsyllus laeviceps	3–4	2.0	Grazhdanov et al., 2002
	15–16	12.0	
	40–44	2.3	
	55–60	0.5	

Sylvilagus floridanus, respectively. Development of the reproductive system, mating, egg maturation and oviposition in these fleas has been shown to be triggered by sex hormones of the pregnant doe rabbits and their kittens (Mead-Briggs, 1964; Rothschild & Ford, 1966, 1969, 1972, 1973; Sobey et al., 1974). The effect of host- and environment-related factors on the reproductive performance of fleas will be discussed later in this book (Chapter 11).

5.2 Larvae

Flea larvae are maggot-like creatures with their morphological characters being surprisingly similar across different species of the order. However, the ecology of larvae is quite variable within the order (e.g. Benton *et al.*, 1979).

The location of larval development varies greatly among flea species. In the majority of species, especially those parasitic on burrow-dwelling hosts, larvae develop mainly off the host in the substrate of host's nest and/or burrow. Moreover, fleas of burrowing mammals usually oviposit in the deepest parts of the host burrows and, thus, larvae develop in the environment where microclimate is relatively stable (e.g. Sokolova *et al.*, 1971). These larvae are usually negatively phototactic and positively geotactic (Sgonina, 1935). In contrast, larvae of vermipsyllid fleas (e.g. *Vermipsylla alakurt*) parasitic on ungulates typically drop from the host body and develop off-host and not in any particular shelter. Similarly, *Tunga* larvae mainly develop in the soil (although larvae of *T. penetrans* were found feeding on host blood near the sites of embedded imagoes: Faust & Maxwell, 1930). Larvae of other species develop on the body of a host, either facultatively as in *Oropsylla silantiewi* (Dubinina & Dubinin, 1951; Zhovty, 1970) or obligatorily as in *Uropsylla tasmanica* (Dunnet & Mardon, 1974).

Among species, variation in larval feeding is startling by the range rather than by diversity of feeding modes. Larvae of most flea species feed on organic debris found in the nest/burrow of the host or in open habitats. They may also feed on flea faeces (e.g. Silverman & Appel, 1994) and even on conspecific or heterospecific eggs, younger larvae and naked pupae (Reitblat & Belokopytova, 1974; Lawrence & Foil, 2002). Moreover, in some species, females have been shown to expel faecal pellets near the clutch which can later serve as a food source for larvae. This behaviour is characteristic, for example, of the cat fleas (Hinkle *et al.*, 1991; Silverman & Appel, 1994) as well as of *Ctenophthalmus nobilis* (Cotton, 1970a) and *Monopsyllus sciurorum* (Larsen, 1995). Furthermore, Hinkle *et al.* (1991) reported that the protein content of flea faeces was actually higher than the bovine blood upon which they fed. Nutritional necessity of faeces of adult fleas for larval development has led to the suggestion that this phenomenon reflects a unique form of parental investment exhibited in some fleas (Hinkle *et al.*, 1991). The amount of proteins in faecal pellets of male fleas is similar to that of females, so male pellets can also be used by larvae (Shryock & Houseman, 2005).

Larvae that develop on a host supposedly feed on derivatives of host's skin and hairs as well as excrements of imago fleas (*O. silantiewi*: Zhovty, 1970). During the winter, *O. silantiewi* oviposits in the pelage of a hibernating host (usually a marmot) and, apparently, the larvae feed not only on host's skin derivatives but also on its blood (Suntsov & Suntsova, 2003). A similar phenomenon was described for larvae of *Oropsylla alaskensis* (Vasiliev & Zhovty, 1971), a flea parasitic on the ground squirrels *Spermophilus undulatus* and *Spermophilus dauricus*.

Females oviposit only after the start of a host's hibernation. Larvae migrate to the snout and oral cavity of the rodent, pierce the tongue, gums and epithelium of the oral cavity, and consume blood. Vasiliev & Zhovty (1971) argued that host blood is the only food source for larval *O. alaskensis* and, thus, their parasitism is obligatory. In contrast, Suntsov & Suntsova (2006) reported that larval *O. silantiewi* fed on blood during especially cold winters and/or at high altitudes. Whether larval parasitism of *Oropsylla* is facultative or obligatory, it undoubtedly hints on the evolutionary potential of fleas to tighten their association with hosts. Moreover, observations on the feeding mode of *Oropsylla* larvae have even led to a new (however, the most bizarre) hypothesis of the origin of plague, a disease that is specifically transmitted by flea vectors (Suntsov & Suntsova, 2003, 2006). According to this hypothesis, larval parasitism of *Oropsylla* facilitated the evolution of *Yersinia pestis* (a causative agent of plague) as a blood parasite of a vertebrate host from ancestors that were parasites of a gastrointestinal tract (as all other *Yersinia* are). Larvae of *Parapsyllus heardi* that parasitize the blue petrel *Halobaena caerulea* and the thin-billed prion *Pachyptila belcheri* on the Kerguelen Islands, and *Glaciopsyllus antarcticus*, an endemic flea from Antarctica found on the southern fulmar *Fulmarus glacialoides*, live on the ventral surface of the body of chicks and feed on blood (Bell et al., 1988; Whitehead et al., 1991; Chastel & Beaucournu, 1992). Larvae of *Euchoplopsyllus glacialis*, a parasite of the Arctic hare *Lepus arcticus*, are also blood feeders (Freeman & Madsen, 1949).

Larvae of the pygiopsyllid *U. tasmanica* are really parasitic. This flea parasitizes various Dasyuromorphia in Tasmania and Australia (Victoria). Larvae burrow into the skin of a host and remain attached until pupation and emergence. Body shape in *U. tasmanica* larva is adapted to parasitism; the anterior segments are expanded, adorned with annular rows of curved spines, and the abdominal segments are reduced (Williams, 1991). This reduction brings the last abdominal spiracle to the posterior end of the larva, supposedly allowing better access to the atmospheric oxygen.

Finally, at the extreme end of the spectrum of larval feeding are aphagous larvae of *T. monositus*. The female of this species is neosomic (see below) and is situated under the skin of a host. The size of the neosome is much larger that that of a newly emerged female, permitting the production of eggs that are larger than a newly emerged imago. The large size of the egg accommodates extra yolk nutrients which provide increased fat body. This, in turn, makes possible (a) aphagous larvae and (b) a shortened larval stage that consists of two instars only (there are three larval instars in other fleas) (Barnes & Radovsky, 1969; Marshall, 1981a).

As is the case with eggs, the survival and rate of development of larvae depend on a variety of both host- and environment-related factors as well as the structure

Table 5.2 *Mean (±S.E.) CO_2 emission at 25°C in* Xenopsylla conformis

Stage	Mean body mass (mg)	CO_2 emission (μl h^{-1} mg^{-1})
Larva	0.19 ± 0.06	0.478 ± 0.039
Pupa	0.20 ± 0.04	0.199 ± 0.023
Newly emerged imago (fed)	0.26 ± 0.03	1.075 ± 0.184
Imago (fed)	0.29 ± 0.02	0.837 ± 0.73

Source: Data from Fielden *et al.* (2001).

of larval community (density and species composition). These factors will be discussed further.

5.3 Pupae

The third-instar larva, on completion of feeding, expels all of its gut content, spins a silken cocoon and becomes camouflaged by adhering to itself particles of the surrounding substrate. Prior to this, genitalia start to appear (Qi, 1990a, b). Silk threads are produced by modified salivary glands (Mironov & Pasyukov, 1987). Due to sharp microclimatic fluctuations, some larvae may leave their cocoons soon after construction and then construct new cocoons (Mironov & Pasyukov, 1987; Krasnov *et al.*, 2001a). This behaviour varies among flea species. For example, *X. conformis* builds these additional cocoons more often than *Xenopsylla ramesis* (Krasnov *et al.*, 2001a).

A cocoon constitutes the protective microenvironment for a flea pupa (Edney, 1947a). As a result, pupae are thought to be resistant to, for example, low relative humidity (at least, in terms of survival: Krasnov *et al.*, 2001a). However, pupae are unable to absorb atmospheric water via their rectal sac at low humidity (Edney, 1947b). As a result, pupae can have significantly higher water loss rates than, for example, adult fleas (see Fielden *et al.* (2002) for *X. conformis*). This relatively high water loss may be compensated by the relatively low metabolic rate of pupae (Fielden *et al.*, 2001) (Table 5.2). Indeed, the quiescent adult of *C. felis* within the cocoon has a lower respiratory demand than the emerged adult, and its survival is considerably longer under low humidity conditions (Silverman & Rust, 1985; Metzger & Rust, 1997). Nevertheless, the pupae of *X. conformis* appeared to be rather sensitive to low humidity (Krasnov *et al.*, 2001a).

Some fleas spin soft cocoons, whereas in other species it is firm and durable (Bacot, 1914); this depends on species-specific width and strength of the silk threads and on species-specific behaviour of coating the internal wall of the cocoon with a substance secreted by the gut epithelium and/or Malpighian tubules (Prokopiev, 1969; Mironov & Pasyukov, 1987). The protective characteristics of soft cocoons are inferior to those of hard cocoons (e.g. in *C. dolichus*: Zolotova, 1968). Soft cocoons are characteristic for the 'body' fleas, whereas firm cocoons are mainly found in the 'nest' fleas (Prokopiev, 1969; Medvedev, 2005).

The texture of substrate that pre-imaginal fleas use to strengthen the cocoon is important for successful development. For example, loess absorbs water better than sand so that the hydrothermic properties of desert loess soils are similar to those of meadow soils (Korovin, 1961). This leads to an increase in humidity that has been shown to be very important in fleas (Sharif, 1949; Smith, 1951; Yinon *et al.*, 1967; Krasnov *et al.*, 2001a). As a result, the duration of pupal stages of some species in sand substrate is longer than that in loess substrate (e.g., X. *conformis* and X. *ramesis*: Krasnov *et al.*, 2002b).

5.4   Imago

Emergence of fleas from cocoons can be triggered by a number of environmental factors such as vibration (Cotton, 1970b), rise of temperature (Hůrka, 1963a, b) and increase in the ambient air CO_2 concentration (Marshall, 1981a). Without external stimuli some fleas can remain in cocoons for long periods of time (more that 1.5 years: Bacot, 1914). In contrast, other fleas emerge from cocoons without any obvious stimulus (Tipton & Méndez, 1966).

Fleas emerge from cocoons using their hind legs and frontal tubercle (Prokopiev, 1969; Amrine & Lewis, 1978). The size and shape of the frontal tubercle tend to be correlated with the structure of the cocoon. Frontal tubercles in fleas that spin soft cocoons (e.g. *Neopsylla*, *Catallagia* and *Ctenophthalmus*) are usually small, while fleas with firm cocoons (e.g. *Ceratophyllus*) possess well-developed frontal tubercles. In some species, the frontal tubercle is absent (e.g. *Acropsylla*), whereas Ischnopsyllidae and Vermipsyllidae are characterized by 'deciduous' frontal tubercles that fall off soon after emergence (Jordan, 1945; Medvedev, 1989a). Nevertheless, de Albuquerque Cardoso & Linardi (2006) ruled out the possibility that the main function of the frontal tubercle is rupturing the pupal capsule, since it is not present in all fleas and may vary even within a genus. Instead, they suggested that it may have a mechano-, thermo- or chemoreceptory function.

There is a great variation in imago body size among flea species. The length of the body varies from about 1.5–2 mm (e.g. *Xenopsylla*) via about 3–4 mm (e.g. *Stenoponia*) to about 7–10 mm (e.g. *Hystrichopsylla kris*: Hastriter & Haas, 2005). Neosomic females of Tungidae and Vermipsyllidae can be even larger.

For example, the body of a female *D. ioffi* with matured eggs may be as long as 16 mm (Ioff et al., 1965). Ecological and evolutionary reasons for interspecific variation in flea body size are largely unknown. For example, a positive relationship between parasite and host body size (so-called Harrison's rule) has been shown for some parasite taxa such as rhizocephalans (Poulin & Hamilton, 1997), chewing lice (Morand et al., 2000) and sucking lice of the genus *Columbicola*, whereas this relationship did not hold for other parasite taxa such as copepods (Poulin, 1995a) and sucking lice of the subfamily Physconelloidinae (Johnson et al., 2005). However, little is known about this pattern in fleas. Kirk (1991) argued that flea body size correlates positively with host body size. However, he analysed only highly host-specific fleas from the British Isles. This limitation could introduce a bias in the analysis. Indeed, comparative analysis of the relationship between body size of 19 flea species from South Africa and 46 flea species from North America and mean body size of their hosts demonstrated that this pattern did not hold (B. R. Krasnov, R. Poulin and S. Morand, unpublished data).

The lifestyle of adult fleas differs greatly among species. Some species spend their life mainly in the host nest or burrow and visit a host for a blood meal only (many bird and rodent species). Other species spend their life mainly on a body of a host but nevertheless are mobile (bat fleas, rabbit fleas, ungulate fleas). In the stick-tight fleas (*Echidnophaga*), females are permanently attached to host by the mouthparts. In chigoes (Tungidae), females are permanently buried under the host's skin. Some researchers have theorized that the sessile lifestyle evolved as a response to the high risk of being detached from a host. For example, Smit (1987) suggested that the sessile lifestyle of Malacopsyllidae is related to their preference for being attached to the ventral side of their armadillo hosts and evolved as an adaptation to withstand brushing against the substrate.

Females of some fleas (Tungidae, Vermipsyllidae, Malacopsyllidae) are neosomic (Audy et al., 1972; Marshall, 1981a; Rothschild, 1992). Neosomy (a transformation of shape with an ability to produce new cuticle without moulting) and the sessile lifestyle in fleas evolved secondarily (evolutionary 'after-thought': Rothschild, 1992) and independently in several families. For example, after emergence, a female *T. penetrans* penetrates the host epidermis and undergoes hypertrophy, becoming a neosome (Audy et al., 1972; Witt et al., 2004). It produces thousands of eggs during a 3-week period and expels them through the posterior abdominal segments extending above the stratum corneum of the host skin (Linardi & Guimarães, 2000). After oviposition, the neosome involutes and the flea dies, being sloughed from the host epidermis by tissue-repair mechanisms (Eisele et al., 2003).

The longevity of the imago depends on a variety of factors such as host species, availability of food during the larval stage, density of larvae, microclimate (temperature and relative humidity) and its fluctuations, substrate, and feeding activity. In general, the longevity of starving fleas is thought to increase at low air temperatures and high relative humidities (Miller & Benton, 1970; Kunitsky et al., 1971a; Larson, 1973; Kosminsky & Udovitskaya, 1975; Talybov, 1975; Ma, 1990, 1993a, 1994a), although there are some conflicting reports (Parman, 1923; Allan, 1956; Suter, 1964). In addition, it has been reported that some fleas can withstand remarkably low air temperatures (e.g. *Ceratophyllus idius*: Larson 1973). Nevertheless, the duration of lifespan seems to be species-specific. For example, under similar air temperatures and relative humidities, starving *N. fasciatus* and *Pulex irritans* survived to 95 and 135 days, respectively, although regularly fed *P. irritans* lived up to 513 days (Bacot, 1914). Starving *Neopsylla bidentatiformis* survived to 227 days (Moskalenko, 1963a). After a single blood meal, *Ctenophthalmus breviatus* and *N. setosa* survived to 715 and 1725 days, respectively (Tiflov & Ioff, 1932). In contrast, starving *X. conformis* and *X. ramesis* survived only 66 and 68 days, respectively (Krasnov et al., 2002b). Starving 'body' fleas usually survive less time than starving 'nest' fleas. For example, under the same air temperatures and relative humidities, newly emerged unfed 'body' *L. segnis* survived one-quarter of the time of newly emerged unfed 'nest' *N. bidentatiformis* (30 versus 120 days, respectively: Moskalenko, 1963a). Newly emerged unfed fleas live longer than fed fleas (Leeson, 1936; Edney, 1945; Krasnov et al., 2002b; Ma, 2002), which may be related to the higher mass-specific metabolic rates of fed fleas (Fielden et al., 2001) (Table 5.2). This demands an increase in oxygen requirements which leads to an increase in spiracular openings and, consequently, to water loss from the tracheal system (Bursell, 1974). In the field, though, flea lifespan is much shorter and varies less among species than in the laboratory. Vashchenok (1988) noted that among most flea species, the variation in the lifespan in the field ranges from several weeks to 2–3 months, although *Xenopsylla* species parasitic on gerbils in Central Asia and fleas of some hibernating hosts live, on average, 8–9 months.

5.5   Seasonality

The survival and reproduction of fleas are dependent on a combination of factors including favourable climatic conditions for development of the immature stages and for adults to survive periods without a blood meal (see details in Chapter 10). This dependence results in seasonal changes of life-history parameters of fleas (abundance, reproduction rate, pattern of parasitism etc.). The annual cycle of a particular flea species in a particular locality corresponds with seasonal climatic fluctuation.

5.5.1 Examples of annual cycles

Two examples of annual cycles are presented below for two fleas parasitic on Wagner's gerbil *D. dasyurus* in the Negev Desert (Krasnov et al., 2002c). *Xenopsylla dipodilli* is an all-seasonal flea. Active imagoes are found on hosts in all months, although prevalence and intensity of infestation are reduced in November–March (Fig. 5.1a). Fleas feed actively all year round (Fig. 5.1b). During early autumn, 95% of fleas possess low and medium fat stores (Fig. 5.1c), whereas this percentage decreases as winter progresses. Weak development of fat tissue is regarded as characteristic of actively reproducing fleas (Vashchenok, 1988). During mid-summer–autumn, most females are parous, whereas their proportion decreases in winter. Reproductively young (immature and nulliparous) females comprise up to 70% of flea in (a) mid-winter when densities, particularly of females, are relatively low and (b) early and mid-spring (Fig. 5.1d). The increased proportion of immatures in mid-winter can be interpreted as evidence for an autumn peak of oviposition. Predominance of immatures and nullipars in early spring also indicates mid-winter reproduction, whereas their occurrence in autumn indicates that reproduction, although it decreases, continues even in the hottest months of summer. The decrease of abundance in winter cannot be attributed to the decrease in reproductive rate (yielding the relatively high proportion of immatures in February–March), but rather to an increase in mortality in winter.

Nosopsyllus iranus is a flea with strict winter activity. Imagoes are found on their hosts only during winter (October–March). The intensity and prevalence of this flea are low in mid-autumn, increase to a maximum in early winter, and then steadily decrease (Fig. 5.2a). The proportion of feeding fleas is always high (Fig. 5.2b). At the start of seasonal reproductive activity, all fleas have medium or highly developed fat stores (Fig. 5.2c). All females are immature and/or nulliparous in October, whereas most become parous for the rest of the winter months (Fig. 5.2d). As numbers decline at the end of the seasonal activity period and the parous rate peaks during March, some females remain immature, showing continued recruitment, but ovarian maturation ceases in nullipars. This suggests that *N. iranus* survive summer as teneral adults and/or pupae within cocoons. They begin to mate and oviposit immediately after emergence and continue reproductive activity until early spring. Female bias in the beginning of the activity period (see Chapter 7; Fig. 7.12c) can be considered as an adaptation which allows a fast increase in population. Alternatively, this bias can be simply a result of higher summer mortality of male pupae than female pupae. Eggs deposited at the end of the activity period evidently develop to the adult or pupae stage and become inactive until next autumn.

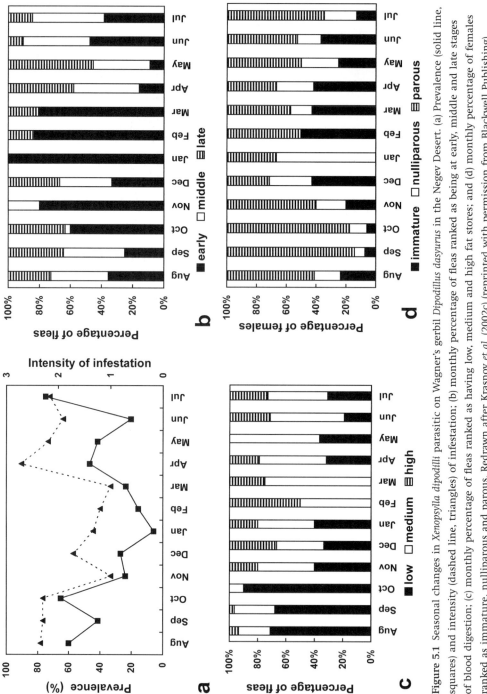

Figure 5.1 Seasonal changes in *Xenopsylla dipodilli* parasitic on Wagner's gerbil *Dipodillus dasyurus* in the Negev Desert. (a) Prevalence (solid line, squares) and intensity (dashed line, triangles) of infestation; (b) monthly percentage of fleas ranked as being at early, middle and late stages of blood digestion; (c) monthly percentage of fleas ranked as having low, medium and high fat stores; and (d) monthly percentage of females ranked as immature, nulliparous and parous. Redrawn after Krasnov et al. (2002c) (reprinted with permission from Blackwell Publishing).

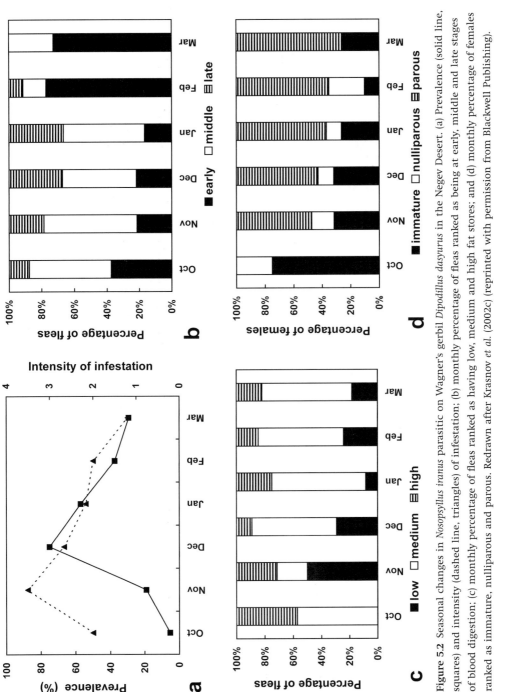

Figure 5.2 Seasonal changes in *Nosopsyllus iranus* parasitic on Wagner's gerbil *Dipodillus dasyurus* in the Negev Desert. (a) Prevalence (solid line, squares) and intensity (dashed line, triangles) of infestation; (b) monthly percentage of fleas ranked as being at early, middle and late stages of blood digestion; (c) monthly percentage of fleas ranked as having low, medium and high fat stores; and (d) monthly percentage of females ranked as immature, nulliparous and parous. Redrawn after Krasnov et al. (2002c) (reprinted with permission from Blackwell Publishing).

5.5.2 Reproductive diapause and overwintering

In regions with sharp seasonality, overwintering fleas, especially those of the genus *Xenopsylla*, demonstrate reproductive diapause (e.g. Novokreshchenova, 1962; Bibikova *et al.*, 1963; Kunitskaya *et al.*, 1971, 1977; Yakunin *et al.*, 1979; Kadatskaya, 1983; Samurov & Ageyev, 1983). Duration of diapause varies geographically ranging from 6–7 months (e.g. in the northern deserts of Kazakhstan: Zolotova & Varshavskaya, 1974; and mountains of the Trans-Caucasus: Teplinskaya *et al.*, 1983) to 2–3 months (e.g. in the Karakum Desert: Novokreshchenova & Kuznetsova, 1964). Reproductive diapause can also occur in summer, during the hottest period (e.g. *X. hirtipes* and *X. conformis*: Kiriakova *et al.*, 1970).

Diapausing fleas of both sexes are characterized by high development of fat tissue (Darskaya, 1955; Vashchenok *et al.*, 1992). This tissue serves as an energetic resource during overwintering or oversummering and allows fleas to survive prolonged starvation, although in winter they continue to feed at a low rate (e.g. *X. hirtipes* and *Xenopsylla nuttalli*: Demin *et al.*, 1970). Furthermore, overwintering fleas are able to synthesize glycerol in late autumn which may act as a cryoprotective agent enabling some of the fleas to survive winter. Apparently, glycerol is synthesized at the expense of glycogen and glucose, and the reverse occurs with the onset of spring and rising temperatures (e.g. *C. idius*: Pigage & Larson, 1983; Schelhaas & Larson, 1989).

Young diapausing females have no sperm in their spermathecae, whereas older females mate prior to diapause *(X. skrjabini*: Yakunin *et al.*, 1979; *Ctenophthalmus strigosus*: Solovieva *et al.*, 1976; *Ctenophthalmus congeneroides*: Litvinova, 2004). The latter seems to start reproduction, but egg production is terminated with the beginning of the cold season (Vashchenok *et al.*, 1992). Nevertheless, oogenesis in these females is not completely terminated but rather decelerated, although the largest and highly developed oocytes are usually resorbed (*X. gerbilli* and *X. hirtipes*: Kunitskaya *et al.*, 1971; Vashchenok *et al.*, 1992). Thus, both virgin and previously mated females can overwinter (e.g. *X. skrjabini*: Kunitsky *et al.*, 1974). Mated individuals compose from 10% (*X. skrjabini*: Tchernova, 1971) to 25–30% (*X. gerbilli*: Vashchenok *et al.*, 1992) of all overwintering females. In contrast, sexually experienced males usually die off immediately after the end of the reproductive period, whereas young virgin males survive the cold season (Vashchenok *et al.*, 1992). The stimulus that triggers diapause in fleas is still unclear. Some evidence suggests that *Xenopsylla* fleas respond by terminating reproduction to photoperiodic changes rather than to changes in air temperature (Starozhitskaya, 1970), although the onset of reproduction in spring seems to be stimulated by an increase in air temperature (*X. gerbilli*: Khrustselevsky *et al.*, 1971). In contrast, feeding and reproductive activity of *Ctenophthalmus* fleas (their annual cycles are similar to those of *Xenopsylla*) seemed to be regulated by air temperature (Kosminsky & Guseva, 1974a).

5.5.3 Classification of annual cycles

In areas with pronounced seasonality in air temperature and rainfall, there are substantial seasonal changes in the abundance and reproductive patterns of fleas, as observed, for example, in Azerbaijan (Kunitsky, 1970; Kadatskaya & Shirova, 1983), Volga–Ural Sands (Samurov, 1985), Canada (Lindsay & Galloway, 1997, 1998) and Europe (Fowler et al., 1983; Peach et al., 1987; Vashchenok & Tretiakov, 2003, 2004, 2005). However, in areas with less pronounced climatic seasonality and in urban areas, there is much less seasonality in the life-history parameters of fleas. For example, Seal & Bhattacharji (1961) found no seasonality in *X. cheopis* in India. However, Schwan & Schwan (1980) and Schwan (1986) found slight seasonal changes in life-history parameters of *Xenopsylla bantorum*, *Xenopsylla debilis* and *Xenopsylla difficilis* in Kenya. They argued that it is difficult to use standard climatological data when attempting to explain seasonal changes in flea populations. Nonetheless, distinct seasonality was reported for fleas of both synantropous hosts in urban areas and wild hosts in regions with weakly expressed seasonal climatic changes (see Makundi & Kilonzo (1994) and Njunwa et al. (1989) for Tanzania; Linardi et al. (1985) for Brazil; Shafi et al. (1988) for Pakistan; Khalid et al. (1992) for Egypt; Krasnov et al. (2002c) for the Negev Desert).

Darskaya (1970) proposed a classification of fleas based on the pattern of their annual cycles as follows (Table 5.3). Fleas with annual cycles of type A are active and reproduce all year round with pre-imagoes and imagoes occurring during the entire year. Imagoes of fleas with annual cycles of type B are active all year, but reproduce in the warm season only. The overwintering stage is represented by imagoes and, in some species, pre-imagoes in cocoons. Fleas with annual cycles of type C have an intermediate position between the previous two types. This type includes fleas of hibernating hosts. They reproduce when their hosts are active as well as being able to produce eggs during the host's hibernation. The type D annual cycle is assigned to fleas parasitic on hosts that reside in shelters or relatively permanent locations only seasonally. Active imagoes occur only during these seasons. Oviposition and pre-imaginal development take place in the locations of this temporary host residence. Imagoes of species with annual cycles of type E are active on hosts and reproduce during only a short period, although their hosts possess permanent shelters all year round. Furthermore, the types of annual cycles described above can be further divided into groups and subgroups (Table 5.3).

Table 5.3 Classification of annual cycles of fleas

Type of cycle	Group	Subgroup	Main hosts	Representatives
(A) Fleas with all-seasonal occurrence of imagoes, reproduction, pre-imaginal development and emergence	(1) Fleas with seasonally stable abundance		Commensal rodents (*Mus, Rattus*) and pets (*Felis, Canis*)	*Ctenocephalides canis, Ctenocephalides felis, Xenopsylla cheopis, Leptopsylla segnis, Nosopsyllus fasciatus*
	(2) Fleas with higher summer compared with winter abundance		Mice (*Apodemus*) and voles (*Microtus, Myodes*)	*Amalaraeus penicilliger, Amphipsylla rossica, Amphipsylla sibirica, Nosopsyllus mokrzeckyi*
(B) Fleas with all-seasonal occurrence of imagoes, but reproducing and developing in the warm period only. Overwintering stage is represented by weakly active imagoes (or pre-imagoes in cocoons in some species)	(1) Fleas with abundance peaking in spring and autumn; sometimes decreasing in summer. Fat reserves of overwintering imagoes are much greater than those of actively reproducing imagoes		Gerbils (*Meriones, Rhombomys*) in deserts; mice (*Apodemus*), voles (*Microtus, Myodes*) and shrews (*Sorex, Neomys*) in the temperate zone	*Xenopsylla* and *Ctenophthalmus* species, *Palaeopsylla soricis, Doratopsylla dasycnema*
	(2) Fleas with abundance peaking in spring and summer. Fat reserves of overwintering imagoes are similar to those of actively reproducing imagoes		Ground squirrels (*Spermophilus*)	*Citellophilus tesquorum, Callopsylla caspia*

(C) Fleas with all-seasonal occurrence of imagoes, reproducing both during period of activity and hibernation of a hibernating host	(1) Fleas with low reproduction during host hibernation and sharp summer decrease of abundance	Ground squirrels (*Spermophilus*)	*Neopsylla setosa*
	(2) Fleas with active reproduction during host hibernation and weak (if any) summer decrease of abundance	Marmots (*Marmota*), dormice (*Eliomys*, *Dryomys*)	*Oropsylla silantiewi*, *Myoxopsylla laverani*
(D) Fleas with occurrence of imagoes only in periods when a host is associated with either a shelter or a certain territory	(1) Fleas with imagoes occurring during warm period only	(a) 'Dispersing' fleas with fast reproduction and development occurring during hosts' nestling period, new imagoes dispersing to the nests of later nestling hosts and often three generation per year	A variety of bird species (e.g. *Parus*, *Tachyneta*)	*Ceratophyllus gallinae*, *Ceratophyllus garei*, *Ceratophyllus petrochelidoni*

(*cont.*)

Table 5.3 (cont.)

Type of cycle	Group	Subgroup	Main hosts	Representatives
		(b) 'Sedentary' fleas with reproduction and development occurring during hosts' nestling period, relatively long pupal stage, rarely dispersing new imagoes and one to two generations per year	Swallows and martins (*Hirundo, Delichon*)	*Ceratophyllus hirundinis, Ceratophyllus delichoni*
		(c) Fleas with reproduction and development occurring during hosts' nestling period, very long pupal stage and non-dispersing new imagoes	Burrowing birds such as sand martins (*Riparia*) and stonechats (*Saxicola*)	*Ceratophyllus styx, Frontopsylla frontalis*
	(2) Fleas with imagoes occurring during cold period only	(a) Semi-sessile fleas with imagoes occurring and reproducing from autumn to spring	Ungulates (Artyodactyla)	*Vermipsylla alakurt, Dorcadia ioffi*
		(b) Fleas with imagoes occurring and reproducing in winter	Bats (Chiroptera)	*Nycteridopsylla pentactena, Nycteridopsylla oligochaeta*

(E) Fleas with occurrence of imagoes and reproduction during relatively short period although a host is associated with a shelter or a certain territory all year round	(1) Fleas with several generations of imagoes coexisting and usually abundant from autumn to spring with autumn peak of abundance	Gerbils (*Meriones*)	*Nosopsyllus laeviceps*, *Nosopsyllus iranus*
	(2) Fleas represented at a given time by imagoes of the same generation		
	(a) Fleas with imagoes occurring and reproducing from autumn to spring	Gerbils (*Meriones*, *Rhombomys*) in deserts; voles (*Microtus*, *Myodes*), shrews (*Sorex*) and moles (*Talpa*) in the temperate zone	*Stenoponia*, *Rhadinopsylla*, *Coptopsylla*, *Wagnerina*, *Paradoxopsyllus*, *Hystrichopsylla talpae*
	(b) Fleas with imagoes occurring and reproducing from mid-summer to late autumn	Voles (*Myodes*)	*Peromyscopsylla silvatica*, *Peromyscopsylla bidentata*
	(c) Fleas with imagoes occurring and reproducing in spring only	Ground squirrels (*Spermophilus*)	*Oropsylla ilovaiskii*, *Frontopsylla semura*

Source: Adapted and modified after Darskaya (1970).

Based on the lines of Darskaya's (1970) classification, Vashchenok (1988) proposed the following annual cycles: (1) adult fleas are active and reproduce all year round; (2) adult fleas are active all year round, but reproduce in the warm season only; (3) adult fleas are active and reproduce in the warm season only; (4) adult fleas are active and reproduce most of the year except for the hottest and driest periods when fleas survive in cocoons; and (5) adult fleas are active and reproduce in the cold season only.

5.5.4 Variation in seasonality: host biology, climate and evolutionary constraints

Whatever classification of flea annual cycles is adopted, it probably reflects ecological differences between flea species and, sometimes, between their hosts, but appears not to involve siphonapteran taxonomic affinities. For example, some evidence suggests that the annual cycle of a flea depends primarily on the ecological properties of its host. Indeed, *X. cheopis* has been shown to reproduce all year round (Seal & Bhattacharji, 1961) and, thus, its annual cycle corresponds to type 1 of Vashchenok's (1988) classification. Other studied *Xenopsylla* fleas have demonstrated seasonal breaks in reproduction and, thus, their cycles correspond to type 2 (Vashchenok, 1988 and references therein). This difference can be easily explained in that *X. cheopis* parasitizes mainly commensal rodents and, thus, climatic fluctuations of its environment are much less pronounced than those of congeneric species that parasitize wild rodents. The same was reported for *Monopsyllus anisus* and *N. fasciatus* parasitic on the Norway rat *Rattus norvegicus* in rural and urban settlements in Siberia (e.g. Zhovty et al., 1983). However, in Japan, these species demonstrated clear, albeit weak, seasonal changes in prevalence and abundance (Nakazawa et al., 1957).

According to observations by Vashchenok (2006) in the Ilmen–Volkhov Lowland (in the north of European Russia), *Palaeopsylla soricis* and *Doratopsylla dasycnema*, both parasitic on shrews, have similar annual cycles. The same is true for *Megabothris turbidus*, *Ctenophthalmus agyrtes* and *Ctenophthalmus uncinatus*, for whom the main host in this region is the pygmy woodmouse *Apodemus uralensis*. However, fleas that exploit ecologically similar hosts may also show strikingly different annual life cycles. For example, *O. alaskensis* feeds and reproduces most actively during deep hibernation of its ground squirrel host (Vasiliev, 1971). Similarly, *N. setosa* also reproduces on hibernating ground squirrels (e.g. the pygmy ground squirrel *Spermophilus pygmaeus*: Myalkovskaya, 1983). In contrast, *C. tesquorum*, another flea of ground squirrels, becomes inactive both in feeding and reproduction on hibernating hosts and starts reproduction only after the host awakes (e.g. Nikulshin, 1980; Nikulshin & Shinkareva, 1983). The same is true for the hedgehog flea *Archaeopsylla erinacei* (Marshall, 1981a). More surprisingly, such contradicting patterns may be demonstrated by fleas belonging to the same taxon. Among bat fleas of the family Ischnopsyllidae, *Ischnopsyllus* species reproduce only on active hosts in summer, whereas *Nycteridopsylla* species breed only on hibernating hosts in winter (Hůrka, 1963a, b; Medvedev, 1989b).

Ecological differences between flea species also do not always lead to different annual cycles. In contrast, ecologically different fleas can have similar seasonal patterns. For example, strictly winter-active species include both 'body' fleas (e.g. *Ischnopsyllus*: Medvedev, 1989b) and 'nest' fleas (e.g. *Rhadinopsylla*: Brinck-Lindroth, 1968; Ulmanen & Myllymäki, 1971; Krasnov et al., 1997). Moreover, 'body' fleas can display either an all-year-round (e.g. *Amphipsylla rossica*: Kosminsky & Guseva, 1974b) or a seasonal (e.g. *N. setosa*: Myalkovskaya, 1983) pattern of reproduction. On the other hand, both summer-active and winter-active species often exploit the same population of the same host species (e.g. Gauzshtein et al. (1967) for fleas on the great gerbil *Rhombomys opimus*). This cycle asynchrony has even been considered as a kind of temporal segregation that has evolved due to interspecific competition (Day & Benton, 1980; Ageyev et al., 1983, 1984).

Within-genus and within-species variation in annual cycles can be related at times to environmental characteristics of a location. Comparison of annual cycles of *Xenopsylla* from different regions demonstrates that there is a trend to expand the reproductive period from summer into winter in southern species and/or populations. In Central Asia, most individuals of summer-hatched *X. skrjabini* and *X. gerbilli* do not reproduce until the following spring, and, thus, overwinter as imagoes (Kiriakova et al., 1970; Kunitskaya et al., 1977). The same is true for *X. skrjabini* from the North Caspian Lowlands (Darskaya, 1970). However, in southern Turkmenistan, *X. gerbilli* and *X. hirtipes* reproduce all year in years with relatively warm winters (Zagniborodova, 1968). There are, however, some exceptions. Abundance of *X. bantorum* in East Africa is higher in the warm and dry season than in the cool and wet season, and this flea presumably pauses or, at least, sharply decreases its reproductive activity in the cool season (Schwan, 1986).

Another example of geographical variation within genera is represented by species of genus *Nosopsyllus* which demonstrate a variety of annual cycles. Comparison of annual cycles among species inhabiting different geographical localities shows that there is a trend to shorten the reproduction period and to shift it towards the cooler season with an increase in summer air temperatures. For example, *N. consimilis* in the North Caucasus and *Nosopsyllus laeviceps* in the North Caspian Lowlands reproduce all year round (Ioff, 1949; Kunitsky, 1970;

Kosminsky et al., 1974; Samurov, 1985), but conspecifics from the South Caucasus and Central Asia reproduce from late summer till late winter only and survive during the hottest period (mid-summer) in cocoons (Kunitsky, 1970) or as imagoes (Samurov & Yakunin, 1979, 1980). *Nosopsyllus iranus* from Azerbaijan is active and reproduces during a relatively short period from mid-autumn till early spring (Kunitsky, 1970), whereas the reproductive period of conspecifics (although belonging to different subspecies, *N. i. theodori*) from the Negev Desert is even shorter (see above).

Different populations of *C. tesquorum* parasitic on the pygmy ground squirrel *S. pygmaeus* and the Caucasian mountain ground squirrel *Spermophilus musicus* in the North Caucasus reproduce in different periods (Belyavtseva, 2002). Reproduction of this species occurs from March to July in the Dagestan Plain and May to August in alpine habitats. Large changes in seasonal patterns within species among localities have been reported for several fleas inhabiting the Tatry Mountains in Poland (Bartkowska, 1973).

These comparisons, as well as the data in Table 5.3, suggest that evolutionary fine-tuning of flea annual cycles is related both to host seasonal behaviour and local environmental conditions. In some genera, the annual cycles seem to be evolutionarily conservative. These genera do not demonstrate geographical variability in their annual cycles that can be explained by climatic differences. Such is the case for all species of the genus *Stenoponia* which demonstrate surprising similarity in their seasonal patterns. They occur as imagoes and reproduce in cold seasons, independent of the climate of the region that they inhabit (Vashchenok, 1988). For example, *Stenoponia sidimi* from the temperate climate zone of the Korean Peninsula, *Stenoponia tripectinata* from the arid Sinai Peninsula and American *Stenoponia americana* and *Stenoponia ponera* are most abundant in November, December and January (Walton & Hong, 1976; Morsy et al., 1993; Hastriter et al., 2006). All *Stenoponia* lay very large eggs and their larval developmental stages are long (M. W. Hastriter, personal communication, 2006). Perhaps these factors dictate similar seasonal patterns. In contrast, other genera such as *Nosopsyllus* and *Xenopsylla* present no evidence of the evolutionary heritability of annual cycles and demonstrate a large variety of seasonal patterns. Another example suggesting that classification by seasonal life-cycle patterns is not related to taxonomy is that the annual cycle of Vashchenok's (1988) type 5 or Darskaya's (1970) type E2a is characteristic for fleas from different genera and families (*Coptopsylla*, *Stenoponia*, *Paradoxopsyllus*, *Rhadinopsylla*, *Wagnerina*, *Jordanopsylla*). However, there are no pulicid species that are active only in the cold season despite the ubiquitous distribution of this family.

5.6　Concluding remark

Despite the strict constraints of a holometabolous life cycle, fleas demonstrate a variety of reproductive and seasonal patterns. Although life cycles and seasonal patterns of many flea species have been described, there is no agreement on how these patterns have evolved. It seems that the evolution of these patterns was driven by joint effects of host ecology and behaviour, local environmental factors and evolutionary constraints.

6

Fleas and humanity

Humans have always coexisted with fleas. This coexistence is asymmetric, usually being favourable for fleas, but unfavourable for humans. Fleas can cause direct medical damage to humans and can serve as vectors for some diseases. They can also cause indirect damage to humans by parasitizing poultry and livestock and, thus, causing economic loss (e.g. Yeruham et al., 1989). Flea damage to human pets (mainly dogs and cats) also represents a serious veterinary problem. The ubiquity of the negative effect of fleas and their role in transmission of diseases have sometimes led to these creatures being blamed even when their negative role has not been explicitly established (e.g. Moynahan, 1987). The negative aspects of fleas as they relate to the economic and medical implications to human society in both urban and rural settings are briefly addressed in this chapter.

6.1   Medical aspects

6.1.1   Dermatological diseases caused by flea parasitism

The most well-known medical condition caused directly by flea parasitism is tungiasis. Tungiasis is a health problem in the tropics and subtropics, especially in underprivileged communities in Latin America, the Caribbean and sub-Saharan Africa (Heukelbach et al., 2001; Eisele et al., 2003; Kehr et al., 2007). This painful parasite-inflicted disease is not exclusive to humans, but also strikes many domestic animals (Heukelbach et al., 2004). It is caused by the sand flea *Tunga penetrans* when the female flea burrows into the skin. Females remain in the skin (the preferred site of attachment is often under the nail bed of the toes, though not exclusively) until their death (in about 5 weeks). After

1–2 weeks post-penetration, the flea becomes neosomic (Geigy & Herbig, 1949; Eisele et al., 2003). Most of the flea body remains buried in the epidermis, except for the posterior-most part. The condition is associated with considerable acute and chronic morbidity, whereas the degree of acute morbidity is directly related to the number of embedded female fleas (Kehr et al., 2007). Tungiasis may lead to pain, pruritus, bacterial infection and autoamputation of toes.

The human flea, *Pulex irritans*, is found worldwide on wildlife (e.g. Durden et al., 2006; Fiorello et al., 2006), livestock (e.g. Christodoulopoulos & Theodoropoulos, 2003; Menier, 2003), and pets (e.g. Franc et al., 1998; Gracia et al., 2000). This flea also inhabits human dwellings, especially in rural settlements. Bites of *P. irritans* can cause severe allergic dermatitis, which is manifested as erythematous oedematous papules with hemorrhagic puncta on the lower extremities, especially on the ankles (see Beck & Clark, 1997 for review). Occasionally, vesicles and bullae appear, as well as larger urticarial lesions. Although *P. irritans* does not represent a serious problem for humans in most developed countries today, its abundance and distribution in human dwellings increases in less-developed regions. For example, the number of settlements with a high rate of *P. irritans* attacks on humans increased in Kazakhstan during the last several decades (Bidashko et al., 2001, 2004). The occurrence of mummified *Pulex* fleas on mummies of domestic animals from prehispanic times in South America (Bouchet et al., 2003; Dittmar de la Cruz et al., 2003) suggests that the history of human suffering from parasitism of this flea is rather long. Numerous cases of human dermatitis caused by bites of other flea species have also been reported (e.g. Haag-Wackernagel & Spiewak, 2004; Yamauchi, 2005).

6.1.2   Fleas as vectors of infectious diseases

The most important negative effect of fleas is their role as vectors of a number of pathogens. Fleas transmit viral, rickettsial and bacterial diseases to humans.

In general, fleas appear to be only poor vectors of several human viral diseases. For example, the tick-borne encephalitis (TBE) virus has been recorded in fleas collected from wild rodents (e.g. Sotnikova & Soldatov, 1969; Tarasevich et al., 1969; Naumov & Gutova, 1984). Moreover, the ability of fleas to transmit the TBE virus among laboratory animals has been shown experimentally with *Amalaraeus penicilliger* (Kulakova, 1962), *Megabothris rectangulatus* (Kulakova, 1962), *Xenopsylla cheopis* (Feoktistov et al., 1968), *Ctenophthalmus congeneroides*, *Neopsylla bidentatiformis* and *Frontopsylla elata* (Tchimanina & Kozlovskaya, 1971a, b). Other studies have also showed that the virus can survive in fleas, but failed to demonstrate that they can serve as vectors for this disease (*Ctenophthalmus assimilis*: Řeháček, 1961; *Nosopsyllus fasciatus* and *X. cheopis*: Smetana, 1965).

Fleas collected from the bank vole *Myodes glareolus* in Siberia were infected with the virus of the lymphocytic choriomenengitis (Fedorov et al., 1959). Bird fleas *Ceratophyllus garei* and *Ceratophyllus gallinae* were able to transmit the virus of Omsk haemorrhagic fever among experimental laboratory mice (Sapegina & Kharitonova, 1969), although the role of fleas as vectors of this disease seems to be minor. *Monopsyllus anisus* and *Leptopsylla segnis* collected from the nests of the fieldmouse *Apodemus agrarius* in China were found to be infected with virus of the haemorrhagic fever with renal syndrome, but the virus was not able to persist in these fleas for more than 48 h (Dong, 1991).

Bacterial diseases that are transmitted by fleas are numerous and diverse (see Vashchenok, 1984, 1988 for reviews). The most important, notorious and legendary of them is the plague. Fleas are specific vectors of this disease with about 250 species documented to have been naturally infected with the plague pathogen (*Yersinia pestis*) worldwide (Gage & Kosoy, 2005). The plague has caused severe epidemics and pandemics during recent human history. This disease is not an extinct medieval monster, and epidemics still happen frequently (e.g. Feng et al., 2004). About 38 000 (2845 of them fatal) human plague cases were recorded in 25 countries during the period 1983–2003 (Anonymous, 2004). An abundant literature deals with the plague and the role of fleas in its transmission. In addition, a variety of morphological, physiological, ecological and epidemiological aspects of the relationship between fleas and the plague pathogen and their implications for human well-being have been studied (see Gage & Kosoy, 2005 for a recent review). Association between fleas as vectors of the plague and their ecological features such as host specificity will be discussed below (Chapter 14).

Other human bacterial diseases that can be transmitted by fleas include listeriosis (e.g. Alekseev et al., 1971) and tularaemia (e.g. Hopla, 1980; Gage et al., 1995). *Listeria* has been shown to persist during relatively long periods in the alimentary canal of experimentally infected *Nosopsyllus consimilis* (Vashchenok & Tchirov, 1976). In infected fleas, bacteria occur along the entire intestine from oesophagus to rectum but most often in the midgut and proventriculus. *Listeria* is capable of penetrating the flea's muscular tissue and, in some cases, enters its body cavity. The occurrence of *Listeria* in the oesophagus suggests the possibility of their transmission via flea bites.

Francisella tularensis, the causative agent of tularaemia, has been documented in several dozens flea species (e.g. 19 species from the territory of the former USSR: Olsufiev, 1975). Some fleas (e.g. *Ctenophthalmus orientalis*, *Ctenophthalmus wagneri*, *Ctenophthalmus secundus*, *N. consimilis*) can maintain and transmit

F. tularensis for long periods (about half a year) after being infected (Volfertz & Kolpakova, 1946; Tiflov, 1959). Nevertheless, the role of fleas in circulation and maintenance of *F. tularensis* in the natural foci of this disease is thought to be minor (Olsufiev & Dunaeva, 1960).

Fleas can be infected by the causative agents of pseudotuberculosis (*Yersinia pseudotuberculosis*) and yersiniosis (*Yersinia enterocolitica*). These bacteria can multiply and persist in a flea (e.g. *X. cheopis*: Blanc & Baltazard, 1944a; Vashchenok, 1979), but the importance of transmission of the disease by fleas to humans is questionable (e.g. *L. segnis*: Yushchenko, 1965). The same is true for the etiological agents of pasteurellosis (*Pasteurella multocida*), salmonelloses (*Salmonella suipestifer* and *Salmonella schotmulleri*), brucellosis (*Brucella melitensis*) and erysepeloid (*Erysepelotrix rhusiopatia*). These bacteria either cannot multiply in fleas (e.g. pasteurellosis in *N. consimilis*: Tiflov, 1964) or cannot be effectively transmitted by them (e.g. brucellosis in *N. setosa* and *Citellophilus trispinus*: Rementsova, 1962; erysepeloid in *Xenopsylla gerbilli*: Punsky & Zagniborodova, 1964). Relationships of fleas and the etiological agents of melioidosis (*Burkholderia pseudomallei*: Blanc & Baltazard, 1941, 1942), pneumococcosis (*Streptococcus pneumoniae*: Vashchenok, 1988) and leptospiroses (*Leptospira grippotyphosa* and *Leptospira sorex*: Soloshenko, 1958, 1962) are poorly known. Results of studies with *X. cheopis*, *Ctenocephalides canis*, *N. fasciatus* and *Megabothris walkeri* did not produce unequivocal conclusions about the ability of fleas to maintain and transmit these pathogens. *Borrelia burgdorferi*, the causative agent of Lyme borreliosis, was isolated from fleas (e.g. from *Ctenophthalmus agyrtes* and *Hystrichopsylla talpae* in the Czech Republic: Jurikova et al., 2002), but the role of fleas in the circulation of this disease is not completely understood.

Recently, several bacteria of the genus *Bartonella* have been recognized as important agents of emerging zoonotic diseases in humans and have been isolated from various mammalian reservoirs (Boulouis et al., 2005). These zoonotic bartonelloses are divided into two groups. The first group is represented by bartonelloses with feline or canine reservoirs. For example, cat scratch disease (CSD) is certainly the most common *Bartonella* zoonosis worldwide. Human cases have been reported in North America, Europe, Australia and in most countries where investigators have looked for such an infection (Breitschwerdt & Kordick, 2000; Sreter-Lancz et al., 2006). The second group of bartonelloses is represented by rodent-borne *Bartonella* (Birtles et al., 1994; Kosoy et al., 1997, 2004; Ellis et al., 1999; Fichet-Calvet et al., 2000; Ying et al., 2002). Fleas are very competent vectors for both feline/canine- and rodent-borne bartonelloses (e.g., Stevenson et al., 2003). For example, in Afghanistan DNA of *Bartonella* was detected in 15.5% of fleas collected from gerbils and in 45% of fleas collected from rats (Marie et al., 2006). In this study, the number of rodents with fleas that were examined was rather low (55 individuals only) and thus the reported prevalence, albeit high, could easily be underestimated.

The most important rickettsiosis transmitted by fleas is murine typhus (also called flea-borne or endemic typhus) caused by *Rickettsia typhi*. This disease is one of the oldest recognized arthropod-borne zoonoses and occurs in many countries on all continents (Parola & Raoult, 2006). Fleas are the primary vector and rodents are the primary reservoir of murine typhus (see Traub *et al.*, 1978; Parola *et al.*, 2005; Parola & Raoult, 2006 for reviews). In contrast with the plague, infection with *R. typhi* does not seem to affect fleas negatively. Fleas remain permanently infected with it and their lifespan is not shortened by the presence of rickettsiae (Farhang-Azad *et al.*, 1984). Natural infection by *R. typhi* was recorded in several flea species, the most important from an epidemiological viewpoint being *X. cheopis, C. canis, Ctenocephalides felis, P. irritans, Echidnophaga gallinacea* and *L. segnis* (Vashchenok, 1988). Transmission occurs via bite, largely when victims rub infected flea faeces into the bite rather than the agent being injected during the bite.

Flea-borne spotted fever group rickettsiosis (also called cat-flea typhus) is an emerging disease caused by *Rickettsia felis*, although it was first detected in *C. felis* as early as in 1918 (Parola & Raoult, 2006). The pathogenicity of *R. felis* in humans has been debated but has been widely accepted since an outbreak of the disease in Mexico in 2000 (Zavala-Velazquez *et al.*, 2000). The disease has been associated with fleas throughout the world including Brazil, Ethiopia, Thailand, Europe, New Zealand, Algeria and the USA (Parola & Raoult, 2006). Fleas that have been found to be naturally infected with *R. felis* include *C. felis, C. canis, P. irritans, Archaeopsylla erinacei* and *Anomiopsyllus nudatus* (Stevenson *et al.*, 2005; Parola *et al.*, 2005; Bitam *et al.*, 2006). Transovarial transmission of both *R. typhi* and *R. felis* in fleas has been reported, suggesting that fleas could act not only as vectors but also as reservoirs of rickettsiae (Farhang-Azad *et al.*, 1985, 1997). Fleas (*X. cheopis, P. irritans*) have been experimentally infected with *Rickettsia prowazekii*, the causative agent of louse-borne (epidemic) typhus, but transmission of the disease by fleas has not been established in most studies (Blanc & Baltazard 1944b; Bozeman *et al.*, 1981).

6.2 Veterinary aspects

The importance of fleas to veterinary medicine is enormous. As with humans, both domestic and wild animals can suffer directly from flea parasitism and from infectious diseases transmitted by fleas. Moreover, the number of animal diseases that require fleas as vectors is higher than that of human diseases. Fleas transmit not only viral, rickettsial and bacterial diseases to animals, as is the case with humans, but also protozoans and helminths. Information on the veterinary importance of fleas is abundant and can be easily found in a variety of veterinary textbooks (e.g. Kettle, 1995; Wall & Shearer, 2001), monographs (e.g. Krämer & Mencke, 2001) and journals. Consequently, a discussion of the veterinary importance of fleas will focus here on the most important issues only.

Direct effect of flea parasitism on animals is related mainly to flea bites and parasitism of the stick-tight and skin-burrowing fleas. Flea-allergy dermatitis is a common disease of cats and dogs. This condition is caused mainly by bites of *C. canis*, *C. felis* and *P. irritans* and, despite recent advances in flea control, flea-allergy dermatitis still continues to be a common problem (Dryden & Blakemore, 1989). Furthermore, these flea species can also infest livestock causing severe dermatological problems (e.g. Araújo et al., 1988; Yeruham & Koren, 2003; Christodoulopoulos et al., 2006; Kaal et al., 2006). *Ceratophyllus gallinae* parasitizes domestic poultry, as well as wild birds. Anaemia in poultry often occurs due to high abundance of *C. gallinae*.

Parasitism of *T. penetrans* (tungiasis) is as dangerous for animals as it is for humans (see above). As mentioned above, tungiasis has been reported from many domestic and wild animals (Heukelbach et al., 2004; Vobis et al., 2005). Animals (e.g. rodents) suffer not only from parasitism of *T. penetrans*, but also from other species of the genus, such as *Tunga monositus* (Hastriter, 1997). Similarly to *Tunga*, stick-tight fleas such as *Vermipsylla* and *Echidnophaga* cause severe disease in domestic and wild animals and birds (e.g. sheep and *Vermipsylla alakurt*: Wang et al., 2004a, b; chickens and *E. gallinacea*: Gustafson et al., 1997).

Viral animal diseases transmitted by fleas include cat leukaemia, caused by feline leukaemia retrovirus (Jarrett, 1975), as well as myxoma and rabbit haemorrhagic disease (see Fenner & Ratcliff, 1965 for review). The latter diseases cause high mortality in rabbits, so the rabbit flea *Spilopsyllus cuniculi* has been introduced in several locations to control rabbit populations. The introduction of rabbit fleas on a small island of the Kerguelen Archipelago in 1987 increased the proportion of rabbits with antibodies from 34% to 85% in 10 years (Chekchak et al., 2000). This use of fleas as agents of biological control is, perhaps, the only case of a positive link between fleas and humans (excluding the use of fleas as a model taxon for ecological and evolutionary studies by scientists).

Animals suffer from the same bacterial (plague, tularaemia, bartonelloses etc.) and rickettsial (murine typhus) infections transmitted by fleas as do humans (e.g. Salkeld & Stapp (2006) for fleas, plague, and carnivore mammals). In addition, fleas transmit feline haemoplasmosis (e.g. Shaw et al., 2004) and other bacterial infections of animals such as, for example, mycoplasmal polyarthritis (Nayak & Bhowmik, 1990).

Among protozoan infections, fleas play an important role in maintenance and transmission of trypanosomes of the subgenus *Herpetosoma* (e.g. Smith *et al.*, 2005). Among 90 species belonging to this subgenus, most are transmitted by fleas. Although these trypanosomes are host-specific in relation to their mammal reservoir (Vashchenok, 1988; but see Hamilton *et al.*, 2005), they are quite opportunistic in relation to a vector and can develop in a variety of flea species (Vashchenok, 1988). Some rodent trypanosomes such as *Trypanosoma lewisi* are non-pathogenic, but Guerrero *et al.* (1997) demonstrated that *T. lewisi* infection increases *Toxoplasma gondii* multiplication in laboratory rats, thus providing a potential synergist for transmission of *T. gondii* to humans.

Some flea species have been shown to be infected with haemogregarines of the genus *Hepatozoon* (e.g. *Hepatozoon erhardovae*) which can be further transferred to a mammalian host (Krampitz 1964, 1981; Gobel & Krampitz, 1982). However, these protozoans usually infect mammals via tick and mite vectors, so the role of fleas in their transmission is not significant. In contrast to ticks and mites, *Hepatozoon* cannot be transmitted via flea bites. A host may be infected when (and if) it consumes an infected flea during grooming.

Ctenocephalides felis and *C. canis* can be intermediate hosts of the tapeworms *Dipylidium caninum* and *Hymenolepis diminuta* (Hinaidy, 1991; Thomas, 1996). When infected fleas are unintentionally eaten by a host (a cat or a dog or even a human for *D. caninum* and a rodent for *H. diminuta*) during grooming or pet handling, the tapeworms enter their definitive host. Helminth eggs pass with the host faeces into the environment where they may be consumed by larval fleas.

6.3 Fleas in human habitats

Although fleas are parasites mainly of wild birds and mammals, humans routinely acquire fleas that invade and adapt to their homes in both rural and urban settlements. Flea assemblages of human habitats are composed of three main components, namely (a) fleas of poultry, livestock and pets; (b) fleas of commensal birds and mammals; and (c) fleas of wild birds and mammals that may invade human habitations (e.g. Sapegina, 1988; Wilson *et al.*, 1991; Visser *et al.*, 2001). Species composition of the first two groups is rather uniform across the world, whereas that of the third group is diverse and depends strongly on host and flea fauna of a geographical region.

The most common fleas of poultry, livestock and pets include *C. gallinae*, *E. gallinacea*, *P. irritans*, *C. felis* and *C. canis*. As mentioned above, *C. gallinae* and *E. gallinacea* parasitize poultry (e.g. Beck, 1999), whereas the remaining two species exploit cats and dogs as well as various domestic mammals, including sheep (e.g. Yeruham *et al.*, 1989), goats (e.g. Yadav *et al.*, 2006), cattle (e.g. Araújo *et al.*,

1988) and donkeys (Yeruham & Koren, 2003). Several flea species that parasitize humans, livestock and domestic animals have limited geographical distribution. For example, *T. penetrans* is limited to human settlements in the tropics (e.g. Heukelbach *et al.*, 2004), whereas *V. alakurt*, a parasite of domestic and wild ungulates, can be found only in Central Asia, Mongolia and China (e.g., Wang *et al.*, 2004a).

Fleas of commensal birds and mammals invade settlements together with their hosts. Commensal birds (sparrows, pigeons, house martins) may harbour *C. gallinae*, *Ceratophyllus fringillae*, *Ceratophyllus columbae*, *Ceratophyllus rusticus* and *Ceratophyllus hirundinis* (e.g. Cyprich *et al.*, 2002; Haag-Wackernagel & Spiewak, 2004; Pilgrim & Galloway, 2004). Rats and mice invade settlements and thus are vehicles for *X. cheopis*, *N. fasciatus* and *L. segnis*. Although the principal hosts of these fleas are feral birds and mammals, they readily and routinely attack domestic animals and even humans (Miyamoto & Hashimoto, 2000; Haag-Wackernagel & Spiewak, 2004; Yamauchi, 2005). For example, Haag-Wackernagel & Spiewak (2004) reported that a couple was repeatedly invaded by *C. columbae* from a pair of feral pigeons whose nest was located in the attic immediately above their apartment, and the fleas found their way along an unsealed heating pipe. The people encountered up to 40 bites per night, so the man gradually developed an allergic urticarial reaction. Another example is represented by the rat flea *X. cheopis*, which readily attacks dogs and cats (e.g. Raszl *et al.*, 1999). Many of these fleas have now ubiquitous geographical distributions. For example, *N. fasciatus* can be found on rats as far north as Iceland (Bengtson *et al.*, 1986) and as far south as New Zealand (Tenquist & Charleston, 2001).

Wild birds and mammals that invade human habitats from the surrounding natural habitats also participate in the 'enrichment' of the settlement flea assemblages (e.g. Zhang *et al.*, 2005). However, the establishment of a 'wild' flea species in a settlement depends on the ability of a wild host not only to invade a settlement but also to find or construct an appropriate shelter with appropriate microclimate conditions allowing successful development of pre-imago fleas. As a result, a variety of flea species can be found in human settlements. They also may switch from their principal 'wild' host to a new commensal or domestic host. For example, house mice *Mus musculus* in the settlements of the coal mines and industrial plants in Poland were exploited not only by the specific *L. segnis*, but also by fleas characteristic for small mammals of the surrounding natural habitats (*Megabothris turbidus*, *C. agyrtes*, *C. assimilis*, *Hystrichopsylla orientalis*, *Peromyscopsylla bidentata*) (Blaski, 1989, 1991). Rats *Rattus rattus* in a human settlement in Vietnam were parasitized not only by the rat flea *X. cheopis*, but also by a 'wild' flea *Aviostivalius klossi* (Suntsov *et al.*, 1992a, b). Commensal rodents (*R. rattus*, *Rattus norvegicus* and *M. musculus*) in Angola harboured not only *X. cheopis*, but also *Ctenophthalmus machadoi* and *Dinopsyllus smiti* (Linardi *et al.*, 1994). Dogs from two rural properties of the county of Piraí in Brazil were found to be infested with *Rhopalopsyllus lutzi* (Scofield *et al.*, 2005), a flea characteristic of wild carnivores (e.g. Rodrigues & Daemon, 2004). These examples clearly demonstrate that the composition of the 'wild' component of a flea assemblage of a human settlement is strongly determined by the regional flea fauna.

Distribution of flea species within a settlement is not uniform. Human houses are inhabited mainly by pet- and commensal-related flea species, whereas fleas that come into a settlement with wild animals are mainly recorded outside houses, e.g. in public gardens and rubbish dumps. For example, 18 of 23 flea species recorded in the city of Novosibirsk (Russia) were found mainly in the forest-park zone (Sapegina, 1988).

6.4 Concluding remarks

Fleas have profound direct and indirect effects on human health and well-being. In particular, fleas are able to transmit various infectious diseases. Nevertheless, in spite of the great efforts invested in the study of the role of fleas in the maintenance and transmission of diseases, we still lack much knowledge on the epidemiology and control of these diseases. One of the reasons for this is that our knowledge is limited by the small number of species that have been tested experimentally for their potential as vectors of human and animal diseases. Although studies of the vector capacities of *P. irritans*, *X. cheopis*, *C. felis* and a few other flea species have been exhaustive, little attention has been directed to the majority of fleas occurring on wild animals. It is essential to fill these gaps in future epidemiological and veterinary studies.

PART II FUNCTIONAL ECOLOGY: HOW DO FLEAS DO WHAT THEY DO?

7

Ecology of sexual dimorphism, gender differences and sex ratio

Males and females play complementary yet distinctly different evolutionary roles. They may differ in size and/or shape, behaviour and response to environmental factors. As a result, the abundance of males and females in a population may be unequal. This inequality in numbers may have important ecological and evolutionary consequences. This chapter starts with the description of differences in size, behaviour and physiology between male and female fleas. I attempt to understand how these differences are reflected in flea ecology. Then the effect of various gender differences on sex ratio in flea populations is considered.

7.1 Sexual dimorphism

7.1.1 Size dimorphism and Rensch's rule

Similar to many arthropods, fleas demonstrate strong female-biased sexual size dimorphism. This is true not only for species with neosomic females that are tens and hundreds of times larger than males (see Chapter 5) but also for species where gender size differences are less pronounced. For example, in six of seven flea species from the Negev Desert studied by Krasnov et al. (2003a), females were significantly larger than males (Fig. 7.1a).

Despite sexual size dimorphism occurring in a great number of animal species, male and female sizes usually covary within a lineage. However, in many cases variation in size is greater in males than in females (e.g. Fairbairn, 1997, 2005; Colwell, 2000). This generates an allometric pattern of sexual size dimorphism known as Rensch's rule. This rule (Rensch, 1960; Abouheif & Fairbairn, 1997; Fairbairn, 1997, 2005; Colwell, 2000) states that in taxa in which females tend to be larger than males, size dimorphism diminishes in larger species,

80 Ecology of sex differences

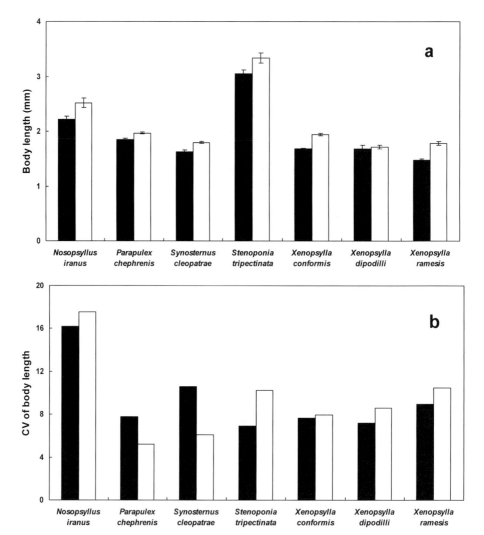

Figure 7.1 (a) Mean (±S.E.) body length (mm) and (b) coefficient of variation (CV) in body length of males (black columns) and females (white columns) of seven flea species from the Negev Desert. Data from Krasnov et al. (2003a).

whereas in taxa where males tend to be larger than females, size dimorphism increases in larger species. Reviews of the quantitative evidence for Rensch's rule (Reiss, 1986, 1989; Abouheif & Fairbairn, 1997; Fairbairn, 1997) indicate that it is a very common allometric trend but that exceptions occur, particularly in taxa in which females are the larger sex.

To test whether sexual size dimorphism in fleas conforms to Rensch's rule, Krasnov et al. (2003a) performed a regression of female body size on male body size using data on seven flea species. The slope of the regression was 0.85 with

95% confidence interval of 0.64–1.10, indicating that this slope did not differ significantly from 1.0 and that allometry of size dimorphism does not occur in fleas. In other words, size dimorphism did not change across species with increasing body size. The absence of allometry of sexual size dimorphism in fleas was probably due to the lack of consistency in female size variation in relation to that of males, which is an important prerequisite for Rensch's rule (Colwell, 2000). Indeed, among seven studied flea species, female size varied less than male size in *Parapulex chephrenis* and *Synosternus cleopatrae*, varied more than male size in *Xenopsylla ramesis*, *Xenopsylla dipodilli*, *Nosopsyllus iranus* and *Stenoponia tripectinata*, and size variation was similar in both genders in *Xenopsylla conformis* (Fig. 7.1b).

Other reasons for the lack of relationship between sexual size dimorphism and body size may be that fleas do not conform to assumptions of the various functional hypotheses explaining the evolution of allometry of sexual size dimorphism (Fairbairn, 1997; Colwell, 2000). In particular, many of the hypotheses explaining Rensch's rule have invoked either sexual selection on males or stabilizing selection on females or both (Fairbairn, 1997). Male fleas have no role in reproduction besides mating and there is no evidence that their mating success depends on their size. Furthermore, fleas do not demonstrate strong stabilizing selection or strong constraints on female size. For example, egg production and egg size in fleas seem to be independent of body size (e.g. Vashchenok, 1988), although these relationships have never been specifically tested. Natural selection for niche differentiation and resource partitioning are also not relevant for fleas, as males and females parasitize the same host individuals and share the same feeding niche.

7.1.2 Shape dimorphism

Very few studies have been undertaken to examine sexual dimorphism in fleas. This is especially true for morphological characters other than body size. Nevertheless, shape differences between male and female fleas have been reported for some species. For example, female *Cediopsylla simplex*, *Ctenocephalides felis* and *Orchopeas howardi* possess significantly more spines in their combs than conspecific males (Amin, 1974, 1982; Amin et al., 1974; Amin & Sewell, 1977). This morphological dimorphism has been used to explain the female-biased sex ratio in these species (Amin & Sewell, 1977). The rationale is that males could be more easily dislodged by host grooming effort as they have fewer spines to anchor themselves to the host's hairs. While this explanation is unsatisfactory (see Marshall, 1981a), it has been observed that *Megabothris acerbus*, a species with no difference between males and females in the number of spines on the

82   Ecology of sex differences

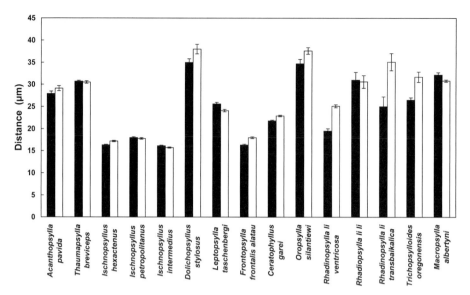

Figure 7.2 Mean (± S.E.) distance (μm) between spines of the pronotal comb in males (black columns) and females (white columns) of 15 flea species. Data from Medvedev (2001).

pronotal comb, occurs on chipmunks in approximately equal numbers (Amin & Sewell, 1977). On the other hand, Hůrka (1963a) reported a higher number of spines in the pronotal and abdominal combs in male *Ischnopsyllus octactenus* compared with those in females, although the sex ratio in this flea was clearly female-biased.

With the exception of genitalia-related characters, the only other morphological feature for which sexual dimorphism has been reported in fleas is the distance between spines of the pronotal comb. In six of 15 species studied by Medvedev (2001), this distance was larger in females than in males, the opposite was the case in two other species, and no difference was found in the remaining seven species (Fig. 7.2). No size correction analysis was performed, so it is unclear whether the reported difference is a true manifestation of sexual dimorphism or it is simply a size-related feature. However, male–female difference in distance between comb spines in *Ischnopsyllus hexactenus* and the lack of this difference in a congeneric species of approximately the same size, *Ischnopsyllus peropolitanus*, suggest that sexual dimorphism in this trait is a real phenomenon, although it is difficult to explain why it occurs in some species but not in others.

7.2   Physiological gender differences

Physiological differences between male and female fleas are associated not only with differences in body size, but also with differences in their

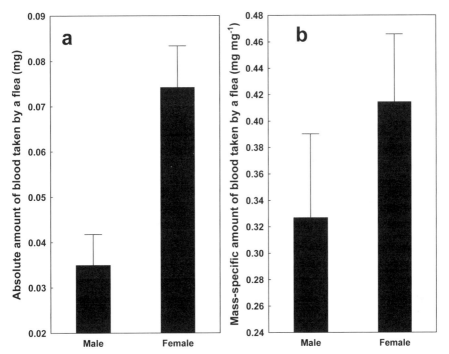

Figure 7.3 Mean (±S.E.) (a) absolute (mg) and (b) mass-specific (per mg of body mass) amount of blood consumed by *Parapulex chephrenis* from the Egyptian spiny mouse *Acomys cahirinus* during 1 h of feeding. Data from Sarfati et al. (2005).

biological roles. As mentioned previously, males have no role in reproduction besides mating. As a result, females are generally more mobile than males and have greater locomotory ability. Between-gender difference in locomotion and associated traits will be discussed in detail in Chapter 8. In addition, male and female fleas differ in their digestion physiology, metabolic rate and water content.

7.2.1   Feeding parameters

Both males and females have the same basic feeding needs. The absolute amount of blood consumed by a female is usually greater than that consumed by a male, all else (host species and feeding time) being equal, due to larger body size. A higher amount of blood taken by a female compared with a male has been reported for such flea species as *X. conformis*, *Xenopsylla cheopis*, *Xenopsylla astia*, *Leptopsylla segnis*, *Nosopsyllus laeviceps*, *Nosopsyllus consimilis*, *Ctenophthalmus golovi*, *Neopsylla setosa*, *Citellophilus tesquorum* and *Coptopsylla lamellifer* (Devi & Prasad, 1985; Vashchenok et al., 1988; Liu et al., 1993; Y. L. Gong et al., 2004). In particular, this has also been shown for *P. chephrenis* (Sarfati et al., 2005) (Fig. 7.3a).

84   Ecology of sex differences

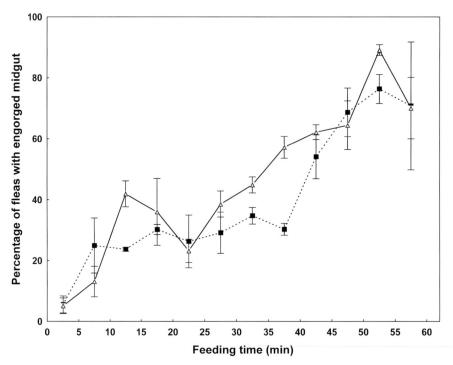

Figure 7.4 Mean (± S.E.) percentage of male (solid line, triangles) and female (dashed line, squares) *Parapulex chephrenis* with engorged midgut after different periods of feeding on the Egyptian spiny mouse *Acomys cahirinus*. Data from Sarfati et al. (2005).

However, when between-sex differences in body size of *P. chephrenis* were taken into account, it appeared that both males and females consume similar relative amounts of blood (Fig. 7.3b). Comparisons of gender differences using absolute and mass-specific blood intake for other species may even provide opposite results (see Chapter 10). Furthermore, no gender difference in the rate of engorgement was found in *P. chephrenis* (Fig. 7.4). Similarly, no gender difference in time to initiation and duration of the first blood meal was reported for *Ctenocephalides canis* (Cadiergues et al., 2001). Nevertheless, females of most species studied by Vashchenok et al. (1988) engorged less blood during the first than during subsequent blood meals, whereas blood portions engorged by males did not increase with age.

Some other gender differences in feeding parameters have been found. In particular, frequency of feeding has been shown to be higher in males than in females in some, albeit not all, flea species (Table 7.1) (see also Y. L. Gong et al., 2004). This can be considered as a compensation for relatively lower chances of the successful attack a host in males due their lower locomotory ability (see Chapter 8). In addition, this can be associated with lower amount and activity

Table 7.1 *Frequency of blood meals in males and females of eight flea species*

Species	Interval between consecutive blood meals (h)		Reference
	Males	Females	
Coptopsylla lamellifer	4.8	6.3	Kunitskaya *et al.*, 1965a
Ctenophthalmus dolichus	5.8	11.4	Kunitskaya *et al.*, 1965a
Frontopsylla elata	6.8	9.6	Guseva & Kunitsky, 1974
Frontopsylla semura	14.1	15.0	Bryukhanova & Surkova, 1970
Nosopsyllus laeviceps	2.5	2.3	Kunitskaya *et al.*, 1965a
Xenopsylla cheopis	9.0	12.7	Kunitskaya *et al.*, 1965a
Xenopsylla gerbillis	6.8	10.3	Kunitskaya *et al.*, 1965a
Xenopsylla hirtipes	4.5	9.0	Kunitskaya *et al.*, 1965a

of the salivary gland lysates (apyrases) in male compared with female fleas (*X. cheopis*, *Thrassis bacchi* and *O. howardi*: Ribeiro *et al.*, 1990). These lysates convert adenosine tri- and diphosphate to adenosine monophosphate and orthophosphate and thus destroy the signal for platelet aggregation.

Newly emerged male and female *P. chephrenis* differed in their rate of blood digestion, as shown by Sarfati *et al.* (2005). In this study, blood digestion status was evaluated following a modified classification of Ioff (1949) as early, middle or late stage (see details in Chapter 10). Overall time of blood digestion was significantly shorter in newly emerged males than in newly emerged females, but adult males and females digested blood at the same rate (Fig. 7.5). The gender difference in newly emerged individuals was due to the duration of the middle, but not early or late, stage of blood digestion (8.15 ± 0.77 h versus 10.60 ± 0.40 h, respectively: Sarfati *et al.*, 2005).

7.2.2   Metabolic rate

The metabolic rate of an individual animal depends on a variety of factors, the most important of which is ambient temperature. Consequently, comparisons of metabolic rate between males and females should be carried out under the same environmental conditions and, if sensitivity to extrinsic factors differs between genders, then the results of such a comparison can be condition-dependent.

Fielden *et al.* (2004) compared metabolic rate between male and female *X. ramesis* under a range of temperatures using CO_2 emission as a measure. They found that adult males and females did not differ in their rate of CO_2 production

86   Ecology of sex differences

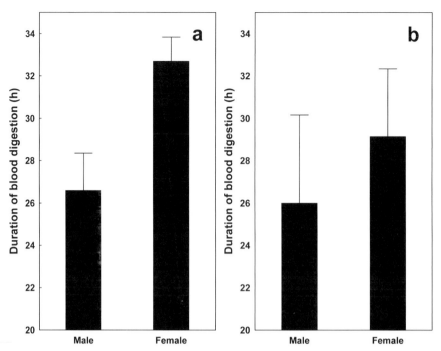

Figure 7.5 Mean (±S.E.) duration (hours) of digestion of blood of the Egyptian spiny mouse *Acomys cahirinus* in (a) newly emerged and (b) adult *Parapulex chephrenis*. Data from Sarfati et al. (2005).

at either relatively low (10 °C) or relatively high (25–30 °C) temperatures. However, at 15 and 20 °C, mass-specific emission of CO_2 was significantly higher in females than in males (Fig. 7.6). Higher metabolic rate in females was reported for six of seven flea species measured at 25 °C by Krasnov et al. (2004b) (Fig. 7.7). The only species that did not demonstrate gender difference in metabolic rate was *S. tripectinata*. Let's keep this species in mind. We will get back to it in the next chapter.

Gender differences in metabolic rate (evaluated via O_2 consumption) were also reported for *X. cheopis*, *Nosopsyllus fasciatus*, *N. setosa*, *C. tesquorum*, *Xenopsylla skrjabini* and *Xenopsylla nuttalli* (Kondrashkina & Dudnikova, 1962, 1968; Kondrashkina & Gerasimova, 1971). In contrast with the results of Fielden et al. (2004), these studies found a higher metabolic rate in adult males compared with adult females in some species (Fig. 7.8).

It is difficult, at first glance, to explain the sharp contradiction between these results and those of Fielden et al. (2004) and Krasnov et al. (2004b), due to the drastic difference in methodology between these studies. However, the majority of experiments by Kondrashkina & Dudnikova (1962, 1968) and

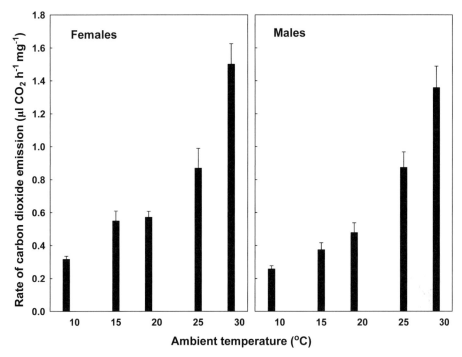

Figure 7.6 Mean (±S.E.) mass-specific production of carbon dioxide (μl per mg body mass per hour) in adult males and females of Xenopsylla ramesis at five different ambient temperatures. Redrawn after Fielden et al. (2004) (reprinted with permission from Elsevier).

Kondrashkina & Gerasimova (1971) used sexually naïve females, whereas Fielden et al. (2004) and Krasnov et al. (2004b) used reproducing females. Therefore, the contradictory results of the two series of studies may be due to differences in metabolic rate between reproducing and non-reproducing females. Indeed, when Kondrashkina & Gerasimova (1971) compared the oxygen consumption of newly emerged females with that of reproductive females of X. skrjabini and X. nuttalli, it appeared that the metabolic rate of the reproductive females significantly exceeded that of the newly emerged ones and was either similar to or considerably exceeded the metabolic rate of males (at 25 °C, 2.95 versus 2.45 μl O_2 mg^{-1} h^{-1} for X. skrjabini and 2.52 versus 1.88 μl O_2 mg^{-1} h^{-1} for X. nuttalli).

Thus, the level of metabolism of reproducing female fleas at favourable temperatures is generally higher than in males. Higher female metabolism is probably associated with a necessity to digest larger amounts of food (see also Sarfati et al., 2005) and with energetic expenses of oogenesis, which may be more energy costly than spermatogenesis.

88 Ecology of sex differences

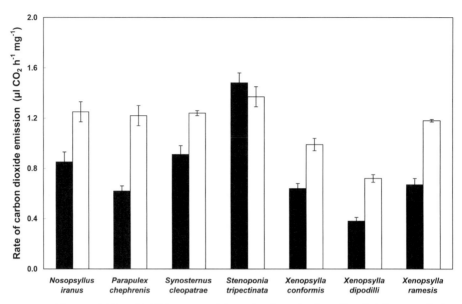

Figure 7.7 Mean (±S.E.) mass-specific production of carbon dioxide (μl per mg body mass per hour) in males (black columns) and females (white columns) of seven flea species at 25 °C. Data from Krasnov et al. (2004b).

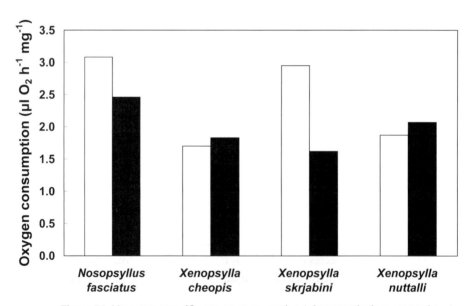

Figure 7.8 Mean mass-specific oxygen consumption (μl per mg body mass per hour) in males (white columns) and females (black columns) of four flea species at 25 °C. Data from Kondrashkina & Dudnikova (1968) and Kondrashkina & Gerasimova (1971).

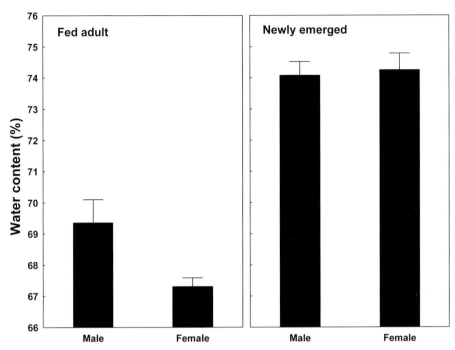

Figure 7.9 Mean (±S.E.) water content (%) in adult and newly emerged males and females of Xenopsylla ramesis. Data from Fielden et al. (2004).

7.2.3 Water content

Careful management of water balance is necessary for fleas because they spend large portions of their life cycle off the host and may thus face periods of starvation when a host is not available. Data on the water content of male and female fleas are scarce. To the best of my knowledge, the only study that documented water content in fleas is that of Fielden et al. (2004). This study demonstrated that newly emerged males and female X. ramesis did not differ in their body water content, whereas adult fed males had a significantly higher water content than adult fed females (Fig. 7.9).

Water can be obtained via food (blood of a host in the case of fleas) or via active water vapour uptake from the air. Active water vapour uptake has been reported in over 60 species of arthropods (O'Donnell & Machin, 1988). Experimental studies have demonstrated water uptake only in the larvae and pre-pupae of X. cheopis and X. brasiliensis (Edney, 1947a; Knülle, 1967; Rudolph & Knülle, 1982; Bernotat-Danielowski & Knülle, 1986), but not the pupae or adults (Edney, 1947a; Knülle, 1967). An absence of a sex-related difference in water content prior to feeding (Fielden et al., 2004), the inability of the imago to uptake water vapour and the similar water loss rate in desiccated male and female fleas (Fielden *et al.*, 2002) suggest that the lower water content of fed females compared to fed males is linked to the processing and/or size of the blood meal. The decrease in water content in females compared to males may reflect a greater accumulation of fat for subsequent egg production. In female mosquitoes, for example, a significant portion of the blood meal is used to synthesize fat tissue and oocyte lipids (Zeigler & Ibrahim, 2001).

7.2.4 Vector competence

Physiological differences between genders may be reflected in gender differences in the ability to serve as vectors of some diseases. For example, such a gender difference was reported for the rate of blockage of flea proventriculus by the plague pathogen *Yersinia pestis*. This blockage is considered to be the main mechanism of plague transmission (see details in Chapter 14). Bazanova & Khabarov (2000), Bazanova *et al.* (2000) and Tokmakova *et al.* (2006) demonstrated this difference for *C. tesquorum*, *X. cheopis* and *Amphipsylla primaris*. However, manifestation of the gender difference in the rate of plague blockage depended on both season and age of the flea. For example, in experiments with *C. tesquorum*, Bazanova & Khabarov (2000) used 'overwintered' fleas that were infected with plague pathogen prior to overwintering and hibernation and 'young' fleas that did not overwinter after being infected. Gender difference in the rate of blockage was found only in the latter category of insects, with males being more often blocked than females. In these fleas, the rate of blockage varied seasonally from 7.0% to 21.7% in males and from 2.0% to 14.0% in females. These results suggest that some unknown physiological changes occurring during hibernation negate gender difference in the effect of the plague pathogen multiplication in the gut of a flea (see also Bazanova & Mayevsky, 1996; Mayevsky *et al.*, 1999).

7.3 Gender differences in behaviour

Fleas may use various signals for host location (Humphries, 1968), dispersal (Darskaya & Besedina, 1961) and hiding (Humphries, 1968). Behaviour of fleas will be discussed in more detail in the subsequent chapters of this book, whereas here I concentrate specifically on gender differences.

In general, starving fleas of many species, in particular those exploiting fossorial hosts, are positively phototactic and negatively geotactic, whereas the opposite is true for fed fleas (e.g., Rothschild & Ford, 1973; see also Chapter 9). The urgency of a blood meal is more critical for females than for males (see Chapter 5). Female fleas are, therefore, expected to be more sensitive to extrinsic stimuli (both originated from the host and from the off-host environment), which should result in more effective host location. Indeed, host-location abilities have been shown to be better in females compared to males of *X. cheopis* (Smith, 1951). However, the ability to distinguish between host species using an odour cue (see Chapter 9) appeared to be similar between males and females of *P. chephrenis* and *X. dipodilli* (Krasnov et al., 2002a). Burdelov et al. (2007) carried out experiments on the relocation of newly emerged and fed individuals of three flea species (*X. conformis*, *X. ramesis* and *P. chephrenis*) in response to light and surface angle. All three species demonstrated positive phototaxis independently of their feeding status, whereas geotactic behaviour varied among species (see further discussion in Chapter 9). No gender difference was found in response to extrinsic stimuli in the two *Xenopsylla* species, whereas the proportion of relocating females of *P. chephrenis* was significantly higher than the proportion of relocating conspecific males.

Moskalenko (1958) observed behaviour of five flea species (*Megabothris calcarifer*, *Ctenophthalmus congeneroides*, *Neopsylla bidentatiformis*, *Frontopsylla elata* and *Rhadinopsylla insolita*) when abandoning the dead body of the fieldmouse *Apodemus agrarius*. It appeared that males left the host corpse earlier than females did. Frequency of host attacks was shown to be higher in females than in males of *N. bidentatiformis* and *C. tesquorum* (Ma, 1994a). Gender differences in activity level were demonstrated for some species (*Ceratophyllus hirundinis*, *Ceratophyllus farreni*, and *Ceratophyllus rusticus*: Marshall, 1981a; Greenwood et al., 1991), but not for others (*X. cheopis*: Clark et al., 1993a).

The above results imply that, in general, behavioural differences between genders are weakly expressed in fleas despite some degree of sexual dimorphism in sensory apparatus. For example, antennae are longer and the number of sensilla on the external margin of the scape and along the dorsal margin of the antennal fossa is greater in males than in females of *Polygenis tripus* (de Albuquerque Cardoso & Linardi, 2006). Nevertheless, sex ratio variation in fleas collected at the same time period from different parts of host burrows (see below) suggests that a weak difference in behaviour may play a role in the spatial distribution of genders.

7.4 Gender differences in responses to environmental factors

Sexual size dimorphism and physiological differences between males and females may be the main cause for differential sensitivity of males and females to environmental factors such as air temperature and relative humidity (RH). Between-gender difference in sensitivity to environmental factors can be illustrated by a comparison of the resistance to starvation in male and female fleas. For example, male *X. conformis* and *X. ramesis* appeared to be less

92   Ecology of sex differences

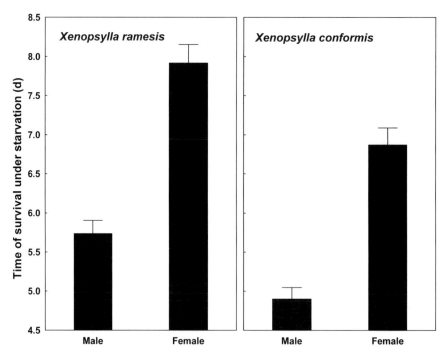

Figure 7.10 Mean (± S.E.) time (days) of survival of male and female *Xenopsylla ramesis* and *Xenopsylla conformis* after a single blood meal at 25 °C. Data from Krasnov et al. (2002d).

resistant to starvation than female conspecifics and survived a significantly shorter time when starved (Krasnov et al., 2001a) (Fig. 7.10). However, this difference disappeared at high air temperature (38 °C). This temperature was probably the threshold for flea survival, so the between-gender differences in resistance to starvation seemed to be levelled. On the other hand, if the level of RH was changed during development of pre-imaginal fleas, newly emerged *X. conformis* (but not *X. ramesis*) demonstrated a significant male bias, suggesting a between-gender difference in the resistance to humidity fluctuations, with females less resistant than males (Krasnov et al., 2001a). Kunitsky et al. (1971a) reported that female *Xenopsylla gerbilli* were more resistant to an increase of moisture in the soil than males. If the soil moisture reached at least 15%, they survived four times longer than males. However, survival of unfed and sexually naïve males and females of *N. laeviceps* did not differ (Amin et al., 1993).

Another example is the developmental rate of female and male eggs in *X. conformis* (Krasnov et al., 2001b). In this species, female eggs developed faster than male eggs at low RH, but not so at high RH (Fig. 7.11). In addition, between-gender

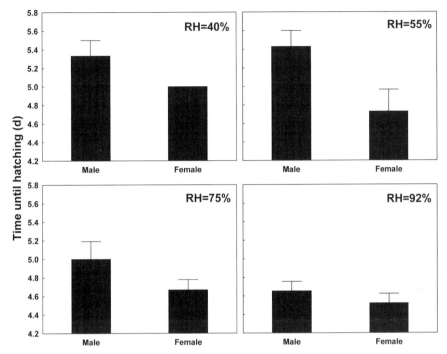

Figure 7.11 Mean (± S.E.) development time (days) of *Xenopsylla conformis* eggs at four levels of relative humidity (RH). Data from Krasnov et al. (2001b).

differences in the duration of larval development of this species were the most evident at lower air temperatures and RH (Krasnov et al., 2001b).

7.5 Sex ratio as an ecological consequence of gender differences

Differential responses to environmental factors may lead to a difference in the relative fitness of male and female offspring under given environmental conditions. Consequently, parents adjust the sex of their offspring according to the environment (Trivers & Willard, 1973). Furthermore, sex ratio can be affected by endocrine disruptors and other chemicals (e.g. Watts et al., 2002; Shutler et al., 2003). Differential fitness is not the only reason for the deviation of the sex ratio from unity. It can also be affected by the probability of mating and the likelihood of inbreeding. Probability of mating, in turn, can be affected by behavioural differences between males and females. As a result of gender differences, it is expected that sex ratios in natural flea populations should deviate from unity.

Indeed, many flea species demonstrate a strong female bias (e.g. Walton & Tun, 1978; Marshall, 1981b; Ryba et al., 1986; Ma, 1993b; Blaski, 2004; Meng et al., 2006), although male bias has been also observed in some species (Skuratowicz, 1960;

Peus, 1970; Manhert, 1972; Haitlinger, 1975). Among 72 flea species belonging to various families, parasitic on various hosts, and inhabiting various geographical regions, 53 species are female-biased, 14 species have a sex ratio close to unity, and only five species show a male bias (Table 7.2). This frequency distribution of species with different sex ratio patterns is similar to that reported by Marshall (1981b). In some cases, fleas with contrasting sex ratios belong to the same family (or even the same genus), but in other instances, fleas from different families have similar sex ratios. This suggests that phylogenetic constraints are not likely to be involved in the determination of sex ratio in flea populations. In most cases, the reasons for biased sex ratio are unknown.

Does the observed sex ratio persist from the egg stage or is it a result of differential mortality of males and females at a later stage? It is believed but, to the best of my knowledge, has not yet been demonstrated that flea gender is determined via Mendelian segregation of sex chromosomes at conception. Thus, males and females should be produced initially in equal numbers (Marshall, 1981a, b). Identification of the primary sex ratio in fleas (as well as in other small holometabolous insects) is difficult because fleas can die at any stage of their life cycle. Consequently, the true primary sex ratio occurs at the stage of the eggs. However, it is practically impossible to sex flea eggs and it is incredibly difficult to sex larvae and pupae. As a result, the earliest the sex ratio of fleas can be identified is in newly emerged adults. Comparison of values of this 'pseudo-primary' sex ratio from the data obtained in laboratory experiments and the values of sex ratios obtained from field samples can help us to understand if an observed sex ratio in a particular flea species is related to differential survival of males and females. For example, in laboratory cultures of *X. conformis* and *X. ramesis* maintained on their natural host (Sundevall's jird *Meriones crassus*), sex ratios (female:male) in newly emerged fleas were 1.1 and 0.9, respectively (B. R. Krasnov, unpublished data). Comparison of these values with those from Table 7.2 as well as the reports of the 1:1 sex ratio in newly emerged fleas from laboratory cultures (e.g. *Oropsylla silantiewi*: Zhovty & Peshkov, 1958) suggests that the main reason for the deviation of the sex ratio from unity observed in the field is related to gender differences in lifespan and/or differential sensitivity of males and females to some extrinsic factors such as environment (e.g. air temperature and RH) or host defensiveness. Gender differences in resistance to starvation seem to be the rule in fleas (see above), although the direction of this difference varies among species. For example, female *X. cheopis* are more resistant to starvation than males (Leeson, 1936; Edney, 1945), whereas the opposite is true in *Caenopsylla laptevi ibera* (Cooke, 1999). Male fleas are also characterized by a shorter lifespan compared to females, as reported for *Callopsylla caspia* and *F. elata* by Talybov (1974) (see also Rothschild & Clay, 1952; Marshall, 1981a, b). As

Table 7.2 *Female : male ratio in 72 flea species (data are pooled across host species, seasons and years)*

Species	Geographical location	Female: male ratio		Reference
		Host body	Host burrow	
Aetheca wagneri	California	1.14		Linsdale & Davis, 1956
Amphalius clarus clarus	North China	1.69	1.58	Ma, 1993b
Amphipsylla daea	North China	1.92		Ma, 1993b
Amphipsylla vinogradovi	North China	3.39	3.31	Ma, 1993b
Anomiopsyllus falsicalifornicus congruens	California	1.54	1.83	Linsdale & Davis, 1956
Atyphloceras echis longipalpus	California	1.00	0.31	Linsdale & Davis, 1956
Atyphloceras multidentatus	California	1.51	1.51	Linsdale & Davis, 1956
Callopsylla dolabris	North China	1.92	2.47	Ma, 1993b
Carteretta carteri	California	1.00	3.00	Linsdale & Davis, 1956
Cediopsylla inequalis	California	1.91		Linsdale & Davis, 1956
Ceratophyllus celsus	Alaska	10.0		Haas et al., 1981
Ceratophyllus idius	Alaska	1.19		Haas et al., 1981
Citellophilus tesquorum sungaris	North China	1.31	1.91	Ma, 1993b
Ctenocephalides canis	California	8.50		Linsdale & Davis, 1956
Ctenocephalides felis	Georgia (USA)	1.14		Durden et al., 2005
Ctenophthalmus agyrtes	Finland	1.22		Ulmanen & Myllymäki, 1971
Ctenophthalmus uncinatus	Finland	1.56		Ulmanen & Myllymäki, 1971
Ctenophyllus armatus	New Mexico	2.22		Morlan, 1955
Dactylopsylla bluei	California	0.20		Linsdale & Davis, 1956
Echidnophaga gallinacea	California	6.03		Linsdale & Davis, 1956
Epitedia stanfordi	New Mexico	1.85		Morlan, 1955
Epitedia wenmanni	West Virginia	1.60		Joy & Briscoe, 1994
Euchoplopsyllus glacialis foxi	California	1.66		Linsdale & Davis, 1956
Eumolpianus fornacis	California	2.60		Linsdale & Davis, 1956
Foxella ignota	California	0.80		Linsdale & Davis, 1956
Frontopsylla aspiniformis	North China	1.88	1.44	Ma, 1993b
Hoplopsyllus anomalus	California	1.33		Linsdale & Davis, 1956
Hystrichopsylla multidentata	North China		2.11	Ma, 1993b
Leptopsylla segnis	California	1.05		Linsdale & Davis, 1956
Malaraeus telchinus	California	1.51		Linsdale & Davis, 1956
Megabothris acerbus	Wisconsin	1.00		Amin, 1976
Megabothris walkeri	Finland	0.92		Ulmanen & Myllymäki, 1971
Meringis cummingi	California	1.07		Linsdale & Davis, 1956
Meringis jamesoni	New Mexico	1.87		Morlan, 1955
Meringis nidi	New Mexico	0.56		Morlan, 1955

(cont.)

Table 7.2 (cont.)

Species	Geographical location	Female: male ratio		Reference
		Host body	Host burrow	
Meringis parkeri	New Mexico	1.56		Morlan, 1955
Meringis rectus	New Mexico	0.89		Morlan, 1955
Nearctopsylla myospalaca	North China	2.03		Ma, 1993b
Neopsylla abagaitui	North China	2.09	1.94	Ma, 1993b
Neopsylla bidentatiformis	North China	1.67	1.64	Ma, 1993b
Neopsylla paranoma	North China	1.15		Ma, 1993b
Nosopsyllus iranus	Negev Desert	1.68		B. R. Krasnov, unpublished data
Nosopsyllus pumilionis	Negev Desert	1.25		B. R. Krasnov, unpublished data
Ochotonobius hirticrus	North China	1.47	1.14	Ma, 1993b
Odontopsyllus dentatus	California	0.87		Linsdale & Davis, 1956
Ophthalmopsylla jettmari	North China	1.80	2.02	Ma, 1993b
Ophthalmopsylla kukuschkini	North China	1.48	1.79	Ma, 1993b
Ophthalmopsylla praefecta praefecta	North China	2.83	2.10	Ma, 1993b
Orchopeas leucopus	Wisconsin	2.41		Amin, 1976
Orchopeas leucopus	West Virginia	1.80		Joy & Briscoe, 1994
Orchopeas sexdentatus	California	1.31	4.67	Linsdale & Davis, 1956
Oropsylla bruneri	Manitoba	1.43		Reichardt & Galloway, 1994
Oropsylla hirsuta	New Mexico	1.77		Morlan, 1955
Oropsylla montana	California	1.30		Linsdale & Davis, 1956
Oropsylla silantiewi	North China	1.34	1.44	Ma, 1993b
Oropsylla tuberculata	New Mexico	1.56		Morlan, 1955
Palaeopsylla remota	Yunnan, China	0.91		Qian et al., 2000
Parapulex chephrenis	Negev Desert	1.02		B. R. Krasnov, unpublished data
Peromyscopsylla hesperomys	California	1.49		Linsdale & Davis, 1956
Pleochaetis exilis	New Mexico	1.13		Morlan, 1955
Polygenis martinezbaezi	New Mexico	3.00		Haas & Wilson, 1998
Pulex irritans	California	3.29		Linsdale & Davis, 1956
Pulex simulans	Georgia (USA)	2.01		Durden et al., 2005
Rhadinopsylla aspalacis	North China	1.18		Ma, 1993b
Rhadinopsylla dahurica vicina	North China	1.30	1.58	Ma, 1993b
Rhadinopsylla multidenticulata	New Mexico	1.32	1.50	Morlan, 1955

Table 7.2 (cont.)

Species	Geographical location	Female: male ratio		Reference
		Host body	Host burrow	
Stenistomera alpina	New Mexico	1.27		Morlan, 1955
Stenoponia singularis	North China	2.25		Ma, 1993b
Stenoponia tripectinata	Negev Desert	1.30		B. R. Krasnov, unpublished data
Synosternus cleopatrae	Negev Desert	0.59		B. R. Krasnov, unpublished data
Xenopsylla conformis	Negev Desert	1.39		B. R. Krasnov, unpublished data
Xenopsylla dipodilli	Negev Desert	0.94		B. R. Krasnov, unpublished data
Xenopsylla ramesis	Negev Desert	1.61		B. R. Krasnov, unpublished data

a result, the proportion of males may decrease in a cohort with an increase of age (e.g. *Synopsyllus fonquerniei*: Klein, 1966).

Furthermore, the data in Table 7.2 were derived by pooling the numbers of male and female fleas across seasons and host species. However, seasonal variation in sex ratio within a flea species can be dramatic (e.g., Morlan, 1955; Haas, 1969; Zhovty, 1970; Jurík, 1974; Haitlinger, 1975; Amin, 1976; Cai et al., 2000), even in the environments where seasonality is weakly expressed (Schwan, 1993; Krasnov et al., 2002c). For example, in *Doratopsylla dasycnema* in the Sudetes Mountains male bias was observed in June, August and October, whereas the sex ratio did not deviate from 1:1 in other months (Haitlinger, 1975). In the same region, *Palaeopsylla soricis* demonstrated male bias in April and October only. In the Tyrol Mountains (Manhert, 1972) and the Sudetes (Haitlinger, 1971, 1973, 1974), female *Ctenophthalmus agyrtes* outnumbered males at the end of spring to the beginning of summer, whereas the opposite was the case at the beginning of autumn. Four flea species parasitic on rodents in the Negev Desert also demonstrated a variety of seasonal patterns in sex ratio (Krasnov et al., 2002c) (Fig. 7.12). Sex ratio in *X. dipodilli* changed significantly between months, averaging ∼1:1 during July–October and March–April, with significant male bias in November–January and female bias in February (Fig. 7.12a). In *X. ramesis*, the sex ratio did not deviate significantly from 1:1 during most of the year, except September and March, when there were excess females and males, respectively (Fig. 7.12b). Sex ratios in winter-active *N. iranus* began the season (October–November) with a strong bias

98   Ecology of sex differences

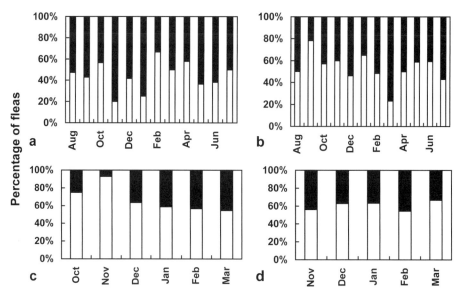

Figure 7.12 Seasonal variation in sex ratio (males — black, females — white) in (a) Xenopsylla dipodilli, (b) Xenopsylla ramesis, (c) Nosopsyllus iranus and (d) Stenoponia tripectinata. Redrawn after Krasnov et al. (2002c) (reprinted with permission from Blackwell Publishing).

to females and then settled at ~1:1 during the next 4 months (Fig. 7.12c). In contrast, females of S. tripectinata consistently outnumbered males by about 3:2 during the entire activity period (Fig. 7.12d).

The sex ratio of a flea species can also vary among different geographical locations as well as among host species (see also Marshall 1981a, b). In the United States, for example, female bias (female:male) in O. howardi was as high as 2.4 in Wisconsin (Amin, 1976) and North Carolina (Shaftesbury, 1934), but was as low as 1.6 in New Work (Layene, 1954) and Indiana (Wilson, 1961). Female bias in Megabothris quirini was found to increase in the northern parts of its geographical range (Gabbutt, 1961). This could be related to a higher ability of females to survive harsh conditions. In New Mexico, the sex ratio in Meringis nidi on Ord's kangaroo rat Dipodomys ordii was male-biased (female:male 0.6), whereas it was female-biased on the banner-tailed kangaroo rat Dipodomys spectabilis (female:male 1.4) (calculated from Morlan, 1955). In the same location, female bias in Orchopeas sexdentatus was higher in individuals collected from the southern plains woodrat Neotoma micropus than in those collected from the white-throated wood rat Neotoma albigula (female:male ratios 1.5 versus 1.17, respectively; calculated from Morlan, 1955). In central California, the sex ratio (female:male) in Malaraeus telchinus was 1.6 in the California vole Microtus

Table 7.3 *Proportion of females in samples of fleas collected from burrows and bodies of the Daurian ground squirrel* Spermophilus dauricus *in southeastern Trans-Baikalia*

Species	Burrow entrance	Nest	Host body
Citellophilus tesquorum	0.57	0.61	0.48
Frontopsylla luculenta	0.63	0.56	0.63
Neopsylla bidentatiformis	0.79	0.64	0.74
Neopsylla abagaitui	0.63	0.91	0.66
Oropsylla asiatica	0.73	0.66	0.48

Source: Data from Bodrova & Zhovty (1961).

californicus, 1.3 in the pinyon mouse *Peromyscus truei* and 0.93 on the western harvest mouse *Reithrodontomys megalotis* (calculated from Linsdale & Davis, 1956).

The sex ratio may differ depending on whether fleas were collected from a host body or host shelter or even from a particular component of a host shelter. For example, Zhovty (1970) reported the sex ratio in *O. silantiewi* collected from bodies and burrows of the grey marmot *Marmota baibacina* in Trans-Baikalia. The mean sex ratio in fleas collected from host bodies was strongly male-biased (female : male 0.7), whereas those in fleas collected from the burrow entrances and nests were closer to unity (0.8 and 0.9, respectively). This can be related to differential microclimatic preferences of males and females as well as to behavioural differences. However, if the sex ratio in fleas collected from host bodies and nests was relatively stable across sampling sites and periods (0.7–1.0 and 0.8–1.1, respectively), the sex ratio in fleas collected from the burrow entrances varied from 0.3 to 2.0, i.e. from being strongly male-biased to being strongly female-biased. Nevertheless, the pattern of sex ratio distribution among microhabitats varies among flea species parasitic on the same host (Table 7.3). For example, Bodrova & Zhovty (1961) reported that the sex ratio in *C. tesquorum* parasitic on the Daurian ground squirrel *Spermophilus dauricus* was similar regardless of where fleas were collected from (burrow entrance, host nests, or host bodies). In contrast, in *N. bidentatiformis* collected from the nests of *S. dauricus*, female bias was much stronger than in conspecifics collected either from bodies or burrow entrances of this host.

One of the reasons for intraspecific variation in the sex ratio of fleas can be the effect of the emergence schedule (Bossard *et al.*, 2000). For example, it has been shown that the peak of emergence of one gender alternates with that of the other gender in *N. bidentatiformis* and *C. tesquorum* (Ma, 1993b). Furthermore, Ma (1993b) demonstrated that snapshots of newly emerged fleas from laboratory

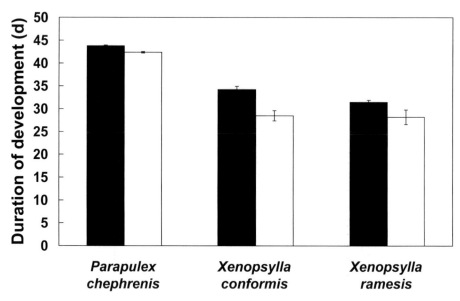

Figure 7.13 Mean (±S.E.) duration (days) of pre-imaginal development of males (black columns) and females (white columns) of three flea species at 25 °C and 92% relative humidity. Data from Krasnov et al. (2001b) and B. R. Krasnov (unpublished data).

cultures demonstrated strong either male or female biases due to the differential emergence schedule (see also Dean & Meola, 2002a). However, if the cumulative number of males and females emerged from pupae over 30 days is considered, it appears that sex ratio demonstrates only a weak female bias (see Fig. 2 in Ma, 1993b).

Differential emergence scale can be associated with between-gender differences in the duration of development. Indeed, females, at least of some species, develop from egg to imago faster than males (Fig. 7.13) (see also Sharif, 1949; Vaughan & Coombs, 1979; Amin et al., 1993). Studying C. felis, Hudson & Prince (1958a), Metzger & Rust (1997), and Kern et al. (1999) also noted that the development time of immature females was shorter than that of males, although this was documented for pupal stages only. It was suggested that the biological significance of this pattern was to prevent inbreeding of fleas from the same cohort by increasing the probability that females will mate with males from other cohorts (Metzger & Rust, 1997).

According to Fisher's (1930) theory of sex allocation, the sex ratio in a population is driven towards the equivalence of investment. In other words, parental investment is equally distributed between male and female progeny. However, if males and females 'cost' different amounts, the production of the more costly

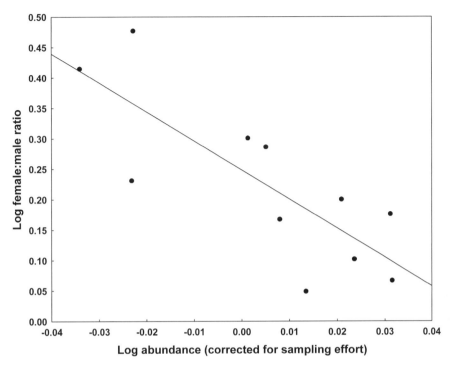

Figure 7.14 Relationship between female bias and total abundance in *Malaraeus telchinus* parasitic on the California vole *Microtus californicus*. Data from Linsdale & Davis (1956). Monthly samples with more than 10 flea individuals were used. The confounding effect of sampling effort was controlled for by substitution of the original values of flea abundance by the residual deviations of a linear regression on the number of examined host individuals in log–log space.

gender is expected to be lower. Female fleas are larger than males (see above) and newly emerged females of some species are characterized by a greater amount of fat tissue compared with conspecific males (Krasnov et al., 2002d). As large size at sexual maturity is associated with certain costs (Ball & Baker, 1996; Anholt & Werner, 1998), female fleas are probably more 'costly' to produce compared with male fleas. If we accept Fisher's (1930) theory about the determination of sex ratio, then we must expect that the sex ratio in fleas should be male-biased, which is far from being a general case. Instead, sex ratio is highly variable. The 'cost-of-production' issue is undoubtedly not a single mechanism determining sex ratio. If we combine Fisherian expectation and considerations related to the probability of mating and chances of extinction, then we can suggest that a trade-off may exist between the high 'cost' of female production and the level of abundance. From an evolutionary perspective, if abundance is low, then the increased number of females (despite their high 'cost') should be advantageous as a single male can inseminate several females. If, however, abundance is high, the chances of extinction are low, the mating probability of every individual is high, and the proportion of females may decrease because they are costly to produce. If the logic of these suggestions is correct, then we may expect that the proportion of females will decrease with an increase in abundance. Indeed, this appeared to be true for *M. telchinus* collected from *M. californicus* in central California (data from Linsdale & Davis, 1956) (Fig. 7.14). However, this single observation should be further validated by studies on other flea and host species in other geographical locations as well as in laboratory and field experiments.

7.6 Concluding remarks

The above examples demonstrate that sex ratio in fleas is far from being uniform both among and within flea species. Furthermore, it appears that the sex ratio in fleas is not associated with a single determinant. It rather is affected by interplay of various ecological and evolutionary factors.

Size and physiological and behavioural differences between male and female fleas lead to their differential response to environmental factors and distribution among microhabitats. This, in turn, affects the sex ratio in natural flea populations. The deviation of the sex ratio from unity is undoubtedly associated with the mating probability of an individual flea and may thus have considerable ecological and evolutionary consequences.

8

Ecology of flea locomotion

The locomotory patterns of fleas reflect their way of life as parasites of fur- or feather-covered hosts. Fleas are able to move through dense host pelage and withstand the host's anti-parasitic grooming. They also are able to jump, to move through the substrate of a host's burrow or nest and to move on vertical surfaces (e.g. fleas parasitic on bats). Here, I briefly review the morphological and physiological aspects of flea locomotory features that facilitate the successful exploitation of hosts.

8.1   On-host locomotion

Flea locomotion in host pelage or feathers differs from that of other mammal and bird ectoparasites. For example, Nycteribiidae (bat flies) have a compressed dorsoventral body and long, spider-like legs (Dick & Patterson, 2006). They are capable of fast sliding movements above the fur of the host. In contrast, the laterally compressed body, high and narrow head capsule and flexible joints of the thorax and abdomen of fleas allow them to move through host pelage by dividing the hair during forward movement.

The flea thorax consists of three separate modified segments (pro-, meso- and metathorax), whereas the abdomen consists of 10 segments. The posterior margins of each segment form collars that overlie the anterior margins of the next segment. As a result, these segments are able to 'squeeze' into each other. In contrast to most winged insects, separation of the mesothorax and metathorax in fleas leads to the absence of a pterothorax which is characteristic of other holometabolous insects. It has been suggested that flea ancestors also did not possess a pterothorax (Medvedev, 2003a, 2005). This lack could be considered as a pre-adaptation to ectoparasitism on fur-covered hosts (see Chapter 4). Separation of the thoracic segments, possession of movable between-thoracic sclerites and highly developed phragmata are features that allow high flexibility of the flea body.

In contrast to other Holometabola, the flea prothorax is not reduced but is tightly connected with the head. The lower part of the prothorax (pleurosternum) is strongly elongated, exceeding at least two times the length of the pronotum. The pleurosternum protrudes anterior to the notum and envelops the posterior part of the head from beneath. As a result, the head and prothorax together constitute a frontal complex which is movable relative to other thoracic segments (Medvedev, 2003a, b). Structures of this complex also include maxillary plates (first segments of the maxilla that possess highly developed collars) and fore coxae. Due to an elongated pleurosternum, fore coxae are situated anterior to the notum. The maxillary plates are broad in the middle, but narrow at the bases and apices. As a result, the anterior frontal complex of a flea is shaped like a keel which divides the host's hairs or feathers or particles of the substrate of its burrow/nest during flea movement. In addition, the frontal and occipital regions of some fleas are covered with basiconic sensilla and large numbers of pores (Amrine & Lewis, 1978; de Albuquerque Cardoso & Linardi, 2006). These pores are openings for the epidermal glands and exude oily substances onto the cuticular surface of a flea facilitating movements among the host hairs (Rothschild & Hinton, 1968; but see Smith & Clay, 1985 and de Albuquerque Cardoso & Linardi, 2006).

8.2 Off-host locomotion

8.2.1 Mechanics of a flea jump

The jump is the most conspicuous characteristic of fleas. For example, *Pulex irritans* can leap a distance of 33 cm, about 200 times the length of their bodies (Rothschild et al., 1972), and the stick-tight flea *Hectopsylla narium* jumps as far as 25 cm (Blank et al., 2007). The recorded heights of flea jumps are no less impressive, attaining, for example, 16.5 cm in *Echidnophaga myrmecobii* (Mules, 1940) and 33 cm in *P. irritans* and *Ctenocephalides felis* (Rothschild et al., 1972, 1975). Jumping allows these wingless blood-sucking insects to attack their hosts successfully, although their major type of locomotion remains walking (Marshall, 1981a).

Enthralled by fleas' jumping abilities, several scientists have asked what are the morphological and physiological mechanisms that allow fleas to accomplish these leaps (e.g. Bennet-Clark & Lucey, 1967; Rothschild et al., 1973; Bossard,

2002; Krasnov et al., 2003a, 2004b). Sometimes this interest has even resulted in humiliating mockeries from the public media (Abrahams, 2004; O'Hare, 2005). Undoubtedly, the major credit in unveiling flea jump mechanisms belongs to Miriam Rothschild and her collaborators who defined this locomotion as a 'flying leap' (Rothschild et al., 1973) and fleas as 'insects which fly with their legs' (Rothschild & Neville, 1967). In their classical series of papers, Rothschild and co-authors (Rothschild et al., 1973, 1975; Rothschild & Schlein, 1975) reported that the main sources of flea saltatorial power are the muscles of the hind legs and a rubber-like protein (resilin) located in the pleural arch. The resilin pad is homologous with the wing-hinge ligaments in flying insects. After being stretched and then released, resilin yields about 97% of its stored energy (Rothschild et al., 1975). An additional advantage of resilin is that the release of energy stored in this elastic structure seems to be a purely physical process and, unlike the chemically controlled release of energy during muscle contraction, does not depend strongly on air temperature (Rothschild et al., 1973).

The jump of the rat flea *Xenopsylla cheopis* was described in detail by Rothschild et al. (1973, 1975). When a flea prepares to jump, it squats down and contracts its body. It orients its hind femurs almost upright, so that only its hind trochanters and tibiae are in contact with the substrate. Contraction of epipleural and trochanteral depressor muscles squeezes the resilin in the pleural arch. Thoracic and coxa-abdominal catches are engaged. Upon the jump, the levator and the ventral longitudinal muscles relax which causes the femur to descend and the catches to be released. The energy stored in the resilin and the arched pleural and coxal wall is freed and rapidly transferred to the legs providing an acceleration of about 150 g in about 1 millisecond. During the jump, the flea arranges its middle or hind legs in an upright position to be able to hook into a host pelage or feathers. During descent, the flea spreads its legs widely sideways and thus controls its landing.

8.2.2   Jumping capacity, sexual size dimorphism and morphology

The ability to make long and high jumps varies greatly among flea species (Cadiergues et al., 2000). Rothschild et al. (1975) argued that this ability and the development of the pleural arch are related. Indeed, pleural arches and resilin protein are well developed in fleas with high jumping capacity such as *P. irritans*, parasitic on medium and large mammals, and *Ceratophyllus styx*, parasitic on birds (Bates, 1962; Rothschild et al., 1975), but are absent or greatly reduced in poor jumpers such as specific bat fleas Ischnopsyllidae and sessile *Tunga* (Traub, 1972a; Rothschild et al., 1973). However, some fleas with extremely low jumping ability have well-developed pleural arches (e.g. *Jordanopsylla allredi*: Hastriter *et al.*, 1998). Nevertheless, the idea that the size of certain parts of the locomotory apparatus can be indicative of jumping capacity, and thus can be used for comparisons among individuals, between genders and among species, has been suggested. Inspired with this idea, Tripet *et al.* (2002a) measured pleural height in preserved flea specimens to be used as indicators of flea mobility for broad comparisons among species of bird fleas from different geographical and host ranges. They found a negative correlation between flea mobility (expressed via measurements of pleural height) and the degree of host colonialism and a positive correlation between flea mobility and their host range. Nonetheless, these conclusions are valid only if pleural height and jumping capacity are indeed correlated.

Furthermore, Rothschild *et al.* (1975) studied gender differences in jumping capacity in *X. cheopis*, *Spilopsyllus cuniculi* and *Nosopsyllus fasciatus*. This study demonstrated that, on average, males jumped a shorter distance than females, which is not surprising due to obvious sexual size dimorphism in fleas, with males smaller than females (see Chapter 7). However, the data were not corrected for body size dimorphism and, thus, it was unclear whether size was the only source of gender difference in locomotory performance or other factors were also involved.

The only experimental study with simultaneous measurements of jumping performance among individuals within species, between genders and among species of fleas was carried out by Krasnov *et al.* (2003a) who searched for correlates between jumping performance and morphometrics of the locomotory apparatus in seven flea species. A flea was allowed to jump, and the jump length was measured. Then, the flea was anaesthetized and measured. Maximal body length was used as a trait describing body size, whereas the pleural height and the length of coxa, femur and tibia of the left hind leg reflected the morphometrics of the locomotory apparatus. Surprisingly, this study demonstrated that morphometrics of the jumping apparatus did not correlate with jumping capacity; no correlation was found between jumping performance and any measure of the locomotory system either among individuals, between genders or among species.

In addition, it was found that interspecific differences in jumping capacity were not related to interspecific differences in body size and locomotory morphometrics (Fig. 8.1; compare with Fig. 7.1a). These results hold also when controlling for the confounding effect of phylogeny. Furthermore, females were generally better jumpers than males, even when accounting for sexual size dimorphism (Fig. 8.1) (Krasnov *et al.*, 2003a). However, males and females of *Stenoponia tripectinata* demonstrated similar jumping ability despite body size

Table 8.1 *Absolute female:male jumping performance ratio and the fraction of performance dimorphism explained by size dimorphism in six flea species*

Species	Female:male performance ratio		Fraction of dimorphism explained by body size
	Absolute	Size-corrected	
Nosopsyllus iranus	1.19	1.16	0.14
Parapulex chephrenis	1.18	1.16	0.10
Synosternus cleopatrae	1.13	1.11	0.16
Xenopsylla conformis	1.48	1.42	0.13
Xenopsylla dipodilli	1.15	1.10	0.34
Xenopsylla ramesis	1.41	1.34	0.16

Source: Modified after Krasnov et al. (2003a) (reprinted with permission from Blackwell Publishing).

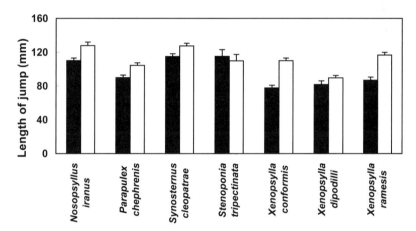

Figure 8.1 Mean (±S.E.) jump length (mm; controlled for body size) of males (black columns) and females (white columns) of seven flea species. Redrawn after Krasnov et al. (2003a) (reprinted with permission from Blackwell Publishing).

difference. In addition, the fraction of the overall differences in jumping ability that was due to size differences between males and females was either low (Table 8.1) or equalled zero (for *S. tripectinata*).

Another way to evaluate the effect of body size differences on jumping ability differences is the regression of female:male ratios of absolute jump length on female:male size ratios. A slope greater than zero would indicate that gender differences in jumping ability can be explained, at least partly, by sexual size dimorphism. Krasnov et al. (2003a) carried out this analysis and found no connection between body size dimorphism and jumping ability differences in studied species.

The occurrence of gender and interspecific differences in flea jumping calls for explanations from morphological, physiological and ecological perspectives. As we have seen, no satisfying morphological explanation of this difference has been found. Physiological and ecological explanations will be discussed in the next two sections.

8.2.3 Jumping performance and metabolic rate

Results of Krasnov et al.'s (2003a) study described in the previous section suggested that jumping ability is determined by factors other than body size or linear metrics of the locomotory system. For example, the amount and/or quality of resilin rather than the pleural height can be a better correlate of jumping ability. In addition, physiological or biochemical differences such as the amount of energy generated by the extensor tibiae muscles (Burrows & Wolf, 2002) or mass-specific activity of enzymes and mitochondria content in muscles (Gäde, 2002) can account for differences in jumping capacity.

Another physiological factor that presumably affects locomotory performance and could explain gender and interspecific differences is metabolic rate. In particular, size-specific differences in metabolic rate between males and females, as well as among species, have been reported for a number of arthropods (Fielden et al., 1999; Rogowitz & Chappell, 2000; see Chapter 7). To understand the mechanism behind intra- and interspecific differences in jumping capacity, Krasnov et al. (2004b) measured resting metabolic rate (RMR) in seven flea species via CO_2 emission (see Lighton, 1991; Fielden et al., 2001, 2004 for methodological details). Both mass-specific and mass-independent RMR were significantly higher in females than in males in all fleas, again except S. tripectinata (see Chapter 7). As we have seen, this species also did not demonstrate gender difference in jumping ability. Furthermore, gender differences in jumping ability between males and females were found to be well explained by gender differences in either mass-specific or mass-independent RMR (Fig. 8.2). In other words, males of most fleas studied were not only poorer jumpers than females, but their RMR was proved to be lower than that of females. Furthermore, higher RMRs were found in species with higher jumping abilities, suggesting a strong connection between metabolic rate differences and jumping ability differences (Fig. 8.3).

As mentioned, fleas originated from winged ancestors presumably similar to the mecopteran families Boreidae or Nannochoristidae (Smit, 1982; Whiting et al., 1997; Lewis, 1998). Although they lost their wings during evolution and returned to a more primitive form of locomotion, fleas retained some flight

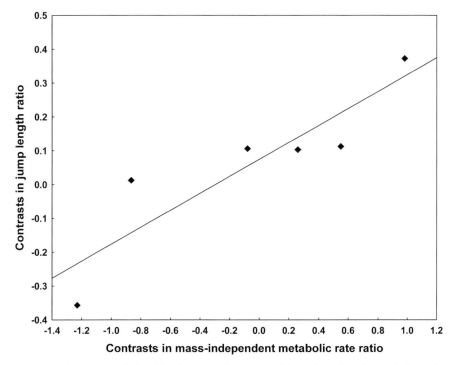

Figure 8.2 Relationships between size-corrected jump length and mass-independent metabolic rates across seven flea species using the method of independent contrasts. Redrawn after Krasnov et al. (2004b) (reprinted with permission from Elsevier).

adaptations which are needed for jumping (Rothschild et al., 1973). In other words, the flight mechanism of flea ancestors has been incorporated into the new jumping mechanism. Therefore, the energetic cost of jumping is presumably high compared with other modes locomotion (Bartholomew & Casey, 1978).

A strong correlation between jumping capacity and RMR in fleas supports the hypothesis about a trade-off between low resting metabolism and efficient metabolism during activity (Reinhold, 1999). This hypothesis states that in species that spend less than half of their daily metabolic energy on resting metabolism, selection will favour mutations that increase RMR but simultaneously decrease metabolic cost of activity; whereas this will not be the case for species where resting metabolism comprises most of the daily energy requirements. Similarly, mutations that decrease RMR but increase the metabolic cost of daily activity will be favoured in species that allocate more than half of their time budget to rest. It was predicted that animals with high energy expenditure on activity would have high RMR. Indeed, flying insects demonstrated higher RMRs than insects with energetically less costly types of locomotion (see Reinhold, 1999; Harrison & Roberts, 2000 for reviews). It is difficult to evaluate

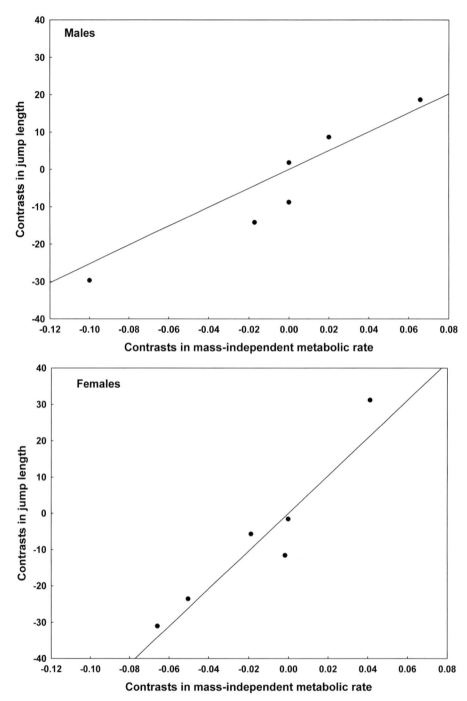

Figure 8.3 Relationship between female : male size-corrected jump length ratios and female : male ratios in mass-independent resting metabolic rates across seven flea species using the method of independent contrasts. Redrawn after Krasnov et al. (2004b) (reprinted with permission from Elsevier).

the proportion of time that a flea allocates to jumping, although this time definitely constitutes less than half of the daily time budget. Nevertheless, fleas demonstrate vigorous jumping activity. For example, newly emerged females of X. cheopis were able to perform up to 250 jumps per 0.5 h (Rothschild et al., 1975). Therefore, the daily metabolic cost of flea jumping can be relatively high, being in between that of flight and that of energetically less demanding locomotory modes such as walking or running.

8.2.4 Ecological correlates of jumping performance

Ecological interpretations of differences in RMR and, as a consequence, in jumping performance between genders and among species of fleas can provide insight into the probable causes of selection for higher or lower levels of RMR that, in turn, were determined by selection for a higher or lower level of jumping capacity.

Gender differences in RMR and jumping can be related to gender differences in reproductive and feeding patterns. As mentioned above, feeding triggers sperm transfer, mating behaviour, egg maturation and oviposition (Chapter 5). The blood meal is more critical for females than for males and, as a result, females need to be more mobile (Iqbal & Humphries, 1970; Rothschild & Ford, 1973; Prasad, 1987; Hsu & Wu, 2001; Dean & Meola, 2002b). As a result of these differences, natural selection would favour females with higher metabolism and, consequently, with increased jumping abilities. Nevertheless, a higher intensity of jumping in males than females (Rothschild et al., 1975) can compensate for shorter jumps in the former. If so, then why do male and female S. tripectinata not differ in their RMR and jumping capacities? Among flea species studied, this species has the shortest period of activity with imagoes being active only for 3–4 months (Krasnov et al., 2002c). Stenoponia tripectinata survive summer in the cocooned stage (teneral adults or pupae in cocoons) and as mature adults that fed and reproduced in the previous season (Krasnov et al., 2002c). The short period of activity during which fleas have to mate and oviposit, and the restricted time that their offspring have to hatch, moult three times and spin a cocoon, requires adults to mate as early in the season as possible. Consequently, both natural and sexual selection would favour more mobile males. As a result, gender differences in both RMR and mobility may disappear. Similar ecological considerations can explain, at least in part, the high ratio of gender differences in the RMR and jumping ability of Xenopsylla conformis and Xenopsylla ramesis that reproduce all year (Krasnov et al., 2002c).

Interspecific differences in RMR and jumping ability seem to be associated with habitat preferences and seasonality. The highest RMR and jump capacity were found either in parasites of psammophylic rodents (*Synosternus cleopatrae*) or species that are active during the short winter season only (*S. tripectinata*) or both (*Nosopsyllus iranus*). The reason of the higher RMR and jumping capacity of these fleas can be that take-off from a mobile sand substrate presumably requires a higher energy investment than that from a hard substrate. Causal relationships between RMR and properties of habitats that determine metabolic costs of activity were reported for different insects. Chironomids from streams had higher metabolic rates than related species from ponds (Walshe, 1948) because the energetic cost of staying in place or swimming against current is higher in streams (Reinhold, 1999). More evidence is provided by *Stenus* beetles with different habitat preferences (Betz & Fuhrmann, 2001). *Stenus comma* which inhabits bare ground has a higher RMR than plant-mounting *Stenus pubescens* presumably because of higher energetic cost of foraging in open sites. Flea studies have confirmed this trend, in that fleas from different habitats differ in RMR and locomotory performance.

8.2.5 Jumping rate and activity

Fleas are able to perform several hundred uninterrupted jumps (Bennet-Clark & Lucey, 1967; Rothschild, 1969; Rothschild et al., 1973, 1975). The rate of flea jumping is affected by various factors. For example, solitary fleas are less active jumpers than fleas maintained in groups (Rothschild et al., 1975; Clark et al., 1993a). High air temperature provokes an increase in jumping rate (Rothschild et al., 1975), whereas water loss inhibits jumping (Humphries, 1968). Moreover, badger fleas *Paraceras melis* maintained in groups demonstrated a cascade jumping response to a single active flea dropped into a container (Cox et al., 1999). Another important parameter affecting jumping rate is blood feeding. Unfed fleas showed a higher jumping intensity than fed ones (Rothschild et al., 1975). Nevertheless, fed *P. melis* separated from a host increased their jumping rate drastically (Cox et al., 1999; Stewart & McDonald, 2003). They also voided their gut content which increased their jumping distance by about 17% (Cox et al., 1999). It was suggested that the voiding of gut content in fleas separated from a host was intended to increase mobility and thus facilitate attempts to return to the host before it moved away. Furthermore, mechanical disturbance (e.g. vibration of a container with fleas) increased flea jumping activity (Rothschild et al., 1975; Cox et al., 1999). The explanation of this was that, after prolonged separation from hosts, fleas relocate to crevices in the substrate or into abandoned burrows seeking protection from desiccation. Vibration can serve as an indicator of a host approaching, stimulating fleas to commence jumping

(Cox et al., 1999). Changes in light intensity provoked jumping of *Ceratophyllus gallinae* (Humphries, 1968, 1969). This can also be interpreted as a cue for a flea to start an active search for a host.

The activity of flea jumping is characterized by diel rhythms. Although only a few flea species have been studied in this relation, it appears that these rhythms of flea host-search activity occur when a host is most readily available. For example, *C. felis* maintained at 12:12 light:dark cycle increased activity just before darkness, i.e. when a host would presumably 'go to bed' (Koehler et al., 1990; Bossard et al., 2000). Good agreement between the circadian rhythm of locomotory activity and potential host availability was found also for *Monopsyllus sciurorum* parasitic on the Eurasian red squirrel *Sciurus vulgaris* (Clark et al., 1997) and several fleas (*Citellophilus tesquorum*, *Frontopsylla luculenta*, *Neopsylla abagaitui* and *Neopsylla bidentatiformis*) parasitic on the Daurian ground squirrel *Spermophilus dauricus* (Pauller & Tchipizubova, 1958).

8.2.6 Walking and climbing

Not all flea species are good jumpers. In general, locomotory patterns of fleas not only reflect their *modus vivendi* but also are complementary with and evolved as a response to the ecology and habits of their preferable hosts (Traub, 1985). For example, Marshall (1981a) argued that flea species that spend their time mainly in nests rather than on hosts, fleas of flying and gliding hosts and fleas of desert rodents are all extremely poor jumpers. Indeed, fleas parasitic on bats almost do not jump at all (Rothschild et al., 1975). However, this generalization is undoubtedly untrue for fleas parasitic on rodents in desert areas. For example, small (about 1.5–1.8 mm long) *X. ramesis* that parasitizes gerbilline rodents (*Gerbillus* and *Meriones*) in the deserts of the Middle East is able to accomplish jumps of about 12 cm in length (B. R. Krasnov, unpublished data).

Most fleas can walk and/or climb and, in some species, walking is the main type of locomotion (Barnes et al., 1977). Others, for example, neosomic females of *Dorcadia*, cannot walk but crawl in a worm-like way, although males of these species are able to walk (Ioff, 1950). Climbing fleas exploit gliding or flying hosts that dwell or roost in caves and hollows. The necessity to climb up the vertical surfaces for host location and attack has led to adaptive changes in the morphology of their legs (Traub, 1985). Indeed, the first segment of the metatarsus in these flea species is unusually long, which is thought to be associated with climbing. Such is the case in *Opisodasys pseudarctomys* and *Tarsopsylla octodicemdentata* (parasitic on flying squirrels), *Choristopsylla tristis* (parasitic on flying phalangers) and *Sternopsylla* (parasitic on cave-roosting bats) (Traub, 1985).

8.3 Concluding remark

Fleas are able not only to move through dense host pelage, but also to jump, to move through the substrate of a host's burrow or nest and to move on vertical surfaces. The diversity of modes of flea locomotion and their efficiency using these modes for host location and exploitation suggest that these locomotory patterns were most likely naturally selected as means to guarantee an evolutionary success of the way of life of the obligatory periodic burrow/nest parasites of avian and mammalian hosts.

9

Ecology of host selection

Selection of an appropriate habitat with necessary and exploitable resources is one of the main tasks for any living organism. If an organism succeeds in fulfilling this task, its reward is translated into an increase or, at least, non-decrease of its fitness. If, however, it fails to find a necessary habitat, its fitness decreases. Continuous fitness decline may finally result in extinction.

The evolutionary motivation of fleas does not differ from that of any other living organism. Fleas have to select carefully their host organisms at both evolutionary and ecological scales. At the evolutionary scale, the selection of an appropriate host or hosts results in a host spectrum for a particular flea species. This spectrum represents a portion of the resource space used by this species, i.e. its fundamental ecological niche. At the ecological scale, an individual flea has to be able to locate and identify an individual of an appropriate host species and to distinguish it from individuals of often similar but less appropriate or even inappropriate species. Furthermore, even if a flea succeeds in finding an appropriate host individual, the selection task is not yet complete. The area of the host body from which (a) a blood meal may be most easily taken and (b) host grooming may be most easily avoided still remains to be found.

In this chapter, strategies of host selection adopted by fleas are discussed. I start with the issues related to differences among hosts from a flea's evolutionary point of view. Then, the patterns and mechanisms of host selection and location by fleas from an ecological perspective are considered. I discuss how fleas select particular parts of the host body and how much time they prefer to spend on a host.

9.1 Evolutionary scale: principal and auxiliary hosts

9.1.1 Definitions

The degree of association between a particular parasite species and a particular host species varies. It is obvious that a strictly host-specific parasite is tightly associated with its principal host, whereas its associations with other host species are probably accidental and can be ignored in most ecological and evolutionary studies. However, host associations of opportunistic parasites represent a more challenging problem. Determining host-selection patterns by generalist parasites is crucial for understanding the causes and consequences of parasite evolution.

The level of specificity in the relationship between a parasite species and its hosts is usually evaluated by comparing the abundance of the parasite in different host species and observing the relative frequency of its association with particular host species in the field. The variation in these parameters among multiple hosts has led to a classification of hosts by the pattern of their relationships with a parasite (e.g. flea) species (Hopkins, 1957; Holland, 1964; Wenzel & Tipton, 1966; Marshall, 1981a). Strict adherence to this classification system is rare and, in some cases, a mammal or a bird may be defined as being a 'true', 'primary', 'accidental' or 'secondary' host for a particular flea based solely on the personal impression of the researcher who studies these associations. However, these terms have been given proper (although not always explicitly clear) definitions. A true host is defined as a single host or a host of primary importance to a flea species. According to Hopkins (1957), a true host is one a flea can reproduce on indefinitely. Holland (1964) argued that a host should be defined as primary if its relationship with a flea is derived from ancient or even original associations. He illustrated the difference between primary and true associations by leporid mammals and spilopsylline fleas, noting that although hares and rabbits are primary hosts for Spilopsyllinae, some ground squirrels and seabirds can be true hosts for some species. Accidental hosts of a flea are those that encounter it by chance (e.g. the squirrel flea *Orchopeas howardi* on the wood duck *Aix sponsa*: Baumgartner & Kane, 1986). However, in many cases hosts that seem to be accidental may be alternative true hosts or may be in the process of becoming true hosts (Sakaguti & Jameson, 1962). Such associations are confirmed by, for example, repeated findings of the shrew fleas on mice and voles and the mouse fleas on arboreal sciurids. For instance, *Doratopsylla dasycnema*, a species characteristic of shrew hosts belonging to the genus *Sorex*, has been repeatedly found on the bank vole *Myodes glareolus* (Nazarova, 1981; Vashchenok & Tretiakov, 2003; Vashchenok, 2006).

A secondary host is an intermediate category for species that are neither accidental nor true hosts. This term was proposed by Smit (1954) for a host that either preys on or co-occurs with a true host of a flea, so it is regularly exposed to and infested by this flea. In some cases, a secondary host may become a true host. For example, *Ceratophyllus lunatus* parasitic on *Mustela* (weasels, minks, ferrets, martens) is the only species of the genus that does not occur on birds and probably evolved by switching from birds to their predators (Hopkins, 1957; Marshall, 1981a; see also King, 1976).

Wenzel & Tipton (1966) also mentioned 'dispersal', 'carrier' or 'sustaining' hosts, whereas Hopkins (1957) and Marshall (1981a) propose to distinguish 'principal' and 'exceptional' hosts among true hosts and 'preferred' and 'normal' hosts among principal hosts. This plethora of terms makes identification of a particular flea–host association rather difficult. As a result, the usefulness of this classification is dubious, and it has rarely been utilized in studies on the ecology and evolution of parasites (in particular, fleas) during last two decades.

Nevertheless, the majority of parasites (including fleas) that exploit more than one host species are unevenly distributed among different host species. Given that the abundance of a consumer in a habitat can be considered as a measure of its efficiency of resource exploitation (Morris, 1987a), the abundance of a parasite on a host species can be considered as a measure of its efficiency of host exploitation. Variation in the abundance of a parasite among its different host species may also reflect parasite specialization. Consequently, the simplest and the most convenient classification is to distinguish between (a) one host species in which the prevalence and intensity or abundance of a parasite are highest (the principal host) and (b) all other hosts in which the prevalence and intensity or abundance of a parasite are lower (auxiliary hosts) (Dogiel et al., 1961; Krasnov et al., 2004c; Poulin, 2005). The principal host may or may not be the original host species in which the parasite first evolved, but it is currently the one used by the majority of individuals in the parasite population.

9.1.2 Flea abundance and the taxonomic distance among host species

Parasite abundance varies greatly not only between the principal host and the auxiliary hosts but also among auxiliary hosts. This uneven distribution of a parasite population may have important ecological and evolutionary implications. For example, if the difference in the abundance of a parasite in different hosts stems from different fitness rewards in these hosts, then different hosts play different roles in the long-term persistence of a parasite population. In such cases, the parasite population would thus depend mainly on one or a few key host species. At the evolutionary scale, differences in the probability of a parasite landing on one host species rather than another would favour adaptations that allow the parasite to exploit successfully the host species that are most likely to be encountered. This can shape the coevolutionary process between hosts and parasites.

The reasons why parasites are distributed unevenly among different host species are sometimes quite obvious. For fleas, this can be because of different reproductive (e.g. Krasnov et al., 2002a, 2004a; see Chapter 11) or exploitation (e.g. Krasnov et al., 2003b; see Chapter 10) performance on different host species. However, this parasite-centred approach does not allow a full understanding of the host parameters that affect the distribution of parasite individuals among their different host species.

The substantial difference in parasite abundance among different auxiliary hosts can be explained by differences in the degree of similarity between the principal hosts and the various auxiliary hosts (Poulin, 2005). For example, different auxiliary hosts can be more similar or less similar to the principal host in their availability to a flea (e.g. they might co-occur or not co-occur in the same habitat) or in their compatibility for a flea (*sensu* Combes, 2001) (e.g. in blood biochemistry or in patterns of behavioural or immune defences).

Phylogenetic relatedness among species is generally a good reflection of their overall life-history and ecological similarity (Brooks & McLennan, 1991; Harvey & Pagel, 1991; Silvertown et al., 1997). In other words, phylogenetically close host species are likely to be more similar in their ecological, physiological and/or immunological characters than phylogenetically more distant host species. The success of colonization of a new host species by a flea would, therefore, depend on the phylogenetic proximity of the new species to the original host and this should be reflected by the abundance of the flea in this new host relative to its abundance on the original host. For example, the stoats *Mustela erminea* introduced into the National Parks of New Zealand were mainly colonized by fleas that are usually parasitic on rats (*Nosopsyllus fasciatus*, *Leptopsylla segnis*), whereas only two of 680 fleas collected from 1501 stoats were bird fleas (*Ceratophyllus gallinae* and *Parapsyllus nestoris*) (King & Moody, 1982).

Krasnov et al. (2004c) tested the hypothesis that taxonomic distance between the auxiliary hosts and the principal host of a flea influences its abundance in the auxiliary hosts using data on 106 fleas parasitic on small Holarctic mammals. The taxonomic distance between the principal host and each auxiliary host was calculated as the path length linking the two host species in a Linnaean taxonomic tree where each branch length was set equal to one unit of distance (= punctuated equilibrium model of speciation). This type of taxonomic distance measure is commonly used in studies that take into account the taxonomic

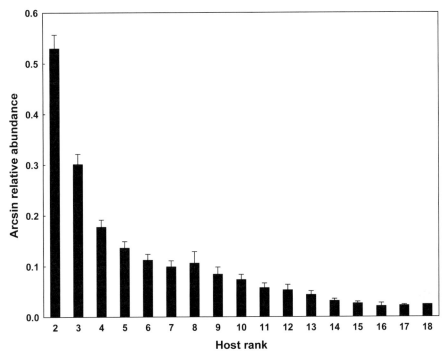

Figure 9.1 Mean (±S.E.) abundance of fleas in their auxiliary host species, as a function of host rank. Host species are ranked from the one in which flea abundance is highest (principal host, rank 1) to the host species in which it is lowest. Abundance values are relative, i.e. expressed as a proportion of the value observed in the principal host. Recalculated and redrawn after Krasnov et al. (2004c) (reprinted with permission from Elsevier).

distinctness of species in an assemblage (Izsák & Papp, 1995; Ricotta, 2004; Poulin, 2005). A comparative analysis demonstrated that the abundance of a flea on its auxiliary hosts decreases with increasing taxonomic distance *ceteris paribus* of these hosts from the principal host (Figs. 9.1 and 9.2).

This means that every time a flea adds a new host to its host spectrum, the taxonomic affinity of this new host matters. It thus appears advantageous for a flea species to exploit taxonomically close host species. If, for example, taxonomically close host species possess similar behavioural or immune defences, a flea could invest less by adapting to a restricted set of host immune defences than it would if its hosts were distantly related and the parasite would be forced to develop multiple adaptations to cope with the array of immune defences of its several hosts (Combes, 1997; Poulin, 1998; Poulin & Mouillot, 2004b). Another explanation of these results might be the greater spatial overlap among related

120   Ecology of host selection

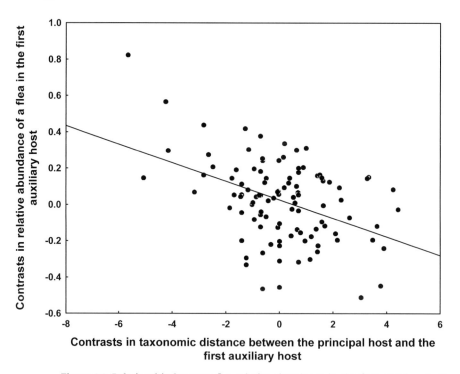

Figure 9.2 Relationship between flea relative abundance in the first auxiliary host and taxonomic distance between the principal host and this auxiliary host using the method of independent contrasts. Redrawn after Krasnov et al. (2004c) (reprinted with permission from Elsevier).

hosts because these hosts may have similar ecological preferences (Brooks & McLennan, 1991). Consequently, their habitat distribution can be similar, so a new host encountered by a flea in the habitat of an original host is possibly a close relative of this original host. Thus, the fact that taxonomically related hosts offer fleas similar immunological and feeding conditions is, perhaps, not the main factor involved. Nevertheless, these two explanations associated with the causes of exploitation of closely related hosts are not mutually exclusive. In other words, these results can be explained in the framework of the host-encounter and host-compatibility filter concept of Euzet & Combes (1980) and Combes (1997, 2001). The host-encounter filter excludes all potential host species that a parasite cannot encounter because of ecological or geographical reasons, whereas the host-compatibility filter excludes all potential host species in which a parasite cannot survive and develop for morphological, physiological or immunological reasons. The possibility of both filters operating simultaneously is higher for closely related than for distant host species. However, the host-compatibility filter probably plays a more important role in the selection by a flea of auxiliary hosts closely related to its principal host than does the host-encounter filter, because in reality taxonomic relatedness between host species does not always determine their similarity in habitat distribution and ecological preferences (Price et al., 2000; Losos et al., 2003).

Nevertheless, the successful colonization of new hosts by fleas is not necessarily restricted to taxonomically related host species. For example, many flea species recorded on the New World Soricidae are representatives of the genera associated mainly with rodents (Morrone & Acosta, 2006). It has already been mentioned that most studies of cophylogeny in host–parasite associations have demonstrated that association by descent (which is indicated by the congruence of host and parasite phylogenies) is not necessarily the norm and that the common history of hosts and parasites is complicated by evolutionary events other than cospeciation, such as host-switching (Paterson et al., 1993; Beveridge & Chilton, 2001; Roy, 2001; Johnson et al., 2002). Furthermore, the evolutionary history of mammal–flea associations has been shown to involve mainly association by colonization with frequent host-switching rather than association by descent (see Chapter 4). However, it seems that in the coevolutionary history of mammal–flea associations host-switching may be somewhat constrained by host taxonomy.

9.1.3 Geographical change of a principal host

Interactions between given species are variable in both space and time (Thompson, 2005). Qualitative and quantitative aspects of interspecific interactions differ among the localities where two or more interacting species are found. One reason for this is that, in the vast majority of cases, hosts and parasites do not have identical geographical ranges. Because some host species that a flea could exploit are not present everywhere throughout the flea's geographical range, we can expect selection to favour local adaptation of the flea to the hosts that are locally available. In other words, a flea has the following evolutionary choice: it can either be host-specific and, consequently, limit its geographical range to the geographical range of its specific host or it can be host-opportunistic and, consequently, change its principal hosts across its geographical range to exploit those hosts that are locally available. For example, *Hoplopsyllus anomalus* is a flea normally associated with ground squirrels. However, in the San Joaquin Valley of California where the giant kangaroo rat *Dipodomys ingens* replaces the niche ordinarily filled by the ground squirrels, this flea is found mainly on *D. ingens* (Tabor et al., 1993). *Ceratophyllus ciliatus* changes its preferred hosts along the Northwest Pacific coast of North America from south to north with Townsend's chipmunk *Tamias townsendii* in Oregon,

Douglas's squirrel *Tamiasciurus douglasii* in southwestern British Columbia and the red squirrel *Tamiasciurus hudsonicus* in southeastern Alaska (Traub et al., 1983; Holland, 1985; Lewis et al., 1988; Haas et al., 2005).

Change of the principal hosts across geographic range appears to be rather common in fleas. If (a) the principal host in a given region is defined as a host in which a flea attains maximal abundance, and (b) between-region change of the identity of the principal host is evaluated as the mean number of the principal hosts per region (when a flea exploits different hosts in each region, it equals 1), then the frequency distribution of this number across flea species allows us to visualize the strength of the tendency of this shift. Among 177 flea species that occurred in at least two of 49 different geographical regions, only 28 exploited the same host species in all regions it inhabited. For example, in all regions where *Xenopsylla hirtipes* was recorded, its principal host was the great gerbil *Rhombomys opimus* (Mikulin (1959a) for the East Balkhash Desert; Zagniborodova (1960) for Turkmenistan; Popova (1968) for the Moyynkum Desert; and Sabilaev et al. (2003) for the Sugaty Valley of the Tien Shan Mountains). Nevertheless, the frequency of the mean number of principal hosts per region is strongly left-skewed (Fig. 9.3). For example, within each of six regions where *Peromyscopsylla hesperomys* was recorded, it attained maximal local abundance on different host species (*Peromyscus nasutus* in New Mexico: Morlan, 1955; *Peromyscus truei* in central California: Linsdale & Davis, 1956; *Onychomys leucogaster* in Idaho: Allred, 1968; *Peromyscus leucopus* in Connecticut: Main, 1983; *Neotoma mexicana* in Colorado: Campos et al., 1985; and *Peromyscus boylii* in southwestern California: Davis et al., 2002).

However, frequency distributions of the mean number of principal host genera, subfamilies and families are much less left-skewed (Fig. 9.3). This suggests that the species that substitutes for the principal host is often phylogenetically (or at least taxonomically) related. As mentioned above, this phylogenetic relatedness among hosts is likely to be associated with their ecological, physiological and/or immunological similarity.

9.2   Ecological scale: host selection

9.2.1   Density dependent host selection

We already know that most fleas are not strictly host-specific and usually parasitize more than one of several co-occurring host species within a habitat. Moreover, an uneven distribution of conspecific fleas among co-occurring hosts with a consistent bias towards the principal host suggests that fleas are somehow

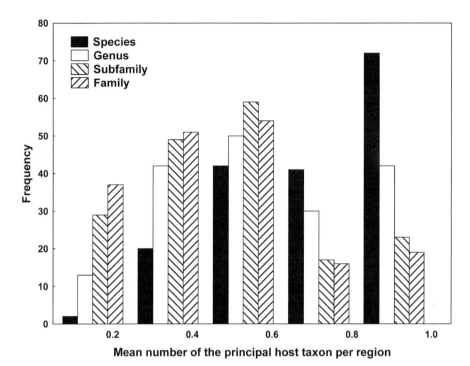

Figure 9.3 Frequency distribution of the mean number of species, genus, subfamily and family identities of the principal hosts across 177 flea species occurring in at least two of 49 different regions. Data from Morlan (1955), Linsdale & Davis (1956), Sinelshchikov (1956), Kozlovskaya (1958), Leonov (1958), Mikulin (1958, 1959a, b), Shwartz et al. (1958), Reshetnikova (1959), Zagniborodova (1960), Kunitsky & Kunitskaya (1962), Alania et al. (1964), Syrvacheva (1964), Brinck (1966), Koshkin (1966), Letov et al. (1966), Paramonov et al. (1966), Pauller et al. (1966), Varma & Page (1966), Vasiliev (1966), Emelianova & Shtilmark (1967), Labunets (1967), Allred (1968), Popova (1968), Violovich (1969), Darskaya et al. (1970), Morozkina et al. (1970, 1971), Elshanskaya & Popov (1972), Smit (1974), Amin (1976), Walton & Hong (1976), Yudin et al. (1976), Novozhilova (1977), Sapegina et al. (1980a, b, 1981a, b), Nazarova (1981), Main (1983), Ageyev & Sludsky (1985), Campos et al. (1985), Starikov & Sapegina (1987), Ravkin & Sapegina (1990), Burdelova (1996), Anderson & Williams (1997), Davis et al. (2002), Stanko et al. (2002), Sabilaev et al. (2003), Nuriev et al. (2004) and Tanitovsky et al. (2004).

able to perceive differences among hosts and to select the most appropriate species.

It is commonly accepted that the behaviour of an individual is greatly influenced by its evolutionary motivation to maximize lifetime fecundity (Lomnicki, 1988). One of the mechanisms to achieve this is to select those habitats (or hosts, in the case of parasites) that guarantee the greatest fitness output. This statement is a keystone of the theory of habitat selection (Rosenzweig, 1981, 1989, 1991) that is based on the mechanism of the ideal free distribution (IFD: Fretwell & Lucas, 1970; see below). Parasites supposedly make the same decisions that every animal has to make regarding resource acquisition and fitness reward. One of the most important differences between environments of free-living animals and parasites is considered to be that the parasite's environment is much more predictable than that of most free-living species (Sukhdeo, 1997).

According to the theory of habitat selection, fitness is a negative function of population density. Consequently, the relationships between fitness and density will be reflected in the distribution and abundance of individuals across habitats (e.g. hosts). Hosts that support different numbers of conspecific fleas probably differ quantitatively or qualitatively, and fleas are likely to be able to perceive this difference.

Morris (1987a, b, 1988, 1990, 2003) proposed a theory that explained the mechanisms of habitat selection and demonstrated that quantitative and qualitative differences between habitats lead to predictable and easily tested differences in population density. He also proposed a technique (the isodar approach) that allows one to infer a mode of habitat selection via census data. The rationale behind the isodar theory and approach is as follows: habitat selection response can be deduced simply by examining patterns of species density in different habitats. Plotting density of a species in one habitat against that in another habitat (N_1 plotted on N_2) produces an isodar, a line at every point at which the fitness of individuals in one habitat is assumed to be equal to that of individuals in another. To the left of the isodar, there are too many individuals in the habitat with the greater fitness relative to those in the other habitat, and the fitness there is depressed. A density-dependent habitat selector should move to the alternative habitat. To the right of the isodar, fitness in the habitat with the lower fitness is depressed. Each mode of habitat selection strategy thus produces its own characteristic isodar, whereas intercepts and slopes of isodars indicate how a consumer perceives the between-habitat difference (Morris, 1988, 2003). To summarize briefly the isodar approach: (1) the slopes of the isodars indicate whether habitats vary qualitatively or quantitatively relative to the foraging of their component species; (2) the intercepts of the isodars can be used to indicate the relative differences in habitat richness perceived by the consumers; and (3) non-significant regressions imply density-independent habitat selection or no habitat selection whatsoever (see Rosenzweig & Abramsky, 1985 and Morris, 1988 for details). Situations where habitats differ only quantitatively result in isodars with a non-zero intercept and slope equal to 1 (parallel population regulation), whereas situations where habitats differ only qualitatively result in isodars with a zero intercept and a slope >1 (divergent population regulation). If habitat 1 is qualitatively and quantitatively more suitable than habitat 2, the resulting isodar has a non-zero intercept and a slope >1 (and divergent population regulation is maintained). If a quantitatively superior habitat is also the one with the lower foraging efficiency, the slope of the fitness function with density is steeper in the habitat with the greater resource, and the fitness curves can cross. This situation produces an isodar with a non-zero intercept and a slope <1 (crossover population regulation). Finally, if habitat 2 is still qualitatively less suitable but quantitatively superior so that its carrying capacity is greater than that of habitat 1, the resulting isodar also has a non-zero intercept and a slope <1 (convergent population regulation), but only for crossovers does the isodar pass through $N_2 = N_1$ before reaching carrying capacity.

A habitat patch for a parasite is its host. Parasite individuals are distributed across host individuals and, thus, the host population can be considered as the habitat for a parasite population. This is especially true for ectoparasites such as fleas because, in contrast to endoparasites, their contact with the host is usually intermittent and an individual flea can move from one host individual to another during its lifetime (see below). Fleas thus represent a convenient model for using the isodar approach to infer a mode of host selection. Furthermore, in contrast to attempts to find the best mathematical fit of data to an observed pattern (e.g. Kingsolver, 1987; Hasibender & Dye, 1988), this approach can explain the evolutionary motivations that drive this pattern (see Kelly & Thompson, 2000).

In the non-sandy plains of the Negev Desert, each of four flea species (*Xenopsylla conformis*, *Xenopsylla ramesis*, *Nosopsyllus iranus* and *Stenoponia tripectinata*) parasitize the two most common co-occurring rodent species, Wagner's gerbil *Dipodillus dasyurus* and Sundevall's jird *Meriones crassus*. In sandy habitats of this desert, one flea species, *Synosternus cleopatrae*, parasitizes two co-occurring gerbils, Anderson's gerbil *Gerbillus andersoni* and the greater Egyptian gerbil *Gerbillus pyramidum*. The two rodent host species in each of the two landscape types were assumed to represent two habitat types for a particular flea species. Could flea distribution between the two hosts be explained in the framework of the theory of habitat selection, which, in this application, can be considered as the theory of host selection? This question can be answered using the isodar approach and thus testing whether fleas use density-dependent or density-independent host selection (Krasnov et al., 2003c).

At low population size, density-dependent host selectors should parasitize the best host only because their fitness output is higher in this case than in case of any other host. With an increase in population size, additional host species would be parasitized too until the overall flea population growth rate is zero (Fretwell & Lucas, 1970; Rosenzweig, 1981; Morris, 1988). Density-independent host selectors should parasitize both hosts at low density, whereas with a density increase the number of parasites on different hosts should correspond to host 'quality', resulting from differential growth rate on the hosts rather than a result of density-dependent regulation (Morris, 1988).

In this study, flea density was calculated per 1 cm^2 of body surface of a rodent. The resulting isodars are presented in Fig. 9.4. Both intercept and slope of the isodar of *X. conformis* were significant and the slope was >1, confirming divergent fitness–density curves (slope >1 and intercept >0) and indicating that *M. crassus* was both a quantitatively and qualitatively superior host for this flea to *D. dasyurus*. Isodars of *X. ramesis* and *S. cleopatrae* had significant slopes (>1 for both species), suggesting divergent population regulation of these fleas. However, intercepts of these isodars did not differ from zero, indicating that *M. crassus* and *G. pyramidum* were qualitatively more suitable hosts for *X. ramesis* and *S. cleopatrae*, respectively, than *D. dasyurus* and *G. andersoni*, respectively, but quantitatively these hosts were identical for the fleas. Finally, the analyses failed to produce significant isodars for *N. iranus* and *S. tripectinata*. Thus, three of five flea species appeared to be able to perceive quantitative (differences in the amount of the resource) or qualitative (differences in the pattern of resource acquisition) between-host differences. *Xenopsylla conformis* and *X. ramesis* perceived a qualitative difference between *M. crassus* and *D. dasyurus*, whereas *S. cleopatrae* perceived a qualitative difference between *G. pyramidum* and *G. andersoni*. In addition, *X. conformis* perceived a quantitative difference between *M. crassus* and *D. dasyurus*. *Nosopsyllus iranus* and *S. tripectinata* did not perceive either qualitative or quantitative between-host differences.

An example of a quantitative difference from a flea 'viewpoint' may be the absolute amount of blood available for a flea imago. In addition, there can be between-host differences in the amount of organic matter in the nest chamber available for flea larvae. This amount can depend on the material of the nest, the time that a host spends in the nest, and the behaviour of a host (e.g. defecation inside or outside the nest). Indeed, the main quantitative difference between *M. crassus* and *D. dasyurus* for *X. conformis* could be caused by the difference in the pattern of their burrow use, although the evidence for this is indirect and circumstantial. Areas inhabited by *X. conformis* (but not *X. ramesis*) are characterized by a sandy-gravel substrate (Krasnov et al., 1997, 1998). Mark–recapture and radio tracking of *M. crassus* in these areas demonstrated that this species was strongly territorially conservative both in relation to its home range within a season and in relation to its burrow within a daily period of activity (Krasnov et al., 1996a). Each individual remained in the close vicinity of a burrow during the entire night for a relatively long period (3–7 days). Thereafter, it moved to another burrow where it also stayed for the same period of time. There were

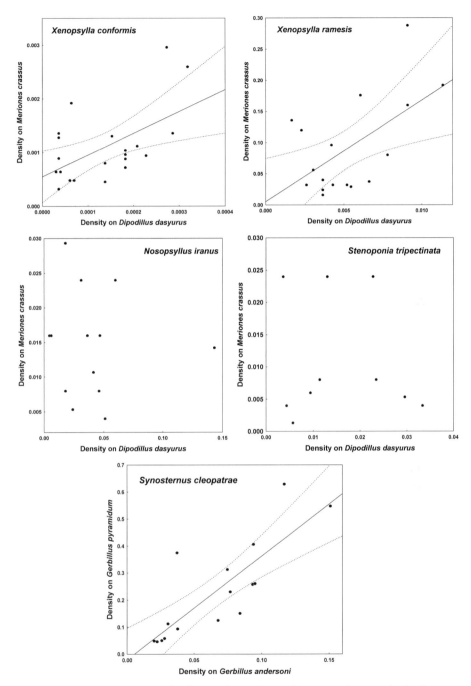

Figure 9.4 Isodars for five flea species each parasitic on two host species in the Negev Desert. Redrawn after Krasnov et al. (2003c) (reprinted with permission from Springer Science and Business Media).

several burrows in each home range, and it appeared that an individual used each of them periodically. This pattern of burrow use presumably increases the amount of organic matter in the burrow (=food resource for the larvae). This is especially important for flea larvae in a sandy-gravel substrate that is poor in organic matter. In contrast, *D. dasyurus* in these areas seemed to be less territorially conservative. Individuals considered to be transients (those that did not construct burrows) comprised a rather high proportion of the entire population (Khokhlova *et al.*, 1994, 2001; Krasnov *et al.*, 2002e).

In contrast to *X. conformis*, closely related *X. ramesis* did not perceive quantitative differences between *M. crassus* and *D. dasyurus*. Areas inhabited by *X. ramesis* are characterized by organic-rich loess substrate (Krasnov *et al.*, 1997, 1998). Both *M. crassus* and *D. dasyurus* here were territorially conservative (Krasnov *et al.*, 1996a; Shenbrot *et al.*, 1997). They both possessed permanent burrows with well-constructed nests composed of dry soft remnants of grasses and sedges (Shenbrot *et al.*, 1997). Thus, there was no sharp between-species difference in the amount of the organic matter in the burrow.

No information on the differences in burrow and nest structure between *G. andersoni* and *G. pyramidum* is available. Nevertheless, the loose sand in their habitats does not allow a great variety of burrow constructions. Consequently, the amount of organic matter available for flea larvae is likely similar in burrows of these two species. In addition, the overall amount of time spent out of the burrow in these two species is similar (as calculated from data published by Kotler *et al.* (1993); see Krasnov *et al.* (2003c) for details) and, thus, there seems to be no quantitative between-host difference for the parasite *S. cleopatrae*.

Examples of qualitative differences include between-host variation in the absolute amount of blood that an individual insect obtains during a blood meal (e.g. Webber & Edman, 1972) and the efficiency of blood digestion by a flea (e.g. Vashchenok, 1976). Hosts also vary in their skin structure (which determines the ease of blood-sucking), fur density (which determines the ease of flea movement) and defensiveness against bites (e.g. grooming activity). Finally, variation in host density can be also considered as a qualitative host difference. This variation represents the difference in the distance between foraging (host body) and breeding (host burrows and/or nests) patches.

The divergent fitness–density curves suggest that there is a divergence in habitat suitability with density and that, at any given density, consumers in one habitat are more efficient at converting resources into offspring than they are in another habitat (Morris, 1987a, b, 1988). The effect of density on reproductive success is greater in those habitats where the foraging is less efficient. In the case presented here, the fitness rewards for individuals of the two *Xenopsylla* species and *S. cleopatrae* parasitizing *M. crassus* or *G. pyramidum*, respectively, are indicated to be higher than for those parasitizing *D. dasyurus* or *G. andersoni*, respectively (which is indeed so; see further details in Chapter 11). Furthermore, at low density, *X. conformis* parasitized the 'best' hosts only, whereas with an increase in flea population size, the additional host was also parasitized. In contrast, *X. ramesis* and *S. cleopatrae* at low density appeared to parasitize both hosts equally because they seemed to be able to achieve similar maximum fitness under such conditions. However, with an increase in the flea population size, parasite pressure on the 'high-quality' host increased at a higher rate than on the 'low-quality' host. In conclusion, *X. conformis* appeared to be a density-dependent host selector that showed sharp selectivity at low density. In contrast, *X. ramesis* and *S. cleopatrae* appeared to select hosts randomly at very low densities but showed density-dependent host selection with a preference for the largest species at high densities. Following Morris (1988), these species can be considered to be 'density-independent host selectors with a direct correspondence of density with host quality'.

Isodar analyses that included *N. iranus* and *S. tripectinata* in a regression design were non-significant. This means that that there was no relationship between the densities of these flea species on their hosts. These flea species can be called 'non-selectors' (Rosenzweig & Abramsky, 1985) or 'density-independent selectors that display no relationship between density and host quality' (Morris, 1988). In other words, these fleas did not perceive the differences between *M. crassus* and *D. dasyurus*. Why are they unable to distinguish between two host species? First, in contrast to *Xenopsylla* and *Synosternus* species, fleas of *Nosopsyllus* and *Stenoponia* genera possess combs (ctenidia) that permit the flea to anchor itself within host fur and to resist the host's grooming effort (Humphries, 1967c; Traub 1972b; see Chapter 13). Consequently, host defensiveness is much less important for them than for fleas without combs. This can explain, at least partly, the absence of qualitative differences between *M. crassus* and *D. dasyurus* in their perception of hosts. Second, *N. iranus* and *S. tripectinata* are winter-active and their activity period is very short (see Chapter 5). The rainfall in the study area occurs in winter and, consequently, the availability of plant material for nest construction as well as seeds for storage by rodent hosts is the highest in winter and early spring. Consequently, the difference between *M. crassus* and *D. dasyurus* burrows might be 'smoothed' under such conditions and, thus, the quantitative between-host differences disappeared. An alternative explanation is that the short period of activity during which fleas feed, copulate, oviposit, hatch, moult and spin a cocoon does not provide fleas with time enough for host selection decisions. This corresponds with Morris's (1988) conclusion that the isodar regression is statistically non-significant in those cases where population size is limited by extrinsic factors below which intraspecific density affects fitness.

9.3 Fleas and the ideal free distribution

The study described in the previous subchapter demonstrated that fleas seem to behave in an IFD-like manner. As we already know, the IFD concept predicts that animals that compete for resources distribute themselves among habitat patches in proportion to the amount of resources available to them, so that resource use per individual will be equal across all patches. In other words, animals are (a) ideal in assessing patch quality and (b) free to enter and use the resources on a regular basis.

Haematophagous parasitic arthropods have been shown to respond generally according to the IFD predictions (e.g. Kelly *et al.*, 1996), although the applicability of the IFD concept to helminth parasites has been questioned (see Sukhdeo & Bansemir, 1996; Sukhdeo *et al.*, 2002). Nevertheless, negative fitness—density relationships have been shown for both endoparasites (Croll *et al.*, 1982) and ectoparasites, including fleas (see Chapter 11). For example, the reproductive success of *C. gallinae* breeding in the nest of the blue tit *Cyanistes caeruleus* was significantly affected by the number of conspecifics in the same nest (Tripet & Richner, 1999).

Kelly & Thompson (2000) developed an IFD model of host choice by blood-sucking insects based on the argument that insect vectors must have evolved to choose the least defensive hosts in order to maximize their feeding success. They argue that an individual blood-sucking insect can improve its feeding success (and, consequently, its fecundity considered as fitness reward) by choosing a host with a higher intrinsic quality, a lower defensiveness and a smaller number of blood-sucking competitors. This model has been applied to parasite individuals distributed both within and between host species. Moreover, the application of the IFD model to parasite distribution may also explain the aggregative distribution of conspecific parasite individuals across a host population (Sutherland, 1983, 1996; Kelly & Thompson, 2000; see also Chapter 15).

The above suggests that fleas can be assumed to be 'free'. Empirical evidence for this assumption will be presented below. However, can they be assumed to be 'ideal'? As already mentioned, an 'ideal' flea should select a host according with the fitness reward it expects to obtain while exploiting it. This issue will be considered in the next subchapter as well as in Chapter 11.

9.3.1 Are fleas 'ideal'? Host search and location

To complete their life cycles, parasites have to find and infect hosts. Processes used by parasites to locate hosts are thought to be similar to the processes used by free-living organisms to exploit their environment (Rea &

Irwin, 1994). Three main categories of signals have been identified that influence the behaviour of parasites (MacInnis, 1976):

(1) signals extrinsic to the host and originating in the environment;
(2) signals extrinsic to the host but of host origin; and
(3) signals intrinsic to the host and originating within the host.

These signals maintain or attract the parasite to the potential host's active space (Rea & Irwin, 1994). Furthermore, signals of host origin should have a special importance for parasites as they can attract a particular parasite to a particular host. Therefore, parasites that have evolved responses to a specific host increase their chance to locate a host by responding to signals of host origin. Behaviour of fleas is realized via direct responses to stimuli received and recognized by sensory structures such as eyes and sensilla (Lewis, 1998).

Signals extrinsic to the host

The main environmental signals that affect flea behaviour include light, air temperature and the angle of the surface. For example, in simulated badger burrows, the badger fleas *Paraceras melis* were negatively geotactic (Cox et al., 1999). Indeed, surface angle may be the most important signal for species that exploit fossorial hosts. The response of fleas to environmental stimuli has been studied on several flea species characteristic for birds (*C. gallinae*: Humphries, 1968; *C. styx*: Humphries, 1969) and large (*P. melis*: Cox et al., 1999), medium (*Ctenocephalides felis*: Osbrink & Rust, 1985), and small (*Parapulex chephrenis*, *X. conformis* and *X. ramesis*: Burdelov et al., 2007) mammals. In particular, studies of the responses of fleas to light and surface angle have demonstrated that different species vary in their behaviour from being positively or negatively photo- and/or geotactic to being indifferent (see review in Marshall, 1981a). This suggests that the response of fleas to these stimuli depends on the patterns of their life history as well as on their level of host specificity. Furthermore, the response to a stimulus alters with the gender, age and condition of an insect. For example, *C. gallinae* is negatively phototactic on emergence, becoming positively phototactic after 48 h (Humphries, 1968). Positively phototactic unfed *Spilopsyllus cuniculi* become negatively phototactic after a blood meal (Rothschild & Ford, 1973). *Paraceras melis* were also strongly positively phototactic when starving (Cox et al., 1999).

Burdelov et al. (2007) studied the response of newly emerged and fed individuals of *X. conformis*, *X. ramesis* (both parasitic on a variety of solitary gerbillines) and *P. chephrenis* (a specific parasite of a social murine) to light and surface angle. They observed flea movements inside either horizontal or tilted cardboard tubes with a different light regime at their ends. Five series of experiments, differing in the position of the tube (tilted or horizontal) and/or the position of the light source were carried out (tilted tube: enlighten end either up or down, both ends either enlighten or darkened). In these experiments, the proportion of relocating *X. conformis* and *X. ramesis* was higher than that of *P. chephrenis* (Fig. 9.5). In the latter species only, adult fleas relocated more frequently then newly emerged fleas.

In general, the majority of fleas moved toward the light source independently of its position in relation to the surface angle, even if it was positioned at the lower end of a tube (Fig. 9.6). When both ends of a tube were darkened, newly emerged *Xenopsylla* moved randomly toward the upper or lower end of a tube, whereas newly emerged *P. chephrenis* moved mainly toward the upper end of a tube (Fig. 9.7b). Adult *P. chephrenis* and *X. conformis* also moved mainly toward the upper end of a tube, whereas adult *X. ramesis* moved mainly toward the lower end. When both ends of a tube were illuminated, newly emerged females of all species, as well as newly emerged female *X. ramesis*, randomly relocated toward the upper or lower end of a tube (Fig. 9.7a). In contrast, newly emerged males and adults of both sexes of *P. chephrenis* and *X. conformis* as well as adult female *X. ramesis* moved mainly toward the upper end of a tube, whereas adult male *X. ramesis* moved mainly down.

Results of this study suggest that (a) light is a more important abiotic signal for flea orientation than surface angle, and (b) there are species-specific differences in flea responses to light and angle stimuli. Light is generally a strong stimulus for fleas. Flea eyes are very simple in structure and represent a transformation of the multifaceted eyes of most insects, replaced with heavily sclerotized, atypical ocelli, or 'eyespots' (Crum *et al.*, 1974; Taylor *et al.*, 2005). Nevertheless, a recent study by Taylor *et al.* (2005) demonstrated that fleas are sensitive to long-wavelength light and, thus, are able to distinguish between light and dark. This ability of fleas is supported by their responses to changes in photoperiod in laboratory experiments (Ma, 1995).

It is quite obvious why starving fleas respond positively to light. This strategy allows them to move to the entrance of the host's burrow, thus increasing the chance of locating and attacking a host (e.g. Darskaya & Besedina, 1961). As mentioned above, some fleas that are positively phototactic when starved become negatively phototactic when satiated (e.g. Humphries, 1968). However, in the experiments of Burdelov *et al.* (2007), fleas responded positively to light independently of the time since a blood meal. This suggests that, at least in some fleas, both satiated and starving individuals are positively phototactic.

Although surface angle has also been shown to be a strong stimulus for fleas (Cox *et al.*, 1999), Burdelov *et al.*'s (2007) study demonstrated that geotactic behaviour varied among flea species and depended on flea feeding status and, sometimes, on gender (see Chapter 7). Nevertheless, the majority of fleas were

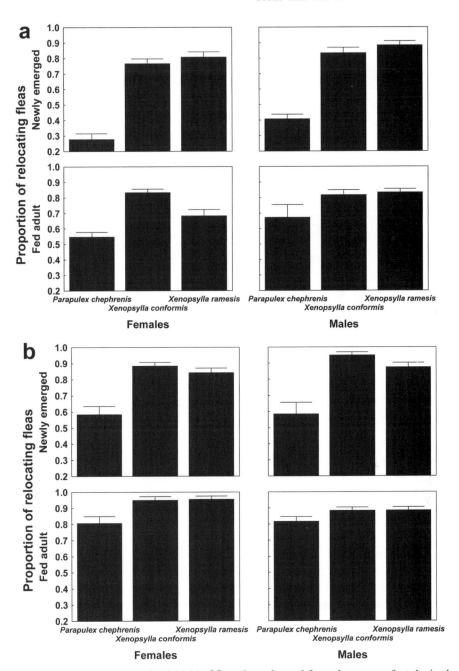

Figure 9.5 Proportion (±S.E.) of fleas that relocated from the centre of a tube in the experiments with tilted tubes and (a) light or (b) dark ends. Redrawn after Burdelov et al. (2007) (reprinted with permission from Springer Science and Business Media).

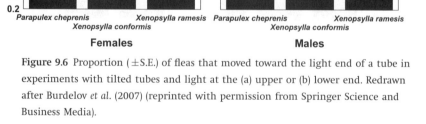

Figure 9.6 Proportion (±S.E.) of fleas that moved toward the light end of a tube in experiments with tilted tubes and light at the (a) upper or (b) lower end. Redrawn after Burdelov *et al.* (2007) (reprinted with permission from Springer Science and Business Media).

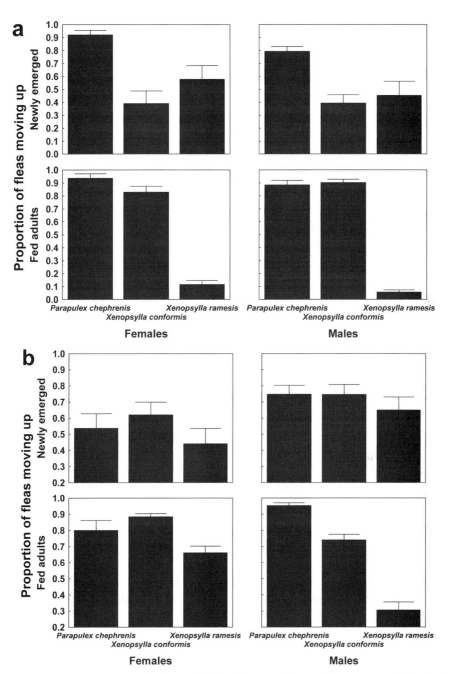

Figure 9.7 Proportion (±S.E.) of fleas that moved toward the upper end of a tube in experiments with tilted tubes and (a) light or (b) dark ends. Redrawn after Burdelov et al. (2007) (reprinted with permission from Springer Science and Business Media).

negatively geotactic, although this was observed only when the light conditions at both ends of a tilted tube were the same. However, positive phototaxis overrode negative geotaxis. When a flea was given the choice to respond either to a light or an angle stimulus, it responded to light even if the location of the light source was opposite to the natural situation.

A surprising result of Burdelov et al.'s (2007) study was positive geotaxis in adult *X. ramesis*. This might be a result of the environmental preferences of this species. Indeed, rodent burrows in habitats occupied by *X. ramesis* are well ventilated, and relative humidity (RH) in these burrows is relatively high (Shenbrot et al., 2002) This is important for *X. ramesis*, which is highly sensitive to low RH (Krasnov et al., 2002d). The preference for higher RH explains the preferential movement of satiated *X. ramesis* downwards, as RH is higher in deeper parts of the host's burrow (Shenbrot et al., 2002).

The intensity of flea responses to cues such as light and surface angle varies depending on a number of factors related to flea conditions (e.g. age, satiation level), host availability and environment (e.g. air temperature, photoperiod) (Kolpakova, 1950; Zagniborodova, 1965; Krivokhatsky, 1984). For example, fleas often abandon deeper parts of a host's burrow and accumulate in or near the entrance of the burrow. In Russian sources, this phenomenon is traditionally called 'flea migration', although it undoubtedly has nothing to do with migration *sensu stricto*, but rather may represent negative geotaxis merely caused by prolonged starvation or search for favourable microclimatic regime. The number of 'migrating' fleas has been shown to vary seasonally. Moreover, interspecific variation in the preferable air temperature and RH results in interspecific variation of seasonal patterns of these 'migrations'. For example, Krivokhatsky (1984) found that among fleas parasitic on *R. opimus* in the eastern Kara-Kum Desert, *Rostropsylla daca* and *Leptopsylla sexdentata* demonstrated peak of 'migration' in January; *Nosopsyllus turkmenicus* in February; *X. hirtipes*, *Coptopsylla bairamaliensis* and *Synosternus longispinus* in March; *X. conformis* and *Nosopsyllus tersus* in April; and *Coptopsylla olgae* in November. Furthermore, no density dependence of this 'migration' activity was found suggesting that the reasons for an increase in positively phototactic/negatively geotactic responses are related to individual behaviour rather than population processes.

Signals intrinsic to the host

How does a flea find a host? It is known that cues used by fleas to find their hosts include vibration, increased concentration of CO_2, increased temperature and silhouette. However, results of experiments that attempt to determine the main cues for locating hosts are contradictory.

Cox et al. (1999) studied the effects of a variety of stimuli on the searching behaviour of *P. melis*. They found that fleas strongly responded to the stimuli of CO_2 (which mimics host respiration), CO_2 with breeze, and a mixture of CO_2, movement and the presence of a dark object. In all cases the fleas responded by jumping toward the source of the stimulus. This study indicated that, when attempting to locate a host, fleas responded most strongly to combinations of stimuli. Furthermore, during experiments with CO_2 stimuli, fleas have been observed to wave their heads backward and forward prior to movement. This is thought to be indicative of the involvement of uni- and bisensory organs (e.g. antennae) in a taxic-orienting response (Cox et al., 1999).

Surprisingly, *P. melis* did not respond to the stimulus of badger hair (Cox et al., 1999). Other studies reported positive responses of fleas to host odour (Vaughan & Mead-Briggs, 1970; Crum et al., 1974; Krasnov et al., 2002a), although this was not found in some experiments (Bates, 1962). Mears et al. (2002) studied host-seeking behaviour in *X. cheopis* by presenting the flea with a choice between a container with an individual rat *Rattus rattus* and an empty container. They found that newly emerged fleas were not influenced by the rat's presence, but older fleas were obviously attracted by the rat's odour. Delay of the host-seeking behaviour in the newly emerged compared with adult fleas was also reported for *C. felis* (Osbrink & Rust, 1985; Dryden & Broce, 1993).

The above-mentioned studies demonstrate that fleas can use odour cues to spot the host. However, several host species usually co-occur in the same location. Fleas capable of responding to a signal from a preferred host increase their chance of locating an appropriate host that, in turn, will probably increase their fitness reward. So, are fleas capable of distinguishing between different host species and, consequently, can they select a 'proper' host? This seems to be the case. For example, Shulov & Naor (1964) reported that, by using an odour cue, *X. cheopis* was more attracted to *Rattus norvegicus* than to the other three rodent species tested. Krasnov et al. (2002a) compared the responses of two fleas, *X. dipodilli* and *P. chephrenis*, simultaneously exposed to the odours of their co-occurring rodent hosts, *D. dasyurus* (a common host of *X. dipodilli*) and *A. cahirinus* (a specific host of *P. chephrenis*). They used a transparent Y-maze that consisted of an adjustment compartment, a central passage, and two choice arms with a stimulus compartment at each distal end. *Dipodillus dasyurus* and *A. cahirinus* were each placed in one of the stimulus compartments. A flea was placed in the adjustment compartment, which was then opened, and observation of flea behaviour was carried out by measuring the latency of movement (time to

138 Ecology of host selection

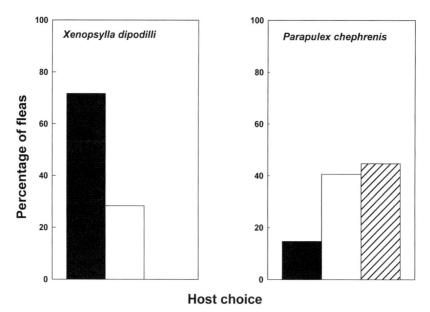

Figure 9.8 Percentage of *Xenopsylla dipodilli* and *Parapulex chephrenis* choosing Wagner's gerbil *Dipodillus dasyurus* (black columns) or the Egyptian spiny mouse *Acomys cahirinus* (white columns) as a host in the Y-maze test, or not making a choice (dashed column). Redrawn after Krasnov *et al.* (2002a) (reprinted with permission from Elsevier).

beginning of movement), latency of choice (time from beginning of movement to choice of a host) and choice of a host species (choice of *D. dasyurus*, choice of *A. cahirinus*, or no choice of host).

It appeared that all *X. dipodilli* chose a host. Furthermore, they chose the stimulus compartment with *D. dasyurus* more often than the compartment with *A. cahirinus* (Fig. 9.8). In contrast to *X. dipodilli*, almost half of *P. chephrenis* were indifferent and did not choose a host at all. However, *P. chephrenis* that did choose selected a compartment with *A. cahirinus* more often than that with *D. dasyurus* (Fig. 9.8). Furthermore, both latency of movement and latency of choice in *X. dipodilli* were shorter than in *P. chephrenis* (Fig. 9.9).

These results demonstrate that fleas, even newly emerged individuals, are able to discriminate between two host species and, therefore, to select an appropriate host species. Indeed, in the study by Krasnov *et al.* (2002a), host odour was the main stimulus to which fleas responded. All other stimuli emitted by both host species that fleas could respond to (CO_2, heat etc.) were most probably similar as both host species are approximately the same size. Furthermore, the two flea species had a different behavioural strategy for host searching: *X. dipodilli* searched actively for a host, whereas *P. chephrenis* seemed to adopt a

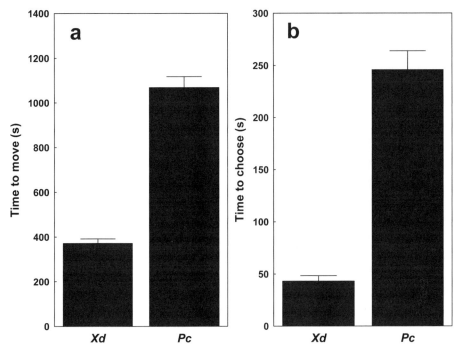

Figure 9.9 Mean (±S.E.) time (seconds) taken by *Xenopsylla dipodilli* (Xd) and *Parapulex chephrenis* (Pc) (a) from the start of the Y-maze test to the first movement and (b) from the start of movement to choosing a host. Redrawn after Krasnov *et al.* (2002a) (reprinted with permission from Elsevier).

strategy resembling that of a sit-and-wait predator. This could depend on properties of host ecology such as the pattern of the burrow usage or the degree of sociality. It could be advantageous to perform a sit-and-wait strategy for fleas parasitizing a host that lives in groups and/or spends considerable time in the same burrow and changes it rarely. If the return of the same or a different host individual to the burrow is predictable and almost guaranteed, a flea may save energy by waiting for the host rather than actively searching for it. Indeed, *A. cahirinus* tends to nest communally (Shargal *et al.*, 2000), and the majority of individuals reside in their home ranges for a long period (Khokhlova *et al.*, 1994, 2001; Shargal *et al.*, 2000). In contrast, a flea whose host is solitary and/or abandons its burrow for long periods of time and changes it often (as is the case with *D. dasyurus*: Shenbrot *et al.*, 1997; Gromov *et al.*, 2000), is better to be an active host-searcher because the probability of an occasional meeting with such a host is low. Other studies support, albeit indirectly, speculation about the link between host-searching strategy and host sociality and/or spatial behaviour. *Xenopsylla gerbilli* parasitizing the group-living *R. opimus* has been reported to cluster at the openings of abandoned burrows and wait for a potential host that they follow when sensed, presumably by odour (Darskaya & Besedina, 1961). Shulov & Naor (1964) reported a lack of movement in the presence of host stimuli in 50% of *X. cheopis* tested. Note that rats (the main hosts of this species) are highly social. Stick-tight vermipsyllids parasitizing ungulates are powerful jumpers and actively seek a host (Marshall, 1981a). In contrast, *Anomiopsyllus amphibolus* lives in the dens of the desert wood rat *Neotoma lepida*, which may have been occupied for hundreds of years, so it is almost guaranteed a host location. Thus, this flea species has no searching activity and almost no jumping capacity (Egoscue, 1976).

Despite their capacity to locate a host and to distinguish between host species, the distance at which fleas can perform these behaviours seems to be rather short. For example, *Ctenophthalmus pseudagyrtes* is able to locate a host successfully if the distance between a flea and its host does not exceed 5 cm (Benton et al., 1959). The same is true for *N. fasciatus* (Iqbal, 1974). Host location distance is even less (about 2 cm) for *Neopsylla bidentatiformis* and *Citellophilus tesquorum* (Ma, 1994b). *Ceratophyllus styx* can perceive the difference between host species at a distance no more than 5 cm (Benton et al., 1959).

9.3.2   Are fleas 'free'? Dispersal and host-to-host transfer

The results of studies on host-searching and distinguishing by fleas suggest that these insects are 'ideal' in that they are capable of assessing the relative 'quality' of different host species and to locate a preferred host. The IFD assumes that individuals have the opportunity to migrate between patch types while they jointly assess density and fitness (Morris, 1988). Consequently, the remaining question related to the assumption of the IFD approach in its application to fleas is: 'are they "free" to enter a habitat patch?' To be 'free', fleas should be able (a) to disperse on their own and/or (b) to transfer between individual hosts belonging to different species.

It appears that individual fleas can disperse rather long distances. For example, free dispersal distance in *C. styx* has been recorded as far as 34 m (Bates, 1962). This flea parasitizes the sand martin *Riparia riparia* which nests in horizontal burrows in vertical sandy cliffs. *Xenopsylla gerbilli* can freely travel up to 2.5 m (Darskaya & Besedina, 1961). Moreover, active movement is the main type of dispersal for some fleas (e.g. *Monopsyllus indages* parasitic on squirrels: Slonov, 1965). Passive dispersal (passing from host to host) is characteristic for other flea species (e.g. *Ceratophyllus petrochelidoni*: Foster & Olkowski, 1968; *Ceratophyllus hirundinis*: Clark et al., 1993b). This can result in dispersal distances that are even greater. For example, individuals of *X. cheopis* dispersed 112 m via their rat hosts (Kuznetsov et al., 1999), whereas *Nosopsyllus laeviceps* dispersed 260–280 m via gerbil hosts (Kuznetsov et al., 1993).

Intraspecific host-to-host transfer has been shown for *Malaraeus telchinus* (Hartwell et al., 1958), *S. cuniculi* (Mead-Briggs, 1964; Williams, 1971; Williams & Paper, 1971) and *C. felis* (Rust, 1994). Interspecific host-to-host transfer is also a rather common phenomenon (Hartwell et al., 1958; Morozov et al., 1972; Rapoport et al., 1976; Marshall, 1981a). Both types of host-to-host transfer have been repeatedly shown to occur either through hosts visiting burrows of other species or through body contact (Haas, 1965; Ryckman, 1971; Nazarova, 1981). For example, Rödl (1979) reported the exchange of fleas between the bank vole *M. glareolus* and the woodmouse *Apodemus flavicollis* which occurred largely through contact between individuals. Other authors also have stated that behavioural interactions mediate flea interchange among small mammal hosts (Hartwell et al., 1958; Sviridov, 1963; Buckle, 1978).

The only experimental evidence for the role of behaviour interactions in flea transfer between host individuals exists for *N. laeviceps* and *Nosopsyllus mokrzeckyi* which parasitize the midday jird *Meriones meridianus* and the house mouse *Mus musculus* in the southern part of Russia. Krasnov & Knyazeva (1983) and Krasnov & Khokhlova (2001) performed interspecific dyadic encounters of the two rodents in a neutral arena and suggested that host-to-host transfer of fleas is facilitated through social contact. The intensity of this exchange was higher between mice and adult male and young jirds than between mice and adult female jirds (Fig. 9.10). The pattern of host behavioural interactions seemed to be the main determinant of the intensity of flea exchange. Adult female jirds were highly aggressive towards mice, whereas interactions between male and young jirds and mice were mainly cohesive, so they resulted in a higher frequency of flea transfer (Fig. 9.11). The proportion of fleas transferred was positively correlated with the number and cumulative duration of tactile contacts between opponents in dyads.

Although rather scarce, data on the interspecific host-to-host transfer of fleas suggest that a 'free' component of IFD assumption does hold for these insects. Consequently, approaches associated with the theory of habitat selection can be applied to study their among-host distribution.

9.4 Distribution of fleas on the body of a host

Distribution of flea individuals within an individual host (on its body) may be associated with (a) selection of body parts preferable for blood-sucking and (b) intraspecific competition among fleas for blood of the host or for those areas of host body where the blood is the most readily available (as determined

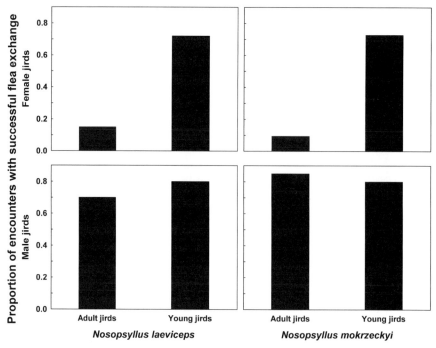

Figure 9.10 Proportion of interspecific dyadic encounters with successful transfer of *Nosopsyllus laeviceps* or *Nosopsyllus mokrzeckyi* between the midday jird *Meriones meridianus* and the house mouse *Mus musculus*. Redrawn after Krasnov & Khokhlova (2001) (reprinted with permission from the Society for Vector Ecology).

by skin thickness, hair density, blood capillary depth). In addition, the concentration of fleas on a particular area of the host body may facilitate meeting of mating partners. While our knowledge on this issue is limited, there are some empirical observations that suggest ecological and behavioural patterns in within-host flea distributions.

9.4.1 Sessile and stick-tight fleas

Host body area specificity is characteristic for permanent (sessile) fleas. It is well known that the chigoe fleas (*Tunga*) are rather body-area-specific. As was mentioned in Chapter 6, the preferred site of attachment of *Tunga penetrans* in humans is often under the nail bed of the toes, but they also occur on the hands, arms, soles, heels and in the genital region (Mashek et al., 1997; McKinney & McDonald, 2001; de Carvalho et al., 2003; Eisele et al., 2003; Muehlen et al., 2003). In pigs, *T. penetrans* usually attach to the feet, scrotum and snout (Cooper, 1976). *Tunga* species parasitic on rodents are usually restricted to the ears of the

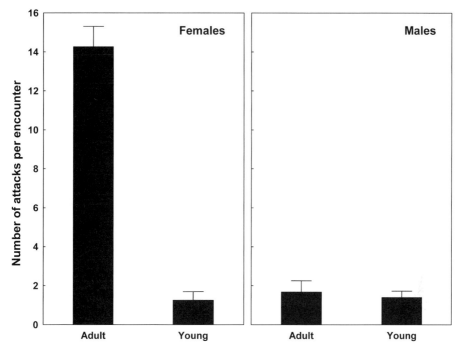

Figure 9.11 Mean (±S.E.) number of attacks initiated by the midday jird *Meriones meridianus* during a 30-min encounter with the house mouse *Mus musculus*. Redrawn after Krasnov & Khokhlova (2001) (reprinted with permission from the Society for Vector Ecology).

host, with *Tunga monositus* preferring the basal part of the upper surface of the pinna, *Tunga caecata* favouring the upper surface and *Tunga caecigena* selecting the edge of the pinna (Jordan, 1962; Barnes & Radovsky, 1969; Lavoipierre et al., 1979; Hastriter, 1997; de Moraes et al., 2003). *Tunga caecata* has also been recorded at the base of the tail (de Moraes et al., 2003). *Rhynchopsyllus pulex* parasitic on bats (e.g. *Molossus molossus*) prefers host body regions with sparse hair such as the ears (Esbérard, 2001). Females of *Neotunga euloidea*, a specific parasite of pangolins (Pholidota), are usually found in the soft skin, which is almost hairless, on the ventral region of the host body (Smit, 1962a; Segerman, 1995).

Females of stick-tight fleas parasitic on birds (*Echidnophaga gallinacea*, *Hectopsylla psittaci*, *Hectopsylla narium*) usually attach to the head of the host in an area that is unreachable during preening (Suter, 1964; Marshall, 1981a; Blank et al., 2007). For example, *H. narium* infesting the nestlings of the burrowing parrot *Cyanoliseus patagonus* has been found inside the nasal cavity and on the relatively dry area under the tongue. The latter area seems to be suboptimal as an infestation occurs mainly towards the end of the breeding season, when the nostrils are already occupied (Blank et al., 2007). In contrast to females, males of these species often remain on other parts of the host body (Suter, 1964). When attacking a mammalian host, female *E. gallinacea* also attach to the head, in particular to the eyelids (Linsdale, 1947) or facial area (Muller et al., 2001). Stick-tight vermipsyllids *Vermipsylla alakurt* and *Dorcadia ioffi* that exploit sheep are usually found on the neck and chest of a host, but when the intensity of infestation is high they also attach to the back, legs and sacrum. *Dorcadia dorcadia* parasitic on the Siberian elk *Cervus elaphus sibiricus* attack the ano-genital region and withers (Zatsarinina, 1972). Neosomic *Vermipsylla* females often attach inside the host's nostrils (Wang et al., 2004a). Similar behaviour is characteristic for three *Ancistropsylla* species parasitic on ungulates in southern and southeastern Asia (Joseph, 1974; Joseph & Mani, 1980; Marshall, 1981a). *Echidnophaga oschanini* parasitic on pikas (e.g. *Ochotona pricei*) attach to the dorsal side of the pinnae, eyelids and nose (Vashchenok, 1967b). However, when these fleas are forced to feed on an unusual host (e.g. a laboratory mouse), they attach to the facial area, around the eyes and on the lower jaw, but rarely to the pinnae (Vashchenok, 1967b). Almost half of individuals of *Chaetopsylla homoea* were recorded on the posterior dorsal area of the body of the steppe polecat *Mustela eversmanni*, although the remaining individuals were found on anterior dorsal and ventral areas of the body (Ma, 1983).

9.4.2   Non-sessile fleas and host grooming

The preferences of non-sessile fleas for certain body areas of a host are poorly known. Linsdale & Tevis (1951) and Linsdale & Davis (1956) reported that *Orchopeas sexdentatus* and *Anomiopsyllus falsicalifornicus* on the dusky-footed wood rat *Neotoma fuscipes* favoured a special 'flea spot' in the middle of the rat's chin. Presumably, it is difficult for *Neotoma* to groom this part of the body. Dubinina & Dubinin (1951) observed that *Oropsylla silantiewi*, *Ochotonobius hirticrus* and *C. tesquorum* were mainly found on the anterior dorsal part of a host body, and *Archaeopsylla sinensis* on the anterior ventral and side parts, whereas species of the genera *Rhadinopsylla*, *Ceratophyllus* and *Nosopsyllus* preferred posterior dorsal areas. In contrast, *Neopsylla* and *Frontopsylla* species were evenly distributed over the entire host body. However, Ma (1983) reported that, in the laboratory, 72.7% of *C. tesquorum* and 75.7% of *N. bidentatiformis* preferred to stay on the posterior dorsal area of a mouse. Although Osbrink & Rust (1985) did not find any significant difference in the mean number of *C. felis* collected from different parts of the body of a euthanized cat, Hsu et al. (2002) argued that cat fleas do prefer specific areas on a cat's body, namely its head and neck. Interestingly, Dubinina & Dubinin (1951) suggested that the moulting pattern of a host may affect flea distribution on its body. They observed that *O. silantiewi*, preferring the dorsal body area of the grey marmot *Marmota baibacina*, relocated to its ventral area during a moult.

Amin (1976) described the distribution of fleas on ether-anaesthetized dogs and reported that the most commonly infested areas of a dog's body were the neck and hindquarters. This distribution was well explained by the grooming pattern of a dog, with fleas occurring preferentially on the body parts that are least subject to autogrooming. Thus, the distribution of fleas on a host body can be explained not only by some intrinsic mechanism in fleas, but also by the direct effect of a host. Indeed, the duration of the scratch grooming of the head and neck with the hind claws takes only about 2% of the entire time of grooming in a cat (Eckstein & Hart, 2000a). However, Tränkle (1989) suggested the reason for aggregation of cat fleas on the head and neck of the host is mating (with subsequent relocation of females to other areas of the cat's body for oviposition) rather than avoidance of the host's grooming effort.

In addition, the preference of fleas to select a particular body area of a host may be related to the 'quality' of this area in terms of quantity of hairs and the degree of secretion of skin glands. For example, Fuller (1974) studied the ability of *X. cheopis* and *C. felis* to settle on the body of a human and to obtain a blood meal. It was found that this ability was associated with the 'net quantity of hair' (cm of hair per cm^2 of skin) rather than with hair density and average length separately. Apocrine secretion affected negatively the rate of flea bites and thus served as a kind of a natural 'repellent'.

9.4.3 *Microclimatic differences, interspecific competition or the effect of the original host?*

The distribution of fleas over a host body can also be related to their microclimatic preferences. For example, Ma (1989) provided indirect evidence of the effect of microclimate on the distribution of *N. bidentatiformis* and *C. tesquorum* across different parts of a host (laboratory mouse) body. Both fleas preferred the anterior area of the mouse over the posterior area both at relatively low (5 °C) and relatively high (25–40 °C) temperatures, but did not show a distinct preference for any part of the host body at 10–20 °C. Furthermore, with an increase in the ambient temperature, fleas abandoned the anterior ventral area of the mouse and relocated to the posterior dorsal area. Vansulin & Volkova (1962) demonstrated that the relative abundance of fleas (mainly *Xenopsylla skrjabini*) on different body areas of *R. opimus* varied seasonally and correlated negatively with hair density. Marshall (1981a) observed that *Archaeopsylla erinacei* preferred the humid ventral area over the dryer dorsal area of a hedgehog host. *Spilopsyllus*

146   Ecology of host selection

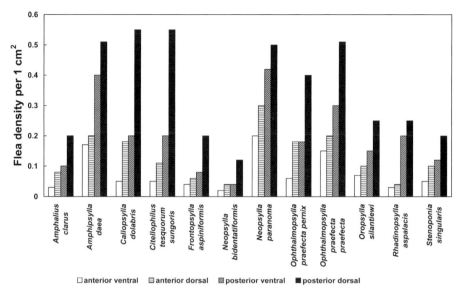

Figure 9.12 Density (per 1 cm^2 of surface) of fleas in different areas of a host's body. Host species are the plateau (Daurian) pika *Ochotona daurica* (for *Amphalius clarus* and *Frontopsylla aspiniformis*), Trans-Baikal zokor *Myospalax psilurus* (for *Amphipsylla daea*, *Rhadinopsylla aspalacis* and *Stenoponia singularis*), Himalayan marmot *Marmota himalayana* (for *Callopsylla dolabris* and *Oropsylla silantiewi*), Daurian ground squirrel *Spermophilus dauricus* (for *Citellophilus tesquorum*), greater long-tailed hamster *Tscherskia triton* (for *Neopsylla bidentatiformis*), plateau zokor *Myospalax fontanierii* (for *Neopsylla paranoma*) and Mongolian five-toed jerboa *Allactaga sibirica* (for *Ophthalmopsylla praefecta pernix* and *Ophthalmopsylla praefecta praefecta*). Data from Ma (1983).

cuniculi preferred the pinnae and periauricular areas of a rabbit (Muller *et al.*, 2001). Mead-Briggs *et al.* (1975) described that, in warm weather, fleas aggregated inside the rabbit pinnae, whereas in winter they moved to the areas of longer fur at the external edge of the ears. Furthermore, if it was very cold, fleas relocated from the ears to the body and concentrated around the anus. Ma (1983) noted that the highest density of 12 flea species was characteristic of the posterior dorsal area of a host (Fig. 9.12) and argued that the unequal distribution of fleas among different areas of the host body is related to between-area variation in the density and length of the hairs.

Co-occurring different flea species may have different preferences for areas on the body of the same host. On rats, most individuals of *X. cheopis* and *N. fasciatus* occur on the hindquarters, whereas the majority of *Xenopsylla astia* and *L. segnis* were found on the head and neck (MacCoy & Mitzmain, 1909; Prasad, 1972; Marshall, 1981a). Differential temperature preference can play a role in this pattern of distribution as the skin surface temperature of the rat's hindquarters is 34.5 °C, whereas that of neck is 35.8 °C (Prasad, 1972). In Australia, *Echidnophaga myrmecobii* occurred on the head and body of the rabbit *Oryctolagus cuniculus*, whereas *Echidnophaga perilis* occurred mainly on the fore and hind feet (Shepherd & Edmonds, 1979). These differential preferences can presumably lead to a decrease in competitive pressure on each flea species. On the other hand, Shepherd & Edmonds (1979) suggested that the differential preferences of the two *Echidnophaga* species are related to the difference in the hair structure of their original marsupial hosts. *Echidnophaga myrmecobii* mainly parasitize hosts with a soft fur (such as possums), whereas hosts of *E. perilis* have short, coarse hairs (such as wombats). As a result, when parasitic on a rabbit, the former selects body parts with soft hair, whereas the latter chooses areas of coarser fur.

9.5 Time spent on- and off-host

9.5.1 'Body' and 'nest' fleas

Fleas are defined as periodic ectoparasites because they spend a considerably longer time on the hosts than is required merely to obtain a blood meal but nevertheless spend a significant amount of time off-host (Lehane, 2005). As mentioned above, the majority of fleas exploit hosts that possess shelters such as burrows and/or nests. A majority of fleas, therefore, alternate periods when they occur on a host body and when they reside in its burrow/nest. Different flea species vary in the proportion of time they spend in each of these two places. Consequently, Ioff (1941) recommended distinguishing between so-called 'body' or 'fur' fleas and 'nest' fleas. As is clear from the suggested terminology, 'body' fleas spend most (if not the entirety) of their life on a host, whereas the contact of a 'nest' flea with its host is intermittent, and the flea stays on a host only to accomplish a blood meal. For example, ischnopsyllid fleas (parasitic on bats) rarely (if at all) abandon their hosts (Hůrka, 1963a, b; Medvedev et al., 1984). In contrast, *Ctenophthalmus* and *Rhadinopsylla* species represent typical 'nest' fleas. For example, Soshina (1973) reported that *Ctenophthalmus wagneri* comprised about 60% of all fleas found on the bodies of *Apodemus* and *Microtus* hosts but more than 80% of all fleas found in the nests. Kozlovskaya & Demidova (1958) reported that among 111 *Frontopsylla luculenta* collected from the fieldmouse *Apodemus agrarius*, only four individuals were taken from the nests, whereas the rest occurred on the host bodies. In contrast, 791 of 1120 *Ctenophthalmus congeneroides* were collected from the nests. Similar results were reported by Smit (1962b) for *Palaeopsylla minor* and *Ctenophthalmus bisoctodentatus* collected from the European mole *Talpa europaea* in the Netherlands. About 99% of individuals of the former species were found on the bodies of the moles, whereas about 73% of individuals of the latter species were found in the nests. However, Khudyakov (1965) observed that in the southern regions of the Russian Far East, the most abundant fleas collected from the nests of several rodent species were represented by the same species as the most abundant fleas collected from the bodies of these rodents.

It has been suggested that many flea species spend more time in the nest than on the body of a host because of the relative stability of the nest environment (Rothschild & Clay, 1952; Hopkins, 1957). Another reason for between-species differences in body versus nest preference is that some species require prolonged contact with a host for their eggs to mature. For example, Starozhitskaya (1968) demonstrated that the eggs in female *X. cheopis* mature mainly when a flea stays on a host, whereas no such relationship was found in *N. fasciatus*. As a result, the mean time of uninterrupted stay on a host body was 48 h for the former species and only 7 h for the latter species.

The idea of the 'body' versus 'nest' flea dichotomy led to vigorous discussions in Russian literature (e.g. Novokreshchenova, 1960; Nelzina et al., 1963; Zhovty, 1963; Vashchenok, 1988), although the intensity of these discussions decreased drastically and eventually diminished in the early 1990s. The termination of this discussion seems to be related to both a decreased importance placed on the issue and a drastic decline of the number of flea researchers. Furthermore, empirical evidence does not always support the attribution of a flea species to one or other of the categories. For example, Medzykhovsky (1971) demonstrated that the proportions of a 'nest' flea *Neopsylla setosa* and a 'body' flea *N. laeviceps* that abandoned the host body during 6 h following the placement on a host did not differ (24.6% versus 25.9% of fleas, respectively). However, in methodologically similar experiments, about 50% of a 'nest' flea *Ctenophthalmus pollex* left the host during 6 h following placement on the host, whereas only 16.3% of a 'body' *C. tesquorum* did so (Medzykhovsky, 1971). It is expected that because the 'body' fleas live within the fur of the host, their sensorial structures would be lost during their evolution (de Albuquerque Cardoso & Linardi, 2006). However, sensorial structures have been found all over the body of a 'body' flea *Polygenis tripus* (de Albuquerque Cardoso & Linardi, 2006).

In addition, the ambiguity of the 'body' versus 'nest' dichotomy is related to circadian (e.g. Dubinina & Dubinin, 1951) and seasonal variation in flea behaviour as well as the effects of the physiological status of fleas (Gauzshtein et al., 1965), ambient temperatures (Darskaya, 1954; Vasiliev & Zhovty, 1961; Zhovty & Vasiliev, 1962a) and host behaviour (Zhovty, 1967). All these factors can influence the relative abundance of fleas on the host and in the burrow (Vashchenok, 1988). Dubinina & Dubinin (1951) reported that when hosts were

Table 9.1 *Mean time of staying on a host in three flea species parasitic on the great gerbil* Rhombomys opimus *in Central Asia*

Flea	Season	Life-cycle phase	Time spent on a host's body (h)
Xenopsylla gerbilli	December–February	Inactivity	19
	March–April	Reproduction start	58.5
	May	Reproductive peak	7
	June	Mass emergence	7
	August	Termination of reproduction	12.7
	October	Decline of activity	27.3
Paradoxopsyllus teretifrons	October	Mass emergence	1.5
	December–February	Reproductive peak	7
	March–April	Imago dying off	26.7
Ctenophthalmus dolichus	October	Mass emergence	26
	December–February	Reproductive peak	4.9
	March–April	Imago dying off	5.4

Source: Data from Gauzshtein *et al.* (1965).

parasitologically examined in the morning or in the evening, the 'nest' species predominated in collections, whereas 'body' fleas were mainly collected from hosts' bodies in the afternoon. Zhovty's (1963) observations of several flea species (*Megabothris calcarifer, C. congeneroides, Amphalius runatus, O. hirticrus, Amphipsylla primaris* and *F. luculenta*) demonstrated that during some seasons fleas prefer to spend most of their time on the body of a host, whereas in other seasons they mostly prefer to stay in its nest. Stark (2002) also noted that the same flea species can behave as a 'body' flea in one season and as a 'nest' flea in another. Gauzshtein *et al.* (1965) used radioactive labelling to demonstrate that the mean time of retention on a host strongly varied among phases of a flea life cycle (Table 9.1).

9.5.2 Reliability of body infestation data

The most commonly used methods for estimating flea population size are calculations of flea abundance (mean number of fleas per examined host) and intensity and prevalence of infestation by fleas (mean number of fleas per infested host and percentage of infested hosts, respectively). However, a large proportion of a flea population occurs off the host body and, consequently, the reliability of these methods has been questioned repeatedly (Muirhead-Thomson, 1968; Lauer & Sonenshine, 1978; Lehmann, 1994). Furthermore, the idea that estimates based on sampling fleas on hosts only are not reliable led to the claims that these estimates cannot be used for either between- or within-species (e.g. among time periods) comparisons (Muirhead-Thomson, 1968; Marshall, 1981a). Alternative methods proposed for the estimation of flea population size have usually not been applied by anybody except the respective authors. For example, the mark–recapture technique of Lauer & Sonenshine (1978) requires radio-labelling, which is difficult in many cases. In addition, radio-labelling has negative effects on flea feeding and mobility (Kharlamov, 1965). A method proposed by Lehmann (1994) to estimate the size of the entire flea population using body infestation data and the reproductive status of female fleas (reinfestation analysis) has rarely been used, probably because of the specificity of the sampling design. Another reason for not using this method is that the large variation in reproductive patterns among different flea species is not taken into account.

Ideally, the sampling and estimation of abundance of periodic ectoparasites such as fleas must include sampling from both the host body and off-host environment (e.g. Haas, 1966). However, the excavation of animal burrows is usually labour-intensive and time-consuming and, therefore, few studies have been undertaken with the aim of comparing the abundance of fleas on host bodies versus host burrows. One of the most frequently cited studies in this relation is that by Stewart & Evans (1941) on fleas parasitic on the California ground squirrel *Spermophilus beecheyi*. They reported a positive correlation between estimates of flea abundance based on host sampling and those based on burrow sampling. Li et al. (2001a) found significant positive correlation between abundances of *C. tesquorum* and *N. bidentatiformis* on the body and in the nests of the Daurian ground squirrel *Spermophilus dauricus*. The same was true for several flea species parasitic on Brandt's vole *Microtus brandti* (Li et al., 2001b). This suggests that flea indices calculated from host body sampling can be reliable indicators of entire population size of fleas (although the conclusions of Stewart & Evans (1941) have been questioned by Lehmann (1994)). Consequently, the majority of later studies that estimated flea numbers have relied exclusively on sampling from the host bodies only (e.g. Schwan, 1975; Krasnov et al., 1997; Lindsay & Galloway, 1997), whereas the correlations between flea estimates from the host bodies and burrows have not been tested in these studies.

It is clear that if the number of fleas that attack a host simultaneously does not depend on the size of a flea subpopulation in a burrow, then the flea numbers based on host trapping and examination would be incorrect estimates of flea population size. If, however, the abundance of fleas on a host body is correlated with the number of fleas in a host burrow, estimates based on body samples can be considered to be reliable estimates of the entire flea population and can be used in between- and within-species comparisons. Krasnov et al. (2004d) tested this correlation by comparing the relationship between abundance on host bodies and abundance in excavated host burrows among different flea species parasitic on rodents using data from published studies that simultaneously sampled fleas from rodents and from their excavated burrows or nests. In addition, to examine whether patterns in the relationship between flea abundance on the host body and in the host burrow differ between fleas considered as either 'body' or 'nest' species, these two groups of fleas were analysed both in combination and separately. In total, data on 55 flea species (31 presumed 'body' species, 15 presumed 'nest' species and nine species with unclear attribution) belonging to four families (Pulicidae, Hystrichopsyllidae, Leptopsyllidae and Ceratophyllidae) were used. These data were controlled for unequal between-study sampling effort as well as for flea phylogeny.

It appeared that the mean number of fleas on host bodies correlated positively with the mean number of fleas in host burrows/nests both when the entire data pool was analysed and for separate subsets of data on 'body' and 'nest' fleas (Fig. 9.13). This positive correlation also held for a within-host (*Microtus californicus*) between-flea comparison.

These results demonstrate that the index of host body infestation by fleas can be used reliably as an indicator of the entire flea population size (see also Haas & Kucera, 2004). Furthermore, this index can be used for interspecific comparisons of flea abundance both between- and within-host species. It is clear, though, that despite the positive correlation between the number of fleas on a host body and the number of fleas in a host burrow/nest, the absolute flea abundance on the body and in the burrow can be profoundly different. For example, Li et al. (2000) reported values of 6.9 of nest/body flea index (mean fleas per nest/mean fleas per host) for 15 fleas from bodies and nests of the Mongolian gerbil *Meriones unguiculatus*, whereas those for four fleas parasitizing the California vole *M. californicus* were >1.0 (Stark, 2002). This is not surprising because the emergence of new fleas from pupae occurs off-host, and thus the number of fleas in the burrow periodically increases. Nevertheless, all fleas must attack the host occasionally to feed. Periods between consecutive feedings vary among flea species, but fleas usually feed one to five times daily (Kosminsky, 1965; Vashchenok, 1988; see also Chapter 10). In addition, temporal variation in the proportion of starving individuals seems to be rather low, although evidence for this is indirect (e.g. Krasnov et al., 2002c). Consequently, a host appears to be a proportional random sampler of flea population.

Finally, the results of the above analysis also suggest that the classification of flea species into 'body' and 'nest' categories is not especially helpful in ecological studies, at least not for interspecific comparisons.

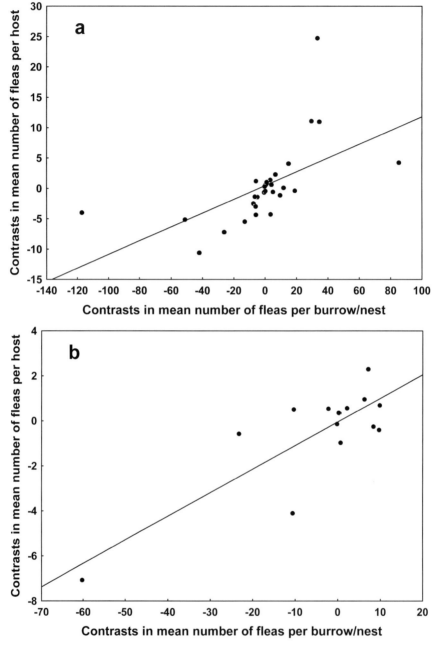

Figure 9.13 Relationship between the mean number of fleas on a host body and the mean number of fleas in a host burrow/nest for (a) 'body' and (b) 'nest' fleas using the method of independent contrasts. Redrawn after Krasnov et al. (2004d) (reprinted with permission from Blackwell Publishing).

9.6 Concluding remarks

The results of studies on the ecology of selection of a particular host and/or a particular part of a host body shows that that fleas, like other animals, behave as if they are able to choose environments in which their reproductive benefit is maximized (see Levins, 1968). Similar evolutionarily motivated behaviour can be expected for other parasites. Indeed, Gurtler et al. (1997) reported that proportion of the kissing bug *Triatoma infestans* feeding on humans as opposed to chickens and dogs decreased with an increase of the density of bugs on the host. However, nature is much more complicated than any of the human-created models such as the IFD. For example, Bansemir & Sukhdeo (1996) and Sukhdeo (2000) reported habitat selection behaviour of the nematode *Heligmosomoides polygyrus* and noted the non-IFD distribution. These helminths can easily relocate within a host's gut, but they often establish in suboptimal locations even when an optimal microhabitat is vacant. Such behaviours cannot be explained in the framework of the classic theory of habitat selection. The consideration of life-history and phylogenetic constraints as well as re-evaluation of the concept of optimality may be required (Sukhdeo et al., 2002).

10

Ecology of haematophagy

Adult fleas are obligatory haematophages. Among arthropods, haematophagy has evolved independently in at least six lineages (Ribeiro, 1995, 1996; Lehane, 2005). Two routes for the evolutionary transition to a parasitic lifestyle have been suggested (Waage, 1979). In the first type, known as type A routes, associations with hosts preceded adaptations for parasitic feeding. This is presumed to be the evolutionary pathway for, among others, psoroptoid acariforms (Fain & Hyland, 1985) and anopluran lice (Kim, 1985a). The second type, known as type F routes, involved adaptations to feeding on a host that preceded the actual association, such as the stylet-like mouthparts of mosquitoes feeding on nectar and hemipterans feeding on plants that can be easily adapted to haematophagy (Radovsky, 1985). In some taxa, evolution of parasitism could be a combination of the two pathways (e.g. dermanyssoid mites: Dowling, 2006). It has been suggested that Siphonaptera might have evolved from free-living mecopteran ancestors mainly along the type F route (Tillyard, 1935; Hinton, 1958, Whiting, 2002a, b; but see Medvedev, 2005 and Chapter 4). In this chapter, the ecology of blood-sucking in fleas is discussed. First, I consider the morphology of flea mouthparts and, based on empirical, albeit extremely scarce, evidence, attempt to find a correlation between their size/shape and the morphology of host skin. Then, I discuss how host-, flea- and environment-related factors can affect the feeding success of a flea.

10.1 Mouthparts and host skin

Haematophagous ectoparasites can feed either directly from blood vessels or from pools of blood created by probing with their mouthparts. The majority of fleas are exclusive vessel-feeders (Marshall, 1981a; Lehane, 2005), with some exceptions such as *Spilopsyllus cuniculi* (Rothschild et al., 1970). Nevertheless, fleas, feeding mainly from blood vessels, may occasionally feed from pools also (Deoras & Prasad, 1967).

The mouthparts of imago fleas represent a modified version of the basic insect pattern (Snodgrass, 1944). The main components are the unpaired mouth stylet (epipharynx) and paired laciniae surrounded by the sheath composed of the maxillary lobes and labium. The latter consists of a premental rod supporting a pair of basally fused and distally free, secondarily subsegmented labial palps (Snodgrass, 1944; Vashchenok, 1988; Michelsen, 1997).

To the best of my knowledge, the relationship between the length of flea mouthparts and host skin structure has never been studied. This is true for both between-host and within-host between-body-area comparisons. Nevertheless, Vashchenok (1988) mentioned that *Vermipsylla alakurt*, a parasite of ungulates, is characterized by a relatively long proboscis (0.823 mm), whereas the proboscis of the rat flea *Xenopsylla cheopis* reaches only 0.412 mm in length. It is true that the body sizes of these two species differ sharply, with *V. alakurt* being much larger; however, it is the absolute rather than the relative length of the proboscis that matters when penetration of the host skin is considered. On the other hand, small *Echidnophaga gallinacea* and *Tunga penetrans* (male/female body lengths 0.97 mm/1.27 mm and 1.06 mm/1.28 mm, respectively: data from Suter, 1964) possess relatively longer proboscises than the larger *Archaeopsylla erinacei*, *X. cheopis* and *Pulex irritans* (male/female body lengths 1.91 mm/2.34 mm, 1.84 mm/2.64 mm, and 1.96 mm/2.94 mm, respectively: data from Suter, 1964). Indeed, the length of proboscis per 1 mm of body length is (male/female) 0.30 mm/0.33 mm in *E. gallinacea* and 0.32 mm/0.28 mm in *T. penetrans*. In contrast, these values are only 0.16 mm/0.17 mm, 0.19 mm/0.16 mm and 0.13 mm/0.10 mm for *A. erinacei*, *X. cheopis* and *P. irritans*, respectively. The relatively longer proboscises of *E. gallinacea* and *T. penetrans* may be explained by the fact that these species use their mouthparts not only for piercing but also for attachment to the host skin (Marshall, 1981a). However, the relatively long proboscis is characteristic not only of the attaching female *T. penetrans*, but also of the non-attaching males (who penetrate the stratum corneum for only a few hours: Witt et al., 2004). In addition, relative proboscis length in non-sessile *Synosternus pallidus* appeared to be only slightly less than that in *E. gallinacea* and *T. penetrans* (0.25 mm per 1 mm of body length: data from Suter, 1964). This suggests that the length of the proboscis is determined not only by its role as a piercing versus an attaching tool, but also by some other factors such as depth of blood vessels or skin thickness of the host. Indeed, *S. pallidus* commonly parasitizes medium-sized mammals such as canids and lagomorphs (e.g. Klein et al., 1975).

The epipharynx is formed by the labrum (upper lip), whereas the paired lacinial stylets arise from levers of stipital derivation (Michelsen, 1997). The anterior edge of the dorsal tubercle of the epipharynx forms a medial ridge that bears a series of nodules furnished with basiconic sensilla (Vashchenok, 1988). The number of nodules is species-specific, being, for example, 14 in *X. cheopis*, 14 in *Cediopsylla simplex*, nine in *Echidnophaga oschanini* and two in *V. alakurt* (Amrine & Lewis, 1978; Vashchenok, 1988). Due to the presence of basiconic sensilla, it has been suggested that these nodules play a certain role in determining the level of submersion of the stylet into the host skin (Wenk, 1953; Amrine & Lewis, 1978).

Laciniae are used to cut the skin of a host, allowing the epipharynx to penetrate a blood vessel. Accordingly, laciniae bear several rows of teeth. Large teeth, each with a backward-projecting cusp, shape the cutting surface of the margins of each lacinia (Amrine & Lewis, 1978). Development of these structures differs to a certain degree among species and is probably correlated with the skin morphology of the main host. Vashchenok (1988) noted that heavily armed laciniae are characteristic of stick-tight fleas such as *V. alakurt*, *Dorcadia ioffi* and *E. oschanini* as well as parasites of rats (e.g. *X. cheopis*) and pikas (e.g. *Ochotonobius hirticrus*). In contrast, laciniae of *Leptopsylla segnis* and species of the genera *Ctenophthalmus* (parasitic on a variety of small mammals) and *Ischnopsyllus* (parasitic on bats) are weakly armed. Fleas with heavily armed laciniae can feed on a wide variety of hosts. For example, *X. cheopis* feeds on many mammals, including humans, birds and even reptiles (Darskaya & Besedina, 1961; Kulakova, 1964; Fox et al., 1966; Vashchenok et al., 1976; Raszl et al., 1998). In contrast, *L. segnis* feeds on humans with great difficulty (Vashchenok, 1988), while the majority of *Ctenophthalmus* species, though they may attempt it, are unable to do this (Ioff, 1941).

The size and shape of flea mouthparts have been suggested to be associated not only with morphological features of a host's skin, but also with morphological features of a host's blood. However, in *X. cheopis*, the diameter of the alimentary canal at the distal portion of the proboscis is about 0.007 mm (Vashchenok, 1988) which is smaller that the diameter of erythrocytes of birds or reptiles. Nevertheless, these fleas are capable of feeding on pigeons and lizards in the laboratory (Vashchenok et al., 1976).

10.2 Measures of feeding success

One reason why a species may occur in a particular habitat (or set of habitats) but be absent from other habitats is that it is adapted to some characteristics of its preferred habitat. These adaptations facilitate the performance of the species and increase its fitness in its preferred habitat. Consequently, comparing the performance of a species in different habitats will help to uncover what determines species–habitat association. When considering haematophagous ectoparasites, this idea can be reformulated as a comparison of the performance of a flea on different hosts. Essentially, this performance involves the ability of a flea to extract a resource (blood) from a host and transform this resource into offspring. Throughout this chapter, I consider the former aspect of performance; the latter aspect will be discussed in the next chapter of this book.

Feeding performance (i.e. feeding success of a flea) can be measured using several parameters, the most important of which are blood meal size, rate of engorgement, rate of digestion, metabolic cost of digestion and number of blood meals necessary for egg maturation and oviposition. In addition, the time that an individual flea survives after a single blood meal may be a helpful measure of the evaluation of the efficiency of the resource (blood) use.

10.2.1 Blood meal size and rate of engorgement

The size of a blood meal is a straightforward measure commonly used in feeding studies of ectoparasitic arthropods (e.g. Fielden *et al.*, 1992, 1999). It is defined as the amount of blood taken in by a flea during a single foraging bout. It is usually measured as the difference in body mass of a flea before and after a timed period of an uninterrupted stay on a host (usually 1 or 2 h: e.g. Krasnov *et al.*, 2003b). This time is usually more than enough to attain satiation in most flea species (see below).

Three main caveats should be noted in relation to the measurement of a blood meal size. First, fleas vary in their body size both among species and between genders within a species, as well as among individuals within a gender. Consequently, the size of a blood meal should be corrected for body size and recalculated, for example, as the amount of blood taken per unit body mass of a starving flea. As a result, although in some species the absolute quantity of blood consumed by a female is greater than that consumed by a male, the gender difference in the blood intake appears sometimes to be absent or even reversed when difference in body size is taken into account (Fig. 10.1) (see also Chapter 7).

Second, some flea species may void their gut content prior to blood digestion in order to increase jumping distance (e.g. *Paraceras melis*: Cox *et al.*, 1999), whereas other species may continuously feed with most of the blood passing through the gut unchanged (e.g. *L. segnis* and *Leptopsylla taschenbergi*: Kosminsky, 1965; *P. irritans*: Shchedrin *et al.*, 1974). Accordingly, there is a school of thought

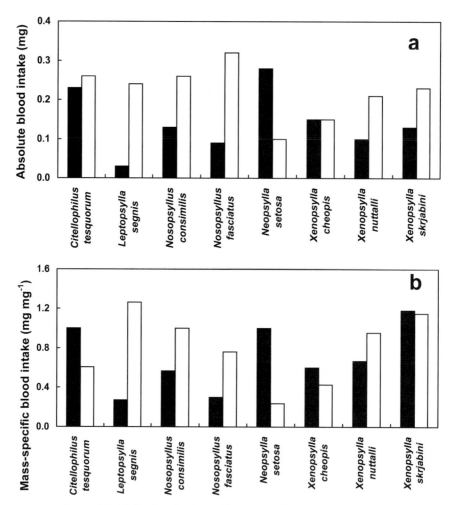

Figure 10.1 (a) Absolute (mg) and (b) mass-specific (mg per mg of body mass) blood intake of male (black) and female (white) fleas. Data from Bryukhanova et al. (1961) and Bibikova & Gerasimova (1967, 1973).

that says that blood meal size, gut capacity or mass change are of little value for such species (Marshall, 1981a). Nevertheless, the difference in body mass prior to and after feeding in these species seems to be a perfectly good estimator of feeding performance, although, indeed, consideration of body mass change as an estimator of the parasitic pressure on a host hardly makes sense.

Third, blood intake is usually greater in older (previously fed) fleas compared with newly emerged fleas and in hungry compared with recently fed fleas (Bryukhanova et al., 1961; Shulov & Naor, 1964; Bibikova & Gerasimova, 1967; 1973; Vashchenok, 1988; see also below). Consequently, measurements of blood meal size should be done for fleas of similar age and similar feeding state.

Rate of engorgement may be evaluated either as the time necessary for a flea to satiate or as the percentage of fleas with a highly engorged midgut after a timed period of feeding. Obviously, there is a positive correlation between the amount of blood taken and time of feeding. For example, female *N. laeviceps* took 40 μg of blood from the Mongolian gerbil *Meriones unguiculatus* during 1 min of feeding, while 5 min of feeding resulted in twice as large (84 μg) a blood meal (Liu *et al.*, 1993). Nevertheless, some fleas need little time to engorge. For example, in *Neopsylla setosa*, *Mesopsylla eucta*, *Ophthalmopsylla volgensis* and *Nosopsyllus mokrzeckyi*, satiation time is 2–5 min only (Kulakova, 1964; Kosminsky, 1965). Other species need twice as much time. This is characteristic for *Xenopsylla gerbilli* (Kulakova, 1964), *Ctenophthalmus dolichus* (Balashov *et al.*, 1965), *L. segnis*, *L. taschenbergi* (Kosminsky, 1965), *Xenopsylla skrjabini* (Vansulin, 1965), *X. cheopis* and *Nosopsyllus consimilis* (Vashchenok, 1988).

10.2.2 Rate of digestion

Digestion rate is another important characteristic of flea feeding success. Ioff (1949) divided the entire digestion process into five stages. Later, Bibikova (1963) and Bibikova & Gerasimova (1967) gave physiological characteristics to the stages of Ioff's (1949) classification, namely: (I) engorgement – stretched midgut with scarlet content; (II) start of digestion – darkening of the midgut content; (III) start of absorption and digestion of haemoglobin – the contour of the midgut is jagged and the content is dark brown or black; (IV) conversion of haemoglobin to haematin – the midgut contains remnants of digested blood; and (V) end of the digestion process and excretion – the midgut is empty. Vashchenok & Solina (1969) suggested that stage V of the above classification is related to starvation and has nothing to do with digestion and, therefore, proposed to distinguish three stages only, namely (I) from blood intake to lysis of the main part of the red blood cells; (II) end of haemolysis and digestion of blood to the final product (haematin); and (III) start of excretion. A modified scheme incorporating the classifications of both Ioff (1949) and Vashchenok & Solina (1969) has been successfully used in some recent studies (e.g. Krasnov *et al.*, 2002c, 2003b). In these studies, the duration of the following stages was measured: (I) early stage – midgut stretched and fully filled with light scarlet or dark red blood; (II) middle stage – jagged midgut with dark brown or black colouring; and (III) late stage – midgut contains only remnants of digested blood or is empty. Presumably, the faster digestion of a blood meal from, for instance, one host species compared with that from another host species may be associated with better performance by the flea on the former and, thus, may be a factor affecting the foraging decision of a flea.

10.2.3 Energetic cost of blood digestion

The rate of digestion, albeit a useful measure of feeding performance, may not be sensitive enough for understanding differences in feeding success of a flea in comparisons among host species or within host species among host individuals, or else within flea species between genders or among age cohorts. This is because the net nutritional and energetic value of food consumed by an animal depends on a variety of factors including ecological, chemical, morphological and/or physiological constraints of the forager (Lee & Houston, 1993; Brown et al., 1994; Kam et al., 1997; Piersma & Drent, 2003; Johnston et al., 2005). An interplay of cost/benefit evaluations concerning each of these factors and their interactions results in a foraging decision. One of the most important factors affecting a foraging decision is the energetic cost of food processing. Of particular importance is the energetic cost of digestion, as different types of food entail different energy costs. These costs would include secretion of enzymes, metabolism of food components, excretion of toxic by-products and the heat increment of feeding (see Clements, 1992 for review).

For a parasite, the differential energy costs of digesting food obtained from different hosts may have important ecological and evolutionary implications. A lower energy cost of resource processing would allow a parasite to allocate more energy to other competing requirements of an organism (host location, mating, oviposition). This would supposedly increase the fitness rewards stemming from selection of an appropriate host and, thus, can shape the coevolutionary process between hosts and parasites. At the evolutionary scale, differences in the energy cost of digestion of a resource extracted by a parasite from one host species rather than another may reflect specific adaptations of a parasite to exploit successfully a particular host species. Specific adaptations to this host species might be favoured by natural selection due to differential fitness rewards between hosts. Thus, energy expended by a parasite to process food obtained from a host may be an indicator of evolutionary success of a parasite in the exploitation of a host species.

To the best of my knowledge, the only study that has measured energetic cost of blood digestion in fleas is that carried out by Sarfati et al. (2005). They measured CO_2 emission by fleas using a flow-through respirometry system and calculated the energy cost to a flea of digesting 1 mg of blood from a host during 1 h of the early, middle or late digestion stages, referring to the energy expended to digest 1 mg of blood as the specific dynamic effect (SDE), as suggested by Withers (1992). Withers (1992: 108) defined SDE as reflecting 'the energetic requirements of many processes that occur as a consequence of food digestion, including mechanical processing, energy exchange through catabolic and anabolic biochemical pathways and amino acid deamination and nitrogen excretion'. First, Sarfati et al. (2005) calculated the average volume of CO_2 emitted per hour per 1 mg of body mass of newly emerged unfed flea. This value was then used to calculate the difference in the mass-specific volume of emitted CO_2 between a digesting flea and an unfed flea for each respirometric replicate. The difference in body mass of a flea before and after feeding (recalculated per 1 mg of body mass before feeding) was considered to be equal to the mass-specific amount of blood consumed. The quotient of mass-specific difference in the volume of emitted CO_2 between digesting and unfed flea and mass-specific amount of consumed blood was considered as a mass-specific indicator of SDE (energy expended in digestion of 1 mg of blood). To convert metabolic rate measured as the rate of CO_2 emission to an energy equivalent, a respiratory quotient was assumed to be equal to 0.8 (giving 24.5 J per 1 ml of CO_2: Schmidt-Nielsen, 1990). This was determined previously for unfed females of the tick *Amblyomma marmoreum* by Lighton et al. (1993) and seemed a reasonable assumption for haematophagous arthropods.

10.2.4 Number of blood meals necessary for oviposition

The blood taken from a host by a flea is further translated into reproductive effort (Lehane, 2005). Consequently, in females ovarian development is physiologically associated with the digestion. As we already know, in fleas oogenesis is tightly linked to feeding (see Chapter 5). Time from the start of the feeding to the first oviposition event varies among flea species (Fig. 10.2). However, it is unclear how many feeding events occur during this time.

Comparison of the numbers of feeding events on different host species or individuals required by a flea to oviposit may help in understanding the success of a flea to transform resources provided by a host into offspring. Krasnov et al. (2004a) estimated this by calculating the number of feedings necessary for 50% of females to develop their eggs (NF_{50}) using the logistic model analogous to the half-maximal response model used in pharmacological research (Neter et al., 1990):

$$P = 100 - \frac{100}{1 + (NF/NF_{50})^b}, \tag{10.1}$$

where P is the percentage of females with fully developed eggs, NF is the number of feedings and b is the slope of the function. The least-squares estimation

162   Ecology of haematophagy

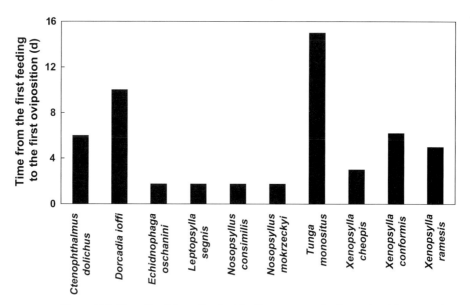

Figure 10.2 Time (days) from the first feeding event to the first oviposition event in eight flea species. Data from Grebenyuk (1951), Kosminsky (1962), Bgytova (1963), Lavoipierre *et al.* (1979), Vashchenok (1988) and Krasnov *et al.* (2004a).

procedures via the quasi-Newton algorithm were applied. The observed relationships between the number of feedings and percentage of females with fully developed eggs fitted well to the model (Fig. 10.3).

10.2.5 Resistance to starvation

The time that a flea survives after a single blood meal on a certain host may be a useful, albeit indirect, indicator of the relative quality of the resource taken from a host. When comparisons among host species or individuals are carried out, it should be remembered that male and female fleas have different survival capacities under starvation, all else being equal (see Chapter 7). In general, fleas can survive for quite a while after a single blood meal (although the survival time is strongly affected by environmental factors — see below). The ability to survive varies greatly among flea species (see Chapter 5).

10.3   Host-related effects

10.3.1 Host species

The physical and chemical properties of a host's blood are known to be important characteristics to which a host-specific flea is adapted (Marshall,

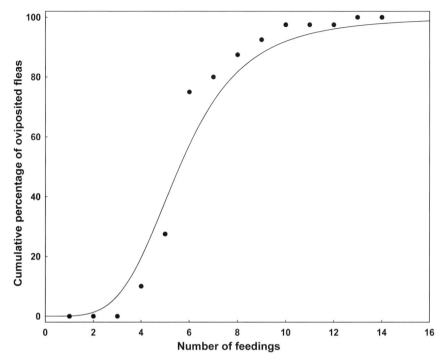

Figure 10.3 Relationship between number of feeding events on Sundevall's jird *Meriones crassus* and cumulative percentage of females with fully developed eggs in *Xenopsylla conformis*. Data from Krasnov et al. (2004a).

1981a; Lehane, 2005). However, studies specifically designed to test the effect of host species on flea feeding performance are somewhat scarce (e.g. Seal & Bhattacharji, 1961; Prasad, 1969; Kamala Bai & Prasad, 1979; Liu et al., 1993). Moreover, in earlier laboratory studies of blood digestion rates, fleas were often fed on laboratory animals rather than on their natural hosts (Table 10.1). Indeed, wild host species were used in only five of the 12 studies of rodent fleas cited in Table 10.1, while the others used mainly laboratory mice, rats, hamsters and guinea pigs.

Nevertheless, the results presented in Table 10.1 clearly show that feeding performance of a flea is strongly affected by the identity of the host species. Moreover, it can be expected that the level of host specificity of a flea species should determine the degree of between-host variation of feeding performance of a flea. In particular, highly host-specific fleas are expected to perform best on their specific host and much worse, if at all, on other host species, because, as we already know, highly specific parasites can become closely adapted to physiological and biochemical traits of a particular host species or group of

Table 10.1 *Median duration of blood digestion by different flea species fed on different host species in laboratory experiments*

Flea species	Host species	Duration of digestion (h)	Reference
Citellophilus tesquorum	*Spermophilus pygmaeus*	6	Bryukhanova & Surkova, 1983
Coptopsylla lamellifer	*Meriones meridianus*	21	Bryukhanova *et al.*, 1983
Ctenophthalmus golovi	*S. pygmaeus*	21.5	Bryukhanova & Surkova, 1983
Leptopsylla segnis	*Mus musculus* (lab)	5	Bryukhanova *et al.*, 1978
Nosopsyllus consimilis	*M. musculus* (lab)	4	Vashchenok, 1967a
Nosopsyllus fasciatus	*M. musculus* (lab)	11	Vashchenok, 1974
	Cavia porcellus (lab)	21	Vashchenok, 1974
	Rattus norvegicus (lab)	21	Vashchenok, 1974
Nosopsyllus laeviceps	*M. musculus* (lab)	4	Vashchenok *et al.*, 1985
Neopsylla setosa	*S. pygmaeus*	21.5	Bryukhanova & Surkova, 1983
Pulex irritans	*Homo sapiens*	5.5	Shchedrin, 1974
Xenopsylla cheopis	*M. musculus* (lab)	12	Vashchenok & Solina, 1969
	Mesocricetus auratus (lab)	17	Vashchenok *et al.*, 1976
	H. sapiens	17	Vashchenok *et al.*, 1976
	C. porcellus (lab)	21.5	Vashchenok *et al.*, 1976
	R. norvegicus (lab)	25	Vashchenok *et al.*, 1976
	Columba livia	25	Vashchenok *et al.*, 1976
	Lacerta viridis	19	Vashchenok *et al.*, 1976
Xenopsylla conformis	*M. meridianus*	11	Bryukhanova *et al.*, 1983
Xenopsylla gerbilli	*M. musculus* (lab)	14	Vashchenok *et al.*, 1985
	Meriones tamariscinus	14	
Xenopsylla skrjabini	*M. musculus* (lab)	16.5	Vashchenok, 1974
	Rhombomys opimus	16.5	Vashchenok, 1974
	C. porcellus (lab)	27	Vashchenok, 1974
	Passer domesticus	22	Vashchenok, 1974
	Phrynocephalus helioscopus	22	Vashchenok, 1974

species (Ward, 1992; Poulin, 2007a). Below, I present the results of several case studies that tested the effect of host identity on feeding success in host-specific and host-opportunistic flea species.

I have already mentioned that *Parapulex chephrenis* is a strictly host-specific flea even though its host occurs in close contact with other potential hosts in rocky habitats of the Negev Desert (Krasnov *et al.*, 1997, 1998). This flea is mainly found on the Egyptian spiny mouse *Acomys cahirinus* but is absent from Wagner's gerbil *Dipodillus dasyurus* and the bushy-tailed gerbil *Sekeetamys calurus*, rodents that coexist with the spiny mouse. Krasnov *et al.* (2003b) and Sarfati *et al.* (2005)

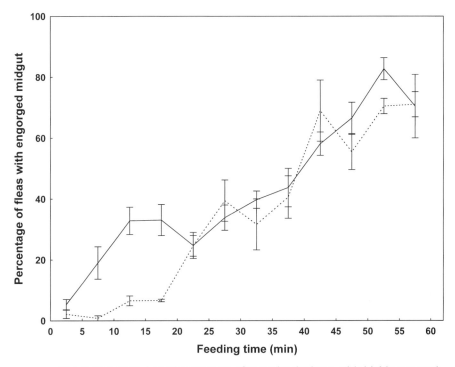

Figure 10.4 Mean (±S.E.) percentage of *Parapulex chephrenis* with highly engorged midgut after different periods of feeding on the Egyptian spiny mouse *Acomys cahirinus* (solid line) or Wagner's gerbil *Dipodillus dasyurus* (dashed line). Redrawn after Krasnov et al. (2003b) (reprinted with permission from Springer Science and Business Media).

studied the feeding rate, rate and energetic cost of digestion and resistance to starvation of *P. chephrenis* when feeding on *A. cahirinus* and *D. dasyurus*, predicting that *P. chephrenis* would fill its gut with blood faster, digest blood for a shorter time with less energy spent for digestion and survive longer when starved while feeding on a specific compared with a non-specific host. All these responses were observed.

For example, 20% of fleas filled their midgut after feeding for 10 min on *A. cahirinus* but this occurred only after 25 min on *D. dasyurus* (Fig. 10.4). In other words, time from contact with the host to the beginning of feeding, which can be considered the latency of foraging decision, was less in *P. chephrenis* feeding on its specific host than on a non-specific host. As Iqbal & Humphries (1983) noted, the relatively long stay on the host may partly be due to the probability of striking a suitable blood vessel and partly due to frequent disturbance by host anti-parasite activity. In their experiments, *Nosopsyllus fasciatus* fed on the host for 2–3 h, although it was physically possible for a flea to obtain a full blood

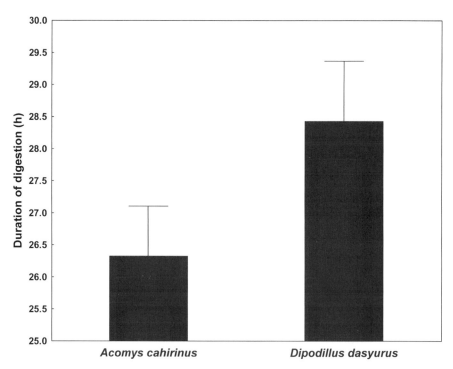

Figure 10.5 Mean (±S.E.) duration (hours) of blood digestion in *Parapulex chephrenis* when feeding on the Egyptian spiny mouse *Acomys cahirinus* or Wagner's gerbil *Dipodillus dasyurus*. Data from Krasnov et al. (2003b).

meal within 10 min. Nonetheless, starving *P. chephrenis* had begun to feed even from a non-specific host, though it took longer to make this foraging decision. The impression was that *P. chephrenis*, when forced to feed on *D. dasyurus*, began feeding after an 'uncertainty' period rather than jumping off the non-specific host and searching for a more suitable host. Interestingly, the time taken for most fleas to reach the fully fed stage was not different. This means that fleas feeding on gerbils started slowly but were then able to feed faster than on spiny mice so that after some time they had caught up with their conspecifics on the specific host. However, in these experiments host animals were not allowed to groom. Consequently, the short latency of feeding may be crucial for a flea when the latency of a host's grooming response aimed to remove fleas is short (e.g. Eckstein & Hart, 2000b). In other words, a flea that begins to feed sooner would be less likely to be removed from a host by grooming before completion of feeding.

The duration of blood digestion was significantly shorter in fleas feeding on *A. cahirinus* than in fleas feeding on *D. dasyurus* (Fig. 10.5). The difference was mainly due to a shorter middle stage of digestion (which includes haemolysis and

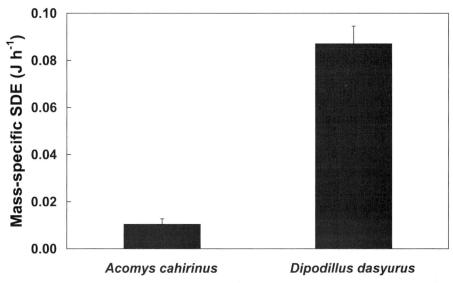

Figure 10.6 Mean (±S.E.) mass-specific dynamic effect (SDE; J per hour) in *Parapulex chephrenis* digesting the blood of the Egyptian spiny mouse *Acomys cahirinus* or Wagner's gerbil *Dipodillus dasyurus* at the middle stage of digestion. Data from Sarfati *et al.* (2005).

digestion of blood to haematin, the final product of blood digestion (Vashchenok, 1988); see above), reflecting between-host differences in the resistance of blood cells (both red and white) to haemolytic activity of the flea digestive system (Vashchenok, 1988; Lehane, 2005). The duration of the early and late stages of digestion that reflect mechanical rather than biochemical processes (midgut filling and release of the undigested remnants and final products, respectively), did not differ between hosts (Krasnov *et al.*, 2003b). The increase in time of the middle stage of digestion when feeding on *D. dasyurus* may indicate that the digestion system of *P. chephrenis* is less adapted to this host than to *A. cahirinus*.

Furthermore, fleas expended significantly more energy digesting the blood of *D. dasyurus* than the blood of *A. cahirinus* (Fig. 10.6). Consequently, *P. chephrenis* not only spent less time digesting the blood of specific as compared with non-specific hosts, but digestion of the blood of the specific host incurred lower energy costs.

Finally, fleas survived for a shorter time when starved after feeding on *D. dasyurus* than after feeding on *A. cahirinus*. This difference was manifested at 25 °C ambient temperature but not at 15 and 20 °C. The effect of the environmental factors on feeding performance of fleas (including resistance to starvation after a blood meal) will be discussed below.

The above results unequivocally demonstrate that a host-specific flea performs on a specific host much better than on a non-specific host. The net outcome of the described responses is that feeding on a specific host should produce a higher fitness reward for *P. chephrenis* than feeding on any other host.

The response of a highly host-specific parasite to the resources extracted from a specific versus a non-specific host is easily predicted. On the other hand, the response of a host generalist to such resources is not so straightforward. Nevertheless, even a host generalist performs differently on different host species. For example, as already mentioned, *Xenopsylla ramesis* exploits Sundevall's jird *Meriones crassus* and Wagner's gerbil *D. dasyurus* in the Negev Desert (see Chapter 9). Narrative description or calculation of standard parasitological indices (e.g. prevalence and intensity of infestation for a particular host species) of its between-host distribution did not provide any hints about host preference or the mechanism underlying this distribution (Theodor & Costa, 1967; Krasnov et al., 1997). However, the feeding performance of this flea on both hosts suggests a clear preference for *M. crassus*, thus conforming to the results of the isodar analyses described in Chapter 9. *Xenopsylla ramesis* required fewer blood meals for the successful start of oviposition when they fed on *M. crassus* than when they fed on *D. dasyurus* (4.2 ± 0.2 versus 5.5 ± 0.4: Krasnov et al., 2004a).

Another host opportunist, *P. irritans*, also demonstrated differential performance in a variety of host species. Zolotova et al. (1971) forced these fleas to feed on laboratory mice, rabbits and guinea pigs. They then deprived fleas of feeding opportunities and measured time to death. *Pulex irritans* actively fed on rabbits, starting the blood meals almost immediately after contact with a host. After a single meal on a rabbit, fleas survived for 1–2 months. Laboratory mice appeared to be less adequate hosts. Fleas started to feed after 15–20 min from the first contact with a mouse and after a meal were usually able to survive for only about 2 weeks. In contrast, the majority of fleas placed on a guinea pig refused to feed, and the several individuals that took a blood meal survived for only 2 days. Bibikova & Gerasimova (1967) found that *X. skrjabini* digested blood of the small five-toed jerboa *Allactaga elater* slower than blood of the great gerbil *Rhombomys opimus*.

In some cases, however, fleas feed successfully on an 'alien' host that is phylogenetically distant from its principal host. For example, Krasnov et al. (2007a) quantified the feeding efficiency of *P. chephrenis* and *X. ramesis* on two rodents, *A. cahirinus* (specific host of *P. chephrenis*) and *M. crassus* (preferred host of *X. ramesis*), and the Egyptian fruit bat *Rousettus aegyptiacus*, an alien host to both fleas. In both fleas, fewer individuals succeed in feeding when offered their non-specific or non-preferred rodent host to feed on compared with those allowed to feed on their preferred or specific rodent host or, surprisingly, on a bat. The proportion of *P. chephrenis* that fed was higher on *A. cahirinus* than on *R. aegyptiacus*. In contrast, similar proportions of *X. ramesis* took blood from *M. crassus* and

Table 10.2 *Proportion of individuals feeding on human blood in seven species of rodent fleas*

Species	Number of fleas used in experiments	Proportion of individuals that took a blood meal (%)
Ctenophthalmus congeneroides	201	0
Frontopsylla elata	100	36.9
Leptopsylla segnis	209	35.4
Megabothris calcarifer	159	65.0
Neopsylla bidentatiformis	546	13.4
Nosopsyllus fasciatus	104	86.5
Rhadinopsylla insolita	20	0

Source: Data from Moskalenko (1958).

R. aegyptiacus. However, each flea species took similar amounts of blood from any of the three host species. Moreover, both fleas digested bat blood significantly faster than blood of either rodent host. In other words, the alien bat host appeared not to be inferior as a source of food to a rodent host phylogenetically close to the flea's principal host. The mechanism behind this could be that a preferred host develops immunological defences against its specific parasite during their common evolutionary history (e.g. Khokhlova et al., 2004a), whereas a parasite should develop specific adaptations to evade or suppress the defences of the immune system of a principal host (Singh & Girschick, 2003; Maizels et al., 2004). Obviously, these means of evasion are costly, and thus are worth mounting only if the benefits of exploiting the host are greater than the costs of these means. Furthermore, immune defence tools are thought to be similar among taxonomically close, compared to taxonomically distant, host species (Poulin & Mouillot, 2004a). Consequently, it may sometimes be advantageous for a parasite to exploit an alien host rather than a non-preferred species related to its principal host, thus possibly trading off the quality of extracted resources against saving energy that would be otherwise spent on immune evasion or immunosuppression.

The 'readiness' of fleas to feed on a non-specific or non-preferred host varies among flea species. For example, the size of a blood meal taken by *N. laeviceps* did not differ depending on host identity (*M. unguiculatus*, *Rattus norvegicus* or laboratory mouse) *ceteris paribus* (Liu et al., 1993). Among seven fleas parasitic on rodents, five species 'agreed' to feed on human blood with the proportion of feeding individuals varying from 13 to 86%, whereas other species refused to feed on a human (Table 10.2) (Moskalenko, 1958; see also Oparina et al., 1989).

This comparison suggests that there exist both intra- and interspecific variability in making foraging decisions.

10.3.2 Host gender and age

Patterns of flea distribution between host genders and age cohorts suggest that the feeding performance of fleas may differ with host gender or age (Morand et al., 2004; Krasnov et al., 2005c, 2006a; Chapter 15; but see Brinkerhoff et al., 2006). In particular, the effect of host gender and age on flea feeding is mainly due to differences in defence functions between genders as well as between age cohorts. For instance, gender differences in immunocompetence may stem from the higher level of androgens found in males that suppress immune function (Folstad & Karter, 1992). The effect of host immunity as well as behavioural defence on feeding and reproductive performance of fleas will be considered in more detail in Chapter 13.

If within-host species parasite feeding performance is related to host defence abilities (Rechav & Dauth, 1987; Varma et al., 1990), then, in general, it can be expected to be higher on male compared to female hosts, and on youngest and oldest compared to 'middle-aged' host individuals (see discussion in Chapters 13 and 15). However, very few experimental studies have specifically aimed at testing the effect of host gender or age on feeding performance of fleas. For example, Haas (1965) studied survival and feeding of *Xenopsylla vexabilis* in relation to gender and age of its host (the Polynesian rat *Rattus exulans*). It was found that fleas survived longer and fed better on adult males followed by adult females and juvenile males.

Nevertheless, experiments carried out in our laboratory demonstrated that better feeding performance on male compared to female hosts and on juveniles compared to adults does seem to be the case. For example, the amount of blood consumed by *P. chephrenis* during 2 h of contact with male *A. cahirinus* was greater than from the same contact time with female *A. cahirinus* (Fig. 10.7a) (B. R. Krasnov, unpublished data). Similarly, *X. conformis* consumed more blood when they fed on juveniles than on adult *M. crassus* (Fig. 10.7b) (Hawlena et al., 2007a). It should be noted that in the latter experiments hosts were prevented from grooming. Therefore, the difference in flea feeding success shown in Fig. 10.7b is most probably related to differences in immunocompetence and/or skin structure between age cohorts of a host.

One of the most famous accounts concerning feeding patterns of fleas as related to host gender and age is that concerning the flea *S. cuniculi* and its rabbit host (e.g. Rothschild & Ford, 1966, 1969, 1972, 1973). This pattern is tightly linked with both flea and host reproductive cycle and will be considered in the next chapter.

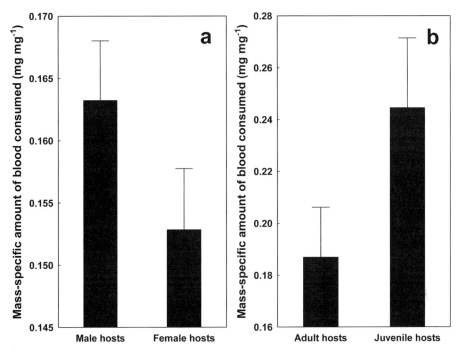

Figure 10.7 Mean (±S.E.) mass-specific amount of blood (mg per mg body mass) consumed by starving (a) *Parapulex chephrenis* when feeding on male and female Egyptian spiny mouse *Acomys cahirinus* and (b) *X. conformis* when feeding on juvenile and adult Sundevall's jird *Meriones crassus*. Data from B. R. Krasnov and L. Ghazaryan (unpublished) and Hawlena et al. (2007a).

10.3.3 Host body conditions

Hosts actively defend themselves against parasites using specific behavioural (e.g. Mooring et al., 2000), physiological (e.g. Banet, 1986) and/or immunological mechanisms (see Chapter 13). This defence against parasites can be costly for a host. Activation of an immune response and maintenance of a competent immune system is an energetically demanding process that requires trade-off decisions among competing energy demands for various activities (Sheldon & Verhulst, 1996; but see Klasing, 1998). The trade-off between the advantage of parasite resistance and its cost could be critical for host individuals that face energy limitations. Consequently, energy-deprived hosts may be less resistant and thus represent better feeding patches for parasites.

Intraspecific host variation in energy reserves can arise for a variety of reasons, such as variation in food availability among individuals (Cumming & Bernard, 1997) or energy constraints related to age (Kam & Degen, 1993) or reproductive status (Zenuto et al., 2002). Low energy intake has been shown to lead to suppression of immune function (Demas & Nelson, 1998). On the other hand, the

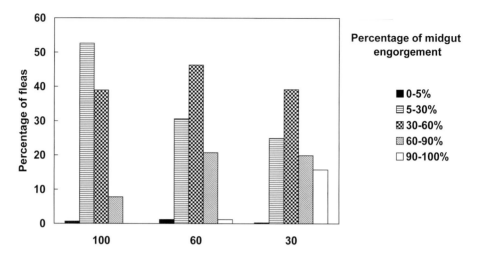

Figure 10.8 Percentage of fleas with different degree of midgut engorgement after 2 days of contact with Sundevall's jirds *Meriones crassus* offered a diet of 100%, 60% or 30% of maintenance energy requirements. Data from Krasnov et al. (2005d).

host-produced 'goods' (*sensu* Combes, 2001) that parasites use (e.g. blood) may be of lower quality in energy-deprived hosts (e.g. decreased plasma glucose during fasting: de Pedro et al., 2003). Thus, hosts in good condition may be a better food source than hosts in poor condition (Lee & Clayton, 1995; Dawson & Bortolotti, 1997). Consequently, parasites themselves face a trade-off between the choice of attacking less well-defended but lower-quality hosts, versus better defended but higher-quality hosts. This may be manifested in different feeding performance depending on host body condition.

Krasnov et al. (2005d) forced *X. ramesis* to feed on *M. crassus* in different nutritional conditions. Rodents were offered diets equivalent to approximately 100%, 60% or 30% of their maintenance energy requirements. The results of these experiments demonstrated not only the effect of host body condition on the quantity and quality of flea offspring (for details see Chapter 11), but also on the feeding performance of fleas (Fig. 10.8). Greater numbers of fleas achieved higher engorgement in hosts experiencing malnutrition. In addition, after feeding, starving fleas survived longer if their rodent host was underfed.

This suggests weakened defence functions of the food-restricted hosts. Weakening of host defence, in turn, favours feeding and, consequently, reproductive success in a parasite (see Chapter 13). In other words, fleas seem to solve the dilemma presented above of choosing between a well-defended but higher-quality resource and a weakly defended but lower-quality resource by choosing the latter. Longer survival of fleas after feeding on a food-restricted host can again be explained by weakened immunoregulatory and effector responses of underfed rodents. Responses stimulated by ectoparasites involve antigen-presenting cells, T-lymphocytes, B-lymphocytes, antibodies, complement, mast cells, circulating granulocytes and cytokines (Jones, 1996). Supposedly lower titres of these components can favour resistance to starvation of an ectoparasite. This can be especially important to fleas because of, at least, two reasons. First, fleas are exposed to strong and/or specific immune attacks because of their intimate association with host blood, the site of major immune defence systems, and skin-associated lymphoid tissues (SALT), the complex of cells responsible for immune response at the cutaneous interface (Streilein, 1990; Wikel, 1996). Second, digestion in fleas is mainly intracellular (but see Filimonova, 1989), and they lack a peritrophic membrane (Vashchenok, 1988) which lines the gut of many arthropods. It separates ingested food from the gut epithelium and thus may restrict penetration of ingested immune effector components (Eiseman & Binnengton, 1994), which is not the case for fleas.

Another type of host body condition that may influence flea feeding performance is hormonal status. As I have already mentioned, the effect of reproduction-related changes in a host's hormonal status is well known for rabbit fleas (although no effect of host sexual hormones on other flea species has been found (Prasad, 1969, 1973; Reichardt & Galloway, 1994; Lindsay & Galloway, 1998)) and will be discussed below. Other hormones may play certain roles in host–flea relationships as well, although our knowledge about this is poor and indirect. *Nosopsyllus laeviceps* has been reported to feed better on active compared to confined hosts *M. unguiculatus* (Liu *et al.*, 1993). *Xenopsylla cheopis* fed better on unrestrained than on restrained hosts (Tarshis, 1956; Bar-Zeev & Sternberg, 1962). Another example was presented by Suter (1964), who noted an increase in the feeding rate of *E. gallinacea* on agitated hosts (which supposedly increased the peripheral blood supply). I failed to find additional examples in the literature. So, the effect of the hormonal status of a host on the performance of ectoparasites is a promising topic that warrants further study.

10.4 Flea-related effects

10.4.1 *Interspecific variation in feeding patterns*

In studying feeding performance of fleas, caution needs to be exercised when making interspecific comparisons. Many feeding patterns such as time between or frequency of blood meals are species-specific, depending sometimes on the pattern of parasitism and the degree of sessility. For example, some species start to feed only after the previous blood meal has been digested (*X. cheopis*: Vashchenok, 1988; *N. setosa*: Novokreshchenova et al., 1968). Other species may feed even before completing digestion of the previously ingested blood (*L. segnis* and *L. taschenbergi*: Kosminsky, 1961, 1965; Bryukhanova, 1966). The frequency of blood meals reported for non-sessile species (including so called 'body fleas') is as high as every 2.3–2.5 h (*N. laeviceps*) or about three to five times per day (*Rhadinopsylla cedestis*, *Coptopsylla lamellifer*, *L. segnis*, *N. consimilis*, *P. irritans*, *Citellophilus tesquorum*, *X. conformis*, *Frontopsylla elata*), whereas other species (*N. fasciatus*, *X. cheopis*, *X. skrjabini*, *X. gerbilli*, *N. setosa*, *Ctenophthalmus golovi*, *Paradoxopsyllus teretifrons*, *Frontopsylla semura*) feed less often – about once or twice per day (Vansulin, 1961; Kunitskaya et al., 1965a; Bryukhanova & Surkova, 1970; Guseva & Kosminsky, 1974; Vashchenok, 1988).

In contrast, it is difficult to talk about the frequency of blood meals in sessile fleas. For example, the feeding process in *E. gallinacea* (Suter, 1964) and *E. oschanini* (Vashchenok, 1967b) is continuous. Nevertheless, blood consumption in sessile *Dorcadia dorcadia* is intermittent, with 0.5 h intervals between consecutive sucking bouts (Zatsarinina, 1972).

10.4.2 Intraspecific variation in feeding patterns

Feeding patterns within flea species can vary between male and female fleas and between newly emerged and adult fleas as well as with ecological situation (e.g. density dependence). Gender differences in feeding parameters have already been discussed in Chapter 7. Here, I consider other reasons for interspecific variation in feeding performance.

Adult (previously fed) fleas consume more blood than newly emerged (never fed) fleas (Bryukhanova et al., 1961; Bibikova & Gerasimova, 1967, 1973; Vashchenok et al., 1988). Adults have also been shown to digest blood faster than newly emerged individuals (Shchedrin, 1974; Bryukhanova et al., 1978, 1983; Bryukhanova & Surkova, 1983; Filimonova, 1986; Krasnov et al., 2003b; Gong et al., 2004a). For example, this was found in *P. chephrenis* when feeding on both specific (*A. cahirinus*) and non-specific (*D. dasyurus*) hosts (Krasnov et al., 2003b). This shorter digestion time in adult fleas might be related to their higher metabolic rate (Fielden et al., 2004) (Fig. 10.9) as well as re-distribution of the fat tissue and an increase in activity of a non-specific esterase after the first blood meal (Xun & Qi, 2004, 2005).

An increase in metabolic expenditure reported for other haematophagous arthropods after a blood meal is typically associated with the processes of blood meal digestion, egg production or cuticle synthesis prior to moulting, as for example in ticks (Fielden et al., 1999) and the reduviid bug *Rhodnius*

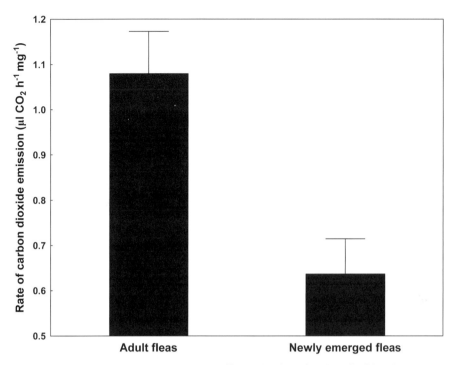

Figure 10.9 Mean (±S.E.) mass-specific production of carbon dioxide (μl per mg body mass per hour) in newly emerged and adult *Xenopsylla ramesis* at 25 °C. Data from Fielden *et al.* (2004).

prolixus (Bradley & Hetz, 2001). Furthermore, newly emerged fleas have a lower fat content than fleas that have had a blood meal (Krasnov *et al.*, 2002c). This may partly explain the increase in mass-specific metabolic rate in adult fleas but not in newly emerged insects. The fat can be used as an energy source which would account for the higher metabolic rate in fed animals. Alternatively, the beginning of blood-feeding may trigger some physiological processes in fleas, and the responses of adult and newly emerged fleas are thus different. For example, Xun & Qi (2004) reported that in newly emerged *Monopsyllus anisus* and *L. segnis* alkaline phosphatase and acid phosphatase were mainly distributed in the midgut, nerve nuclei, testes, ejaculatory ducts, oviducts and spermathecal glands, while adenosine triphosphatase was distributed in all tissues. After a blood meal and digestion, the activity of all three enzymes increased in the midgut. In contrast, in engorged adults, apart from the activity of alkaline phosphatase which decreased 72 h after a blood meal, there was no significant difference in the increasing degree of the activity of the three enzymes at other times after blood intake.

In fleas, feeding may also weaken (but not eliminate) the circadian rhythm of activity and change the overall time allocated for activity (*Monopsyllus sciurorum*: Clark *et al.*, 1999).

As mentioned in the previous chapter, Kelly & Thompson (2000) suggested that an individual blood-sucking insect can improve its feeding success by choosing a host that supports a small number of competitors, all else being equal. In other words, the suitability of a host is assumed to be density dependent, where higher densities of parasites will lower host suitability, suggesting that there exists intraspecific competition among blood-sucking arthropods. Intraspecific competition among ectoparasites should thus result in a decrease in feeding success of an individual with an increase in the number of conspecifics. Indeed, density-dependent feeding success has been reported for various haematophagous insects (Webber & Edman, 1972; Kelly *et al.*, 1996; Schofield & Torr, 2002; see Lehane, 2005 for review).

Intraspecific competition can be both exploitative and interfering. However, where fleas are concerned, intraspecific competition, at least between imagoes, is likely to be interfering. It does not seem feasible that the blood in a host can be a limiting factor (Khokhlova *et al.*, 2002). However, as we already know, the limiting factors may be those areas of a host body where blood is most readily available (e.g. thinnest skin or closest position of capillary to body surface). In addition, interference among fleas can be mediated via the host. For example, if there is a threshold of host sensitivity to parasite attacks, then its defence systems (behavioural or immune) may be activated once exploiters attain a certain level of abundance (Mooring, 1995; de Lope *et al.*, 1998; Shudo & Iwasa, 2001). From this point on, the host defence may be the main force limiting parasite success. As a result, feeding success of an individual parasite should decrease with increased density. On the other hand, if the cost of suppressing the feeding of a great number of parasites is too high, then the strength of the host's response will decrease with an increase in the number of attackers (e.g. Khokhlova *et al.*, 2004a). In particular, this may happen if a great number of co-occurring attackers suppress the defence system of a host by a cumulative effect of factors contained in their saliva (Roehrig *et al.*, 1992; Ribeiro, 1995; Wikel, 1996; Gillespie *et al.*, 2000). As a result, in this case feeding success can be expected to increase with an increase in the number of co-feeding parasites. A density increase may also improve feeding success of an individual flea via unknown behavioural effects. For example, this pattern was observed by Bar-Zeev & Sternberg (1962) in *X. cheopis* fed through a membrane.

Krasnov *et al.* (2007b) tested these two alternative predictions by studying the feeding success of *X. conformis* and *X. ramesis* during a 1-h feed on either *M. crassus* or *D. dasyurus* at different densities (from five to 50 fleas per individual host). They found that *X. conformis* consumed significantly less blood on *D. dasyurus* at

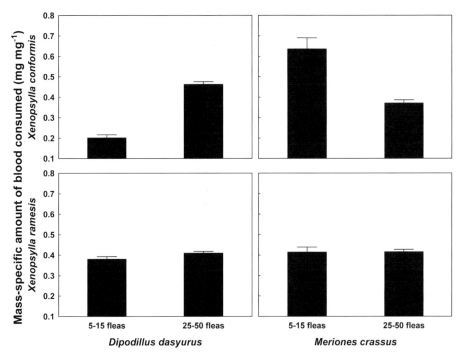

Figure 10.10 Mean (±S.E.) mass-specific amount of blood (mg per mg body mass) consumed by *Xenopsylla conformis* and *Xenopsylla ramesis* when feeding on Wagner's gerbil *Dipodillus dasyurus* and Sundevall's jird *Meriones crassus* at different densities. Data from Krasnov et al. (2007b).

low densities (5–15 fleas per host) than at high densities (25–50 fleas per host), whereas when feeding on *M. crassus*, fleas consumed significantly more blood at low densities (Fig. 10.10). Mean blood intake of *X. ramesis* parasitizing either *D. dasyurus* or *M. crassus* was similar and was not affected by flea density. Evaluation of feeding success using another measure, the proportion of fleas with highly engorged midguts, provided similar results.

Differences in flea responses to density may be associated with their different strategies of host selection described in the previous chapter (at low density *X. conformis* demonstrated sharp host selectivity, whereas *X. ramesis* chose hosts randomly). The difference in response to density in *X. conformis* when feeding on different hosts might be due to the differential effect of fleas on the host's energy balance and differences in the pattern of mounting an immune response against fleas. In particular, the effect of flea parasitism on *D. dasyurus* is supposedly linked with stimulation of an immune response (Khokhlova et al., 2002; see Chapter 12). Furthermore, *D. dasyurus* demonstrated 'post-invasive' immunity against fleas and mounted immune responses only after flea attacks (Khokhlova et al., 2004b; see Chapter 13). Thus, when a *D. dasyurus* individual is attacked by a relatively low number of fleas, it presumably mounts an immune response that suppresses the feeding success of fleas. When, however, the number of fleas is high, the energy available to the host may be insufficient to mount an effective immune response, so the feeding success of the fleas increases with density. This suggests apparent host-mediated facilitation. In contrast to *D. dasyurus*, the energy requirements of *M. crassus* are not affected by flea parasitism (I. S. Khokhlova, unpublished data; see Chapter 12). Consequently, the magnitude of the immune response in this species is unlikely to be affected by the number of haematophagous attackers. However, the parasites compete with each other, for example, for areas of the host body where the blood is more readily and/or easily available. As a result, feeding success per flea decreases with an increase in density. Thus, this study showed that density dependence of feeding success (a) varies among flea–host associations and (b) indicates intraspecific competition in some cases, but facilitation via hosts in other cases.

10.5   Environment-related effects

As has been mentioned several times, fleas spend a large part of their life cycle off-host. Consequently, they are subject to the influence of a variety of environmental factors (temperature, relative humidity, light regime etc.) that inevitably affect various aspects of flea ecology, including feeding patterns.

Being ectotherms, fleas demonstrate a strong metabolic response to ambient temperature. For instance, the metabolism of *X. ramesis* was shown to be temperature-dependent (see Fig. 7.6) with an average thermal coefficient (Q_{10}) of 2.57 for females and 2.55 for males over the temperature range of 10–30 °C (Fielden et al., 2004). Consequently, fleas are expected to respond to changes in ambient temperature both behaviourally (e.g. changing frequency of blood meals) and physiologically (e.g. changing duration of blood digestion). For example, proportion of previously fed male *Citellophilus tesquorum* that took a blood meal increased from 36% to 77% with an increase of air temperature from 5 °C to 20 °C and then steadily decreased to 60% with further increase in air temperature to 40 °C, whereas these proportions for newly emerged males were 1%, 20% and 5%, respectively (Gong et al., 2004a). In the experiments carried out by Bryukhanova (1966), female *C. tesquorum* took a blood meal from a host (the pygmy ground squirrel *Spermophilus pygmaeus*) on average 1.5 times per day at 4–6 °C, 2.5 times per day at 17–24 °C and 4.2 times per day at 34–35 °C, whereas frequency of feeding of *N. setosa* under these temperature changes was 2.3, 1.1 and 0.3 times per day, respectively. The sharper decrease of feeding frequency with a decrease in air temperature in *N. setosa* could be related to its pattern of parasitism. Fleas of this species spend much less time on the body of a host

Table 10.3 *Proportion of individuals of five flea species taking a blood meal after 6 h from the first contact with a host at different ambient temperatures*

Species	Temperature (°C)			
	17–18	24–25	28–29	34–35
Leptopsylla segnis	0.0	60.0	62.5	7.7
Megabothris calcarifer	0.0	9.1	–	–
Neopsylla bidentatiformis	8.7	42.0	23.1	20.0
Nosopsyllus fasciatus	0.0	14.3	45.1	27.2
Xenopsylla cheopis	0.0	0.0	60.0	33.3

Source: Data from Moskalenko (1966).

compared with *C. tesquorum* and thus are more strongly exposed to the effect of air temperature.

The latency of blood-sucking has been shown to decrease with an increase in air temperature (Moskalenko, 1966) (Table 10.3). It should be noted, however, that the measure of feeding implied in this study was somewhat strange and indirect. Frequency of feeding is also temperature-dependent, peaking at some optimal species-specific temperature and decreasing at both lower and higher temperatures (Bar-Zeev & Sternberg, 1962; Novokreshchenova et al., 1968).

Duration of blood digestion is another feeding parameter strongly influenced by ambient temperature. This was reported for *Ceratophyllus avicitelli* and *F. frontalis* by Bibikov & Bibikova (1955), *X. cheopis* and *Ctenophthalmus dolichus* by Balashov et al. (1961) and *X. conformis*, *C. tesquorum* and *N. setosa* by Bryukhanova et al. (1961) and Bryukhanova (1966). *Xenopsylla skrjabini* and *Xenopsylla nuttalli* digested blood of the great gerbils *R. opimus* four times more slowly at 10 °C than at 28 °C (Bibikova & Gerasimova, 1967; see also Vansulin, 1965). A higher rate of digestion at higher air temperatures in *X. gerbilli* and *N. laeviceps* was shown by Murzakhmetova (1958) (Fig. 10.11).

In general, the effect of air temperature on feeding performance of a flea species is manifested in a seasonal pattern of flea feeding activity. For example, although all-seasonal *X. ramesis* feeds all year round, its feeding activity is relatively high from February till September and is relatively low in October–January (Krasnov et al., 2002c).

The effect of light regime on feeding patterns varies among flea species. Although this issue is extremely poorly studied, available data suggest that fleas parasitic on wild hosts with distinct circadian rhythms (*C. tesquorum*, *X. conformis*) are influenced more strongly by the light regime than fleas exploiting commensal hosts in which the circadian rhythm is weaker (*X. cheopis*, *L. segnis*, *N. fasciatus*) (Novokreshchenova et al., 1968) (Fig. 10.12).

180 Ecology of haematophagy

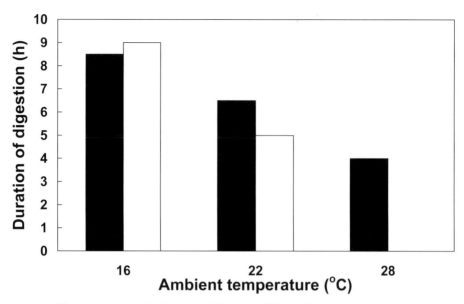

Figure 10.11 Duration (hours) of digestion of blood of a rodent host (laboratory mouse, great gerbil *Rhombomys opimus* or midday jird *Meriones meridianus* gerbil) by *Xenopsylla gerbilli* (black columns) and *Nosopsyllus laeviceps* (white columns) at different ambient temperatures. Data from Murzakhmetova (1958).

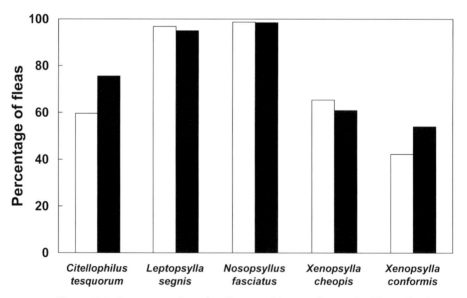

Figure 10.12 Percentage of starving fleas attaching to a host and taking a blood meal in the light (white columns) and in the dark (black columns). Data from Novokreshchenova et al. (1968).

10.6 Concluding remarks

Differential feeding performance on different host species is characteristic not only of host-specific fleas, but also of fleas that, at first glance, seem to be host generalists. In terms of the feeding performance, fleas seem to be able to distinguish between male and female hosts as well as between young and adult hosts. Furthermore, the feeding success of fleas appears to depend on the physiological and nutritional conditions of a host individual. Perception of among-host differences may be one of the mechanisms behind the unequal distribution of conspecific fleas among different host individuals and species. Studies of the feeding performance of fleas should take into account inter- and intraspecific variation in the feeding patterns of fleas as well as the possibility of strong dependence of these patterns on environmental factors.

11

Ecology of reproduction and pre-imaginal development

The main evolutionary motivation of every species is to increase its reproductive success, a goal that is achieved via various adaptations to specific ecological conditions, biotic and abiotic. Consequently, investigating reproductive patterns in different situations and disentangling factors that affect reproduction traits is an integral part of any ecological and evolutionary study. We already know that a flea life cycle is typical for a holometabolous insect. Consequently, individuals at different stages of this cycle have neither the same requirements nor the same possibilities and constraints. Furthermore, the effect of an abiotic (e.g. ambient temperature) or biotic (e.g. host identity) factor on flea individuals belonging to one stage of the cycle may be transformed or compensated by the effect of the same or another factor on individuals belonging to another stage of the cycle. Therefore, flea reproductive success should be measured both at each stage of the life cycle and over the entire cycle. This chapter discusses how various factors affect flea reproductive success. I start with issues related to measurement of reproductive outcome considering both quantity and quality of offspring. Then, the effect of both biotic (hosts and other fleas) and abiotic (external environment) interactions will be discussed.

11.1 Measures of reproductive success

At first glance, measuring reproductive success seems quite straightforward. Indeed, the number of offspring per female may represent the simplest measure. However, several caveats should be mentioned. The first is related to the time period for which offspring production should be considered. As was already mentioned in Chapter 5, fleas (as well as many other animals) can either produce eggs one by one more or less continuously, or partition their reproductive effort into discrete clutches and/or breeding periods. Furthermore, the rate of egg production as well as the clutch size and time interval between clutches vary in dependence of the age of an individual flea (e.g. Vashchenok, 2001; see also Chapter 5). The reader should therefore bear in mind that these parameters represent components of fecundity rather than the entire lifetime fecundity (Poulin, 2007a). In practice, the lifetime reproductive success of a flea is rarely known (see Chapter 5 for several exceptions). In most cases, reproductive success has been evaluated by the number of eggs, larvae, pupae and/or newly emerged imagoes produced by a female flea after timed contact with a host (Buxton, 1948; Alekseev, 1961; Vashchenok, 1993; Krasnov et al., 2002a, 2004a).

Second, from an evolutionary perspective, it is the *net* result of the reproductive effort that matters. In other words, the most important outcome in the reproduction of a flea is how many second-generation imagoes emerged and how many of them were able to produce offspring of the third generation and so on, whereas the egg-productive ability of a parent female is of secondary importance. However, from an ecological perspective, the numbers of eggs, larvae and pupae produced by a female are no less important than the number of the imagoes in successive generations because these numbers are associated with (a) parental investment and (b) quantity and quality of the offspring.

Quality of the offspring can be evaluated, for example, by assessing their survivability at each pre-imaginal stage. In other words, a higher survival rate at a given stage indicates higher quality of offspring. Hatching of each egg, pupation of each larva and emergence of a new imago from each pupa indicates that reproductive effort of a female has resulted in offspring of the highest quality, given their 100% fertility. This looks very simple.

The reality, however, is more complicated. This is because the quality of offspring is associated not only with their survivability, but also with their ability to compete with the offspring of other females, both con- and heterospecific. In particular, this is related to larvae which, in most flea species, are not parasitic (see Chapter 5). As we already know, larvae of the majority of fleas feed on all kinds of organic matter found in the burrow or nest of a host (see Chapter 5). The amount of the organic matter can be a limiting factor for the larvae, so they are expected to compete for it (e.g. Day & Benton, 1980; Krasnov et al., 2005e). In addition, cannibalism (both intra- and interspecific) in pre-imaginal fleas was shown to be a common occurrence, with older larvae readily cannibalizing younger larvae and naked pupae (Lawrence & Foil, 2002). Consequently, larvae that hatch earlier may have a certain competitive advantage over larvae that hatch later. An imago that emerges earlier has a better chance of locating

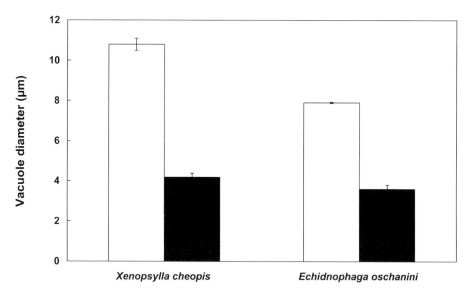

Figure 11.1 Mean (±S.E.) diameter (μm) of fat vacuoles in the trophocytes of newly emerged (prior to the first feeding) (white columns) and adult (after more than 10 days of feeding) (black columns) *Xenopsylla cheopis* and *Echidnophaga oschanini*. Data from Vashchenok (1988).

and attacking a host and producing more new generations during the breeding season than a late-emerging imago. All this suggests that offspring quality may be evaluated via rate of development, with a shorter development time being an indicator of a higher quality.

Another indicator of the quality of offspring is the resistance to starvation in newly emerged imagoes. Like any other insect, fleas possess a fat body composed of trophocytes. The fat body is considered a multifunctional organ, because it acts in various metabolic processes with a high biosynthetic activity during the entire life of an insect (Levenbook, 1985). When a flea emerges from a cocoon, it already possesses energetic storage in the fat tissue. Vacuoles of the trophocytes in the fat body of unfed newly emerged individuals are larger than those of previously fed imagoes (Fig. 11.1) (Vashchenok, 1966b, 1967a, 1988; Vashchenok & Solina, 1972). The energy storage provided by the fat tissue allows the newly emerged flea to survive until it has an opportunity to attack a host. The host may not always return in a regular or predictable manner to its nest or resting area where the immature fleas develop. The ability of a newly emerged imago to survive unpredictable and sometimes lengthy periods without a blood meal is thus extremely important.

11.2 Host-related effects

11.2.1 Host species

The unequal distribution of conspecific fleas among different host species has been already discussed in previous chapters. It is clear that the level of abundance of a particular flea on a particular host is a result of two opposing forces, namely flea performance on this host and host defensiveness against this flea. Here I focus on the former.

A strong effect of the host species on the reproductive success of host-specific fleas is not surprising. One of my favourite examples, mentioned already in this book, involves two rodent species that co-occur in the rocky habitats of the Negev Desert, Wagner's gerbil *Dipodillus dasyurus* and the Egyptian spiny mouse *Acomys cahirinus*. Although these two rodents live in close contact and frequently visit each other's shelters (rocky niches and crevices), *D. dasyurus* is parasitized almost exclusively by *Xenopsylla dipodilli*, whereas *A. cahirinus* is parasitized almost exclusively by *Parapulex chephrenis*. *Xenopsylla dipodilli* can be found occasionally on other gerbil species (*Sekeetamys calurus*, *Meriones crassus*), but not on spiny mice, whereas *P. chephrenis* can be found on the closely related *Acomys russatus*, but not on gerbils (Theodor & Costa 1967; Krasnov *et al.*, 1997, 1998, 1999). Furthermore, when *A. cahirinus* densities peak and individuals migrate to the non-rocky habitats, they are accompanied by *P. chephrenis*. These strong host preferences were reflected by the reproductive performance of female fleas in an experiment in which they were forced to feed upon the two hosts. Both species produced significantly more eggs when they fed upon their specific host than when they fed upon the other host species (Krasnov *et al.*, 2002a) (Fig. 11.2).

In the particular case of the host-specific *P. chephrenis*, then, the exploitation of a non-specific host has led to a decrease in quantity of offspring. But did this also affect their quality? The answer seems to be 'no'. The eggs produced by *P. chephrenis* when they were fed on a non-specific host were as viable as eggs produced by conspecifics fed on a specific host (B. R. Krasnov and L. Ghazaryan, unpublished data). Furthermore, pre-imaginal development of these fleas was slightly, albeit significantly, faster than that of offspring produced by females fed on *A. cahirinus* (42.6 ± 0.2 days versus 43.1 ± 0.1 days) and they survived slightly longer under starvation (16.2 ± 0.2 days versus 15.3 ± 0.4 days) (B. R. Krasnov and L. Ghazaryan, unpublished data). In addition, fleas fed on a non-specific host produced slightly more female offspring than fleas fed on a specific host (Fig. 11.3). Whatever the mechanism of this female bias was (primary sex ratio or differential survival of males and females), it may be considered as compensation for the lower fertility of parent females fed on *D. dasyurus*.

186   Ecology of reproduction and development

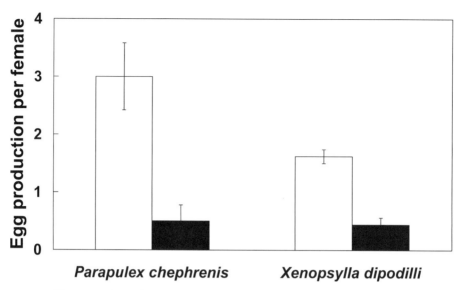

Figure 11.2 Mean (±S.E.) number of eggs produced by females of *Parapulex chephrenis* and *Xenopsylla dipodilli* after 8 days of feeding on the specific (white columns) and non-specific (black columns) host. Data from Krasnov *et al.* (2002a).

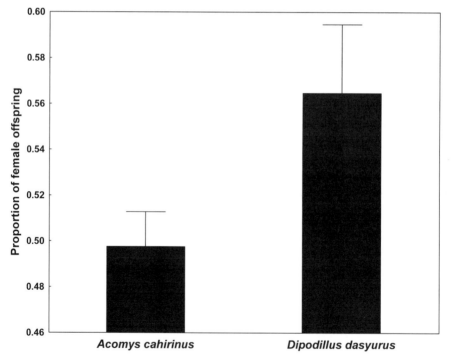

Figure 11.3 Mean (±S.E.) proportion of female offspring born to *Parapulex chephrenis* when feeding on the Egyptian spiny mouse *Acomys cahirinus* and Wagner's gerbil *Dipodillus dasyurus*. Data from B. R. Krasnov and L. Ghazaryan (unpublished).

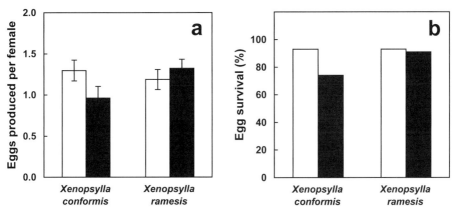

Figure 11.4 (a) Mean (±S.E.) egg production per female and (b) egg survival (%) in *Xenopsylla conformis* and *Xenopsylla ramesis* when feeding on Sundevall's jird *Meriones crassus* (white columns) and Wagner's gerbil *Dipodillus dasyurus* (black columns). Data from Krasnov et al. (2004a).

What about host-opportunistic fleas? Inequality of their distribution among host species suggests a differential reproductive response. Another good example is *Xenopsylla conformis* and *Xenopsylla ramesis* exploiting *M. crassus* and *D. dasyurus*. Let's recall the results of the application of isodar analysis to the between-host distribution of these fleas from Chapter 9. In brief, the isodars indicated that both fleas were able to perceive quantitative and/or qualitative differences between hosts. The isodar for *X. conformis* host selection suggested the perception of *M. crassus* as a both quantitatively and qualitatively superior host compared with *D. dasyurus*. The isodar for *X. ramesis* suggested that *M. crassus* was perceived as a qualitatively more suitable host than *D. dasyurus*, but quantitatively these hosts were identical for this flea. This difference in host-selection strategy in the two fleas could be manifested in the differences in their reproductive patterns when exploiting the two host species. Sharp host selectivity even at low density in *X. conformis* suggested that the egg production of individuals exploiting *M. crassus* will always be greater than that of individuals exploiting *D. dasyurus*. Random host selection at low densities and preference for *M. crassus* at high densities in *X. ramesis* suggested that the number of eggs produced by individuals exploiting either host can be similar. However, if the food resource for larvae (the amount of organic matter in the burrow) is a limiting factor at high densities, then a shorter development time may be advantageous for an individual flea.

The results of the study of Krasnov et al. (2004a) supported these predictions. Female *X. conformis* fed on *M. crassus* produced significantly more eggs than those fed on *D. dasyurus*, and the survival of the eggs was higher if the host was *M. crassus* (Fig. 11.4). In contrast, mean number as well as viability of eggs produced by female *X. ramesis* did not differ between host species (Fig. 11.4). However, time to hatching was similar for eggs produced by female *X. conformis* fed on both host species (6.74 ± 0.08 days for females fed on *M. crassus* and 6.70 ± 0.11 days for females fed on *D. dasyurus*). Larvae of *X. ramesis* hatched significantly faster if parent females were fed on *M. crassus* than if they were fed on *D. dasyurus* (6.01 ± 0.13 days and 7.25 ± 0.07 days, respectively).

The reproductive patterns of *X. conformis* and *X. ramesis* thus proved to be consistent with their strategy of host selection. The following scenarios can be envisaged. In *X. conformis*, exploitation of *M. crassus* provided a greater fitness reward than exploitation of *D. dasyurus*. Behaving in an IFD-like way, individual fleas maximize their reproductive success when flea population size is low by parasitizing *M. crassus* only. However, when the flea population grows, the amount of organic matter in the host burrow becomes a limiting factor for larval growth and development. Consequently, between-larva competition for food resources (see Chapter 15) intensifies (see below) and their survival decreases, thus decreasing per capita fitness of *X. conformis*. When the increased density reduces fitness rewards in fleas parasitic on *M. crassus* due to intensification of between-larva competition to the level of fitness rewards of fleas parasitic on *D. dasyurus*, some fleas will select *D. dasyurus*. Survival of their larvae is likely to be higher because of their lower density in burrows of *D. dasyurus*. Therefore, the lower egg production of fleas parasitic on *D. dasyurus* will be compensated by higher survival of the larvae and the fitness reward will thus be equalized between hosts at different within-host densities. From this point on, any further increase in flea population size will be divided equally between both hosts (Morris, 1988).

Xenopsylla ramesis achieves similar maximum fitness at low population densities by exploiting either host and thus selects its hosts randomly. As in *X. conformis*, the amount of larval food resources becomes a limiting factor with an increase in flea population size. Under these conditions, larvae that hatch earlier (i.e. from females exploiting *M. crassus*) will most probably have an advantage over larvae that hatch later (i.e. from females exploiting *D. dasyurus*), simply because the former rapidly consume most of the available food. The cumulative time to larva hatching consists of a period of egg development inside the female and a period of egg development after oviposition. Both these periods are shorter when a flea feeds on *M. crassus* than on *D. dasyurus* (recall that fewer blood meals are necessary for oviposition; Chapter 10). Therefore, parasitism on *M. crassus* compared to that on *D. dasyurus* produces a 'delayed' fitness advantage which is manifested only at high flea densities. In addition, shorter development time allows an increasing number of flea generations per breeding season.

The feasibility of these scenarios is indirectly supported by observations on the distribution of *M. crassus* and *D. dasyurus*. In areas inhabited by *X. conformis* (but not by *X. ramesis*), rodent densities as well as spatial overlapping between individuals of the two species are relatively low, and burrows of *M. crassus* and *D. dasyurus* are almost completely separated (Krasnov *et al.*, 1996b). Consequently, processes of between-larva competition for food presumably occur independently and are spatially separated in fleas parasitic on *M. crassus* and in fleas parasitic on *D. dasyurus*. In contrast, in areas inhabited by *X. ramesis* (but not *X. conformis*), the densities of both hosts are relatively high and their distributions overlap spatially, so that *D. dasyurus* is repeatedly recorded in burrows of *M. crassus* (Krasnov *et al.*, 1996a, b; Shenbrot *et al.*, 1997). This suggests spatial co-occurrence of larvae of fleas fed on *M. crassus* and larvae of fleas fed on *D. dasyurus* and thus competition for food between these groups (see Chapter 16). Consequently, host-dependent differences in the time of larval hatching supposedly affect the competitive outcome in *X. ramesis*, but not in *X. conformis*.

Xenopsylla conformis and *X. ramesis* are not unique in exhibiting differing reproductive patterns on different host species. Decreased reproduction in fleas fed on one particular host but not when fed on another has been reported for other flea species as well. The rat fleas *Xenopsylla cheopis* and *Xenopsylla astia* did not reproduce when fed on humans (Seal & Bhattacharji, 1961). Fecundity and egg hatchability in *X. cheopis* were higher when the fleas were fed on the black rat *Rattus rattus* than on the bandicoot rat *Bandicota bengalensis* (Prasad, 1969). *Xenopsylla nuttalli* and *Xenopsylla skrjabini* produced more offspring when exploiting a specific host, the great gerbil *Rhombomys opimus*, as opposed to the non-specific hosts, the pygmy ground squirrel *Spermophilus pygmaeus*, yellow ground squirrel *Spermophilus fulvus* and small five-toed jerboa *Allactaga elater* (Gerasimova, 1973). *Xenopsylla gerbilli* fed on *R. opimus* at air temperature of 20–24 °C produced up to 6.2 offspring per female per day, whereas the mean number of offspring produced by females fed on laboratory mice under the same air temperature did not exceed 2.5 per female per day (Zolotova *et al.*, 1979). Although egg hatching and adult emergence in *Ctenocephalides felis* fed on human blood did not differ significantly from egg hatching and adult emergence in conspecifics fed on cats (Pullen & Meola, 1995), a higher proportion of fleas attained maturity on cats compared with calves (Williams, 1993). Although dog fleas *Ctenocephalides canis* fed on cats laid eggs, their offspring did not develop beyond the first larval stages (Baker & Elharam, 1992). Variation in reproductive success in dependence on host species has also been reported for *Citellophilus tesquorum* (Tchumakova *et al.*, 1981).

Sometimes, reproductive performance on a highly unusual host can be even higher than that on the usual preferred host. For example, *X. cheopis* and

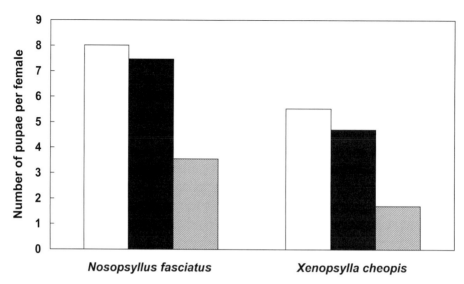

Figure 11.5 Number of pupae produced by female *Xenopsylla cheopis* and *Nosopsyllus fasciatus* per day when feeding on the golden hamster (white columns), laboratory rat (black columns) or laboratory mouse (hatched columns). Data from Samarina et al. (1968).

Nosopsyllus fasciatus produced more offspring when feeding on golden hamsters than on laboratory rats and mice, even though the latter are much closer phylogenetically to the natural hosts of these fleas (Samarina et al., 1968) (Fig. 11.5). An explanation for this pattern may be analogous to the explanation of the better feeding performance of rodent fleas on an unusual bat host (see Chapter 10).

Furthermore, differential reproductive success as a function of host species can be seen in fleas even at scales finer than the species. For example, Hudson & Prince (1958b) in California collected *Pulex irritans* from human habitations and from the abandoned den of the kit fox *Vulpex macrotis* and studied the reproductive performance of fleas from the two populations after feeding them on blood from various hosts. It appears that fleas that presumably exploited humans performed well when feeding on human blood and produced, on average, 8–10 pupae per female per week, but their performance when feeding on dog blood was poor (0.4–3.5 pupae per female per week depending on the rearing method employed). In contrast, individuals from the fox 'strain' produced only a few offspring (0.0–0.2 pupae per female per week) when they were forced to feed on human blood, but their reproductive success was high (up to 11 pupae per female per week) when they were fed on the blood of a dog. Adaptations by local populations of the same parasite species to different host species are known from other parasite taxa as well (Lively, 1989; Ballabeni & Ward, 1993; Ebert, 1994; Elmes et al., 1999).

The identity of host species can influence the reproductive success in fleas not only via the effect on the parent individuals but also via a direct effect on larvae. This is because faeces of imago fleas are an important component of larval diet in many species (Moser et al., 1991; Silverman & Appel, 1994; Correia et al., 2003). These faeces may contain blood at different stages of digestion, including semi-digested blood (e.g. in *Echidnophaga gallinacea*: Suter, 1964; *Echidnophaga oschanini*: Vashchenok, 1967b). Linardi et al. (1997) added dried blood from various hosts to larval medium and found that larvae of *C. felis* fed with diets containing mouse or *Mastomys* blood developed faster than larvae fed with diets containing blood of a dog or a pigeon. However, the survival of the larvae demonstrated the opposite. Larvae offered dog or pigeon blood survived better than those offered mouse or *Mastomys* blood.

11.2.2 Host gender and age

The idea that host gender and age may affect reproductive performance of fleas stems from strong gender- and age-biases of flea infestation reported from field studies on a variety of hosts (for details see Chapter 15). Although the investigation of flea reproductive success in male and female as well as in young, adult and geriatric hosts is obviously the next step in our attempt to understand the mechanisms of these biases, surprisingly few studies of these issues have been undertaken except for several famous studies on the rabbit fleas *Spilopsyllus cuniculi* and *Cediopsylla simplex* (Rothschild, 1965a, b; Rothschild & Ford, 1966, 1969, 1972, 1973; Sobey et al., 1974). The patterns revealed in the latter studies, however, are related to the effect of changes in the hormonal status of hosts rather than to pure gender or age effects and will be considered in the next section.

The rat flea *X. cheopis* appears to be the favourite model species for studies of host gender and age effects on flea reproduction. For example, Buxton (1948) studied the effect of host age on flea fertility and found that *X. cheopis* reared on juvenile mice (from 2 to 10 days old) were less fertile than conspecifics reared on adult mice. Experiments with *R. rattus* carried out by Mears et al. (2002) supported these findings. Fecundity of fleas was lower on juvenile rats, although no effect of host gender was found. The results of Trukhachev's (1971) study of the effect of age and gender of the laboratory mouse serving as a host for *X. cheopis* on quantity and quality (in terms of resistance to starvation after emergence) of offspring were different. Fleas produced more offspring when fed on adult females as opposed to adult and juvenile males (Fig. 11.6a). However, fleas of the new generation were less resistant to starvation if their parents fed on adult female hosts (Fig. 11.6b).

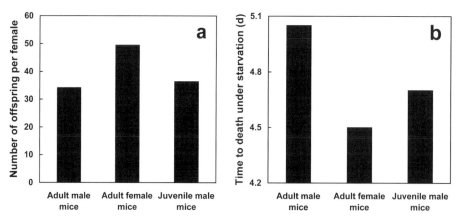

Figure 11.6 (a) Number of fleas of the new generation produced by *Xenopsylla cheopis* when feeding on adult female and adult and juvenile male laboratory mouse and (b) their resistance to starvation. Data from Trukhachev (1971).

Lehmann (1992) studied reproductive activity of *Synosternus cleopatrae* infesting Anderson's gerbil *Gerbillus andersoni* in sandy areas of the Negev Desert. In this study, fleas were collected from host individuals captured in the field. The reproductive activity of fleas was evaluated as (a) proportion of reproductive females (based on counts of gravid and non-gravid females) and (b) reproductive intensity (based on measurements of the oocyte size). It appeared that host gender had a significant effect on flea reproduction, with the proportion of reproductive fleas and their reproductive intensity being higher on male than on female hosts. No effect of age was found. In addition, effects of the host's reproductive status on flea reproductive parameters were also absent, suggesting that the 'hormonal' explanation of the sex-biased infestation should be ruled out.

Despite the scarcity of information on the effect of host gender and age factors on flea reproduction, we can conclude that gender- and age-related biases in flea infestation patterns are related not only to behavioural differences among hosts of different cohorts, but also to differential reproductive performance of fleas on these hosts. Mechanisms for these biases may involve not only encounter, but also Combes's (2001) compatibility filter concept. These mechanisms will be considered in detail in Chapter 15.

11.2.3 Host body conditions

Physiological variation among host individuals may be crucial for a flea in terms of a host-selection decision. Within the same host gender or age cohort, some individuals may represent patches of better quality than other individuals.

This results in inequality of flea distribution among individual hosts of the same species and will be discussed later. Physiological changes in host organisms may be predictable, such as those related to reproduction, or unpredictable, such as those related to nutrition. The border between these categories is conditional because in some cases changes associated, for example, with malnutrition may be predictable and occur in certain seasons.

As mentioned above, the most extraordinary and well-known pattern of flea reproduction related to predictable changes in the organisms of the hosts is the association between the reproduction of the rabbits *Oryctolagus cuniculus* and *Sylvilagus floridanus* and their respective fleas *S. cuniculi* and *C. simplex*. This fascinating story, based on the studies of Mead-Briggs & Rudge (1960), Mead-Briggs (1964), Rothschild (1965a, b), Mead-Briggs & Vaughan (1969), Mead-Briggs *et al.* (1975), Rothschild & Ford (1966, 1969, 1972, 1973) and Sobey *et al.* (1974), has been repeatedly retold in various sources (e.g. Marshall, 1981a). In brief, fleas usually concentrate around the ears of an adult rabbit. After mating, rabbit ear temperature rises, which supposedly triggers and facilitates voracious blood consumption by fleas. Ovulation in a female rabbit is followed by the release of sexual steroids into the bloodstream. These hormones are inevitably consumed by fleas with the rabbit's blood. It seems that the increase in blood hormone concentration affects fleas, as they attach more tightly to the ears of a rabbit. Moreover, a cohort of young female rabbits that are pregnant for the first or the second time is the most heavily infested by fleas. About 10 days prior to delivery of a rabbit's litter, the concentration of adrenocorticotrophic hormone in the blood of the pregnant doe increases. This triggers the development of flea ovaries, maturation of their oocytes and deposition of yolk. Male fleas seem also be affected by the hormonal changes in their host as their testicles start to develop. At this time, changes in the digestive system of fleas also occur (hypertrophy of salivary glands, increase in the rate of feeding and defecation etc.). The fleas attain full maturation about a day prior to a rabbit's delivery. After the kittens are born, the majority of fleas relocate on them, whereas the fleas remaining on a doe undergo regression of the reproductive system and decrease of the gut size. Fleas that have relocated to the kittens continue to feed actively and start mating and oviposition. It should be noted that this is the only time when fleas copulate. Oviposition starts immediately on the first day of a stay on a kitten with egg production steadily decreasing over the following days and terminating after about a week. The week-long stay on a kitten by a female flea is interrupted by leaving the host to lay the eggs in a burrow and then returning to the host. Hormonal changes in the blood of a kitten seem to occur when it is a week or so old. At that time, fleas terminate their breeding activity, leave the kitten and return to a doe. Some fleas (mainly males) die off, whereas the reproductive system in the survivors regresses. Thus, the reproduction of rabbit fleas is strongly controlled by the reproductive hormones of their hosts. Nevertheless, host hormone levels are apparently not important for other flea species and their reproductive periods have nothing to do with the breeding of their hosts (Rothschild, 1965a; Prasad, 1969, 1973, 1976; Prasad & Kamala Bai, 1976; Marshall, 1981a; Lindsay & Galloway, 1998; Pigage et al., 2005). This is true not only for fleas exploiting other than lagomorph hosts (e.g. *X. cheopis* and *X. astia*: Prasad, 1973, 1976), but also for non-spilopsylline fleas parasitic on rabbits and hares such as *Echidnophaga myrmecobii* and *Echidnophaga perilis* (Shepherd & Edmonds, 1979), *Xenopsylla cunicularis* (Cooke, 1990b) and *Ctenocephalides felis damarensis* (Louw et al., 1993). However, Brinck & Löfqvist (1973) argued that the reproduction of *Archaeopsylla erinacei* may be controlled by the hormonal status of the hedgehog, although Marshall (1981a) disproved this hypothesis.

Variation in host body conditions may result from variation in food availability among hosts. Can this variation affect reproductive performance of fleas? We have already seen that *X. ramesis* fed better on food-restricted compared with normally fed *M. crassus* (Krasnov et al., 2005d; Chapter 10). This feeding success was translated into greater egg production. Fleas that parasitized control animals produced significantly fewer eggs than those that parasitized food-restricted animals. Moreover, egg production of fleas fed on rodents with 60% of maintenance energy intake was significantly lower than that of fleas fed on rodents with 30% of maintenance energy intake (Fig. 11.7).

Furthermore, food availability of rodent hosts affected not only quantity, but also quality of flea offspring in terms of both survival and development time. More than twice the number of eggs from fleas on food-restricted rodents survived than those on control rodents, whereas the highest survival of larvae (but lowest survival of newly emerged imagoes) was recorded in fleas fed on rodents with only 30% of maintenance energy intake (Fig. 11.8). Time to hatching and time to pupation of larvae were the longest if the host's diet comprised 30% of energy requirements for maintenance, whereas resistance to starvation was the weakest in offspring of fleas that parasitized *M. crassus* offered the minimal amount of food (Fig. 11.9).

The described patterns can be related to a decline in immune function in hosts with limited food intake (see further discussion in Chapter 13). The increase in flea fitness because of host food limitation was due to both a higher number of eggs produced and a higher survival of eggs and of larvae hatched from these eggs. It should be noted that flea egg production increased even when hosts were limited to 60% of maintenance energy intake and did not lose their body mass.

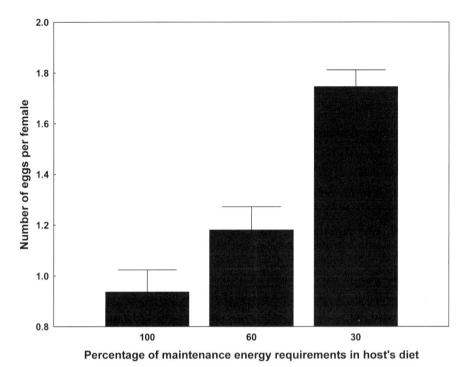

Figure 11.7 Mean (±S.E.) number of eggs produced by female *Xenopsylla ramesis* after a 4-day contact with Sundevall's jirds *Meriones crassus* offered a diet of 100%, 60% or 30% of maintenance energy requirements. Redrawn after Krasnov *et al.* (2005d) (reprinted with permission from Blackwell Publishing).

Longer developmental time and earlier death due to starvation of fleas whose parents parasitized underfed hosts suggest that although fleas can benefit by choosing to exploit a weakened host in terms of quantity of offspring and one of their quality components (higher survival ability), they can also suffer in terms of other quality components of offspring (longer development). Field observations on the distribution of parasites among hosts with different body conditions indicate that the former option is usually chosen (but see Dawson & Bortolotti, 1997; Liker *et al.*, 2001). For example, higher abundances of amblyceran and ischnoceran lice were found in the Galapagos hawks *Buteo galapagoensis* with poorer body condition (Whiteman & Parker, 2004). A negative correlation was found between body condition of capybaras *Hydrochoerus hydrochaeris* and their intensity of infestation by some helminth species in Venezuela (Salas & Herrera, 2004). However, it is difficult to distinguish between a cause and a consequence in observational studies of host body condition and parasite abundance. Indeed, manipulation of parasite numbers demonstrated that, at least in some host–parasite systems, it was parasitism that caused the decrease in host

196 Ecology of reproduction and development

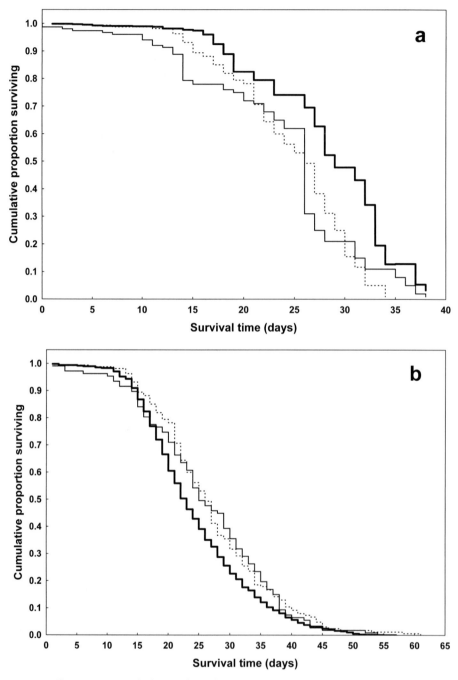

Figure 11.8 Cumulative survival of (a) larvae and (b) newly emerged imagoes of *Xenopsylla ramesis* from parents fed on Sundevall's jirds *Meriones crassus* offered a diet of 100% (solid line), 60% (dashed line) and 30% (bold line) of maintenance energy requirements. Data from Krasnov et al. (2005d).

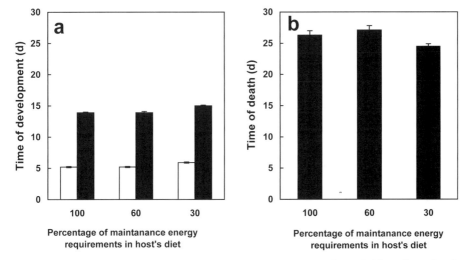

Figure 11.9 Mean (±S.E.) time (days) (a) of development of eggs (white columns) and larvae (black columns) and (b) to death under starvation of newly emerged *Xenopsylla ramesis* from parents fed on Sundevall's jirds *Meriones crassus* offered a diet of 100%, 60% and 30% of maintenance energy requirements. Data from Krasnov et al. (2005d).

body condition rather than inferior host body condition that attracted parasites (Neuhaus, 2003).

Other studies that tested the effect of host body conditions on the reproductive performance of fleas provided contrasting results. Ma (2000) measured reproductive success and resistance to starvation in *Neopsylla bidentatiformis* and *C. tesquorum* on properly fed and food-restricted laboratory rats and reported a negative effect of feeding on food-restricted host on both parameters. Tschirren et al. (2007a) manipulated both access to food in the broods of the great tit *Parus major* and exposure of the nestlings to *Ceratophyllus gallinae*. It was found that the food supplementation of the nestlings significantly influenced the parasites' reproductive success. Female fleas laid significantly more eggs when feeding on food-supplemented hosts. These opposing results could be due to differences in the experimental design of the studies: Krasnov et al. (2005d) compared the reproductive success of parasites feeding on food deprived and control animals in the laboratory using their natural host species, while Ma (2000) used an unnatural host and Tschirren et al. (2007a) used food supplementation to improve the condition of the hosts in the wild. In addition, contrasting results of Krasnov et al. (2005d) and Tschirren et al. (2007a) might reflect differences in the coevolutionary history of the two host–parasite systems. The association between pulicids (e.g. *Xenopsylla*) and rodents has a much longer history than that between ceratophyllids (e.g. *Ceratophyllus*) and birds. Ceratophyllidae is phylogenetically the youngest siphonapteran family, so flea–bird associations are probably younger than flea–rodent ones (see Chapter 4).

Previous experience of a host with fleas has been shown to have a profound effect on flea reproduction. This is related to so-called acquired resistance against parasites that an individual host can develop during its lifetime (Rechav et al., 1989; Fielden et al., 1992; see also Chapter 13). Walker et al. (2003) studied the effect of previous experience of P. major with C. gallinae in Switzerland and found that exposure of neonates to fleas early in the nestling period reduced the reproductive output of fleas late in the nestling cycle. However, the effect of the induced nestling response was seasonal. Exposure to fleas during the initial days post-hatching reduced the hatching success of flea eggs and the number of larvae produced at the end of the nestling cycle in early tit broods, whereas no effect was found in late broods. In addition, there was no effect of the induced neonate response on the size of the larval population at the end of the nesting period, suggesting that other factors, such as the environmental temperature and density dependent larval competition (see below), may be more important in determining the size of future parasite populations.

The study of Tschirren et al. (2007a) on the same host–parasite system and in the same geographical region as those of Walker et al. (2003) supported findings of the latter and added an important extension. It appeared that previous parasite exposure of P. major did indeed affect the reproductive success of C. gallinae. However, the impact of this induced host response on flea reproduction depended on the birds' natural level of immunocompetence, assessed by the phytohaemagglutinin (PHA) skin test (see Chapter 13). Flea fecundity significantly decreased with increasing PHA response of the nestlings in previously parasite-exposed broods, although no relationship between flea fitness and host immunocompetence was found in previously unexposed broods. These results demonstrate that the negative effect of an induced host response on parasite fecundity is not unconditional, but depends on a variety of factors, many of which are largely unknown.

Surprisingly, the effect of previous exposure to fleas of A. cahirinus on reproductive output of P. chephrenis was the opposite of that expected. Fleas that exploited immune-naïve mice produced significantly fewer eggs than fleas that exploited flea-experienced mice (Fig. 11.10a) (B. R. Krasnov & L. Ghazaryan, unpublished data). However, the offspring developed significantly more slowly if their parents were fed on flea-experienced compared with previously unexposed rodents (Fig. 11.10b). This suggests that, at least in some flea–host associations, previous parasitological experience of a host may affect flea fitness negatively in terms of the quality but not the quantity of the offspring.

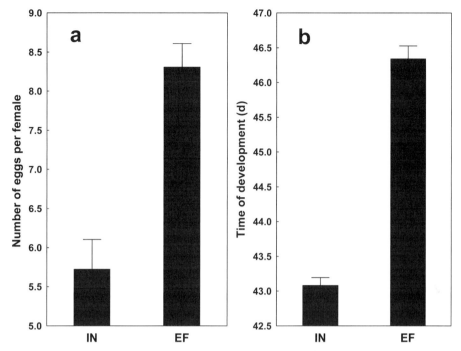

Figure 11.10 Mean (±S.E.) (a) number of eggs produced by a female *Parapulex chephrenis* after 3 days' contact with immune-naïve (IN) and flea-exposed (EF) Egyptian spiny mice *Acomys cahirinus* and (b) development time (days) of the new generation born to these females. Data from B. R. Krasnov & L. Ghazaryan (unpublished).

11.3 Flea-related effects

11.3.1 Flea species

We have already discussed variation in egg production, oviposition rate and lifetime fecundity among flea species (see Chapter 5). Unfortunately, no study has yet aimed specifically at finding any ecological or evolutionary rule governing this variation. It can be suggested, for example, that the egg production of fleas may be correlated with their body size as is the case in other parasites (e.g. Skorping *et al.*, 1991). Nevertheless, Cooke (1990a) observed that although female *S. cuniculi* and *X. cunicularis* are of approximately the same size, their egg production abilities substantially differ (10 versus four eggs per female per day, respectively). However, the eggs of *S. cuniculi* are relatively small (420 by 250 μm), whereas those of *X. cunicularis* are larger (450 by 320 μm) (Cooke, 1990a). Thus, there may be a trade-off between egg number and egg size (e.g. Poulin, 1995a). Egg production or egg size (which determines larval size) may be

200   Ecology of reproduction and development

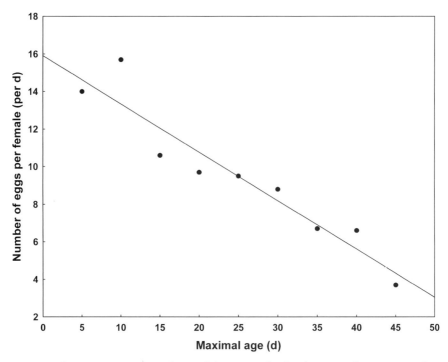

Figure 11.11 Age dependence of the egg production in *Leptopsylla segnis*. Data from Vashchenok (2001).

correlated with some features of hosts such as body size, burrow structure and behaviour. These issues are still a waiting further investigation.

11.3.2   Flea age

In general, flea fecundity decreases with age (Zolotova et al., 1979; Dryden, 1989; Cooke, 1990a; see also Chapter 5). For example, Vashchenok (2001) studied egg production in female *Leptopsylla segnis* that were allowed continuous access to a host (laboratory mouse) for 40 days. It appeared that young females produced, in general, more eggs than older females (Fig. 11.11), although the peak of the fecundity occurred when a flea was 6–10 days old.

Age-related changes in egg production in other flea species are characterized by more strongly expressed unimodality (e.g. Korneeva & Sadovenko, 1990) (Fig. 11.12) or even multimodality (e.g. Kunitskaya et al., 1974). This was reported for *X. skrjabini* by Gerasimova (1973) and Korneeva & Sadovenko (1990), for *X. nuttalli* by Gerasimova (1973), for *X. gerbilli* by Bibikova et al. (1971) and Kunitskaya et al. (1974) and for *X. conformis* by Grazhdanov et al. (2002). In contrast, Kunitsky (1970) reported that egg production in *Nosopsyllus laeviceps* (in terms of

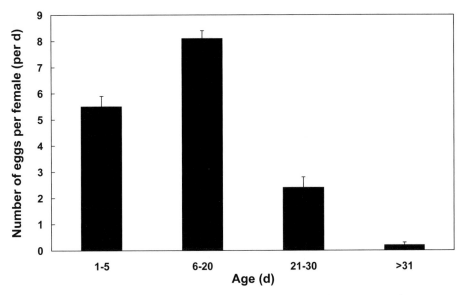

Figure 11.12 Mean (±S.E.) number of eggs produced per female per day by *Xenopsylla skrjabini* females of different age. Data from Korneeva & Sadovenko (1990).

clutch size) increased with flea age. Unfortunately, the methods applied in these studies do not often allow us to distinguish between the age effect per se and the effects of other factors, such as the time of contact with the host.

The effect of flea age on quality of offspring is even less well known. Nevertheless, Tchumakova & Kozlov (1983) noted that in *Nosopsyllus consimilis* viability of eggs produced by young females was lower than that of eggs produced by older females (42% versus 80% of hatchings).

11.3.3 Flea density

As mentioned above, if fleas behave in an IFD-like manner (see Chapter 12), then the effect of flea density on reproductive performance is expected to be negative. A negative fitness–density relationship may stem from intraspecific competition (Tripet & Richner, 1999a, b) which, in the case of imagoes of ectoparasitic insects, seems to be interfering rather than exploitative (see Chapter 10). We have already seen that interference can arise due to competition for those areas of a host body where blood is most readily available or that it may be mediated via the host that can activate its defence system after a certain threshold of parasite attacks is attained (e.g. Shudo & Iwasa, 2001). Both these factors can vary among host species (e.g. Khokhlova *et al.*, 2004b). In addition, host body size can also play a role. For example, for the same number of parasites

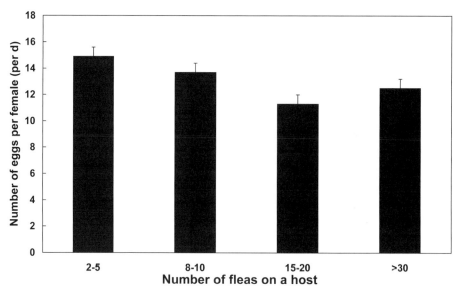

Figure 11.13 Mean (±S.E.) number of eggs produced per female per day by *Leptopsylla segnis* females in relation to their density on a host. Data from Vashchenok (1995).

the degree of crowding will be different in small-bodied and large-bodied hosts and, consequently, the effect of density will be manifested more strongly in a small host than in a large host. In contrast, intraspecific competition of flea larvae can be both interfering (e.g. cannibalism: Fox, 1975 and references therein) and exploitative (e.g. Braks et al., 2004).

Studies of the relationship between the density of fleas on a host and their reproductive success are rare. Vashchenok (1995) studied egg production of *L. segnis* in relation to the number of fleas simultaneously fed on a laboratory mouse. In these experiments, he allowed fleas to stay on a restricted host (a mouse that was prevented from grooming) for at least 4 days and counted the number of eggs produced per female per day. Egg production appeared to decrease slightly but significantly with an increase in flea density from 2–5 to 15–20 (Fig. 11.13). However, further increases in density did not result in further decreases in fecundity. In contrast, under very high densities, egg production tended to increase (Fig. 11.13). Vashchenok (1995) suggested that the factors that regulate density of fleas operate only when the density fluctuates within some natural limits, whereas in cases of 'hyperinvasion' fleas are unable to regulate their density. He also hinted that the response of a host may be one of these factors, but did not provide any further explanation.

Tripet & Richner (1999a) found that the reproductive success of *C. gallinae* breeding in the nest of the blue tit *Cyanistes caeruleus* in Switzerland was affected by the number of founder fleas in this nest. They carried out two experiments. The first was aimed at establishing the importance of density dependence and examined the effect of founder density on the reproductive rate of fleas. They found that the number of young larvae and the final number of newly emerged adults produced per female flea during host hatching and nestling periods increased with an increase in flea density, but the slope of the population growth decreased. In other words, there was a negative relationship between density and population growth rate rather then reproductive success per se. However, when the data from this experiment were recalculated as the number of offspring produced per female flea per nest, this parameter appeared to decrease with increasing founder density (Tripet et al., 2002c). The second experiment was designed to investigate density dependence of larval production and adult flea survival and attempted to differentiate between various processes that could lead to density dependence. The results of this experiment demonstrated that the number of eggs laid by females during a host incubation period did not decrease with flea density. Instead, fleas tended to produce fewer eggs under low density. Tripet & Richer (1999a) associated this with low mating opportunities and reduced sexual selection. Nevertheless, the number of larvae and newly emerged fleas decreased with increasing flea density.

These results point to two important issues. First, intraspecific competition between larvae seems to be the main process underlying density dependence of flea population growth. Second, host biology may affect density dependent processes in fleas. Indeed, competition between larvae was manifested during the host's incubation period, but decreased during the nestling period. Tripet & Richner (1999a), however, attempted to avoid implicating the host's life cycle in their explanation of this pattern, implicating instead the effect of the flea life cycle. It could be expected, however, that competition would be better expressed when the number of larvae increased, i.e. during the host's nestling period. The reason for the opposite finding, as suggested by Tripet & Richner (1999a), may lie in the fact that larvae use both organic debris of the host's nest itself and the blood faeces excreted by adult fleas. Thus, they may first use the limited organic resources provided by a nest and compete for them. Then, once these resources are consumed, the number of larvae is proportional to the number of adults that provide food.

Tripet et al. (2002c) continued to investigate the effect of density on flea reproduction, this time considering the quality of the offspring. As an indicator of quality, body size and time of dispersal (earlier or later) of the newly emerged fleas were used. Tripet et al. (2002c) hypothesized that intraspecific competition negatively affects the phenotypic quality of the fleas in a new generation and thus their capacity for overwintering as well as optimal timing of dispersal. As birds often avoid heavily infested nests, it was assumed that a host's nest choice behaviour may determine the direction of the evolution of a density-dependent phenotypic response to crowding in fleas. If so, early dispersal would be optimal at high density, whereas late dispersal (if at all) would be optimal at low density. The results of this study demonstrated that the density of flea offspring in the nests was negatively correlated with the proportion of early dispersers and thus ruled out the working hypothesis. Nevertheless, flea density negatively affected the body size of the newly emerged individuals. The latter, in turn, was positively correlated with potential reproductive ability. Indeed, the number of developed ovarioles, although not the number of oocytes per ovariole, decreased with an increase in female body size. In addition, the larger individuals usually dispersed later. Furthermore, intraspecific larval competition in the laboratory experiments has been shown to affect strongly the development of offspring, their imaginal body size and their chances of overwintering successfully. The results of this study combined with those from Tripet & Richner (1999a) suggest that, despite the high cost of the intraspecific larval competition in terms of offspring size and overwintering ability, this competition affects only a small fraction of individuals, namely those that live as early-instar larvae and experience competition either during the host's incubation period or after the host nestlings fledge and adult fleas disperse. The supply of adult flea faeces as larval food is likely to be short (if it exists at all) during both these periods. This can explain the negative relationships between the number and size of early dispersers and density because early dispersers presumably finish their larval development under the fierce competition that follows host departure. Tripet *et al.* (2002c) argued that competitively inferior (poorly fed) individuals suffer from diminished water and/or energetic resources and thus should emerge and start their host search earlier. In other words, fleas adjust their dispersal behaviour according to their phenotypic quality which, in turn, is determined by the level of larval competition. This strategy may maximize their chance of transmission.

Khokhlova *et al.* (2007) studied the effect of flea density on egg production in *X. conformis* and *X. ramesis* when exploiting *M. crassus*. The number of eggs produced by both species at high densities was significantly lower than at low densities (Fig. 11.14). In this case, density dependence of flea egg production might be an outcome of a density-dependent immune response of a host due to an increase in immune responses when the number of attackers increases (Randolph, 1994; see also Chapter 13).

Flea density affected not only quantity but also quality of the offspring (B. R. Krasnov & A. Hovhanyan, unpublished data). This effect was, however, manifested differently in different species. In particular, survival of pre-imaginal fleas decreased with an increase in parent density in *X. ramesis*, but not in *X. conformis*

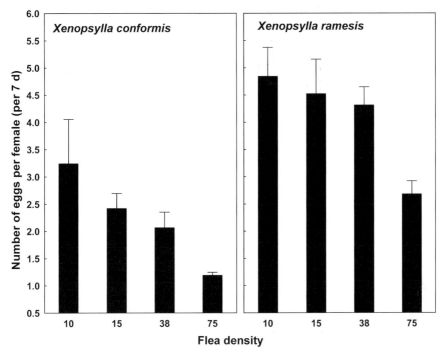

Figure 11.14 Mean (±S.E.) number of eggs produced per female *Xenopsylla conformis* and *Xenopsylla ramesis* during 7 days of oviposition when feeding on Sundevall's jird *Meriones crassus* at different densities. Data from I. S. Khokhlova, A. Hovhanyan and B. R. Krasnov (unpublished).

(Fig. 11.15). Time of development of the offspring increased sharply at the highest density in *X. ramesis*, but peaked at median density in *X. conformis* (Fig. 11.16). Between-species differences in this pattern cannot be associated with the effect of larval competition because, in contrast to parent fleas, all larvae in these experiments experienced the same density. Perhaps different flea species merely respond differently to crowding.

The effect of larval competition, nevertheless, was supported by the results of other experiments (B. R. Krasnov and A. Hovhanyan, unpublished data). In general, larval mortality in both species was higher at higher densities (Fig. 11.17).

In addition, the effects of density can be mediated by environmental factors. For example, relative humidity (RH) strongly affects survival of fleas at various stages of their life cycle (Heeb et al., 2000; Krasnov et al., 2001a, b; see below). As a result, the outcome of intraspecific competition may be, in turn, dependent on humidity. For example, Krasnov et al. (2005e) studied the performance of larvae

206 Ecology of reproduction and development

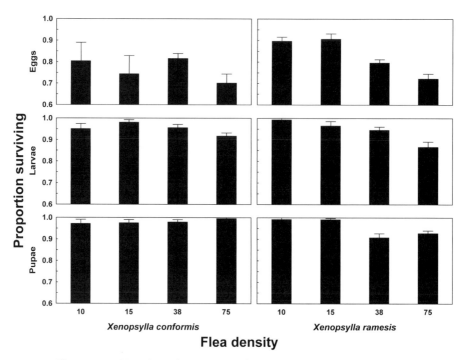

Figure 11.15 Mean (±S.E.) proportion of surviving eggs, larvae and pupae of *Xenopsylla conformis* and *Xenopsylla ramesis* produced by females fed on Sundevall's jird *Meriones crassus* at different densities. Data from B. R. Krasnov and A. Hovhanyan (unpublished).

of *X. conformis* and *X. ramesis* in terms of their developmental success in mixed-species and single-species treatments under different food abundance and RH. In particular, at low food availability (thus fiercer competition), larvae survived better if RH was higher (Fig. 11.18).

The results of the experiments described above suggest that density dependence of reproductive performance in fleas is a general rule. Furthermore, the manifestation of the effect of density on reproductive success of a flea is a result of a complicated interactions between host-, flea- and environment-related effects.

11.4 Environment-related effects

11.4.1 Air temperature and relative humidity

Microclimatic factors such as air temperature and RH affect various parameters of flea reproduction. For example, egg production and rate of oviposition in *N. laeviceps* increased with an increase in air temperature (Kunitsky,

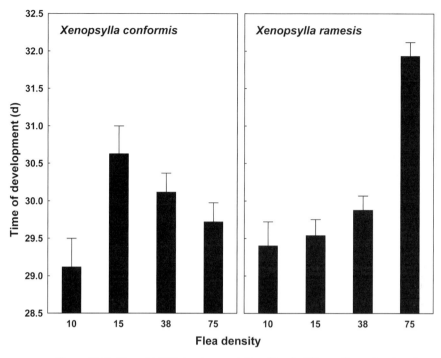

Figure 11.16 Mean (±S.E.) development time (days) of *Xenopsylla conformis* and *Xenopsylla ramesis* produced by females fed on Sundevall's jird *Meriones crassus* at different densities. Data from B. R. Krasnov and A. Hovhanyan (unpublished).

1961, 1970; Kunitsky et al., 1963). *Xenopsylla cheopis* produced, on average, 1.71 pupae per female at 22–24 °C, 3.45 at 26–27 °C and 5.18 at 28–30 °C (Samarina et al., 1968). However, further increase in air temperature led to a decrease of fecundity to 2.01 pupae per female at 32 °C, a further decrease to only 0.26 pupae per female at 34 °C and a drop to zero at 36 °C (Samarina et al., 1968). Similar results were reported for this species by Vashchenok (1988). Gerasimova (1973) showed that the maturation of the oocytes and oviposition of *X. nuttalli* and *X. skrjabini* at 10–15 °C compared with 20 °C slowed down. In another species, *Nosopsyllus tersus*, increase in air temperature from 11–13 °C to 19–20 °C was accompanied by increases in both rate of oviposition (from a 12.5-h interval between consecutive clutches to 7.3 h) and clutch size (from 2–3 to 4 eggs per clutch) (Kunitskaya et al., 1965b). Similar results were found for *Frontopsylla elata* (Guseva & Kosminsky, 1974). Decrease and even termination of egg production at low air temperature was also recorded in *Ctenophthalmus wladimiri*, *Ctenophthalmus teres* and *Ctenophthalmus wagneri* (Kosminsky & Guseva, 1974a, 1975a, b). Interestingly, the effect of low air temperature on the rate of oviposition was expressed more strongly in older than in young *C. wagneri* (Kosminsky & Guseva,

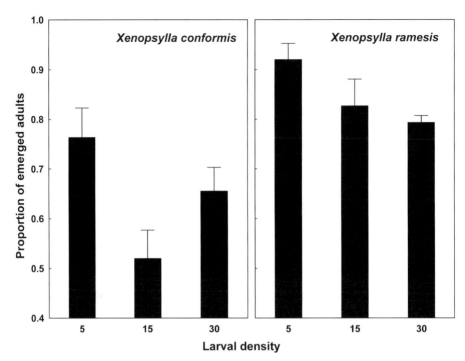

Figure 11.17 Mean (±S.E.) emergence success (number of newly emerged adults per larva) of *Xenopsylla conformis* and *Xenopsylla ramesis* developing under different larval densities. Data from B. R. Krasnov and A. Hovhanyan (unpublished).

1975b). However, no effect of air temperature on egg production was found in *Leptopsylla taschenbergi* (Kosminsky, 1960), *Paradoxopsyllus teretifrons*, *Paradoxopsyllus repandus*, *Rhadinopsylla cedestis* (Kunitskaya et al., 1965b, 1969) and *Amphipsylla rossica* (Kosminsky & Guseva, 1974b). In addition, in some flea species such as *L. segnis*, *Monopsyllus anisus* and *Megabothris calcarifer*, fecundity decreased rather than increased with an increase in air temperature (Moskalenko, 1963b) (Table 11.1). Mean number of clutches per female per day in *Frontopsylla semura* was 2.2 at 22–23 °C and 1.8 at 7–10 °C (Bryukhanova & Surkova, 1970).

The effect of RH on reproductive output has been less well studied than the effect of air temperature. Nevertheless, some information is available. For example, female *X. skrjabini* produced 6.5 times more eggs at 80% than at 100% RH (Yakunin & Kunitskaya, 1980). Vashchenok (1993) also found a strong effect of RH on egg production in *L. segnis*. In contrast to *X. skrjabini*, females of this flea produce significantly more eggs at high RH (77–95%; i.e. imitating the host's burrow) than at low RH (7–22%; i.e. imitating external environment).

These examples suggest that there is no general direction for the relationship between reproductive output of fleas and air temperature or RH, but rather

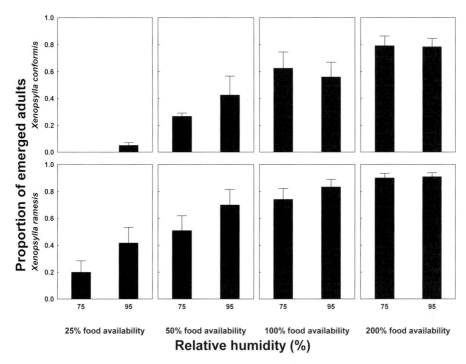

Figure 11.18 The effect of relative humidity and food availability on the mean (±S.E.) proportion of *Xenopsylla conformis* and *Xenopsylla ramesis* larvae that survived until emergence. Data from Krasnov *et al.* (2005e).

that each species has its own preferred temperature and/or humidity range. Reproductive success decreases when air temperature and/or humidity is either lower or higher than preferable. Furthermore, the geographical distribution of a flea and its species-specific pattern of seasonality (see Chapter 5) are probably associated with species-specific preferred microclimatic conditions. For example, *F. semura*, which exhibited a decrease in the rate of oviposition with an increase in air temperature is characterized by winter periods of activity and breeding (Ioff & Tiflov, 1954; Bryukhanova & Myalkovskaya, 1974).

The effect of microclimate on survival and development of pre-imaginal fleas has been known for a long time (e.g. Bacot, 1914; Uvarov, 1931; Tiflov & Ioff, 1932; Edney, 1945, 1947a, b). Much attention has been paid to fleas that have medical and/or veterinary importance. For example, it was found that air temperature influenced the development time and emergence of *X. cheopis*, *C. felis*, *C. gallinae* and *E. gallinacea* (Suter, 1964; Panchenko, 1971; Margalit & Shulov, 1972; Silverman & Rust, 1983; Metzger & Rust, 1997; Heeb *et al.*, 2000). In the cat flea, the pre-emerged adult stage is capable of surviving prolonged periods during the absence of hosts or during unfavourable environmental conditions

Table 11.1 *Reproductive success and duration of development in four flea species at two different air temperatures*

Species	Number of new imagoes per parent female		Minimal time of larval hatching		Minimal time of imago emergence	
	13–16 °C	19–20 °C	13–16 °C	19–20 °C	13–16 °C	19–20 °C
Leptopsylla segnis	27.0	17.4	18	8	49	25
Monopsyllus anisus	16.0	11.7	13	12	58	33
Megabothris calcarifer	1.3	0.0	20	–	72	–
Neopsylla bidentatiformis	36.2	55.6	14	12	64	35

Source: Data from Moskalenko (1963b).

such as winter or midsummer. The quiescent adult within the cocoon has a lower respiratory demand than the emerged adult, and its survival is considerably longer under low humidity conditions (Silverman & Rust, 1985; Metzger & Rust, 1997). It has been speculated, though never experimentally verified, that the prolonged survival of quiescent adults within the cocoon is in part due to a reduction in respiratory water loss because less time is spent with the spiracles open (Silverman & Rust, 1985). Larvae, in contrast, cannot close their spiracles, and thus are extremely sensitive to low humidity (Mellanby, 1933) although Bahmanyar & Cavanaugh (1976) demonstrated that *X. cheopis* can complete its life cycle at 60% and *Xenopsylla brasiliensis* at 51% RH. Bruce (1948) reported that the survival of *C. felis* larvae was relatively high at 21–32 °C and declined at higher temperatures whereas no larval survival occurred at RH less than 45% or higher than 95%. Larvae and pupae of the cat flea did not survive at air temperatures higher than 35 °C even at the optimal RH (Silverman *et al.*, 1981; Silverman & Rust, 1983). Outdoor survival of cat flea larvae was greatest at moderate temperatures and humidities (Kern *et al.*, 1999). In general, low humidity reduces the lifespan of all stages, slows down larval activity, inhibits silk production causing the cocoon to be small and soft and affects adult emergence and size (Sharif, 1949; Smith, 1951; Yinon *et al.*, 1967; Heeb *et al.*, 2000; Krasnov *et al.*, 2001a, b, 2002d).

Development time of pre-imagoes is affected by air temperature and humidity as strongly as their survival. For example, eggs of *C. felis* hatched in 48 h and larvae spun cocoons in less than 10 days at 26.7 °C, whereas at 15.5 °C these periods were extended to 6 days and more than 26 days, respectively (Silverman *et al.*, 1981). Imagoes of this species emerged from cocoons in 12–20 days at 26.7 °C but stayed within a cocoon for more than 3 months at 15.5 °C (Silverman *et al.*, 1981). Larval development of *X. cheopis* at 20–35 °C and 95–100% RH lasted, on average, about 20 days, whereas it was shorter (about 15 days) when RH decreased to 75–80% (Panchenko, 1971). Similar trends were found in the larvae of two other rat fleas, *N. fasciatus* and *M. anisus* (Panchenko, 1971).

The effects of air temperature and humidity on survival and development patterns in pre-imagoes of fleas parasitic on wild hosts have also been studied. Gerasimova (1970, 1973) showed that eggs and larvae of *X. nuttalli* and *X. skrjabini* did not survive at 10 °C. Immature *C. tesquorum* survived at an air temperature no lower than 9 °C (Karandina & Darskaya, 1974). Significantly more imagoes of *S. cuniculi* emerged when their larvae were reared at 27 °C than at 25 °C (Vaughan & Coombs, 1979). A shorter duration of development at higher air temperatures was found for *N. laeviceps* (Kunitsky *et al.*, 1963; Amin *et al.*, 1993), *X. skrjabini* (Vansulin, 1965), *X. gerbilli* (Zolotova, 1968), *Ctenophthalmus dolichus* (Zolotova & Afanasieva, 1969), *Coptopsylla lamellifer* (Sokolova & Popova, 1969), *Neopsylla setosa* (Darskaya & Karandina, 1974), *C. wladimiri* (Talybov, 1976) and *C. teres* (Yurgenson & Maximov, 1981). The same was true for *L. segnis*, *M. anisus* and *N. bidentatiformis* (Moskalenko, 1963b), despite decreased production of the offspring (see Table 11.1), suggesting that different mechanisms are behind the effect of air temperature on egg production on the one hand, and pre-imaginal development of the offspring on the other. This is likely as the former is associated with parent fleas (either females or males or both), whereas the latter represents the direct effect of the environment on the offspring themselves.

Kosminsky *et al.* (1970) studied the effect of air temperature (0–2, 4–5, 7–10, 18–23, 25 and 30 °C) and RH (60%, 75%, 90% and 100%) on survival and duration of development in *C. wladimiri* (Table 11.2). It appeared that the minimal air temperature for egg survival was 4–5 °C, whereas that for larvae was 18–23 °C. Furthermore, both eggs and larvae responded strongly to changes in RH and did not survive at low humidity even if the air temperature was favourable. Duration of development decreased with increase in temperature. No apparent effect of RH on this parameter was found, although extremely high humidity accompanied by extremely high temperature lengthened the duration of flea development (at 30 °C it was longer at 100% relative than at 90% RH).

Kunitsky (1970) found that larvae of *N. laeviceps* were able to develop at a broad range of air temperatures (from 6–10°C to 31–32 °C) with the lowest mortality at 80–85% RH. Higher humidities negatively affected larval survival at high temperatures only, whereas lower humidities decreased larval survival at any temperature. Furthermore, egg and larval development were faster at

Table 11.2 *Mean survival and development of pre-imaginal* Ctenophthalmus wladimiri *at different air temperatures and relative humidities*

Temperature (°C)	Relative humidity (%)	Survival			Duration of development (days)		
		Eggs	Larvae	Pupae	Eggs	Larvae	Pupae
0–2	60	0.0	–	–	–	–	–
	75	0.0	–	–	–	–	–
	90	0.0	–	–	–	–	–
	100	0.0	–	–	–	–	–
4–5	60	0.0	–	–	–	–	–
	75	5.0	0.0	–	49.4	–	–
	90	57.0	0.0	–	39.0	–	–
	100	0.0	–	–	–	–	–
7–10	60	0.0	–	–	–	–	–
	75	7.0	0.0	–	41.0	–	–
	90	30.0	0.0	–	40.6	–	–
	100	54.0	0.0	–	42.7	–	–
18–23	60	10.0	0.0	–	5.4	–	–
	75	30.6	0.0	–	5.0	–	–
	90	49.3	21.6	81.2	5.9	14.7	15.7
	100	52.2	43.6	90.2	5.3	15.3	14.4
25	60	3.8	0.0	–	4.0	–	–
	75	35.0	0.0	–	4.3	–	–
	90	68.3	34.0	82.0	4.2	15.0	12.4
	100	78.2	38.0	94.1	4.0	13.6	13.1
30	60	0.0	–	–	–	–	–
	75	42.5	0.0	–	3.3	–	–
	90	63.3	29.0	70	3.3	10.3	10.0
	100	94.3	50.0	97.0	3.3	13.3	10.5

Source: Data from Kosminsky *et al.* (1970).

higher temperatures. Higher humidity, especially at lower temperatures, led to shortening of the development time in larvae, but did not affect the development of eggs.

Gerasimova (1970) showed that an increase in air temperature from 15 to 32 °C was associated with faster development of all pre-imaginal stages of *X. nuttalli*. However, the effect of RH on the rate of development in this species was pronounced only in larvae and at lower (15–20 °C) air temperatures. Cooke & Skewes (1988) demonstrated that normal development of *S. cuniculi* from eggs to imagoes only occurred if air temperature was 15–30 °C and RH was 70–95%.

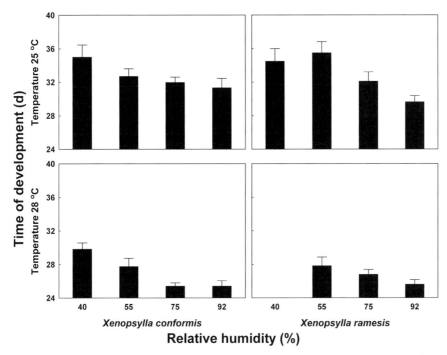

Figure 11.19 Mean (±S.E.) development time (days) of *Xenopsylla conformis* and *Xenopsylla ramesis* at two different air temperatures and four different relative humidities. Data from Krasnov *et al.* (2002b).

Krasnov *et al.* (2001a, b) examined the survival and rate of development of immature *X. conformis* and *X. ramesis* in relation of air temperature and RH. These studies supported previous findings about strong effect of these factors on flea developmental success. In general, the quality of pre-imaginal *X. conformis* and *X. ramesis* (in terms of their survival and development time) increased at higher (but not too high) ambient temperature and RH (see, for example, Fig. 11.19 for development time). However, the effect of RH appeared to be somewhat stronger than that of air temperature. In addition, there was a difference in the responses of different stages. In particular, eggs and larvae responded to both factors, whereas no effect of air temperature on pupae was found.

Interestingly, occasional findings of fleas with abnormalities of chaetotaxy were sometimes explained by development of pre-imaginal fleas under suboptimal microclimatic conditions. For example, Smit (1977) suggested that an abnormally high number of lateral plantar bristles in *E. gallinacea* found occasionally in Africa might arise because these individuals developed under a 'foreign environment' (see also Hastriter, 2000b).

Summarizing the results of these case studies on the effect of air temperature and RH on survival and development of pre-imaginal fleas, we should note three important issues. First, not only the separate effects of each factor but also their interactions are important (see also Bibikova (1965) for *Citellophilus trispinus*). The effect of the interaction between air temperature and RH was most strongly pronounced in the larval fleas. This is most probably related to their ability to absorb water via the rectal sac (Edney, 1947a; Knülle, 1967), which depends on the water vapour pressure depending, in turn, on air temperature and RH. Second, different pre-imaginal stages respond differently to microclimatic factors. Third, the trend in the effect of microclimate on the duration of development is similar in different flea species.

To conclude, the effects of air temperature and RH on reproductive parameters such as egg production and rate of oviposition vary among flea species and are associated with seasonal patterns of their activity. In contrast, the effect of these factors on survival and duration of development of pre-imagoes is surprisingly similar in all flea species studied. This suggests that the effect of the environment on parent fleas is ecologically based, whereas the effect of the environment on pre-imagoes is physiologically based. Consequently, the responses of parent fleas to changes in air temperature and/or RH were probably selected under pressure of various ecological factors such as seasonal availability of host, seasonal changes in host status (e.g. hibernation) and competition among flea species, while the responses of the offspring fleas were probably selected under strong pressure of the constraints determined by physiological and biochemical processes. Nevertheless, an ecological or geographical component was also important for the selection of preferable microclimatic conditions in immature fleas. For example, optimal RH for immature *X. skrjabini* has been shown to be higher than that for immature *C. lamellifer* (Sokolova & Popova, 1969; Gerasimova, 1970).

11.4.2 Substrate texture

Texture of the substrate is another environmental factor that may affect reproductive success in fleas. First, different substrates have different water-storage capacities which is important for larval development. Second, a flea cocoon is covered by adhering particles of the nest substrate. The size and physical properties of the particles of different substrates differ in their abilities to absorb water which can affect the survival and development of pupae. For example, Launay (1989) suggested that one of the factors limiting geographical and habitat distribution of *Xenopsylla cunicularis* is the nature of the substrate, which should have a sandy texture.

To the best of my knowledge the only experimental study that was specifically aimed at testing the effect of substrate on fleas was the study of Krasnov et al. (2002b) on performance of pre-imaginal X. conformis and X. ramesis in sand and loess substrate. It was found that, in general, the texture of substrate did not affect the survival and development rate of eggs, but strongly affected the survival of larvae and the rate of development of both larvae and pupae. The effects of the substrate differed, however, between developmental stages as well as between fleas (see details in Chapter 18). In particular, larvae of both species developed faster in sand than in loess, but the opposite was true for the pupae.

11.4.3 Light regime, photoperiod and host odour

Some fleas are known to lay their eggs with a certain circadian rhythm. For example, E. gallinacea and E. oschanini oviposit mainly at dusk and at night (Suter, 1964; Vashchenok, 1967b). A female E. oschanini laid, on average, 2.0 eggs per hour in lighted conditions and 12.6 eggs per hour in darkness (Vashchenok, 1967b, 1988). A similar pattern was found in Vermipsylla alakurt and Dorcadia ioffi (Ioff, 1950) and C. felis (Kern et al., 1992). The time of peak oviposition corresponds to the host's resting period. Experimental manipulations with the light regime (light versus dark) by Vashchenok (1993) demonstrated that L. segnis lays more eggs in darkness. Darkness and higher RH (which also increases the rate of oviposition) mimic the host's burrow. This pattern was thought to have evolved as an adaptation to increase the chances of survival of immature fleas (Vashchenok, 1993).

As we already know, microclimate dependence of reproductive parameters in some flea species corresponds to seasonal pattern of their activity and reproduction. It can therefore be expected that flea reproduction may also be influenced by a season-specific photoperiod. This effect was studied in C. felis by Metzger & Rust (1992). They measured egg production under light:dark photoperiods of 8:16, 12:12 and 16:8 h that mimicked winter, spring/autumn and summer, respectively, but did not find significant differences in either daily or total flea egg production among photoperiods. However, the photoperiod at which the eggs were produced affected the duration of development of immature stages as well as subsequent adult emergence from cocoons.

Finally, the odour of a host has been shown to increase egg production in L. segnis (Vashchenok, 1993). Furthermore, the highest number of eggs produced by a female flea has been recorded at combined conditions of high RH, darkness and the presence of bedding material from the cage of a mouse. This, again, was suggested to be a regulating mechanism for oviposition that guarantees the offspring will develop under favourable conditions in terms of microclimate, food availability for larvae and chances of the new generation imago to locate and attack a host.

11.5 Concluding remarks

As I mentioned in the conclusion to the previous chapter, a narrative description or calculation of standard parasitological indices of flea distribution can only suggest, but not definitely reveal, mechanisms underlying host preferences and distribution. In contrast, experimental studies of flea reproduction allow us to detect the mode by which a flea perceives differences between host species and to explain between-host flea distribution by subtle differences in fitness consequences of host selection. Furthermore, a variety of factors associated with hosts, fleas themselves and the environment affect reproductive output. All these factors should be taken into account when fitness-related traits are considered. On the other hand, many ecological and evolutionary issues related to flea reproduction remain to be tested. At present, we know quite a bit about reproductive patterns in fleas, but the processes underlying these patterns largely remain to be discovered.

12
Ecology of flea virulence

A parasite is commonly defined as an organism that lives in or on host and from which it derives food and other biological supplies (Kim, 1985b). In addition, it reduces host fitness (Watt *et al.*, 1995; Clayton & Moore, 1997), including causing an increase of host mortality and/or morbidity (Shaw & Moss, 1990; Delahay *et al.*, 1995). A concept of harm or any other negative effect on a host is an integral part of the ecology of parasitism. Consequently, one of the main asymmetries in host–parasite relationships is that fitness of a host without a parasite is maximal *ceteris paribus*, whereas fitness of a parasite without a host is zero (Combes, 2001; Poulin, 2007a). Another asymmetry is that the fitness of a host is determined by a variety of extrinsic and intrinsic factors, but fitness of an infested host is also affected by competition with the parasite, which may use part of the resources that a host would otherwise allocate for an increase in its own fitness (Combes, 2001; Poulin, 2007a). Thus, a parasite may reduce fitness in a host. The component of loss of fitness resulting from a parasite, i.e. parasite-induced loss of fitness, is virulence (Combes, 2001). Virulence is a parameter characterizing the parasite, but it can be measured via host fitness or its elements (Combes, 2001).

Although both the parasite and the host have the same evolutionary aim of maximizing fitness, these asymmetries determine the different ecological tasks of a parasite and a host when these two entities are associated. From the parasite's perspective, the tasks include location of an appropriate host, its successful exploitation and the guarantee of transmission of offspring (see Chapters 9–11). From the host's perspective, the tasks are defence against parasites (avoidance or elimination) and minimizing damage (see Chapter 13). However, both the parasite and the host are constrained in fulfilling these tasks. It is generally of no interest to the parasite what happens to the host as a consequence of exploitation when an increased exploitation improves the parasite's fitness (Poulin, 2007a). However, an exaggerated exploitation can impair the parasite's fitness because the host is not only the parasite's resource but also its habitat and, in many cases, its tool of dispersal (Combes, 2001). As a result, it is important to the parasite that the host and/or its offspring live long enough to ensure the parasite's successful transmission. Thus, on the one hand, an unlimited increase of exploitation may be disadvantageous to the parasite. On the other hand, although the well-being of the parasite is obviously of no interest to the host, anti-parasitic defence is costly (see below). As a result, the host should struggle with the parasite only if the fitness increment due to parasite elimination is higher than the fitness decline due to defence cost.

The effects of parasites on the host(s) can be direct, such as using energy and nutrients of hosts (Khokhlova *et al.*, 2002) and indirect, such as increasing activity of the immune system (Wedekind, 1992; Lochmiller & Deerenberg, 2000), modifying behaviour (Barnard *et al.*, 1998; Kavaliers *et al.*, 1998; ter Hofstede & Fenton, 2005) and decreasing food intake (Milinski, 1990; Kavaliers & Colwell, 1995; Tripet *et al.*, 2002b). Parasites can also be vectors of various pathogens (see Chapter 6). However, the effects of a parasite as a vector of pathogens represent relationships of quite a different type, and will not be considered here. In this chapter, I start with the effect of fleas on various fitness-related host traits, that is, examine how flea parasitism affects various components of host fitness. Then I consider the net results of these effects on host fitness, that is, evaluate the level of flea virulence.

12.1 Host metabolic rate

An animal obtains energy from its food intake, which it digests and absorbs. Part of this energy is lost through faeces, urine and combustible gases, whereas the remaining (apparent metabolizable energy) is available for an animal's maintenance and production (Degen, 1997). Fleas compete with the host for this energy. From a flea perspective, net energy that it obtains from the host is the energy of consumed blood minus energy spent for blood digestion and energy spent for both avoiding the host grooming and coping with the host immune system. From a host's perspective, the loss of energy from flea parasitism equals the energy of blood extracted by a flea plus energy spent for both the haemopoietic compensation for the lost blood and for behavioural and/or immune responses. Consequently, flea parasitism is expected to have energetic consequences on the host. To examine this, Khokhlova *et al.* (2002) studied the effect of parasitism of *Xenopsylla ramesis* on Wagner's gerbil *Dipodillus dasyurus* and predicted that the energy requirements of parasitized gerbils would be higher than those of non-parasitized gerbils.

Average daily metabolic rate (ADMR) was used as a measure of metabolizable energy requirements for maintenance under laboratory conditions. ADMR is the metabolizable energy intake required by a caged, laboratory animal to maintain constant body energy content. It includes basal metabolic rate, heat increment of feeding for maintenance, some minimal locomotory costs and possibly some thermoregulatory costs (Degen, 1997; Degen et al., 1998). The gerbils were divided randomly into three groups in which each group received a different amount of food: 85%, 100% and 115% of their energy requirements for maintenance. Two 2-week trials were done for each animal — one with parasite pressure (parasitized) and one without (non-parasitized). Fifty starving fleas were placed on each gerbil to be parasitized for 3 h daily. In these experiments, parasitized rodents were not allowed to groom during flea feeding. For each gerbil, dry matter intake (DMI) was calculated as the difference between dry matter offered and dry matter remaining, whereas apparent dry matter digestibility was calculated from dry matter intake and faecal output. Digestible energy intake (DEI) was calculated as the difference between gross energy intake and faecal energy output, while metabolizable energy intake (MEI) was assumed to be 98% of DEI (Grodzinski & Wunder, 1975). Daily energy requirements for maintenance (ADMR) were estimated separately for parasitized and non-parasitized gerbils from the linear regression of body mass change of the gerbil on its MEI, with ADMR taken at the point where there was no change in body mass.

The linear relationships of body mass (m_b) change on MEI for non-parasitized and parasitized D. dasyurus (Fig. 12.1) were: m_b change (% d^{-1}) = $-3.95 + 0.59$ MEI (kJ $g^{-0.54}$ d^{-1}) and m_b change (% d^{-1}) = $-4.34 + 0.56$ MEI (kJ $g^{-0.54}$ d^{-1}), respectively. Therefore, ADMR of non-parasitized gerbils was 6.69 kJ $g^{-0.54} d^{-1}$ and body mass loss at zero MEI was 3.95% d^{-1}. For parasitized animals, ADMR was 7.75 kJ $g^{-0.54} d^{-1}$ and body mass loss at zero MEI was 4.34% d^{-1}. In other words, the difference between the intercepts of the regression lines reflected difference in energy expenditure between the two groups of gerbils with ADMR of the parasitized gerbils being about 16% higher than that of non-parasitized gerbils. At zero MEIs, the parasitized gerbils lost body mass at a faster rate than the non-parasitized gerbils. Non-parasitized gerbils maintained steady-state body mass when offered food at maintenance (100% of energy requirements) or above maintenance (115% of energy requirements) energy levels and lost body mass when offered food at submaintenance (85% of energy requirements) energy levels. In contrast, parasitized gerbils maintained steady-state body mass only when offered food at the highest level (115% of energy requirements) and lost body mass when offered food at 100% and 85% levels of energy requirements.

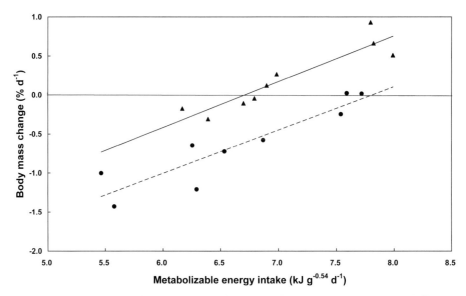

Figure 12.1 Relationships between body mass change (% per day) and metabolizable energy intake (kJ per $g^{-0.54}$ per day) in infested with *Xenopsylla ramesis* (circles, dashed line) and uninfested (triangles, solid line) Wagner's gerbils *Dipodillus dasyurus* consuming different levels of energy. Redrawn after Khokhlova et al. (2002) (reprinted with permission from Blackwell Publishing).

The slopes of the regression equations of body mass change on MEI were similar in both groups (Fig. 12.1). This indicated that the efficiency of utilization of metabolizable energy in the two groups was similar, as there is generally a relationship between change in body mass and change in energy content of the body (Degen et al., 1998). This conclusion was further supported by the findings that dry matter and energy digestibilities were not different between parasitized and non-parasitized gerbils. This suggests that the utilization of energy to combat parasitism and for maintenance is similar (Kam & Degen, 1997).

Although 50 *X. ramesis* having one blood meal per day caused *D. dasyurus* to increase its maintenance energy requirements by only 16%, this increase in energy requirements could be vital under desert conditions where food supplies are often scarce. Furthermore, in this study, each flea consumed about 0.074 mg of blood which amounted to 34.3% of its unfed body mass, but the total blood lost by the host was only 0.17% of its blood volume. Thus, it is possible that the major effects of fleas on the energy expenditure of the host could be through means other than blood loss, such as costly immune responses (see below).

The study of Khokhlova et al. (2002) was the first to determine the energy cost of flea parasitism. However, this study dealt only with a single flea and a single host. Hawlena et al. (2006a) used the same methodology and looked at what happened with another flea species (*Synosternus cleopatrae*) and another

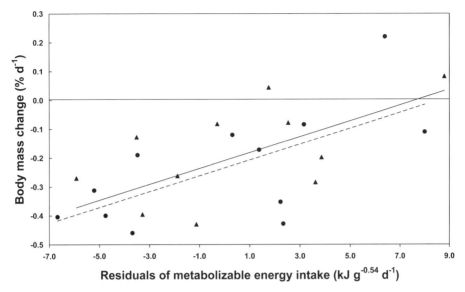

Figure 12.2 Relationships between body mass change (% per day) and the residual metabolizable energy intake (kJ per $g^{-0.54}$ per day) (extracted from the regression between metabolizable energy intake (dependent variable) and mean body mass (independent variable)) in Anderson's gerbils *Gerbillus andersoni* infested with *Synosternus cleopatrae* (circles, dashed line) and uninfested (triangles, solid line) consuming different levels of energy. Redrawn after Hawlena et al. (2006a) (reprinted with permission from Blackwell Publishing).

host species (*Gerbillus andersoni*) that inhabit the same region as those used by Khokhlova et al. (2002). Surprisingly, in contrast to the findings in *D. dasyurus*, there was no measurable difference between energy requirements of parasitized and non-parasitized *G. andersoni* (Fig. 12.2). The energetic cost of fleas to *G. andersoni* was only 3% of its total energy requirements.

This sharp difference in energetic response of different host species to flea parasitism can be related to differences in the strategy of immune defence against fleas. *Gerbillus andersoni* is characterized by continuous immunological 'readiness' and, thus, it demonstrates a 'pre-invasive' (constitutive) response that is always present and is capable of defence without previous flea contact, whereas *D. dasyurus* possesses a 'post-invasive' (induced) immune strategy in which the response is employed only after an attack of a flea (Khokhlova et al., 2004b). The reasons for this difference in strategies will be discussed below (see Chapter 13). Non-parasitized *G. andersoni* (pre-invasive immune response) required a higher energy maintenance expenditure than *D. dasyurus* (post-invasive immune response) reflecting the cost of maintaining the immune 'readiness'. This supports the idea that costs of mounting an immune response and maintaining a competent immune system are high in vertebrate hosts (e.g. Lochmiller &

Ecology of flea virulence

Table 12.1 *Results of investigations of the effect of flea parasitism on host's energy requirements: A — Is this flea a characteristic parasite for this host? (Y/N); B — Has the negative effect of flea parasitism on host's energy requirements been found? (Y/N)*

Host species	Flea species	A	B	Reference
Dipodillus dasyurus	*Xenopsylla ramesis*	Y	Y	Khokhlova et al., 2002
	Xenopsylla conformis	Y	Y	I. S. Khokhlova, B. R. Krasnov and A. A. Degen, unpublished data
	Parapulex chephrenis	N	Y	I. S. Khokhlova, M. Sarfati, B. R. Krasnov and A. A. Degen, unpublished data
Meriones crassus	*Xenopsylla ramesis*	Y	N	I. S. Khokhlova, B. R. Krasnov and A. A. Degen, unpublished data
	Xenopsylla conformis	Y	N	I. S. Khokhlova, B. R. Krasnov and A. A. Degen, unpublished data
Gerbillus andersoni	*Synosternus cleopatrae*	Y	N	Hawlena et al., 2006a
Acomys cahirinus	*Parapulex chephrenis*	Y	Y	I. S. Khokhlova, M. Sarfati, B. R. Krasnov and A. A. Degen, unpublished data
	Xenopsylla ramesis	N	Y	I. S. Khokhlova, M. Sarfati, B. R. Krasnov and A. A. Degen, unpublished data

Deerenberg, 2000; see also Chapter 13) in contrast to what was thought earlier (Klasing, 1998). These costs may include the additional energy invested in the cumulative mass of immune products (cells and molecules), the metabolic requirements of immune cells and the indirect consequences of mounting an antigen-induced immune response (e.g. acute inflammatory response).

Additional studies of various flea–host associations showed that the negative effect of fleas on energy balance of a host is, although common, not a universal phenomenon (Table 12.1). Moreover, there is evidence that the level of evolutionary 'familiarity' between a particular flea and a particular host does not play an important role in this relation (note a host *Acomys cahirinus* and fleas *Parapulex chephrenis* and *X. ramesis* in Table 12.1), whereas host identity appears to be very important. For example, the negative effect of flea parasitism on energy requirements of the host was always revealed in *D. dasyurus* and *A. cahirinus* (Table 12.1).

The only study of the effect of fleas on metabolic rate in birds was done by Nilsson (2003) in Sweden. The mass-specific resting metabolic rate of the marsh tit *Poecile palustris* nestlings from flea-infested nests was significantly higher than in nestlings from control nests. Studies on the relationship between parasitism and energy balance of a host are scarce (see review in Degen, 2006) and more investigations are needed.

12.2   Host body mass and growth rate

An effect of parasite infestation and/or its intensity on host body mass has been found in some parasite–host associations (Shields & Crook, 1987; Puchala, 2004; Gwinner & Berger, 2005; Holmstad et al., 2005) but not in others (Bennett et al., 1988; Christian & Bedford, 1995; Pacejka et al., 1998; Henry et al., 2004). Fleas have been also investigated in this relation because blood loss due to flea parasitism and the energy cost of behavioural and immune defences against fleas could be reflected in the body mass loss of parasitized hosts.

In a field study of flea infestation of the black-tailed prairie dogs *Cynomys ludovicianus* in Colorado, Brinkerhoff et al. (2006) found that heavier rodents had fewer fleas *Oropsylla hirsuta* and *Oropsylla tuberculata*. In other field studies that used change in body mass of parasitized hosts as an indicator of flea effect (Krasnov et al., 1997; Hawlena et al., 2005; Boughton et al., 2006), no significant difference between infested and uninfested hosts was found. It should be noted that comparisons of body mass between infested and uninfested hosts in field surveys is difficult to interpret as it is unclear as to what is the cause and what is the consequence. Is it flea parasitism that causes the difference in body mass between hosts or is it the difference in body mass between hosts that causes fleas to exploit hosts with higher or lower body mass? In contrast, laboratory and field experiments may provide much clearer answers.

One of the reasons for the lack of an effect of flea parasitism on host body mass found in the above-cited field surveys could be that natural levels of flea infestation were too low for the manifestation of a negative effect on host body mass. Indeed, when Hawlena et al. (2006b) measured body mass change in *G. andersoni* under parasitism by *S. cleopatrae* at numbers of fleas on a rodent much greater than the natural infestation level, the parasitized animals lost body mass at a higher rate than non-parasitized control individuals (Fig. 12.3a). These results suggest that fleas have the potential to damage their host when occurring at very high densities.

A density-dependent macroparasite effect is included in almost every model of epidemiology (Roberts et al., 1995) and is supported by empirical studies that manipulated parasite numbers (Lehmann, 1993; Gulland, 1995). Moreover, most empirical studies show that natural infestation levels are costly to the host and removal of ectoparasites improves host body conditions, survival and reproductive success (e.g. Brown & Brown, 1986; Møller, 1991). Nevertheless, there can be a different situation when equilibrium natural densities of fleas are at a point in which the negative effect on a host is small enough to be below measurement accuracy. Possible mechanisms regulating ectoparasite numbers may be density-dependent population processes of the parasites themselves (e.g.

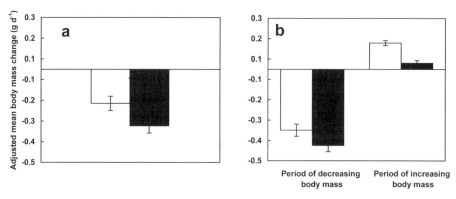

Figure 12.3 Mean (±S.E.) daily body mass change (g) of (a) adult and (b) juvenile Anderson's gerbils *Gerbillus andersoni* infested with *Synosternus cleopatrae* (black columns) and flea-free (white columns). The latter was measured over 4 days of decreasing body mass and 23 days of increasing body mass (see text for explanations). Redrawn after Hawlena *et al.* (2006b) (reprinted with permission from Cambridge University Press).

Tripet *et al.*, 2002c) or host-mediated behavioural (Moore, 2002) and physiological responses (e.g. Walker *et al.*, 2003). Whatever these regulation processes are, in their absence density-dependent negative effects of parasites would be strong.

Another reason for the lack of difference in body mass of infested and uninfested adult hosts in field surveys is perhaps related to the host-age dependence of the effect of flea parasitism. If this is the case, then this effect should be readily noticeable in growing young host individuals due to their higher mass-specific energy requirements. Also, the effect of fleas may be manifested in the dynamics of body growth rather than in body mass per se in that energy obtained by a young animal is allocated not only to maintenance but also to somatic growth and maturation. Moreover, young animals have a larger surface-to-volume ratio and thus higher energy requirements for maintaining unit body mass (Kleiber, 1961). Thus, a young individual is more energy-constrained than an adult individual.

In experiments with many bird species (but a limited range of flea species), fleas have been shown to affect negatively the growth rate of nestlings. For example, Richner *et al.* (1993) heat-treated nests of the great tit *Parus major* in Switzerland to kill all ectoparasites and randomly infested some of the nests with 60 adult *Ceratophyllus gallinae*. At the age of 14 and 17 days, nestlings in infested nests had lower body mass and wing length than control nestlings. Parasitism of *C. gallinae* (sometimes coupled with the effect of other ectoparasites) has been shown to reduce body mass and/or growth rate in nestlings of the great tit

P. major in Sweden and Switzerland (Dufva & Allander, 1996; Allander, 1998; Heeb *et al.*, 2000; Tschirren *et al.*, 2003; Fitze *et al.*, 2004a, b), blue tit *C. caeruleus* in Switzerland (Tripet *et al.*, 2002b), marsh tit *P. palustris* in Sweden (Nilsson, 2003) and the swallow *Tachycineta bicolor* in Canada (Thomas & Shutler, 2001). Similarly *Echidnophaga gallinacea* reduced the growth rate of nestlings of the Florida scrub jay *Aphelocoma coerulescens* in Florida (Boughton *et al.*, 2006), and *Xenopsylla gratiosa* reduced growth rate of nestlings of the European storm petrel *Hydrobates pelagicus* at Benidorm Island in Spain (Merino *et al.*, 1999).

In other experiments, often with the same flea and bird species, there was no effect of flea parasitism on body mass of nestlings (Dufva, 1996; Tripet & Richner, 1997a; Brinkhof *et al.*, 1999; Bouslama *et al.*, 2001, 2002; Shutler *et al.*, 2004). For example, high flea numbers were not associated with smaller nestlings in *T. bicolor* in Canada (Shutler *et al.*, 2004). Furthermore, Kedra *et al.* (1996) in Poland found that fledglings of the pied flycatcher *Ficedula hypoleuca* and the great tit *P. major* in flea-infested nests were heavier than in flea-free nests. Mazgajski *et al.* (1997) showed that the presence of flea larvae or imagoes in the nests of the European starling *Sturnus vulgaris* in Poland did not affect nestling weight or body size. However, when broods of different sizes were analysed, there was a strong negative influence on the weight and body size of nestlings in bigger broods (four nestlings).

The lack of effect of flea parasitism on body size of young birds or increase in body mass in flea-infested nests can be attributed, at least in part, to parental compensation of flea parasitism (e.g. Christe *et al.*, 1996a; Richner, 1996, 1998; Tripet & Richner, 1997a; Bouslama *et al.*, 2002; see below). In addition, the effect of fleas on body condition and survival of young birds may be mediated by environmental effects, such as variation in weather conditions (Dufva & Allander, 1996; Allander, 1998; but see Fitze *et al.*, 2004b). For example, Dufva & Allander (1996) investigated the effect of *C. gallinae* on the reproductive success of *P. major* in Sweden, manipulating flea levels by either spraying nests with pyrethrin or by introducing 100 adult fleas during the egg-laying period. They found that the effect of fleas on growth of nestlings was significant in only two of five years of a study, namely in years with lower temperatures and higher precipitation. Another factor that might mask manifestation of flea effect on young birds is a decrease of brood size in parasitized nests (e.g. Fitze *et al.*, 2004b; see below). Smaller broods are usually associated with larger nestlings, thus compensating for the flea effect per individual nestling (Kedra *et al.*, 1996).

The impact of flea parasitism on body mass and other traits of nestlings might also be masked by differential response to fleas by male and female nestlings. In particular, this is because of the relationships between sexual hormones and the immune system, with androgens generally suppressing immune function (e.g.

Folstad & Karter, 1992; see Chapter 13). In a field study in Switzerland, Tschirren et al. (2003) manipulated the level of flea infestation in nests of *P. major* and tested the effect of flea parasitism on male and female nestlings separately. They found that males were significantly smaller and lighter in parasitized than in non-parasitized nests, whereas in females there was no difference in morphological traits such as body mass and metatarsus length due to parasitism.

The reduced body mass and growth rate of young animals due to flea parasitism was found not only in avian but also in mammalian hosts. For example, Van Vuren (1996) demonstrated that the growth rate of yearling yellow-bellied marmots *Marmota flaviventris* in Colorado was correlated negatively with the number of fleas *Thrassis stanfordi*. Hawlena et al. (2006b) measured body mass change in juvenile *G. andersoni* under continuous parasitism of 30 *S. cleopatrae* during 25 consecutive days. Since the energy requirements of growing juveniles increased with time, Hawlena et al. (2006b) varied food offered the gerbils during the experiment. During the first 4 days, each animal was offered 1.6 g of millet seeds a day. This amount did not meet energy requirements for maintenance and the gerbils lost body mass (period of decreasing body mass). During the following 4, 6, and 12 days they were offered 2, 2.2 and 2.4 g of millet seeds a day, respectively (period of increasing body mass). Parasitized juveniles lost body mass faster and gained body mass slower than non-parasitized juveniles, with no significant difference between the periods of decreasing and increasing body mass (Fig. 12.3b).

The study by Neuhaus (2003) suggested that the effect of flea parasitism on body mass or growth rate of young is a particular case of a more general trend of the strong manifestation of the effect of fleas during critical and the most energy-demanding period of the host's life. He studied the effect of ectoparasites, mainly fleas (although the identities of species were not mentioned), in the Columbian ground squirrel *Spermophilus columbianus* in Canada. All individuals in the study area carried some fleas and prevalence was high. Animals were trapped before mating and ectoparasites were removed from half the females using a commercially available insecticide. Treatment was then applied weekly until the end of lactation. Body mass did not differ significantly between treated and untreated females when emerging from spring hibernation or just after parturition, but was higher in treated than untreated females at weaning. During lactation, that is, from parturition to juvenile emergence, treated females gained 19 g while untreated females lost 40 g. Furthermore, treated females gained more mass than untreated females from emergence in spring to the emergence of young. There was no difference between treated and untreated groups in body mass per young at birth, but because of the larger litters in treated than untreated females (see below), the mass of litter was higher in treated females.

The contrasting results cited above suggest that the effect of fleas on body parameters and growth rate is mediated by a variety of factors, both extrinsic and intrinsic. This effect is manifested differently in different periods of a host's life cycle. An additional reason for general ambiguity of our knowledge on the relationship between flea parasitism and body mass is that only a limited number of host–flea associations have been studied.

12.3   Host haematological parameters

As the haematophagy of fleas is one of their most conspicuous characters, it has often been hypothesized that blood loss of a host leading to anaemia should be the main effect of fleas. Indeed, severe anaemia caused by extremely high levels of flea infestation is well known from veterinary practice (e.g. Kusiluka et al., 1995; Yeruham & Koren, 2003). However, surprisingly, there have been only few studies examining this in wild animals.

Richner et al. (1993) demonstrated an increase in haematocrit of great tit nestlings from nests infested with hen fleas in Switzerland. In a field study of G. andersoni in the Negev Desert, Lehmann (1992) found that infestation by S. cleopatrae correlated negatively with red blood cell count and haemoglobin concentration but did not affect white blood cell counts. However, laboratory experiments with this host and this flea by Hawlena et al. (2006a) did not support Lehmann's (1992) findings. No significant difference in various biochemical (e.g. AST (L-aspartate aminotransferase), ALT (L-alanine aminotransferase), total protein, albumin and urea) or haematological (e.g. blood cell counts) variables between G. andersoni parasitized by S. cleopatrae and non-parasitized G. andersoni were observed.

Goüy de Bellocq et al. (2006a) studied the effect of fleas on blood parameters in Sundevall's jird Meriones crassus in the laboratory. Haematocrit and leukocyte concentration were measured in all rodents on the first day of the experiment. Then, half of the rodents were exposed to parasitism by X. ramesis. After 15 days, both blood parameters were remeasured. No differences in haematocrit and leukocyte concentration between experimental and control rodents were found either prior to or after flea exposure. The same was true for the comparison between the two parameters on the first or 16th days of experiment. However, the variance in difference of haematocrit or leukocyte concentration between the first and 16th days was significantly higher in the experimental than in the control group.

Boughton et al. (2006) reported a severe negative effect on blood parameters in *A. coerulescens* when parasitized by *E. gallinacea* in Florida. Haematocrit of infested jays was sharply impacted and was negatively correlated with the extent of infestation. Leukocyte counts were higher in infested jays; however, plasma immunoglobulin levels were lower. Physiological stress levels, measured as plasma corticosterone, increased more rapidly in infested than uninfested jays and were positively correlated with heterophil/lymphocyte ratios.

In Poland, Słomczyński et al. (2006) replaced natural (infested with a variety of ectoparasites including fleas) nests of *C. caeruleus* with clean artificial nests twice during the nestling stage. This treatment caused an increase of $7-10.5$ g l^{-1} in haemoglobin of 12-day-old nestlings compared to control nestlings, suggesting that nestlings developing in parasite-pathogen-free nests improved their health status.

Again, no general trend in effect of fleas on host haematology can be concluded from empirical studies. Two important issues should be pointed out. First, changes in the haematological background of a host might be related not only to the effect of fleas per se, but also to the immune responses of a host. Although these processes can be related, their ecological consequences can be quite different. Second, field measurements and findings of the difference in haematological (as well as many other) parameters between flea-infested and flea-free hosts cannot demonstrate unequivocally a flea effect. This is because a host is subject to attacks of a large variety of ecto- and endoparasitic metazoans as well as viruses, fungi, bacteria, protozoans etc. Levels of infestation by some of these natural attackers can be manipulated, whereas this is not the case for others. Furthermore, some parasites and pathogens occurring in the field are usually undetected and, consequently, ignored. However, differences found between infested and uninfested hosts could be either merely coincidental and unrelated to fleas or mediated by fleas but caused directly by other parasites (e.g. Watkins et al., 2006).

12.4 Host features related to sexual selection

It is well known that visual signals play an important role in sexual selection and mate competition. However, producing and carrying the signal is costly (e.g. Zahavi, 1977; Hamilton & Zuk, 1982). Parasites hijack resources that a host could otherwise allocate for various fitness-related aims, including signalling, and thus may affect the expression of signals. Moreover, the expression of a particular signal may be used by potential mates of a given individual for an assessment of its ability to resist parasites. Indeed, lower expression of sexually oriented signals in parasitized individuals has been reported for fish (e.g.

Milinski & Bakker, 1990) and birds (e.g. Thompson *et al.*, 1997; Hörak *et al.*, 2001; Roulin *et al.*, 2001).

Experimental evidence for the effect of parasites on the expression of such signals is scarce as only a few studies have been carried out. Some of these studies have dealt with flea parasitism. Fitze & Richner (2002) studied the effects of *C. gallinae* on carotenoid- and melanin-based traits in breeding great tits in Switzerland. Nests were either infested with fleas or cleaned of all ectoparasites. The colour of carotenoid-based yellow plumage and the colour and size of melanin-based black breast stripes were assessed both in the year of experimental parasite infestation and during the following breeding season after the annual moult. An effect of flea parasitism on the size of the melanin-based trait but not on the colour of either carotenoid- or melanin-based traits was found. Exposure to fleas during breeding significantly affected the area of the breast stripe in the subsequent year in both male and female birds. After the moult, the size of the breast stripe decreased in infested birds but increased in uninfested birds. This is one of a few pieces of experimental evidence that flea parasitism can induce changes in signalling traits of their hosts.

An additional signal that can play an important role in sexual selection is asymmetry. The degree of fluctuating asymmetry is thought to reflect the ability of an individual to cope with different types of stress (Parsons, 1990). The effect of flea parasitism on the level of fluctuating symmetry in their hosts has not been studied, but can be expected as it has been reported for other haematophagous ectoparasites. Møller (1993) experimentally manipulated the load of the tropical fowl mite *Ornithonyssus bursa* in the nests of the barn swallow *Hirundo rustica* in Denmark by either increasing or reducing the number of mites, or keeping nests as controls. The degree of fluctuating asymmetry for male tail length measured in the subsequent year after the swallows had grown new tail ornaments was larger at increasing levels of parasitism.

12.5   Host behaviour

Many parasites, especially those possessing complex life cycles that require transmission of an individual parasite between different hosts, alter host behaviour in exactly the same way as would be expected if a host was to behave in such a way as to guarantee successful transmission of a parasite. Numerous examples of such alterations have been reported (see Combes, 2001; Moore, 2002; Poulin, 2007a). Can we say anything about relationships between flea parasitism and host behaviour? It is highly unlikely that fleas manipulate host behaviour to ensure transmission. First, this is because of their *modus vivendi* as they occur mainly in the nests and/or burrows of the host which it has to visit at least from time to time. Second, this is because of their *modus operandi*, i.e. their ability to actively search for, locate and attack a host (see Chapter 9).

Nevertheless, flea parasitism does affect the behaviour of their hosts. In general, hosts have two (or three; see Chapter 13) lines of defence against parasites (Combes, 2001, 2005). The first line attempts to avoid the parasites by choosing appropriate behaviour, whereas the second line attempts to remove the parasites. In case of ectoparasites, this latter line consists of behavioural and immunological defence mechanisms. Obviously, flea parasitism may affect the first, purely behavioural, defence line and the behavioural component of the second defence line. These changes of host behaviour triggered by flea parasitism are associated with host resistance and will be considered in the next chapter.

Another type of behaviour alteration in a host under flea parasitism is associated with attempts by the host to compensate for the harm caused by fleas. In birds, the detrimental effect of fleas on the incubating female or nestlings (the cohort most easily exploited by nest-dwelling bird fleas) can be compensated by increased provisioning by male or both parents. Fitze et al. (2004b) demonstrated that in the presence of *C. gallinae*, the incubation and nestling periods of *P. major* in Switzerland were prolonged, thus leading to an increased parental effort.

Christe et al. (1996a) compared the foraging behaviour of parent *P. major* in clean and infested with *C. gallinae* nests in Switzerland and found that males increased the frequency of feeding trips by about 50%, whereas there was no change in females. Tripet & Richner (1997a) reported that parent *C. caeruleus* in nests infested by *C. gallinae* increased their rate of food provisioning by 29%. This issue was more extensively studied by Tripet et al. (2002b) who asked whether flea load affects the number of male feeds delivered to incubating females and the rate of food provisioning by parents to their nestlings. In Switzerland, they video-recorded the changes in frequency and duration of provisioning bouts in *C. caeruleus* from nests experimentally infested with low and high densities of *C. gallinae*. Male birds were not affected by flea density; they did not increase their frequency of food provisioning to incubating females or to nestlings in heavily infested nests. However, females increased the frequency of provisioning bouts. Similar results were reported by Bouslama et al. (2002) for blue tits in Algeria. As we already know, fleas, especially at high density, may reduce nestling body mass at an early age (e.g. Richner et al., 1993; Christe et al., 1996a; but see Thomas & Shutler, 2001). These costs can be partly or completely compensated for by an increase in feeding effort by adult birds. As was mentioned above, parent compensation may be one of the reasons for the lack of the relationship between flea load and nestling body mass reported in some studies (e.g. Shutler et al., 2004). Additional foraging efforts by female *C. caeruleus* were compensated for, in turn, by decreased duration of foraging bouts (Tripet et al., 2002b). Consequently, their time and energy budgets were not affected by flea density, and thus the cost of an increase in provisioning effort was probably not high.

The allocation of compensatory feeding effort between adult males and females differs among bird species and is probably determined by species-specific behavioural patterns. For example, an increase of foraging effort by male *P. major* might be related to a risk of being abandoned by a female (Christe et al., 1996a) because females usually divorce males if breeding success is low (Lindén, 1991). However, Fitze et al., (2004b) did not find an effect of fleas on the divorce rate in *P. major* following infestation. The divorce-risk explanation is also weakened by the fact that the divorce behaviour is also characteristic of *C. caeruleus* in which males did not provision more infested broods (see above).

Interestingly, nestlings parasitized by fleas stimulate their parents to increase provisioning effort by increasing begging behaviour. Christe et al. (1996a) reported that in Switzerland nestlings of the great tit in parasite-free nests called for their parents 18 min each hour, whereas nestlings in flea-infested nests did so for 39 min each hour. In contrast, Thomas & Shutler (2001) in Canada found no relationship between begging of *T. bicolor* nestlings and flea and blowfly loads in the nests. Furthermore, parental feeding rate was not affected by parasitism; these nestlings did not have higher mortality or reduced body size and body condition. The conclusion from this study was that nestling tree swallows were able to buffer the effects of naturally occurring ectoparasite loads without significant help from their parents.

Natal dispersal behaviour in birds was reported to be affected by flea parasitism. Moreover, contrasting patterns have been reported. For example, in Nebraska, nestlings of the cliff swallow *Petrochelidon pyrrhonota* that were heavily parasitized by *Ceratophyllus celsus* dispersed to non-natal colonies to breed in the subsequent year, whereas nestlings that were lightly parasitized returned to their natal colony to breed (Brown & Brown, 1992). In Switzerland, great tit nestlings infested by *C. gallinae* dispersed shorter natal distances than nestlings from uninfested broods (Heeb et al., 1999; Tschirren et al., 2007b). However, Fitze et al. (2004b) examining the same bird and flea species in the same region found that infested females dispersed longer distances than uninfested females, but that fleas did not affect male dispersal distance. In these cases, it is difficult to distinguish between the direct flea effect and host behavioural response to avoid fleas, that is, it is unclear whether flea-induced variation in dispersal distance should be considered as an effect of parasitism or as a behavioural defence mechanism of a host. For example, Heeb et al. (1999) suggested that by travelling shorter dispersal distances from infested broods, birds remain closer to their natal territory where they are likely to be better adapted or show higher tolerance to local parasites. This suggestion was recently supported by a study by Tschirren et al. (2007b) on the association between flea presence, dispersal distance and future reproductive success in *P. major* nestlings. They showed that relatively short dispersal distances in the presence of *C. gallinae* were indeed adaptive. The reproductive success of individuals originating from flea-infested nests decreased with increasing natal dispersal distance, while the opposite was the case for birds originating from flea-free nests.

12.6 Host survival

Host survival under parasitism is an important indicator of the parasite's virulence. Obviously, the detrimental effects of fleas on hosts described above are expected to be translated into lower chances of survival of infested individuals in comparison with uninfested individuals. A summary of some studies aimed at testing for the effect of fleas on survival of their hosts is given in Table 12.2.

One can observe from Table 12.2 that most studies concerned avian hosts, and only a few studies dealt with mammals. Furthermore, the effect of fleas on survival of adult hosts was found in some studies, but not in other studies. For example, Winkel (1975a, b) in Germany did not observe any effect of *C. gallinae* on survival in *P. major*, *Parus ater* and *C. caeruleus*. In contrast, Van Vuren (1996) demonstrated that in Canada highly parasitized *M. flaviventris* had lower overwinter survival. Lehmann (1992) analysed the survival probability of *G. andersoni* captured in the field in the Negev Desert as affected by *S. cleopatrae* infestation. He used logistic regressions and found that higher infestation predicted lower survival. Then, gerbil infestation level was manipulated by removal of all ectoparasites, including fleas, from individual hosts, to test whether the initial infestation level of the host and the number of fleas removed was related to survival. These analyses showed higher host survival as the number of fleas removed increased and lower survival as the initial number of fleas on the host increased.

In contrast to adult survival, a negative effect of fleas on the survival of young animals was detected in most investigations. For example, Fitze et al. (2004a, b) reported a negative effect of fleas on nestling mortality in *P. major* but explained that this occurrence was due mainly to complete nest failures rather than individual nestling mortality. Nevertheless, during ontogenesis, an individual often passes through very significant changes in morphology, physiology, ecology and behaviour. Some of these changes can be translated into changes in the exposure to infective stages of parasites, resistant ability and susceptibility to parasites (Gregory et al., 1992). In particular, acquired resistance to parasites is, by

Table 12.2 Results of investigations into the effect of flea parasitism on the host's survival

Flea	Host	Reduction of adult survival? (Y/N)	Reduction of offspring survival? (Y/N)	Reference
Ceratophyllus celsus	Petrochelidon pyrrhonota	N	N	Brown & Brown, 1992
C. celsus (?)	P. pyrrhonota	Y	—	Brown et al., 1995
Ceratophyllus gallinae	Cyanistes caeruleus	N	Y	Winkel, 1975a, b
	C. caeruleus	—	N	Tripet & Richner, 1997a
	Ficedula hypoleuca	—	Y	Eeva et al., 1994
	F. hypoleuca	—	Y	Merino & Potti, 1996
	Parus ater	N	Y	Winkel, 1975a, b
	Parus major	—	Y	Richner et al., 1993
	P. major	—	N	Eeva et al., 1994
	P. major	N	Y	Winkel, 1975a, b
	P. major	—	Y	Fitze et al., 2004a, b
	P. major	—	Y	Heeb et al., 2000
	P. major	—	N	Allander, 1998
	P. major	—	Y	Richner et al., 1993
C. gallinae (?)	C. caeruleus	—	N	Bouslama et al., 2001
Oropsylla stanfordi	Marmota flaviventris	Y	?	Van Vuren, 1996
Synosternus cleopatrae	Gerbillus andersoni	N	—	Lehmann, 1992
	G. andersoni	N	Y	Hawlena et al., 2006c
?	Spermophilus columbianus	—	Y	Neuhaus, 2003

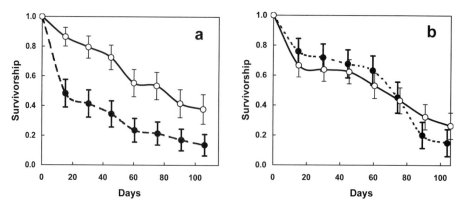

Figure 12.4 Survival curves of pyrethrin-treated (open circles, solid line) and untreated (closed circles, dashed line) (a) juvenile and (b) adult Anderson's gerbils *Gerbillus andersoni* from the day of pyrethrin application. Survivorship is the probability that a gerbil survives till day t. Vertical lines represent standard errors of the life-table estimation. Redrawn after Hawlena et al. (2006c) (reprinted with permission from Springer Science and Business Media).

definition, low in young host individuals. This may be one of the reasons for their high mortality under parasitism.

Hawlena et al. (2006c) studied age-related differences in survival under parasitism by ectoparasitic arthropods, the most abundant of which was the flea *S. cleopatrae*, in a free-living population of *G. andersoni* in the Negev Desert. Ectoparasite infestation on gerbils was manipulated in rodents from two 1.5-ha plots. Both juvenile and adult gerbils were assigned randomly to two treatments: (a) ectoparasites removed with pyrethrin and (b) untreated controls. A proportional hazards model was applied to test the effect of parasite removal on survival of individual hosts. It was found that average survival time of insecticide-treated juveniles was twice as high as untreated juveniles (61.4 ± 7.0 days versus 32.3 ± 7.1 days) (Fig. 12.4a). However, average survival times of insecticide-treated and untreated adults were similar (51.2 ± 7.0 days versus 54.7 ± 7.6 days) (Fig. 12.4b).

Studies of the same flea–host associations but from different regions or from the same region but different time periods gave contrasting results on survivorship (Table 12.2). This can be due to differences between studies in methods of manipulation and observation and/or methods of the evaluation of survivorship. In addition, the effect of fleas on host survival and parent compensation for the flea effect could be mediated by environmental factors. For example, parent and/or juvenile hosts could compensate for the negative effects of flea parasitism by increasing food intake (e.g. Tripet & Richner, 1997a) in years with abundant resources, whereas in years with scarce resource, this option would be much more difficult (Dufva & Allander, 1996).

12.7 Host fitness

From an evolutionary perspective, the most important issue related to the effect of flea parasitism on the host is whether fleas influence host fitness. In other words, whether the virulence of fleas, i.e. flea-induced loss of host fitness, is the net result of all the effects described above.

Loss of host fitness may result from a number of causes such as a decrease in the number of matings (including a decrease of mating attractiveness), shortening of the breeding period, decrease of litter/clutch size, loss of litters/clutches, mortality of offspring, shortening of lifespan etc. As we have already seen, flea parasitism may cause each of these misfortunes to happen. For example, the negative effect of fleas on the size of the black breast stripe in *P. major* (Fitze & Richner, 2002) may prevent an infested individual from achieving a successful mating. Oppliger et al. (1994) reported that in Switzerland great tits in flea-infested nests delayed egg laying by 11 days compared with birds in parasite-free nests. The reproductive success of parasitized tits was reduced by about 40%, as there was less opportunity for additional clutches and less time for rearing broods for birds from flea-infested nests (see Oppliger et al., 1994 for details). Interestingly, in the experiments of Fitze et al. (2004b) with the same flea–bird association from the same region as in Oppliger et al. (1994), the laying date of the first egg was not significantly different between infested and uninfested broods. However, during the nestling period, the proportion of deserted nestlings in flea-infested nests was significantly higher than in uninfested nests (Fitze et al., 2004b). The first and subsequent clutches of *P. major* nestlings infested by fleas during their youth were smaller than those of the nestlings reared in clean nests (Fitze et al., 2004a).

A negative effect of fleas on host fitness may also be caused by distortion of the host sex ratio. Heeb et al. (1999) in Switzerland tested whether *C. gallinae* affects gender-related recruitment of great tit fledglings. Using sex-specific DNA markers, they showed that flea infestation led to a higher proportion of male fledgling recruitment despite higher susceptibility of male nestlings to flea parasitism (Tschirren et al., 2003). However, the proportion of males in the brood at fledging differed neither between infested and uninfested broods nor from parity. Nevertheless, the proportion of male recruits differed in relation to flea infestation. A greater proportion of male fledglings was recruited from flea-infested broods, whereas recruitment was not gender-biased in uninfested broods. Thus, flea infestations resulted in bias in the proportion of male fledglings recruited, and could result in gender-related differences in dispersal (see above) and/or mortality of fledglings after leaving the nest.

A direct effect of flea parasitism on fitness of mammals has also been reported. Neuhaus (2003) in Canada observed that insecticide-treated female *S. columbianus* weaned 5.25 young whereas untreated females weaned 3.6 young per litter and the number of young surviving to yearling age was higher for treated than untreated females (3.5 versus 2.0 young per litter). The flea-induced juvenile mortality described by Hawlena *et al.* (2006c) may have important ecological consequences for the population of *G. andersoni* in the Negev Desert. When survival times of pyrethrin-treated and untreated juvenile gerbils were converted to survival probability, parasitism by *S. cleopatrae* led to a 48% reduction in survival, which influences the number of reproductive adults in the following year.

Some patterns that seem related to the influence of parasites can, in reality, be related to the host response aimed at minimizing the negative effects. For example, changes in the litter or clutch size under flea parasitism may result from manipulation of the litter/clutch size by the host itself. The duration of the life cycle of fleas is comparable to the duration of the nestling period of an avian host. Richner & Heeb (1995) suggested that in such a case, larger host broods should be favoured as parasites become increasingly diluted with an increase in brood size. Earlier experiments (Richner *et al.*, 1993) showed that great tits tended to lay more eggs in flea-infested than in flea-free nests. Moreover, female *P. major* that were experimentally infested with fleas after laying the second egg produced larger clutches than uninfested birds, although this was true in some years but not in others (Richner, 1998). Analogously, results of the experiments of Heeb *et al.* (1999) on sex-biased recruitment in tits under flea infestation were interpreted as modification of their sex-allocation decisions by hosts as a response to parasite infestations.

Fitze *et al.* (2004b) argued that the parasite-induced effect on the host's reproductive success is not only due to infestation, but is also related to the host's allocation of resources between current and future reproduction under parasite pressure. They noted correctly that understanding the parasite influence on host fitness requires the determination of the effects of parasites over the entire lifespan of a host. If parents adjust their reproductive effort to the reproductive value of their offspring (Møller, 1997), then hosts should reduce their investment in the presence of parasites and, thus, enhance future reproduction (Stearns, 1992). To assess the lifetime reproductive success of great tits parasitized by *C. gallinae*, Fitze *et al.* (2004b) manipulated the flea load in the bird nests over 4 years and evaluated the components of current and future reproduction on survival, divorce and breeding dispersal. Fleas led to a decline in current reproductive success due to an increase in the probability of nest failure, decrease of clutch size and decrease in the number of fledglings. An increase in parental effort by longer incubation and nestling periods compensated partly for parasitism, but current reproductive success was still reduced. From the perspective of future reproduction, females from infested nests dispersed over longer distances between breeding attempts, but the divorce rate and the probability of breeding locally in the future were not affected. In general, fleas reduced the lifetime reproductive success of birds estimated as the number of surviving and non-emigrating offspring over the entire period of observations. Two important conclusions resulted from this study. First, fleas have a direct negative effect not only on current reproductive parameters but also on lifetime reproductive parameters. Second, the trade-off between current and future reproduction is not affected by fleas as parasitized parents adjusted only to current but not future reproductive investment.

12.8 Concluding remarks

Flea parasitism has various negative effects on their hosts. Often, these effects are manifested differently, even within the same flea–host association. The reason for this variation may be the difficulty in determining the direct effect of fleas. The observed responses represent an interplay between flea effects and host compensatory and/or defence efforts. This interplay is further mediated by environmental factors.

Nevertheless, fleas generally harm their hosts. This suggests that the classical concept of a relationship between a parasite and a host evolving from an initial period of adjustment when the parasite causes harm to the host to a final period of coexistence when the parasite becomes benevolent to the host is incorrect (Poulin, 2007a).

As we have seen from the examples in this chapter, the majority of studies related to the effect of fleas on their hosts have been done using a model system of birds (mainly parids) and *C. gallinae* and artificial nest-boxes. Two points should be mentioned in this relation. First, tits and a hen flea are only a small sample from a huge variety of host–flea associations. Moreover, this association is likely to be evolutionarily young (see Chapter 4). Therefore, trends found to be true for this association are not necessarily true for other flea–host associations (e.g. Richner et al., 1993 versus Thomas & Shutler, 2001). Second, using nest-boxes in field experiments could introduce a bias in the results (Møller, 1989). For example, Wesołowski & Stańska (2001) studied flea infestation in the nests of tits and flycatchers in natural holes in Poland. They demonstrated that, in contrast to the studies that used nest-boxes, both the prevalence and intensity of infestation of nests by fleas were very low. However, in the same forest, flycatchers breeding in nest-boxes, and tits breeding in natural holes but within the area where nest-boxes were introduced, had significantly higher infestation rates. These observations indicated that the high flea loads reported for tits and flycatchers may be a product of biased sampling (data collection in nest-box areas) rather than reflecting natural conditions.

13

Ecology of host defence

A host is not a passive victim of a parasite but rather defends itself actively against the detrimental effects of parasitism. I have already mentioned two strategies of defence that can be implemented by a host: (a) it may attempt to avoid the parasite by choosing an appropriate habitat or patch; and (b) it may attempt to kill the parasite (Combes, 2001, 2005). The first strategy is purely behavioural, whereas the second strategy comprises behavioural, physiological and immunological defence tools. Moreover, anti-parasitic behaviour may include attempts to kill the parasite not only by using the host's own instruments such as beaks, teeth and claws but also by using extrinsic materials such as plants with insecticide or repellent properties. In other words, these are two lines of defence against parasites suggested by Combes (2001, 2005). In case of fleas, I advocate modifying this scheme and distinguishing three lines of defence. The first line is to avoid an encounter with fleas (implementing appropriate behaviour), the second line is to get rid of fleas by repelling or killing them (again, implementing appropriate behaviour), whereas the third line is to minimize the harm done by fleas (implementing the immune system). The boundary between the second and the third line is thus instrumental (what is the tool that a host uses) rather than ideological (what is the aim that a host attempts to achieve). For example, the immune response may not only decrease the amount of blood taken by fleas, but may also decrease flea reproductive success, thus diminishing the number of fleas that can attack the host in future.

In general, host resistance is defined sometimes as virulence against a parasite, since it induces in the parasite a host-mediated loss of fitness and even host-induced mortality (Combes, 2001). Therefore, analogously with virulence (see Chapter 12), resistance is a parameter used to characterize the host, but it should be measured in terms of parasite fitness. However, if the virulence of a parasite should be and usually is, indeed, measured via its host, resistance, although it should be measured via the parasite, is measured mainly via the host too. In other words, many resistance-related studies are strongly host-focused.

In this chapter I discuss how hosts defend themselves against fleas and how fleas cope with these defences. The discussion starts with flea-avoidance behaviour and then issues related to anti-flea behaviour and immunity are addressed.

13.1   First line of defence: avoidance

13.1.1   Avoidance in space

The simplest way for a host to minimize potential harm from a parasite is merely to avoid encounters with it. However, behavioural adaptations of hosts allowing them to avoid parasites are scarce. Combes (2001) suggested three reasons for this scarcity. First, the encounter between a parasite and a host is much more important for parasite than host because without an encounter, the fitness of a parasite is zero. Consequently, a parasite fights for its life. A host, however, often risks only diminished health or higher exposure to predators. Second, a host often has a chance rid itself of a parasite using the further lines of defence (e.g. immunity). Third, selection of behavioural traits to avoid a parasite is costly and parasite-avoidance behaviour may come into conflict with other fitness-related traits. Let's add two more reasons. It is not always possible to find a place free from parasites, especially if a host and a parasite have similar environmental preferences at least at certain stages of their life cycles. This is the case with fleas and, for instance, burrowing small mammals. Finally, and perhaps most importantly, successful parasite-avoidance behaviour requires the host to be able to predict the temporal and/or spatial distribution of parasites and to evaluate the risk of being attacked.

Despite these limitations, several examples of behavioural responses by animals to avoid flea attack have been reported. Oppliger et al. (1994) and Christe et al. (1994) examined the effects of the hen flea *Ceratophyllus gallinae* on nest-site choice in the great tit *Parus major* in Switzerland. They tested whether birds prefer a clean nest-box to a box containing fleas. After a breeding season, they provided tits with a pair of nest-boxes located at a short (0.3–1 m) distance from each other; one of the boxes contained an old, but flea-free nest while another was infested with 20 fleas. When the birds started nesting, 18 of 23 tit pairs preferred parasite-free boxes.

In Sweden, Merila & Allander (1995) also offered *P. major* a choice between a clean nest-box and a nest-box containing an old nest of either *P. major* (area A) or the collared flycatcher *Ficedula albicollis* (area B). Great tits preferred to nest in empty boxes in area A but not in area B. Nevertheless, birds that were forced to start their breeding in boxes with old nests chose those boxes where *C. gallinae* numbers were low.

Experiments of Mappes *et al.* (1994) showed that the pied flycatcher *Ficedula hypoleuca* in Finland preferred nest-boxes with old nest material from the previous year over clean nest-boxes. This choice of the 'dirty' nest-boxes is rather surprising. However, it is possible that the benefit from a 'prepared lodging' may be greater than the cost of defending against fleas as this energetic cost may be lower than that of constructing a new nest. This can be especially important in northern regions where the breeding period is relatively short.

Olsson & Allander (1995) asked whether *F. hypoleuca* and three tit species (*P. major*, *Cyanistes caeruleus* and *Poecile palustris*) in Sweden discriminate between empty nest-boxes and nest-boxes containing either clean or *C. gallinae*-infested old nests. Results supported Orell *et al.* (1993) and Mappes *et al.* (1994) in that flycatchers preferred nest-boxes containing old nests, regardless of the occurrence of fleas, but did not support results of Oppliger *et al.* (1994) in that tits did not discriminate between the three types of boxes. However, Rytkönen *et al.* (1998) reported that in Great Britain *P. major* avoided nest-boxes infested with fleas. In Canada, the tree swallows *Tachycineta bicolor* preferred empty and clean nest-boxes or those where the old material had been microwaved over nest-boxes with old, untouched material (Rendell & Verbeek, 1996). However, empty nest-boxes and those with microwaved nest material were more spacious, so the results suggested that either swallows avoided potentially high numbers of parasites in nests with old material or they preferred larger cavities or both.

In an elegant 2-year experiment in Canada, O'Brien & Dawson (2005) tested whether birds are capable of assessing early cues that predict future risk of flea parasitism and adjust their primary reproductive investment in response to the anticipated energetic costs of combating fleas. They presented a visual cue of *Ceratophyllus* fleas on the outer surface of nest-boxes of *T. bicolor* without direct exposure of birds to the parasites. This experimental design examined behavioural responses without physiological effects of fleas on bird reproduction. During one of the study years, birds preferentially occupied control boxes; however, across two study years, birds nesting in treatment nest-boxes produced smaller clutches than birds in control nest-boxes. This difference indicated that for these birds, the perception of future risk from parasites might be sufficient to induce a reduction in reproductive investment early, before flea numbers increased.

It is apparent that birds have the ability to discriminate between infested and uninfested nest sites. This ability seems to be based on the occurrence of the old nest material that is usually an indicator of a risk of fleas. Birds thus are able to select their strategy and can therefore either avoid fleas by choosing a clean site or confront fleas by reusing the old nest. Weak manifestation of selectivity and a 'wrong', at first glance, choice do not, however, vote for the bird's lack of 'intellect'. Rather, this selection may depend on a variety of factors such as the level of detrimental effect of a parasite and length of a breeding period of a bird in a given geographical area. Sometimes, it may be advantageous for a bird to select an old nest even with fleas over the flea-free nesting site because (a) nest reuse may be less costly than nest-building (Cavitt *et al.*, 1999), and (b) old nests might serve as cues for nest-site selection (Thompson & Neill, 1991; Mappes *et al.*, 1994; Olsson & Allander, 1995).

Another way to avoid flea parasitism is to relocate to a clean or, at least, cleaner nest or burrow. For example, European badgers *Meles meles* switched their preferred sleeping sites in response to ectoparasite (mainly the flea *Paraceras melis* and the biting louse *Trichodectes melis*) accumulation (Reichman & Smith, 1990; Butler & Roper, 1996; Roper *et al.*, 2001). Roper *et al.* (2002) in South Africa studied Brants' whistling rats *Parotomys brantsii* which inhabit complex burrow systems containing several nest chambers. The effect of ectoparasites, mainly fleas *Xenopsylla eridos*, on the choice of alternative nest chambers was investigated before and after treating them to remove ectoparasites. Prior to anti-flea treatment, animals used several different nest chambers within a single burrow, moving from one chamber to another every 1.6 day on average, and most individuals slept in more than one burrow. Anti-flea treatment reduced the rate at which animals switched from one nest chamber to another within a burrow. Thus, whistling rats reduced the rate of flea accumulation by switching periodically from one nest chamber to another.

13.1.2 Avoidance in time

Beside selection of a resting or reproduction-related location where attack by fleas is least probable, a host can select a time of use of this location when the chances of encountering the fleas are low. Great tits in Switzerland delayed egg-laying by 11 days in a territory containing flea-infested nest-boxes, which is in agreement with the hypothesis that birds take the probability of flea attacks into account when starting to breed (Oppliger *et al.*, 1994). The rationale of this hypothesis is that if a nest is not occupied, then emerged fleas will either die or emigrate and, eventually, the flea load will become much lower.

This seems to be an ad-hoc explanation as 11 days is a short timespan compared with the period that *C. gallinae* wait in nests for birds to come to breed. It is difficult to imagine that if there are two flea-infested nests, and birds start to breed in one of these nests immediately with the beginning of reproductive period and in another after 11 days, the difference in the flea effect on the two broods will be considerable. It is also doubtful whether the benefit of 11 days' delay in egg-laying, even if this reduced flea parasitism, would be higher than the cost of shortening the breeding period. No experimental testing of this hypothesis has been carried out. However, the possibility of a strategy to adjust the timing of reproduction or any other activity in such a way that the chances of an encounter with fleas would be minimal cannot be ruled out.

13.2   Second line of defence: repelling or killing fleas

13.2.1   Managing nest infestation

If a host can not avoid an encounter with fleas, it should attempt to rid itself of them. As we know, fleas alternate periods when they occur on the body of a host and when they occur in its nest or burrow. Consequently, a host should attempt to eliminate fleas that either reside in its nest/burrow or/and on its body.

The simplest way to get rid of those flea species that spent most of their time in the nest/burrow and visit host body for the sake of a blood meal is nest sanitation. In Switzerland, Christe *et al.* (1996b), defining nest sanitation as a period of active search with the head dug into the nest material, manipulated the density of *C. gallinae* in nests of *P. major* during the nestling period. Males did not perform nest sanitation activities regardless of the infestation status of a nest. During daytime, females in infested nests allocated more time to nest sanitation than those in flea-free nests. At night, nest sanitation increased significantly from 8% of total time in flea-free nests to 27% in infested nests with mean duration of a nest sanitation bout being significantly longer in flea-infested nests (41.3 s versus 28.4 s). Tripet *et al.* (2002b) carried out a similar study with the blue tit *C. caeruleus* during both the incubation and nestling periods. Female blue tits with higher flea densities increased the frequency but decreased the duration of nest sanitation bouts. It remained unknown whether during nest sanitation tits caught fleas or whether fleas were killed during a bird's grooming. Nest sanitation could simply result in chasing fleas away and thus preventing them from biting the incubating female and nestlings. Sanitation behaviour might also force flea larvae to pupate in cooler parts of the nest where their development would be slowed down (Tripet & Richner, 1999b) as well as destroy flea cocoons. Whatever the exact mechanism was, nest sanitation might reduce flea reproduction.

Nest-sanitation behaviour is also characteristic for some mammalian hosts. For example, European badgers renew their bedding material periodically, 'airing' the bedding by bringing it to the surface and exposing it to sunlight (Neal & Roper, 1991). Removing old bedding from burrows and replacing it with new material in *P. brantsii* has been suggested to be a strategy of ectoparasite (mainly, fleas) elimination (Jackson, 2000).

Birds are known to incorporate fresh herbs into their otherwise dry nest material. The idea that this green plant material (preferably from plants rich in volatile compounds) is used as a bactericide, insecticide or repellent dates back to 1922 (Widmann, 1922) and is still implied as an explanation (Banbura *et al.*, 1995; Roulin *et al.*, 1997; Petit *et al.*, 2002). Analogous behaviour has been recorded for some mammals. For example, badgers incorporate into their nests plant material that may have biocidal or fumigant properties (Hart, 1990).

Very few experimental studies have been done to investigate this phenomenon. For example, wild carrots put into the nests of European starling *Sturnus vulgaris* in Philadelphia during the nestling period inhibited the development of mites (Clark & Mason, 1988). In Germany, Gwinner & Berger (2005) exchanged natural starling nests with experimental nests with or without fresh green material from six plant species and found that, although nestlings from nests with herbs fledged with higher body mass than nestlings from nests without herbs, herbs had no effect on nest infestation by mites. The effect of fresh green material was thus associated with its bactericidal rather than insecticidal activity. Lambrechts & Dos Santos (2000) proposed a new 'potpourri' hypothesis which stated that birds introduce various fresh plants into their nest to produce a cocktail of strong odours which increases the efficiency of defence against one or more parasite and/or pathogen species. This assumes that a mixture of different strong-smelling odours produced by the different herb species is more efficient in defence than the odour of each of the plant species separately.

Hemmes *et al.* (2002) tested the hypothesis that dusky-footed wood rats *Neotoma fuscipes* in California placed bay leaves (*Umbellularia californica*) in or near their sleeping nests as a fumigant against fleas. Bay leaves were nibbled by *N. fuscipes* in a fashion consistent with the release of fumigating volatiles. Samples of whole or torn (to simulate nibbling effects) bay leaves were incubated with larvae of *Ctenocephalides felis* for 72 h. Torn leaves reduced larval survival to 26%, providing support for 'nest fumigation' hypothesis. It should be noted that *N. fuscipes*, as a rule, is not naturally infested with *C. felis* and that this flea was used in the experiments merely because it is the only species commercially available in large numbers. Nevertheless, it is likely that torn (or nibbled) bay leaves have the same effect on the larvae of fleas that are characteristic for *N. fuscipes* (mainly, *Orchopeas* and *Anomiopsyllus*). In contrast, results of Dawson (2004) did not support the 'nest fumigation' or 'nest protection' hypothesis. He experimented with *T. bicolor* in Canada by adding green leaves of yarrow *Achillea millefolium* to nests. Nests with yarrow had significantly higher levels of infestation by *Ceratophyllus idius* than controls. However, experiments of Shutler & Campbell (2007) with yarrow, *T. bicolor* and *C. idius* in Canada provided strong support for the 'nest protection' hypothesis. An average of 773 fleas was found in 44 control nests versus 419 in 23 nests where yarrow leaves were introduced. Experimental evidence on the potential effect of some plant material introduced by a host to its nest or burrow against fleas is equivocal. In addition, it is thought that introducing green material into nests by birds is aimed at attracting mates rather than to repelling or killing parasites (e.g. Fauth et al., 1991; Brouwer & Komdeur, 2004). This issue is worth further investigations.

Finally, Haftorn (1994) described a peculiar tremble–thrusting behaviour performed by females of some tit species. During this action a female stands in the nest cup and pokes her bill deep into the sides or bottom of the cup. At each poke, she makes series of rapid and vigorous twisting movements of a head. She often starts poking close to the nest rim and successively moves deeper down into the nest cup, eventually attaining an extreme vertical position. During vertical poking, she virtually penetrates the nest with the bill. It was suggested that tremble–thrusting keeps ectoparasites, especially fleas, away from the nest cup.

13.2.2 *Managing body infestation*

One of the most frequently and regularly performed defence behaviours in mammals and birds aimed at removing and/or killing ectoparasites, including fleas, is grooming (Hart, 1990; Clayton & Cotgreave, 1994; Moore, 2002). Auto- and allogrooming serve a number of purposes such as body temperature regulation (e.g. Nijssen, 1985), forming and maintaining social relationships (Palombit et al., 2001), reduction of tension (Waeber & Hemelrijk, 2003) and communication (e.g. Ferkin et al., 1996) in animals living in groups. An anti-parasitic effect is characteristic mainly of autogrooming, although allogrooming can also play a role in the removal or scaring away of ectoparasites (B.-G. Li et al., 2002).

On the one hand, grooming and scratching can be a response to the irritating bites of ectoparasites (e.g. Alexander, 1986). Indeed, excessive grooming has been be used to identify a flea-infested cat (Scheidt, 1988). On the other hand,

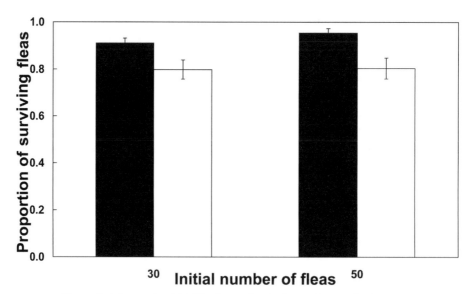

Figure 13.1 Proportion (±S.E.) of *Parapulex chephrenis* that survived on the grooming-restricted (black columns) and free grooming (white columns) Egyptian spiny mice *Acomys cahirinus*. Data from B. R. Krasnov and M. Leiderman (unpublished).

anti-parasitic grooming is considered to be efficient and even to be a major host-originated factor of ectoparasite mortality (Marshall, 1981a; Hinkle *et al.*, 1998).

In general, the efficiency with which hosts remove fleas by grooming is related to flea and host species as well as some factors such as flea density and host age. Host individuals and species with higher manipulating ability seem to be more effective groomers than hosts with lower ability. Wade & Georgi (1988) demonstrated that some cats were able to remove 50% of *C. felis* in a week by grooming. A high ability of cats to remove fleas was also shown by Eckstein & Hart (2000b). Collars which prevented grooming were fitted on nine cats in a flea-infested household and, 3 weeks later, were compared with nine control cats in the same household. Flea numbers dropped in the control cats but not in the collared cats, on which fleas were about twice as numerous as in the control cats. In other mammalian species, the anti-flea effect of grooming was not particularly high. For example, the Egyptian spiny mouse *Acomys cahirinus* that were allowed to groom succeeded in removing a low proportion of *Parapulex chephrenis* (B. R. Krasnov and M. Leiderman, unpublished data) (Fig. 13.1). Nikitina & Nikolaeva (1979) infested the common vole *Microtus arvalis*, bank vole *Myodes glareolus* and fieldmouse *Apodemus agrarius* with *Nosopsyllus fasciatus*, *Ctenophthalmus uncinatus* and *Ctenophthalmus agyrtes* and found that *A. agrarius* was the most efficient groomer, whereas *M. arvalis* was the least efficient groomer. Zhovty & Vasiliev (1962b) mentioned among-host variation in grooming efficiency against any flea species and demonstrated high grooming abilities in the dwarf hamster *Phodopus sungorus* and low grooming abilities in the Daurian ground squirrel *Spermophilus dauricus*.

Hinkle et al. (1998) studied grooming efficiency of cats against fleas by infesting cats with various numbers of *C. felis*, then collecting the cat faeces and extracting the fleas to determine how many had been groomed off. Inter-individual variation in grooming efficiency was found. The best groomer removed 17.6% of its fleas daily, whereas the poorest groomer removed only 4.1%. Cats were most efficient at grooming fleas at infestations of either <50 or >150 fleas. Hinkle et al. (1998) suggested that at high flea densities, the irritation from bites stimulated intense grooming, whereas at low infestation levels, removal of even small numbers was a substantial proportion of the entire population. At median infestation levels, the stimulus to groom was reduced.

Host age and, sometimes, host gender have been shown to influence grooming efficiency. Several authors have explained higher level of infestation in young animals as being due to less efficient grooming (Buxton, 1948; Brinck-Lindroth, 1968; Cotton, 1970b). For example, Osbrink & Rust (1985) demonstrated that younger cats were less effective groomers than older cats in terms of grooming efficiency and frequency. Chandy & Prasad (1987) studied the effect of grooming behaviour in laboratory rats and mice and reported that flea mortality was higher on adult and female than on young and male rodents.

Lower grooming efficiency of younger animals is, however, not always the case. Hawlena et al. (2007b) compared the efficiency of anti-flea grooming in adult and juvenile (1 month old) Sundevall's jirds *Meriones crassus* that were experimentally infested with *Xenopsylla conformis*. Of 37 adult and 37 juvenile rodents, 24 of each were infested with fleas (parasitized groups); the remaining 13 adult and 13 juvenile animals served as control groups. Each experiment trial lasted for two consecutive nights; during the first night no fleas were introduced, whereas during the second night rodents from the parasitized groups were infested with fleas. Initial flea densities per unit of host's body surface (calculated as the number of fleas divided by the host body mass to the power of 0.67: Hawlena et al., 2005) were equal for adult and juvenile hosts (4.63 fleas $g^{-0.67}$) and within their natural infestation range. Behaviour of the rodents was recorded during the night hours. At the end of each hourly observation one rodent was randomly chosen, fleas were brushed off its body and its cage was checked for fleas. Mortality probability per flea was estimated as the ratio of the

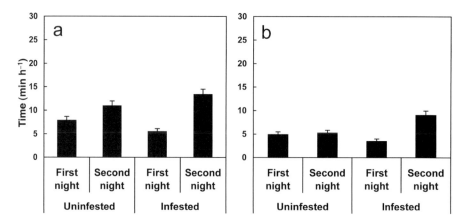

Figure 13.2 Mean (±S.E.) time (minutes per hour) spent grooming by flea-free and flea-infested (a) adult and (b) juvenile Sundevall's jirds *Meriones crassus*. Rodents were infested only on the second experimental night. Data from Hawlena et al. (2007b).

number of missing fleas (initial minus final flea numbers) to the initial number of fleas. Both infested and uninfested rodents spent more time grooming during the second night of the experiment, but the increase was significantly higher in the former than in the latter (Fig. 13.2).

Furthermore, adults spent more time grooming than juveniles (Fig. 13.2). Total time spent grooming correlated strongly with the mortality probability per flea which indicated that grooming was efficient at killing fleas (Fig. 13.3). Juveniles killed fleas at higher rates than did adults (Fig. 13.3). However, the overall (within a trial) mortality probabilities per flea on adult and juvenile rodents were similar (0.28 ± 0.04 versus 0.34 ± 0.05). As a result, flea densities on juveniles and adults at the end of a trial were similar (3.1 ± 0.25 fleas $g^{-0.67}$ and 3.3 ± 0.19 fleas $g^{-0.67}$). Although adults spent more time grooming, they did not remove more fleas and, thus, were less efficient in killing fleas than juveniles. These results support the idea that ectoparasites evoke grooming activity. Furthermore, in this flea–host association, grooming was the major cause for flea mortality and explained 57% and 70% of the variation in adult and juvenile hosts, respectively. Anti-parasitic behaviour of the host may therefore explain why fleas survived only 7–14 days when feeding on hosts, whereas they survived for months with no food when separated from the host (Tripet & Richner, 1999a, b).

Age-related differences in grooming activity may be related to the body-size principle (Mooring et al., 2000, 2002, 2004). Smaller individuals should groom harder in order not to allow too many parasites per unit body surface. Although

Second line of defence: repelling fleas 249

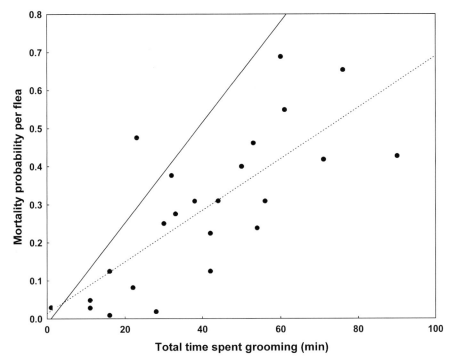

Figure 13.3 Relationships between total time spent grooming (minutes) and mortality probability per flea *Xenopsylla conformis* in adult (dashed line) and juvenile (solid line) Sundevall's jirds *Meriones crassus*. Redrawn after Hawlena et al. (2007b) (reprinted with permission from Blackwell Publishing).

this differential grooming with age has been reported for sheep, goats and antelopes (Mooring et al., 2000; Hart & Pryor, 2004), juvenile gerbils actually spent less time grooming than adults while imposing a similar risk of death on fleas (Hawlena et al., 2007b). The discrepancy between rodents and ungulates may be due to age-specific differences in grooming efficiency that exist in the former but not in the latter. For example, the fur of juvenile rodents is sparser than that of adults (Webb et al., 1990), and therefore fleas are more susceptible to grooming in juveniles. This allows juveniles to spend less time grooming yet still be efficient in killing or scaring away fleas. Nevertheless, as juvenile rodents suffer more mortality than adults under the same flea density (see above; Hawlena et al., 2006c), natural selection should favour juveniles that groom themselves until flea density falls below the average density in adults. The fact that juvenile rodents spent less time grooming than adults and sustained similar flea densities may indicate that juvenile hosts are constrained by other factors.

Table 13.1 *Anti-flea grooming efficiency (% fleas removed during 24 h) in five rodent species at different air temperatures*

Host	Flea	Ambient temperature (°C)		
		10	20	30
Spermophilus undulatus	*Oropsylla silantiewi*	4.0	8.0	9.0
	Neopsylla bidentatiformis	4.0	6.0	6.0
	Citellophilus tesquorum	14.0	10.0	5.0
	Ctenophthalmus assimilis	15.0	6.7	23.8
Phodopus sungorus	*Monopsyllus anisus*	16.0	18.0	38.0
	Frontopsylla luculenta	63.0	27.0	9.0
	N. bidentatiformis	14.0	5.0	16.0
	Neopsylla pleskei	14.0	10.0	14.5
	C. tesquorum	39.7	11.0	12.0
	C. assimilis	42.5	20.0	23.8
	Xenopsylla cheopis	16.0	18.0	12.7
Meriones unguiculatus	*C. tesquorum*	19.0	4.0	0.0
	N. bidentatiformis	12.0	2.0	10.0
	N. pleskei	11.5	9.0	8.0
	F. luculenta	24.0	6.0	31.0
	X. cheopis	6.0	16.0	13.0
Mus musculus (lab)	*M. anisus*	5.0	17.0	37.0
Rattus norvegicus (lab)	*X. cheopis*	4.0	12.0	2.8

Source: Data from Zhovty & Vasiliev (1962b).

Among external factors that could affect the efficiency of the anti-flea grooming, only air temperature has been studied (Zhovty & Vasiliev, 1962b). The direction of this effect varied in relation to both host and flea species (Table 13.1). Some hosts decreased the efficiency of their grooming against one flea under higher temperature but increased this efficiency against another flea at the same conditions. In other flea–host associations, the grooming efficiency of a host either peaked or declined at median air temperatures. The reasons for this variation are still unclear.

13.2.3 Neurophysiology of anti-parasitic grooming

There are two main non-mutually exclusive models for the neurophysiological regulation of anti-parasitic grooming. The first model, programmed grooming, assumes a type of central programming that periodically evokes a bout of grooming in order to remove ectoparasites before they are able to attach (Hart et al., 1992; Mooring, 1995). According to this model, a host is expected to groom on a regular basis even in an ectoparasite-free environment (Hart et al., 1992). Moreover, the host who grooms the most will have the lowest number of parasites, and thus negative relationships are expected between grooming rate and ectoparasite load. Finally, it is assumed that programmed grooming evolved to balance the costs of parasites against the costs of grooming. These costs may vary among host individuals depending on body size, reproductive status and ecological conditions, and therefore the programmed grooming rate is expected to be affected by these variables (Hart et al. 1992). The second model, stimulus-driven grooming, postulates a peripheral mechanism that is a direct response to cutaneous irritation from ectoparasite bites (Wikel, 1984). According to this model, in the absence of a stimulus an animal is not expected to groom (Alexander, 1986). Similarly, the animal with the lowest number of ectoparasites experiences the fewest stimuli, and thus will groom the least (Hart et al., 1992). Finally, differences in stimulus-driven grooming rate between individuals should reflect only variations in infestation level and thus should not be related to body size, reproductive status or ecological conditions.

Numerous observations on ungulates infested mainly with ticks have suggested that programmed grooming predominates in the natural environment (Mooring et al., 2004). Hawlena et al. (H. Hawlena, B. R. Krasnov & Z. Abramsky, unpublished data) attempted to disentangle the two mechanisms experimentally using M. crassus and X. conformis as a model host–ectoparasite association. The behaviour of adult and juvenile flea-infested and flea-free rodents was recorded and the duration and frequency of scratch-grooming (fast movement of the extremities, usually the hindpaws, toward a single region of the body; suggested to be regulated by stimulus-driven control) and scan-grooming (fast paw movements and mouthing which has a clear cephalocaudal sequence, starting with nose wipes by the forepaws and ending with ano-genital and tail grooming; suggested to be regulated by programmed control) were quantified. It was expected that in the case of programmed grooming: (a) flea-free rodents groom; (b) flea infestation increases the grooming frequency but not the duration; and (c) juvenile rodents groom more and sustain lower flea densities than adults because the expected gain in removing fleas would outweigh the costs of grooming in juveniles but not in adults (the body-size principle). Expectations from the stimulus-driven grooming model were that: (a) no grooming activity occurs in flea-free rodents; (b) flea infestation increases grooming duration, directed to specific body parts experiencing flea bites; and (c) juvenile and adult rodents invest similar time in grooming. The first two predictions of the programmed grooming hypothesis were confirmed. Rodents groomed even prior to flea infestation. Introduction of fleas increased the frequency of scan-grooming bouts without affecting duration. This suggested that programmed grooming rather than stimulus-driven grooming prevailed in this rodent–flea association. However, in contrast to the programmed grooming model, fleas induced further bouts of scratch-grooming, and the body-size principle was not confirmed in that juvenile rodents scratch- and scan-groomed less and for less time than adults.

The contradiction of this result to data on size-related differences in the grooming rates in sheep, goats and antelopes, which had strongly supported the body-size principle (Mooring & Hart, 1997; Mooring et al., 2002; Hart & Pryor, 2004), remains to be investigated. It may be associated with lower difference in surface-to-mass ratio between juvenile and adult rodents or higher constraints by feeding or sleeping hours in juvenile rodents compared with those in ungulates. Nevertheless, it seems that anti-parasitic grooming is well programmed and regulated in a similar ways across different mammalian species, although interspecific variation in host morphology (i.e. fur density, differences in the size of the grooming-operating organs, manipulation abilities) may require a more species-specific consideration in generating the predictions from the programmed grooming model.

13.2.4 Trade-offs of the behavioural defence against fleas

Despite its benefits, anti-parasitic behaviour has considerable costs. Grooming may incur energetic costs (Giorgi et al., 2001), distract from vigilance for predators (Mooring & Hart, 1995), facilitate microparasite and pathogen transmission (Heitman et al., 2003), entail thermoregulatory costs of winter hair loss in cold environments (Mooring & Samuel, 1999), cause loss of water via saliva (Ritter & Epstein, 1974) and even damage dental elements (McKenzie, 1990). The most important cost of grooming is that it requires a substantial amount of time that could otherwise be used for other fitness-related activities. For example, Kern (1993) found that cats spent an average of 20.5% of their time budget in grooming, although Rust (1992) estimated that cats spent only 5% of their time grooming.

The time budgets of infested and uninfested animals indicated that they often face a trade-off between different activities. In Switzerland, female great tits in nests infested by *C. gallinae* reduced their sleeping time significantly (73.5% versus 48.1%) (Christe et al., 1996b), with the time deducted from sleeping being allotted mainly to nest sanitation rather than to grooming. In contrast, female *C. caeruleus* with a high density of *C. gallinae* did not allocate more time to anti-parasite

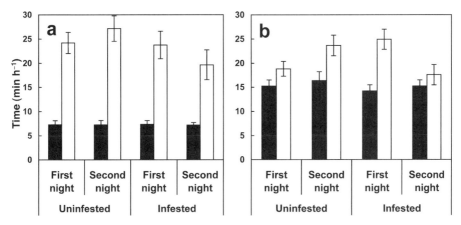

Figure 13.4 Mean (±S.E.) time (minutes per hour) spent feeding (black columns) and resting (white columns) by flea-free and flea-infested (a) adult and (b) juvenile Sundevall's jirds *Meriones crassus* during the first and second experimental nights. Rodents were infested only on the second experimental night. Data from Hawlena et al. (2007b).

behaviour than conspecifics with low density of fleas (Tripet et al., 2002b). In the experiments of Hawlena et al. (2007b) with *M. crassus* and *X. conformis*, the increased time in grooming by infested rodents (Fig. 13.2) came at the expense of resting but not feeding (Fig. 13.4). *Acomys cahirinus* infested by *P. chephrenis* allocated 62.5% more time to grooming, 25% less time to eating and 21% less time to sleeping than uninfested animals (Fig. 13.5) (M. Leiderman and B. R. Krasnov, unpublished data).

These examples show that behavioural defences against fleas are generally costly in terms of time. Time cost may further be translated into other costs. For example, some hosts sacrifice resting for anti-flea defence. However, rest deprivation may have both short-term (reduced foraging efficiency at the activity period) and long-term (decrease in survival and overall reproduction value) negative effects on a host (see Christe et al., 1996b). Sleep deprivation may also suppress the immune function (Öztürk et al., 1999). In other cases, hosts may not be able to increase the amount of time allocated to anti-flea defence. For example, during incubation female *C. caeruleus* were constrained in their nest-cleaning behaviour by the amount of disturbance that eggs can sustain (Tripet et al., 2002b). During the nestling period, time that birds allocated to anti-flea behaviour competed strongly with time allocated to feeding offspring (Christe et al., 1996b). Furthermore, responses to flea parasitism in terms of time budget can also be mediated by environmental conditions. Hosts occupying

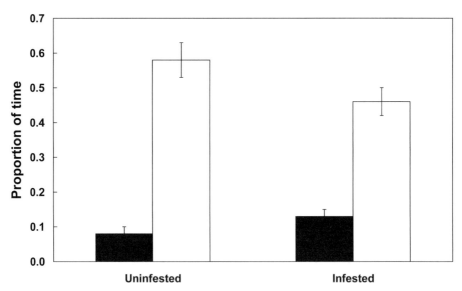

Figure 13.5 Mean (±S.E.) proportion of time spent grooming (black columns) and resting (white columns) by flea-free and flea-infested Egyptian spiny mice *Acomys cahirinus*. Data from M. Leiderman and B. R. Krasnov (unpublished).

high-quality habitats (e.g. food-rich) may not increase the time allocated to defence against parasites if food can compensate for the extra costs (Tripet & Richner, 1997a). In low-quality habitats, however, an increase in anti-parasite defences may reduce the costs of parasitism more efficiently (Christe *et al.*, 1996b; Tripet *et al.*, 2002b).

13.2.5 *Withstanding host grooming*

Nikitina & Nikolaeva (1981) infested four rodent species (*Mus musculus*, *Lagurus lagurus*, *M. arvalis* and *M. glareolus*) with five species of fleas (*Leptopsylla segnis*, *N. fasciatus*, *Amalaraeus penicilliger*, *C. uncinatus* and *Ctenophthalmus assimilis*) and observed the rodents' grooming efficiency. Rodents successfully removed all other fleas except for *L. segnis*, suggesting variation among flea species in their ability to withstand host grooming, which in turn stems from the among-flea variation in behaviour and morphology. From a behavioural perspective, the distribution of fleas across the body of a host is often characterized by preference for those body areas that are the least reachable by the paws or teeth of a host (see Chapter 9). From a morphological perspective, it was previously mentioned that helmets, ctenidia, spines and setae allow fleas to attach to the host's hairs and to resist the host's grooming (see Chapter 4). The most fascinating and comprehensive descriptions of the weaponry that fleas use to enhance their attachment to the host body and avoid being dislodged either by host grooming effort or the host environment can be found in Traub (1980, 1985).

Indeed, *L. segnis*, which rodents in the experiments of Nikitina & Nikolaeva (1981) were unable to remove, is usually found on the skull vertex and the middle of the back of a rodent, i.e. areas that are difficult to reach by a rodent's limbs (Traub, 1980). Moreover, *Leptopsylla* species possess a 'crowns of thorns' (Traub, 1980, 1985) which allows them to hook onto the hairs of the host and remain in this position for several hours or even days (Farhang-Azad *et al.*, 1983; Traub, 1985).

Zhovty & Vasiliev (1962b) also demonstrated that grooming efficiency depends mainly on some characteristics of a flea species. They reported that the Mongolian marmot *Marmota sibirica* during 24 h successfully removed 37.6% of *Neopsylla pleskei*, 20.0% of *Oropsylla silantiewi* and *Citellophilus tesquorum*, 16.5% of *Neopsylla bidentatiformis*, 15.0% of *Frontopsylla luculenta*, 4.0% of *A. penicilliger*, and only 2.9% of *C. assimilis*. Zhovty & Vasiliev (1962b) suggested that the 'aggressiveness' of flea species is what matters, as high 'aggressiveness' triggers a vigorous host response. They argued that the ability of the flea to withstand grooming and the grooming capabilities of the host species play only secondary roles.

Finally, there is no common opinion on the relationships between the grooming efficiency of a host and the level of specificity of a flea to that host. Dubinina & Dubinin (1951) infested five rodent species with specific and non-specific fleas, and argued that hosts were more efficient in removal of the latter than the former, although their results are not especially convincing (Table 13.2). Furthermore, the experiments carried out by Zhovty & Vasiliev (1962b) did not support this conclusion but suggested quite the opposite (Table 13.2).

13.3 Third line of defence: immune response against fleas

The immune system is aimed at discriminating between 'self' and 'non-self', and at minimizing the consequences of contact with foreign molecules introduced into the host by feeding parasites. The immune defence mechanisms of vertebrates comprise two components, innate and acquired (adaptive) immunities that develop during evolutionary (innate immunity) or ontogenetic (acquired or adaptive immunity) time (Janeway *et al.*, 1999). Generally, innate immunity is non-specific, whereas acquired immunity is an induced response characterized by 'specificity' for the sensitizing immunogen and memory. It is acquired immunity that is believed to play a major role in the host developing resistance to parasites (Wakelin, 1996).

Table 13.2 *Anti-flea grooming efficiency (% fleas removed during 24 h) in nine rodent species in relation to flea specificity (S — specific, NS — non-specific)*

Host	Flea	Specificity	Efficiency	Reference
Marmota sibirica	*Oropsylla silantiewi*	S	48.0	Dubunina &
	Frontopsylla luculenta	NS	26.0	Dubinin, 1951
	Citellophilus tesquorum	S	40.0	
	Amphalius runatus	NS	20.0	
	Ceratophyllus gallinae	NS	36.0	
	Neopsylla bidentatiformis	NS	30.0	
Spermophilus dauricus	*O. silantiewi*	NS	31.0	Dubunina &
	F. luculenta	S	26.0	Dubinin, 1951
	C. tesquorum	S	43.0	
	A. runatus	NS	36.0	
	C. gallinae	NS	27.0	
	N. bidentatiformis	S	30.0	
Ochotona daurica	*O. silantiewi*	NS	21.0	Dubunina &
	F. luculenta	S	41.0	Dubinin, 1951
	C. tesquorum	NS	38.0	
	A. runatus	S	29.0	
	C. gallinae	NS	37.0	
	N. bidentatiformis	NS	27.0	
Microtus gregalis	*O. silantiewi*	NS	16.0	Dubunina &
	F. luculenta	S	32.0	Dubinin, 1951
	C. tesquorum	NS	41.0	
	A. runatus	NS	43.0	
	C. gallinae	NS	34.0	
	N. bidentatiformis	S	45.0	
Cavia porcellus (lab)	*O. silantiewi*	NS	23.0	Dubunina &
	F. luculenta	NS	23.0	Dubinin, 1951
	C. tesquorum	NS	36.0	
	A. runatus	NS	46.0	
	C. gallinae	NS	23.0	
	N. bidentatiformis	NS	29.0	
Spermophilus undulatus	*O. silantiewi*	NS	7.0	Zhovty &
	C. tesquorum	S	9.7	Vasiliev, 1962b
	Leptopsylla segnis	NS	14.3	
Meriones unguiculatus	*O. silantiewi*	NS	3.0	Zhovty &
	C. tesquorum	NS	7.7	Vasiliev, 1962b
	F. luculenta	S	20.3	
	Neopsylla pleskei	S	8.3	
	Xenopsylla cheopis	NS	12.0	
Rattus norvegicus	*X. cheopis*	S	12.0	Zhovty & Vasiliev, 1962b
Mus musculus	*L. segnis*	S	65.0	Zhovty & Vasiliev, 1962b

13.3.1 Flea antigens and host response

Ectoparasitic arthropods are often thought of as crawling or flying hypodermic needles that suck blood and inject disease causing agents. However, the saliva of blood-feeding arthropods contains factors that not only help them evade host haemostatic defences (Ribeiro, 1995; Wikel, 1996) but also have potent immunogens that influence the immune responses of the host (e.g. Roehrig et al., 1992). The development of immune responses to ectoparasitic arthropods such as ticks, mites, chiggers, fleas, mosquitoes and lice is well documented (Ribeiro, 1987, 1995; Rechav, 1992; Wikel, 1996; Wikel & Alarcon-Chaidez, 2001). However, most studies of immune responses to ectoparasites have involved livestock or laboratory animals; relatively little is known about immune responses to ectoparasites in wild animals.

The skin reaction (usually itch) is the first response to the bite of any haematophagous insect. This response is caused by an antigen contained in the saliva of an insect (see Lehane, 2005 for review). Furthermore, there are several allergens in the saliva of each insect species (e.g. Lee et al., 1999 for the cat flea), and different hosts respond to different antigens contained in the saliva of the same insect. For example, major allergens in the saliva of C. felis for humans and dogs are represented by different proteins (Trudeau et al., 1993; McDermott et al., 2000). Nevertheless, host reactions against fleas appear to be similar among hosts (Hudson et al., 1960a, b; Benjamini et al., 1963; Larrivee et al., 1964; Dryden & Blakemore, 1989). Larrivee et al. (1964) exposed guinea pigs to bites of C. felis and described a sequence of allergic skin reactions as follows: (a) induction stage with no abnormal macro- or microscopic changes in the skin; (b) stage of delayed hypersensitivity with intense mononuclear infiltration at the vicinity of the bite site (24 h after the bite); (c) stage of both immediate and delayed hypersensitivity, with immediate reactions appearing 20 min after the bite and characterized by an eosinophilic infiltration and the delayed reaction appearing within 24 h and characterized by mononuclear infiltration; (d) stage of immediate hypersensitivity only, the reaction appearing 20 min after the bite and characterized by an infiltration of eosinophils at the bite site; and (e) stage of non-reactivity resulting from desensitization, when no skin reactivity and cellular abnormalities occur (see also descriptions in Jones (1996) and Lehane (2005)). In guinea pigs exposed to bites of Pulex irritans, Pulex simulans and C. felis, the strongest response occurred during the second week after flea feeding began (Hudson et al., 1960a). Despite the similarity in the sequence of skin reactions of different hosts to the same flea or the same host to different fleas (compare Hudson et al. (1960a) and Larrivee et al. (1964)), the details and the strength of the response may differ. For example, the sensitivity of humans was greater to bites of *P. irritans* than to *P. simulans* or *C. felis* when measured by either immediate or delayed reactions (Hudson et al., 1960a).

On the one hand, immunological, physiological and histological particulars of host responses against fleas are not within the scope of this book. On the other hand, ecological and evolutionary patterns of host resistance cannot be fully understood without knowledge of their mechanisms. This knowledge is, however, extremely scarce and mostly concerns the response to fleas in pets and livestock (e.g. Halliwell & Longino, 1985; Halliwell & Schemmer, 1987; Greene et al., 1993). In addition, a detailed description of the histological changes at the attachment site of *Tunga monositus* in mice has been made by Lavoipierre et al. (1979), and skin reactions to feeding of *Xenopsylla cheopis* has been described for guinea pigs (Johnston & Brown, 1985; Brown, 1989) and rats (Vaughan et al., 1989).

Dogs that had clinical signs of hypersensitivity to *C. felis* had high levels of immunoglobulins IgE and IgG, whereas flea-exposed dogs with no signs of allergy had IgE levels close to background and low IgG levels (Halliwell & Longino, 1985). Injection of dogs with whole-flea extracts led to an increase of both IgE and IgG (Schemmer & Halliwell, 1987), whereas cats injected with extracts from the gut membrane of *C. felis* demonstrated elevated anti-flea antibodies in sera (Opdebeeck & Slacek, 1993). Antigens from *C. felis* induced high titres of IgE in laboratory mice (Zhao et al., 2006). However, Khokhlova et al. (2004a) found no difference between flea-infested and flea-free *M. crassus* in the concentration of immunoglobulins and circulating immune complexes in contrast to the number of white blood cells, leukocyte blast transformation index and phagocytic activity of leukocytes. This suggested that the immune response to flea parasitism in this rodent was linked mainly to cell-mediated immunity. The same has been shown to be true for the immunity against ticks. For example, there was no correlation between rabbit serum antibodies to soluble antigens from tick salivary gland extracts and protective immunity (Heller-Haupt et al., 1996). Yet production of anti-flea antibodies and transfer of resistance to fleas with immune serum have also been reported (Greene et al., 1993; Heath et al., 1994; Jones, 1996). Thus, humoral factors can also have a role in host resistance against fleas, especially given their intracellular digestion and the lack of a peritrophic membrane.

13.3.2 Acquired resistance

Cellular changes in the skin of a guinea pig had no effect on flea feeding success (Johnston & Brown, 1985). This can be related to the fact that the association between *X. cheopis* and the guinea pig does not occur in nature. However, basophilic response in rats to *X. cheopis* (Vaughan et al. 1989) (natural hosts of this flea) also did not suppress flea feeding and survival. Nevertheless, an individual host attacked repeatedly by an ectoparasite is known to develop acquired resistance against this ectoparasite (Willadsen, 1980; Rechav, 1992). This resistance is manifested by decreased feeding and reproduction of the ectoparasite (e.g. Fielden et al., 1992). In other words, ectoparasite arthropods downregulate host innate and specific acquired immune defences, inducing host responses that impair their own ability to feed (Rechav et al., 1989). For example, a study of acquired resistance in guinea pigs to tick larvae showed that repeated infestation of the host resulted in a sharp reduction in body mass of engorged larvae (Fielden et al., 1992). As early as 1939, Trager (1939) found that one infestation with larvae of the tick *Dermacentor variabilis* induced resistance in guinea pigs, while two or three infestations were needed to induce effective immunity in a natural host, the deer mouse *Peromyscus leucopus*. Similar observations for other species of ticks that infest laboratory and natural hosts were made by Rechav & Dauth (1987) and Rechav & Fielden (1995, 1997). Acquired resistance to fleas has been studied much less than acquired resistance to ticks.

When fleas *P. chephrenis* fed on its specific host *A. cahirinus*, they consumed more blood when they fed on immune-naïve rodents than on rodents previously exposed to parasitism (B. R. Krasnov and L. Ghazaryan, unpublished data) (Fig. 13.6a). This suggests that the host's response may have a negative effect on flea feeding success. However, when the host was a non-specific host, Wagner's gerbil *Dipodillus dasyurus*, no difference in feeding success between fleas fed on immune-naïve and flea-experienced rodents were found (B. R. Krasnov and L. Ghazaryan, unpublished data) (Fig. 13.6b). Consequently, the efficiency of the acquired resistance against fleas may depend on the flea species and be related to the degree of 'tightness' of a particular flea–host association. Indeed, differential immune responses of rodents to natural and unnatural flea species have been described (e.g. Vaughan et al., 1989).

It has been reported that natural hosts of some tick species do not acquire immunity against these ectoparasites (e.g. Rechav & Fielden, 1997). This is probably due to the presence of various substances in the tick saliva which allow successful evasion of the host immune response (Ribeiro, 1987). In Australia, Studdert & Arundel (1988) reported a severe allergic reaction in cats which hunted rabbits infested with *Spilopsyllus cuniculi*. The severity of these symptoms indicated that cats had a much higher response to the rabbit flea than to the cat flea with which they were normally infested. The pattern in *P. chephrenis* was opposite. However, it appeared that it was more energy-consuming for *P. chephrenis* to digest blood from flea-experienced than from immune-naïve *D. dasyurus* (Fig. 13.6c), while no difference in managing a blood meal from a specific host

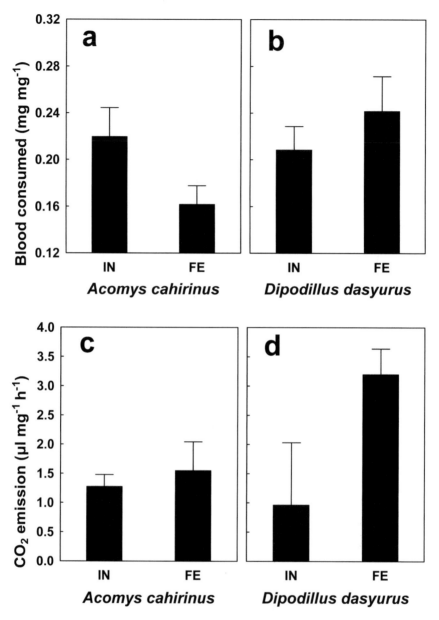

Figure 13.6 Mean (±S.E.) (a, b) mass-specific amount of blood consumed (mg per mg body mass) while feeding and (c, d) mass-specific emission of carbon dioxide (μl per mg body mass per hour) while digesting blood in *Parapulex chephrenis* feeding on immune-naïve (IN) and flea-experienced (FE) Egyptian spiny mice *Acomys cahirinus* (a, c) and Wagner's gerbils *Dipodillus dasyurus* (b, d). Data from B. R. Krasnov and L. Ghazaryan (unpublished).

(*A. cahirinus*) in dependence of its immune status was found (Fig. 13.6d) (B. R. Krasnov and L. Ghazaryan, unpublished data). In other words, it is not enough to measure a single fitness-related variable in a parasite to understand the link between acquired immunity and the degree of coevolutionary 'familiarity' of a host and a flea.

13.3.3 Cross-reactivity

Clear manifestation of the effect of acquired resistance of a host on different parasites in relation to a host's 'familiarity' with these parasites may be distorted by a phenomenon commonly known as cross-resistance or cross-reactivity. In cases when an investigation is focused on processes occurring in a host rather than on responses of a parasite in terms of its feeding and/or reproductive success, the term 'cross-reactivity' seems to be preferable. In contrast, when responses of a parasite are studied, 'cross-resistance' seems to be more appropriate. However, 'cross-resistance' and 'cross-reactivity' are two sides of the same coin.

Although haematophagy has evolved independently in different taxa of arthropods, and thus chemical mediators that are contained in their saliva are different (Ribeiro, 1995; Jones, 1996), salivary anti-clotting, anti-platelet and vasodilatory substances can be similar within a parasite taxon (Ribeiro *et al.*, 1990; Mans *et al.*, 2002; but see Warburg *et al.*, 1994). As a result of this within-taxon similarity, cross-resistance of a host against closely related parasites often occurs and has been frequently reported, for example in ticks (McTier *et al.*, 1981; Njau & Nyindo, 1987; but see Rechav *et al.*, 1989).

The first report suggesting the existence of cross-reactivity against different flea species and some role of taxonomic relatedness in these species was published by Boycott (1913). It was noted that a human sensitive to bites of *P. irritans* did not respond to *X. cheopis*, whereas a human sensitive to *C. felis* did not respond to *N. fasciatus*. However, when the former subject was repeatedly bitten by *X. cheopis*, he/she developed responses not only to this flea but also to *N. fasciatus*, whereas when the latter subject was repeatedly bitten by *N. fasciatus*, he/she developed responses not only to it but also to *X. cheopis*. Thus, cross-reactivity between *X. cheopis* and *N. fasciatus* was obvious, while no cross-reactivity between each of these species and *P. irritans* has been proved. Later, cross-reactivity between *P. irritans*, *P. simulans* and *C. felis* has been shown for humans and guinea pigs (Hudson *et al.*, 1960a, b), and also between *C. felis* and *X. cheopis* (Feingold *et al.*, 1968). From these results, Feingold *et al.* (1968) expected cross-reactivity between *X. cheopis* and *P. irritans*, although Boycott (1913) failed to find this.

Polygenis bolhsi and *X. cheopis* are another pair of flea species for which cross-reactivity has been established (Hecht, 1943).

I am aware of only two studies that have aimed specifically to either support or reject the existence of immune cross-reactions of a wild host against different flea species. Khokhlova et al. (2004a) studied the immune responses of *M. crassus* to three flea species belonging to the same family. The rodents used in their experiments were maintained in an outdoor enclosure and parasitized by *X. conformis* and *Xenopsylla ramesis*. The immune response was quantified using the leukocyte blast transformation test which measures the in-vitro reactivity of mononuclear cells to sensitizing (in-vivo encountered) antigens. Cell growth was quantified by means of the glucose consumption technique. Blood taken from each individual animal was tested in five variants, namely: (a) untreated control culture; (b) phytohaemagglutinin (PHA) treated culture; (c–d) antigens of 'familiar' fleas *X. conformis*- and *X. ramesis*-treated cultures; and (e) antigen of 'unfamiliar' flea *Synosternus cleopatrae*-treated culture. Antigens were whole-body extracts from fleas. The transformation index of leukocytes with *S. cleopatrae* antigen did not differ from those with antigens from both *X. conformis* and *X. ramesis* and was higher than both spontaneous glucose consumption and response to PHA (Fig. 13.7).

Similar experiments were also carried out with two other rodent species, *Gerbillus andersoni* and *D. dasyurus* (Khokhlova et al., 2004b; I. S. Khokhlova, M. Spinu, B. R. Krasnov and A. A. Degen, unpublished data). Fleas that represented 'familiar' and 'unfamiliar' species differed between these rodents with *G. andersoni* being 'familiar' with *S. cleopatrae* and 'unfamiliar' with *X. ramesis* and *P. chephrenis*, and *D. dasyurus* being 'familiar' with *Xenopsylla dipodilli* (the antigen from which has not been tested in this study) and 'unfamiliar' with all three flea antigens that have been used. The results were similar to those obtained for *M. crassus*, namely the response, in general, did not differ between flea species due to 'familiarity', although the response of *G. andersoni* to the antigen from a 'familiar' flea was weaker than those to 'unfamiliar' fleas (Fig. 13.8).

These data supported the existence of cross-reactivity between several flea species in wild rodents. The reason for this may be monophyly of fleas (see Chapter 4) as well as opportunistic feeding in many species (see Chapter 10). In addition, all species in the above studies belong to the same family Pulicidae. It is possible that the manifestation of cross-reactivity between distantly related species would be weaker. However, cross-reactivity between distantly related parasite taxa has also been reported. Rabbits infested with mites *Prosoptes cuniculi* produced antibodies reactive with both mite and tick extracts, whereas mite-free rabbits did not (den Hollander & Allen, 1986). However, cross-reactivity between fleas and mosquitoes has not been established (Feingold et al., 1968).

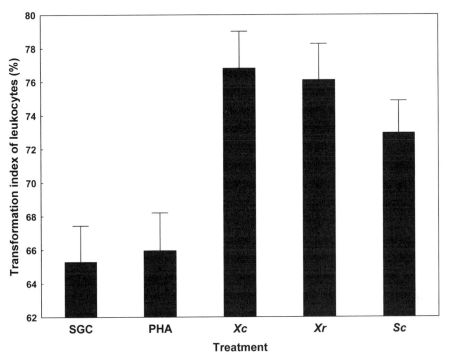

Figure 13.7 Mean (±S.E.) transformation index of leukocytes (%) of Sundevall's jird *Meriones crassus*. SGC — spontaneous glucose consumption; PHA — glucose consumption under phytohaemagglutinin treatment; *Xc*, *Xr* and *Sc* — glucose consumption under treatment with antigens from *Xenopsylla conformis*, *Xenopsylla ramesis* and *Synosternus cleopatrae*, respectively. Animals were parasitized by *X. conformis* and *X. ramesis*, but were never parasitized by *S. cleopatrae*. Data from Khokhlova et al. (2004a).

13.3.4 Immunocompetence

Immunocompetence is the general capacity of an organism to mount an immune response against pathogens and parasites (Schmid-Hempel, 2003; Schmid-Hempel & Ebert, 2003). Within a host population, immunocompetence varies among individuals depending on age, gender and nutritional and/or reproductive status (Lochmiller et al., 1993; Nelson & Demas, 1996; Humphreys & Grencis, 2002). As a result, parasite load varies among hosts characterized by different levels of immunocompetence. For example, immunocompetence in the greater mouse-eared bats *Myotis myotis* in Switzerland was negatively correlated with the number of mites *Spinturnix myoti* that they were harbouring (Christe et al., 2000).

264 Ecology of host defence

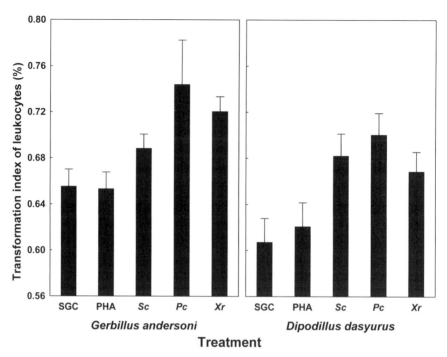

Figure 13.8 Mean (± S.E.) transformation index of leukocytes (%) of Anderson's gerbils *Gerbillus andersoni* and Wagner's gerbil *Dipodillus dasyurus*. SGC – spontaneous glucose consumption; PHA – glucose consumption under phytohaemagglutinin treatment; Sc, Pc and Xr – glucose consumption under treatment with antigen from *Synosternus cleopatrae*, *Parapulex chephrenis* and *Xenopsylla ramesis*, respectively. *Gerbillus andersoni* was parasitized by *S. cleopatrae*, but was never parasitized by *P. chephrenis* and *X. ramesis*. *Dipodillus dasyurus* was parasitized by *X. dipodilli*, but was never parasitized by either of the three tested flea species. Data from Khokhlova et al. (2004b) and I. S. Khokhlova, M. Spinu, B. R. Krasnov and A. A. Degen (unpublished).

Techniques for the assessment of immunocompetence can be divided into non-functional and functional. Non-functional techniques consist of measurements of immuno-related structures including leukocyte concentration and mass of lymphoid organs such as the spleen. An individual host with a relatively high leukocyte concentration or a large spleen presumably possesses better immune defences than an individual with a low leukocyte concentration and/or with a smaller spleen. A functional way to assess immunocompetence is to challenge individual hosts with antigens that trigger cellular or humoral immune responses. The phytohaemagglutinin injection assay (PHA test) is commonly used to measure immunocompetence in birds and mammals (Smits et al., 1999). The test is done by subcutaneous injection of vegetal lectin, a phytohaemagglutinin, which induces local T-cell stimulation and proliferation that causes swelling. The degree of swelling gives a measure of the potential proliferative response of circulating T cells. The PHA test has become popular in immunoecological field studies because it does not need sophisticated equipment, and the response is seen soon after the PHA injection (Smits et al., 1999).

Brinkhof et al. (1999) studied the effect of infestation by C. gallinae on the immunocompetence of P. major nestlings in Switzerland by using the PHA test. They carried out cross-fostering experiments and tested full siblings reared in flea-infested and flea-free nests. The results of this study showed that the nestlings' cell-mediated immunity was not affected by flea infestation. Therefore the PHA response has been suggested to be an inherent and objective measure of immunocompetence, being independent of experimentally created levels of parasitism (see also Saino et al., 1998). Another reason for the lack of the effect of fleas on the PHA response (Brinkhof et al. 1999) might be merely a low degree of effect of fleas on these hosts. However, in concomitant infections, one parasite influences the outcome of an infection of another through depression or stimulation of the immune response (Cox, 2001). As the host mounts an immune reaction against the PHA, individuals suffering from a highly virulent parasite may be expected to mount a different response against the PHA than parasite-free individuals. Indeed, the experiments of Goüy de Bellocq et al. (2006a) with M. crassus infested with X. ramesis provided support for this idea. The response to the PHA injection was significantly lower in parasitized than in non-parasitized animals (Fig. 13.9). The attenuating effect of parasites on the PHA response was found in birds as well; the response of the red jungle-fowl Gallus gallus parasitized by the intestinal nematode Ascaridia galli was lower than in control birds (Johnsen & Zuk, 1999).

Two possible mechanisms explain the decrease in the PHA response in flea-infested rodents. First, antigens produced by fleas had a depressing effect on immune cells, and thus the recruitment of T cells involved in the PHA response was low. Second, the energy required to withstand flea infestation was allocated to other anti-parasitic defences or other types of immune responses and, consequently, less energy was available to mount an immune response against the new antigen (PHA). These two explanations are not necessarily mutually exclusive. The depressive effect on T cells could be an adaptive response to a lack of sufficient energy to mount a PHA response, although the in-vitro stimulation of leukocytes by PHA was higher for flea-infested than uninfested M. crassus (Khokhlova et al., 2004a). However, the measurements of the immune response in this in-vitro experiment were done in a culture medium with glucose present, and thus energy was not a limiting factor. Furthermore, although no effect of X. ramesis on energy requirements of M. crassus was found

266 Ecology of host defence

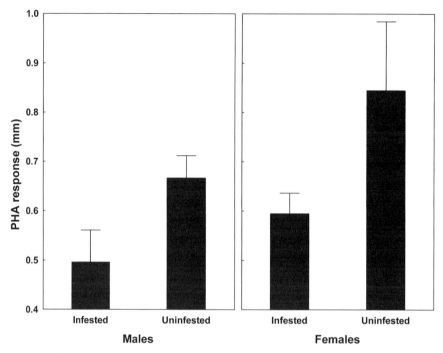

Figure 13.9 Mean (±S.E.) swelling response (mm) to PHA injection in male and female Sundevall's jirds *Meriones crassus* infested with *Xenopsylla ramesis* and in uninfested control *M. crassus*. Redrawn after Göuy de Bellocq *et al.* (2006a) (reprinted with permission from Blackwell Publishing).

(see Table 12.1), higher feeding and reproductive success of *X. ramesis* on food-restricted *M. crassus* (Krasnov *et al.*, 2005d; see Chapters 10 and 11) suggested that the latter preferred to surrender to fleas rather than to spend much energy on defence (see below). Consequently, energy allocation among different functions, or between different types of immune responses, seems to be the more likely explanation of the depressing effect of flea infestation on the PHA response (see also Kyriazakis *et al.*, 1998).

13.3.5 Host gender and immune response

Gender differences in immunocompetence have been reported for a variety of animals, with males being generally less immunocompetent than females (e.g. Olsen & Kovacs, 1996; Poulin, 1996a). This difference is supposedly related to the higher levels of androgens in males which suppress the immune function (Folstad & Karter, 1992). However, evidence for the relationship between testosterone and immune function is equivocal (Castro et al., 2001; Rolff, 2002; Schmid-Hempel, 2003, Vainikka et al., 2004). For example, Rolff (2002) proposed an alternative hypothesis explaining sexual differences in immunocompetence as being due to a higher investment by females in immune defence. Nevertheless, testosterone treatments reduced both innate and acquired resistance of the bank vole M. glareolus and the woodmouse Apodemus sylvaticus to parasitism of the tick Ixodes ricinus (Hughes & Randolph, 2001). Greives et al. (2006) measured endogenous levels of testosterone and two components of innate immunity (total levels of non-specific immunoglobulin and complement levels) in a wild breeding population of the dark-eyed junco Junco hyemalis in Virginia. Testosterone levels were significantly negatively correlated with both immune-related variables, suggesting that elevated testosterone levels may compromise innate immune function.

In an in-vitro study of M. crassus parasitized by fleas, Khokhlova et al. (2004a) found higher levels of circulating immune complexes in females than in males, but other measured immunological parameters did not differ between genders. This indicated a higher synthesis of antibodies and clearance of the antigen through complexation in females than in males, which supports the hypothesis of gender differences in immunocompetence and reduced humoral response in males (e.g. Zuk & McKean, 1996). In contrast, the level of circulating immune complexes did not differ between males and females of flea-infested D. dasyurus and G. andersoni (Khokhlova et al., 2004b). However, gender differences in phagocytic activity of leukocytes was found in G. andersoni but not in D. dasyurus and, surprisingly, this parameter was significantly higher in males than in females. These results contradict the 'immunohandicap' hypothesis of lower immunocompetence in males (Folstad & Karter, 1992) as well as the alternative hypothesis of Rolff (2002) that females invest more than males in immune defences. Phagocytosis by leukocytes presumably imposes high energetic costs (Blount et al., 2003).

Similarly, in-vivo studies also did not provide clear-cut results. Tschirren et al. (2003) manipulated infestation of P. major nestlings with C. gallinae in Switzerland and assessed their immunocompetence. They found strong gender differences in the response to the PHA assay, with males showing a reduced cellular immunity, suggesting that gender differences in immunocompetence develop early in life. Goüy de Bellocq et al. (2006a) reported gender differences in the PHA response in non-parasitized M. crassus (higher in females), but this difference disappeared after these rodents were exposed to parasitism by X. ramesis (Fig. 13.9). Furthermore, no gender difference in the PHA response was found in nine other rodent species parasitized by fleas in the Negev Desert (Goüy de Bellocq et al., 2006b).

268 Ecology of host defence

Contrasting results described above suggest that (a) androgens may suppress some but not all components of the immune defence against fleas (Khokhlova *et al.*, 2004a); (b) manifestation of gender difference in the immune defence may be species-specific (for host and/or flea) and/or mediated via environment (see Chapter 15); and (c) further investigations of the gender differences in anti-flea immunity are needed.

13.3.6 *Immune responses and maternal effects*

Parasite circulating antigens, immunoglobulins, immune cells, cytokines and other cell-related products can be transferred from mothers to offspring. In mammals, this transfer occurs during pregnancy and/or lactation (Carlier & Truyens, 1995). In birds, this transfer may be realized via the egg yolk (Buechler *et al.*, 2002). Maternal transfer of immunity can induce a long-term modulation of the offspring's capacity to mount immune responses against parasites. Maternal transfer of immunity has been suggested for a number of protozoans and helminths (e.g. Heckmann *et al.*, 1967; Shubber *et al.*, 1981; Kristan, 2002). It is difficult to measure directly the maternal transfer of immunity in terms of immune parameters of placental–foetal circulation or milk or yolk. Nevertheless, the occurrence of maternal transfer of immunity could be inferred from the comparison between immune responses of individuals that have been previously parasitized, for instance by fleas, and those of individuals that have never been parasitized but are the offspring of parasitized mothers.

In experiments with great tits and hen fleas in Switzerland by Heeb *et al.* (1998), female birds were exposed to fleas during egg production, whereas other females were kept flea-free. Fleas were killed at the start of incubation and then the nests of both previously parasitized and control females were re-infested. Nestlings born from mothers who were uninfested during the egg production grew faster and attained higher body mass at day 16 after hatching than nestlings from parasitized mothers. Moreover, when birds were recaptured the following year, it appeared that significantly more young birds were recruited from infested than from control females.

Buechler *et al.* (2002) also investigated maternal transfer of immunity in this host–flea system, in this same region. During egg-laying, female tits were either infested by *C. gallinae* or remained uninfested. Then the newborn nestlings were cross-fostered between the two treatments. The result was that, within the same nest, nestlings from the flea-infested mothers grew faster than nestlings of flea-free mothers. This study also tested for the transfer of parasite-induced immunoglobulins via the egg. Females were kept free of fleas until they laid the first egg and were then either infested by fleas or kept uninfested until the end of egg production. The concentration of immunoglobulins increased from the first to the eighth egg produced by infested tits, but not in eggs of uninfested tits. Consequently, the first experiment showed that fleas could induce maternal transfer of anti-flea immunity at egg-laying, while the second experiment suggested that this transfer was associated with immunoglobulins conveyed via the egg.

Khokhlova et al. (2004a) compared immunological parameters between M. crassus that were maintained in an outdoor enclosure and parasitized by X. conformis and X. ramesis and those that were maintained in a flea-free animal room but born from mothers previously infested by these fleas. Although transformation indices of leukocytes were generally lower in 'laboratory' than in 'enclosure' rodents (compare Fig. 13.7 and 13.10), their responses to antigens from X. conformis and X. ramesis were higher than those to PHA, as occurred with 'enclosure' individuals (Fig. 13.10). However, in contrast to 'enclosure' rodents, 'laboratory' rodents demonstrated no difference in transformation index between S. cleopatrae antigen and either the PHA treatment or controls (Fig. 13.10).

Significant immune responses to X. conformis and X. ramesis in rodents born in the animal room suggested that they received protection against these fleas from their mothers, although the protective effect of maternal antibodies transfer to offspring is limited (e.g. Knopf & Coghlan, 1989). Indeed, responses to antigens of both Xenopsylla species in 'laboratory' rodents were lower than those in the 'enclosure' animals, suggesting that the protective level of maternal immunity was probably lower than the acquired immunity against the same flea species. However, the relatively short lifespan of immune cells that could be transferred from mothers to offspring suggests a higher probability of finding them in juveniles rather than in young adults, as was the case in this study.

Maternal protection of offspring against parasitism may be associated not only with the transfer of immunity but also with adjustments of the offspring hormonal levels during their prenatal development. Tschirren et al. (2004) hypothesized that when female great tits anticipate high levels of flea parasitism, their eggs contain a lower level of androgens. Eggs from flea-infested and uninfested tits were collected and concentrations of three androgens were measured in the yolk. Eggs of infested females contained significantly lower amounts of two of the three androgens, supporting the hypothesis. However, when Tschirren et al. (2005) experimentally manipulated yolk testosterone and exposed nestlings to fleas or kept them uninfested, it was found that high levels of yolk testosterone promoted growth of the nestling's body mass similarly in flea-infested and flea-free nests, and neither affected the level of nestling's immunocompetence.

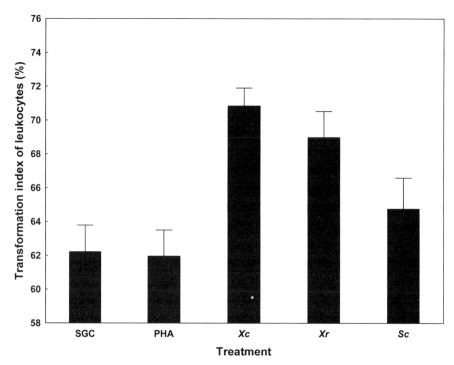

Figure 13.10 Mean (± S.E.) transformation index of leukocytes (%) of Sundevall's jirds *Meriones crassus*. SGC — spontaneous glucose consumption; PHA — glucose consumption under phytohaemagglutinin treatment; Xc, Xr and Sc — glucose consumption under treatment with antigen from *Xenopsylla conformis*, *Xenopsylla ramesis* and *Synosternus cleopatrae*, respectively. Animals were never parasitized by fleas, but their mothers were parasitized by *X. conformis* and *X. ramesis*. Data from Khokhlova *et al.* (2004a).

13.3.7 Immune responses and infestation patterns

Mounting immune responses and investing in immune defences should depend on the pattern of parasite pressure (see Combes, 2001). In particular, the frequency and probability of parasite attacks may strongly affect the patterns of immune defence (Martin *et al.*, 2001; Tella *et al.*, 2002). Selection of mechanisms of resistance in hosts is expensive, and thus of little advantage if encounters with the parasite are rare (Poulin *et al.*, 1994). Consequently, if the frequency and/or probability of attacks by parasites are low, then a host can limit its allocation of energy for immune responses by the development of the responses only after being attacked by the parasite ('post-invasive'). If, however, frequency and/or probability of parasitism are high, a continuous maintenance of a certain level of immune 'readiness' in the host is advantageous, although the investment in immune defence can be high (Jokela et al., 2000). In addition, the long and continuous association between a particular host and a particular parasite can induce host genotypical changes via selection. In particular, these changes could affect the major histocompatibility complex which is the region of the genome that controls immune responses (Gruen & Weissman, 1997). As a result of any of these processes, the host can maintain some protection against a parasite whose attack is highly probable, even though the host has never been attacked (Jokela et al., 2000).

Another component of parasitism that can affect the pattern of development and persistence of defensive responses in hosts is the variety of parasite challenges. Maintaining several different means of defence is probably more costly than sustaining one specific type of defence (L. H. Taylor et al., 1998). Consequently, in spite of the occurrence of cross-resistance, host species that are exploited by a small number of specific parasites can acquire specific immune resistance against these parasites but not against other parasites (Rechav et al., 1989). In contrast, hosts with a diverse parasite spectrum can develop multiple immune responses against a variety of parasite species. As a result, mounting immune responses to non-familiar parasites should be expected in a 'parasite-rich' rather than in a 'parasite-poor' host.

Two gerbils, D. dasyurus and G. andersoni, inhabiting the Negev Desert differ sharply in their natural species-richness of flea assemblages and prevalence of infestation. The former occupies a variety of habitats and is parasitized naturally by several flea species (X. dipodilli, X. conformis, X. ramesis, Nosopsyllus iranus, Stenoponia tripectinata, Coptopsylla africana, Rhadinopsylla masculana) (Krasnov et al., 1997, 1998, 1999). Prevalence of infestation of D. dasyurus by fleas ranged between 20% and 65% among habitats (Krasnov et al., 1998). Intensity of D. dasyurus infestation by fleas also differed among habitats ranging from a low of 2.0 fleas per infested individual to a high of 6.3 fleas per infested individual (Krasnov et al., 1998). In contrast, G. andersoni is a specialist sand-dweller and is parasitized mainly by a single flea species S. cleopatrae (Lehmann, 1992; Krasnov et al., 1999; Hawlena et al., 2005). Prevalence of G. andersoni infestation by fleas was 95–100%, whereas intensity of infestation averaged 12.2 fleas per infested individual (Hawlena et al., 2005, 2006c).

The study of Khokhlova et al. (2004b) mentioned above dealt with in-vitro immunological parameters and the pattern of immune responses to an antigen from an 'unfamiliar' flea species (X. ramesis) in these two gerbils comparing uninfested individuals and individuals experimentally infested with X. dipodilli (D. dasyurus) and S. cleopatrae (G. andersoni). In contrast to expectations, the level of circulating immune complexes and concentration of immunoglobulins did not differ between parasitized and control rodents in either species as well as

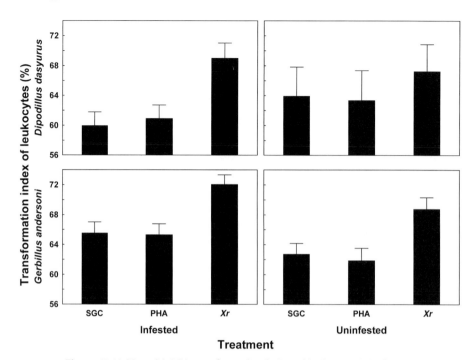

Figure 13.11 Mean (±S.E.) transformation index of leukocytes (%) of Wagner's gerbil *Dipodillus dasyurus* and Anderson's gerbils *Gerbillus andersoni*. SGC – spontaneous glucose consumption; PHA – glucose consumption under phytohaemagglutinin treatment; Xr – glucose consumption under treatment with antigen from *Xenopsylla ramesis*. Animals were never parasitized by this flea. Redrawn after Khokhlova *et al.* (2004b) (reprinted with permission from Springer Science and Business Media).

between species. However, the number of white blood cells was significantly lower in control than parasitized *D. dasyurus*, but did not differ between control and parasitized *G. andersoni*. This again hints that immune responses to flea parasitism were linked mainly to cell-mediated rather than humoral immunity. Responses to antigen from *X. ramesis* were higher than both spontaneous glucose consumption and response to phytohaemagglutinin in parasitized and control *G. andersoni* and parasitized *D. dasyurus* (Fig. 13.11). However, there was no significant difference in the spontaneous index of glucose consumption and responses to both phytohaemagglutinin and flea antigen in non-parasitized *D. dasyurus*.

These results demonstrated that *D. dasyurus* was characterized mainly by 'post-invasive' immune responses, whereas even non-parasitized *G. andersoni* showed immune responses suggesting that the expected probability of attack by fleas plays an important role in determining the immune strategy of the host. A high expected probability of flea attacks could impose strong natural selection for immune defences in *G. andersoni*. However, this was manifested mainly in the persistence of pre-invasive immune responses rather than in magnitude of post-invasive immune responses, which was not greater in *G. andersoni* than in parasitized *D. dasyurus*. Perhaps it is too costly for *G. andersoni* to be capable both of maintaining pre-invasive immune ability and of mounting stronger immune responses. Mechanisms of the pre-invasive immune 'readiness' remain to be tested. Between-host difference related to the second component of parasitism, natural diversity of flea assemblages, has not been found. As mentioned previously, it should be advantageous to have the ability to cope with multiple challenges for a host inhabiting parasite-rich habitats. Nevertheless, both species demonstrated similar immune responses to antigen from an 'unfamiliar' flea species. The reason for this is probably due to the similar saliva proteins in closely related *X. ramesis*, *X. dipodilli* and *S. cleopatrae*, resulting in cross-immunity between them (see above).

Goüy de Bellocq et al. (2006b) investigated patterns of the PHA response in 10 rodent species from the Negev Desert. These rodents differed markedly in species-richness of their flea assemblages as well as in flea prevalence and abundance. The PHA response was measured 6, 24 and 48 h post-injection in the footpad. Two types of PHA responses were found. One was rapid (peaked ~6 h after injection) and characteristic of rodents that have either species-poor flea assemblages or that are rarely attacked by fleas (Fig. 13.12a). The second type of response was delayed (peaked 24 h after injection) and was typical of rodents that have either species-rich flea assemblages and/or high abundance and prevalence of fleas (Fig. 13.12b). Furthermore, rodents that responded promptly had a lower maximum response than rodents with a delayed response.

These results suggest that (a) the time of maximal response to PHA injection is a species-specific character; and (b) a trade-off exists between strength and rapidity of the immune response (see Navarro et al., 2003). The latter stems from trade-offs between defence against parasites and other concurrent needs of the organism (Sheldon & Verhulst, 1996; see below). Rodents mounting a prompt PHA response had a lower PHA response peak than rodents with a delayed response, indicating that a host may not be able to develop both strong and rapid responses to parasite infection. Therefore, the strength and latency should be optimized according to the pattern of parasite pressure. The latter can be evaluated in terms of host species-specific probability of a parasite attack that is assessed via prevalence of infestation, parasite abundance and/or diversity of immunological challenges (species-richness of parasite assemblages). For example, on the one hand, if parasite pressure is relatively low (e.g. low prevalence, abundance and species-richness of parasites), then it may be advantageous for a rodent to mount a weak but fast immune response. On the other hand, if a

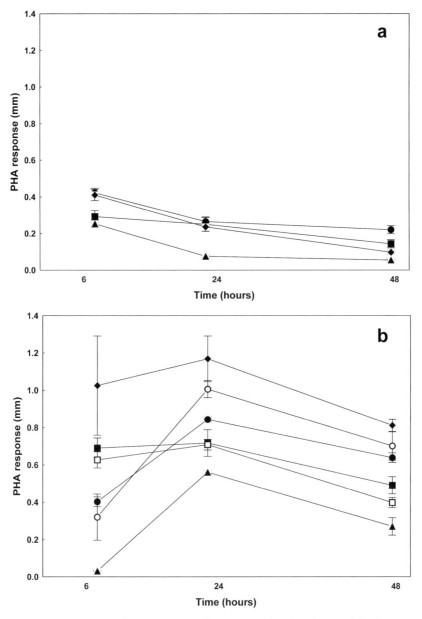

Figure 13.12 Mean (± S.E.) swelling response (mm) to the PHA injection at 6, 24 and 48 hours after PHA injection in 10 rodent species: (a) the Egyptian spiny mouse *Acomys cahirinus* (circles), golden spiny mouse *Acomys russatus* (squares), house mouse *Mus musculus* (triangles), pygmy gerbil *Gerbillus henleyi* (diamonds), (b) Anderson's gerbil *Gerbillus andersoni* (closed squares), Wagner's gerbil *Dipodillus dasyurus* (closed circles), greater Egyptian gerbil *Gerbillus pyramidum* (diamonds), lesser Egyptian gerbil *Gerbillus gerbillus* (open circles), Sundevall's jird *Meriones crassus* (open squares) and fat sand rat *Psammomys obesus* (triangles). Redrawn after Göuy de Bellocq *et al.* (2006b) (reprinted with permission from Elsevier).

host is frequently attacked by many parasites of different species, then it may be advantageous to mount a strong but slower immune response. Indeed, bird species parasitized by many flea species mount stronger immune responses than bird species parasitized by few flea species (Møller et al., 2005).

13.3.8 Costs and trade-offs of the immune defence against fleas

As has already been mentioned, activation of an immune response and even maintenance of a competent immune system is an energetically demanding process which requires trade-off decisions among competing energy demands for maintenance, growth, reproduction, thermoregulation and immunity (Sheldon & Verhulst, 1996). In other words, trade-offs occur between defence against parasites and the other needs of the organism, so that the use of the immune system cannot be sustained simultaneously with other energy-demanding activities (e.g. Nordling et al., 1998). Empirical evidence suggests that such costs can be high (e.g. Moret & Schmid-Hempel, 2000). As a result, many hosts generally have low circulating titres of immune effectors such as leukocytes and immunoglobulins (Klein, 1990).

The costs of immunity against parasites and trade-off decisions have been widely discussed (see Degen, 2006 for recent review). Here, I will restrict myself to studies that consider costs of host immunity against fleas. Earlier, I mentioned that Krasnov et al. (2005d) demonstrated an increase in egg production by X. ramesis exploiting underfed M. crassus (Chapter 11). I discussed the results of this study from the flea's perspective. Let's now look at these results from the host's perspective.

Let's recall that experimental M. crassus were offered food equivalent to approximately 100% (control group), 60% (T1 group) or 30% (T2 group) of maintenance energy requirements. Animals from the control group maintained body mass throughout the experimental period and their body mass after the first week was $98.9 \pm 2.1\%$ of initial body mass. The same was true for T1 animals and their body mass after the first week of experiment was $97.3 \pm 1.7\%$ of the initial body mass. However, body mass of rodents from the T2 group decreased after the first week to $81.1 \pm 1.8\%$ of initial body mass. Fleas that parasitized control animals produced significantly fewer eggs than those that parasitized rodents with 60% and 30% of maintenance energy intake, while egg production of fleas fed on rodents with 60% of maintenance energy intake was significantly lower than that of fleas fed on rodents with 30% of maintenance energy intake (see Fig. 11.7). The most surprising result of this study was that fleas increased their egg production exploiting rodents from the T1 group, that is, those that did not decrease body mass although underfed. It is possible that body content of the rodents changed and body energy was reduced without a change in body mass (Degen, 1997; Kam et al., 1997). This confirms that host resistance against parasites is both energetically and nutritionally demanding (Lochmiller & Deerenberg, 2000).

From an ecological viewpoint, short-term suppression of the immune system may be advantageous to a host because it enables reallocation of resources to functions that support immediate survival during periods of food limitation (Apanius, 1998). However, if food limitation occurs in a predictable manner (e.g. seasonally), it can be advantageous to suppress other functions (e.g. reproduction) rather than immune function. Indeed, the 'winter immunoenhancement' hypothesis was suggested by Nelson & Demas (1996) to explain the increase in immune parameters during the winter reproduction break in small mammals from temperate environments (Lochmiller & Dabbert, 1993; Lochmiller et al., 1993).

Given the high cost of immune defence and the multiple trade-offs, a relationship between flea density and the strength of a host immune response should be expected. If a host immune response is an efficient tool for overcoming parasite pressure, an increase in parasite load will lead to an increased response (de Lope et al., 1998). Using similar logic, when parasite pressure is low, the response of a host should be relatively low (Combes, 2001). However, due to the cost of using the immune system, a relatively low response by a host can also be expected when the cost of eliminating parasites is higher than the cost of limiting its pressure to a 'tolerable' level (Combes, 2001). Consequently, the responses of the host should peak at intermediate levels of parasite load, and thus the curve describing the relationship between parasite load and host response level should be hump-shaped. Khokhlova et al. (2004a) tested this prediction with *M. crassus* parasitized by different numbers of *X. conformis* and *X. ramesis*. Among a variety of immunological parameters measured, only the phagocytic activity of leukocytes was affected by flea burden, decreasing significantly with an increase in the number of fleas (Fig. 13.13).

This suggests that even a weak attack of fleas triggered the immune system of *M. crassus*. However, this system could not overcome the attack by large number of fleas, perhaps due to the cost of the immune system and/or the additive immunosuppression effect of a high number of fleas.

Finally, there can be a trade-off not only between immunity and other needs of an organism, but also between immunity and other tools of anti-parasitic defence. Goüy de Bellocq et al. (2006a) assessed the behavioural defence of *M. crassus* by counting *X. ramesis* remaining in rodent cages and on rodent bodies 4 days after flea introduction. In addition, they tested the relationships between grooming success and host immune responses (leukocyte concentration and the

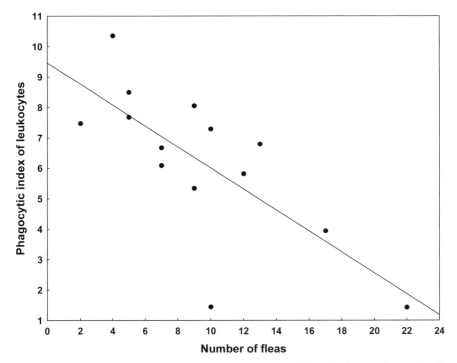

Figure 13.13 Relationship between phagocytic activity of leukocytes in Sundevall's jird *Meriones crassus* and flea burden. Redrawn after Khokhlova et al. (2004a) (reprinted with permission from the Company of Biologists).

PHA response), but did not find any trend. Although ectoparasite removal is one of the most important functions of grooming, grooming has other functions as well (see above). The multiple functions of grooming could mask the potential relationship between immune and behavioural defences. Further experiments combining behavioural observations with immune measurements are required to investigate this issue.

13.3.9 Coping with the host immune response

As we have already seen, when fleas attack a host, the latter attempts to defend itself. The host's aim is thus either to eliminate the fleas or, at least, to impair their feeding and/or reproductive abilities. Obviously, defence activity is very important from the immunoecological perspective. However, from the parasitological perspective, it is much more important how fleas respond to host defence, and what is the net effect of host's defensiveness on fleas.

Several studies have demonstrated the effect of the host immune response on flea feeding and/or reproductive success (see Chapters 9 and 10). For example,

278   Ecology of host defence

C. gallinae feeding on previously flea-exposed great tits had lower hatching success and fewer larvae than hen fleas fed on immune-naïve hosts (Walker *et al.*, 2003). Negative relationships between immunocompetence of great tit nestlings and hen flea fecundity were found by Tschirren *et al.* (2007a).

However, there was no effect on quantity (but not quality) of flea offspring between fleas parasitizing immune-naïve and previously flea-exposed *A. cahirinus* (B. R. Krasnov and L. Ghazaryan, unpublished data; see Chapter 11). In experiments by Goüy de Bellocq *et al.* (2006a) with *X. conformis* and *M. crassus*, fleas consumed more blood when they fed on previously parasitized than on non-parasitized animals. Among fleas that fed on previously parasitized animals, blood consumption was positively correlated with the initial leukocyte concentration of the rodents and negatively correlated with difference in leukocyte concentration between the first and 16th days of flea infestation, while there was no correlation between blood consumption in fleas that fed on control animals and any other immunological variable of the hosts, including the level of immunocompetence measured via the PHA test. Furthermore, mean egg production and hatching success of fleas were not related to either initial haematocrit, leukocyte concentration or the PHA response of the rodent. However, the two flea-related parameters were negatively correlated with differences in leukocyte concentration in their hosts between the first and 16th day of infestation (Fig. 13.14). In particular, this showed that the strength of the PHA response which is commonly used to infer the outcome of host–parasite arms race does not reflect the overall immunocompetence of individual hosts.

Nevertheless, correlation between changes in rodent leukocyte concentration after 16 days of flea parasitism and flea feeding and fitness variables implies that the host's immune responses affected the reproductive performance of the fleas. Taken together, (a) the relationship between difference in leukocyte concentration and flea reproductive/ feeding variables and (b) intrahost heterogeneity in the leukocyte response to flea parasitism suggest that fleas could select host individuals with the lowest level of immunocompetence and thus minimize the negative impact of the host's immune responses. This is similar to the mechanisms of parasite aggregation inferred by the 'tasty chick' hypothesis (Christe *et al.*, 1998) which states that the last-hatched chick in an asynchronously hatching brood would have the least efficient immune system, resulting in subsequent parasite aggregation on this chick.

13.4   Concluding remarks

Scheidt (1988) noted that the amount of money and energy spent on studies related to the immune responses of animals to fleas is much greater

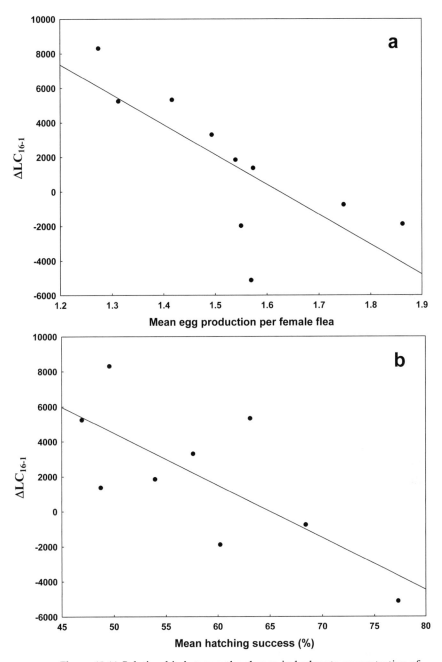

Figure 13.14 Relationship between the change in leukocyte concentration of Sundevall's jird *Meriones crassus* before and after 16 days (ΔLC_{16-1}: cells mm^{-3}) of infestation by *Xenopsylla ramesis* and (a) mean egg production per female flea and (b) mean hatching success (%) of flea eggs. Redrawn after Göuy de Bellocq et al. (2006a) (reprinted with permission from Blackwell Publishing).

than those spent on any other issue in veterinary medicine (see also Jones, 1996). Nonetheless, little progress has been made during the last 20 years. Many questions remain unanswered, and studies have often provided contradictory results. In general, some results suggest that the immune defence against fleas is not especially effective, and fleas cope successfully with the host's immune responses. This can be one of the reasons why hosts often give up immune defence and surrender to fleas.

On the other hand, similarly to the questions related to flea virulence (Chapter 12) and apart from veterinary research, a very limited number of host–flea systems have been studied in association with the host defence issues. These studies have mainly investigated the *C. gallinae–P. major* model in Switzerland (by the team of Heinz Richner at the University of Bern) and the rodent–flea associations in the Negev Desert of Israel (by my team at the Ben-Gurion University of the Negev). Most of these studies have focused on the hosts, while only a few have considered the resistance of a host from both the host and flea aspects. Much further effort is required to understand better the intricate flea–host relationships.

PART III EVOLUTIONARY ECOLOGY: WHY DO FLEAS DO WHAT THEY DO?

14

Ecology and evolution of host specificity

Host specificity is one of the fundamental properties of any parasitic organism. Highly host-specific parasites exploit a single host species, whereas host-opportunistic parasites use hosts belonging to several different species. Consequently, from an ecological perspective, host specificity is nothing other than the breadth of ecological niche of a parasite, representing the diversity of resources used by it (Futuyma & Moreno, 1988). From an evolutionary perspective, host specificity reflects the parasite's historical associations with its hosts (Brooks & McLennan, 1991; Page, 2003). In this chapter I discuss patterns of host specificity in fleas, starting with how to measure it and reviewing the patterns of host specificity found in fleas. I go on to consider the relationships between degree of host specificity and evolutionary success in fleas and discuss the evolutionary forces that determine selection for either high or low host specificity. Finally, I address some applicative questions that involve investigations of flea host specificity.

14.1 Measures of host specificity

Intuitively host specificity can be defined merely as the number of host species that are used by a parasite population or species. However, even host-opportunistic (=host generalist) parasites do not use all the variety of their hosts equally, but rather some host species are used more intensely than others. Accordingly, several attempts to introduce some ecological information (e.g. prevalence and/or abundance of a parasite in different host species) into a single index of host specificity have been made. For example, Rohde (1994) proposed an index of specificity, based on the number of parasite individuals found in each host species. However, this index was criticized because it appeared to be unreliable for comparison of parasite species using different numbers of host species (Poulin, 2007a).

From an evolutionary perspective, the host specificity of a flea is not merely a function of how many host species it can exploit, but also of how closely related these host species are to each other. For example, consider two flea species each exploiting the same number of host species; if one of these fleas exploits only congeneric hosts whereas the other exploits hosts belonging to different families, then the host specificity of the former should be considered higher than that of the latter. Therefore, the study of host specificity should take into account phylogenetic (or at least taxonomic) relationships among all host species of a parasite. Recently, Poulin & Mouillot (2003, 2004a, b) have proposed a new measure of host specificity that takes into account host relationships, focusing on the average taxonomic distinctness of all host species used by a parasite species (see details below). This index (S_{TD}) and its variance (VarS_{TD}) are analogous to the indices of taxonomic distinctness and taxonomic asymmetry (Δ^+ and Λ^+) proposed by Clarke & Warwick (1998, 1999, 2001) and Warwick & Clarke (2001). An attempt to combine both ecological and phylogenetic information into a single index of host specificity has also been made (Poulin & Mouillot, 2005a).

In this chapter, I consider flea host specificity using two measures. The simplest measure is the number of host species on which the flea species is found. One of the main flaws in using this measure to estimate host specificity is sampling effort. Indeed, high host specificity can be an artefact of inadequate sampling (Poulin, 2007a), and the division between highly specific and relatively non-specific parasites may really be a division between rare and common species (Klompen et al., 1996). Corrections for sampling effort are therefore necessary in any broad survey of host specificity. In the following text, whenever host specificity is considered as the number of host species, the confounding effect of sampling effort has been controlled for (except Krasnov et al., 2005a).

Another measure of host specificity used for fleas is the specificity index S_{TD}, and its variance VarS_{TD} (Poulin & Mouillot, 2003). It should be noted that the indices S_{TD} and VarS_{TD} have only been applied to date to fleas parasitic on mammals. Patterns of host specificity in bird fleas remain to be studied. The index S_{TD} measures the average taxonomic distinctness of all host species used by a flea species and represents the mean number of steps up the taxonomic hierarchy of the hosts that must be taken to reach a taxon common to two host species, computed across all possible pairs of host species (see Poulin & Mouillot 2003 for details). For any given host species pair, the number of steps corresponds to half the path length connecting two species in the taxonomic tree, with equal step lengths of 1 being postulated between each level in the taxonomic hierarchy. The greater the taxonomic distinctness between host species, the higher the number of steps needed, and the higher the value of the index S_{TD}: thus it is actually inversely proportional to specificity. A high index value means that on average the hosts of a flea species are not closely related.

Using the taxonomic classification of Wilson & Reeder (2005), mammal host species were fitted into a taxonomic structure with five hierarchical levels above species, i.e. genus, subfamily, family, order and class (Mammalia). The maximum value that the S_{TD} index can take (when all host species belong to different orders) is thus 5, and its lowest value (when all host species are congeners) is 1. Since the index cannot be computed for parasites exploiting a single host species, an S_{TD} value of 0 is assigned to these flea species, to reflect their strict host specificity.

The variance in S_{TD}, $VarS_{TD}$, provides information on any asymmetries in the taxonomic distribution of host species (Poulin & Mouillot, 2003); it can only be computed when a flea exploits three or more host species (it always equals zero with two host species). High values of $VarS_{TD}$ usually mean that one main branch in the taxonomic tree of host species contributes proportionally more species to the list of host species than other branches. In many (although not all) cases, S_{TD} and $VarS_{TD}$ are sensitive to the number of host species in a host spectrum. In such cases the respective corrections have been made (e.g., Krasnov et al., 2004e).

14.2 Variation in host specificity among flea species

14.2.1 Patterns of host specificity

Distribution of host specificity values among parasite species of many taxa is typically right-skewed, suggesting that there are many host-specific species, whereas only a few species are true host-generalists. For example, among helminths parasitic in small mammals in Central Asia and the Iberian peninsula, between one-third and a half of known parasite species in the region are strictly host-specific and found in only one host species, whereas the majority of other helminth species use five or more host species, and only very few species use 10 or more host species (Poulin et al., 2006a). The same is true for chewing lice parasitic on rodents worldwide, namely the vast majority of species occur on a single host species, or less frequently on two hosts (Poulin et al., 2006a). Given that the data for lice came from the entire global data set, whereas those for helminths were extracted from regional surveys, these results suggest that lice are, in general, more host-specific than helminths. This greater specificity may be a consequence of the mode of transmission, with lice being contact-transmitted parasites, whereas most helminth species are transmitted via ingestion. Fleas, however, do not strongly depend on host-to-host contact to be transmitted between host species. Although they can be transmitted via contact (Krasnov & Khokhlova, 2001; see Chapter 9), to a great extent their main method of host location and attack is active host selection (Krasnov et al., 2003c; see Chapter 9). As a result, among fleas, the distribution of numbers of host species used is less right-skewed than for other parasite taxa (Fig. 14.1). This pattern remains true when the data for fleas from several regions are pooled (Fig. 14.2). Although many flea species are found on only one or two host species, there is a substantial number of flea species that can exploit several host species. For example, this has been reported for fleas of China (Tian, 1995; Guo & Xu, 1999) and Mexico (Acosta, 2005).

This pattern is based on host specificity measured as the number of host species used. Other measures of host specificity could produce different patterns. For example, applying a measure of the average taxonomic distinctness of host species, S_{TD}, generates a roughly symmetrical distribution of host specificity values (Poulin et al., 2006a) (Fig. 14.3). The distribution peaks between values of 2.5 and 4, suggesting that among fleas that exploit more than one host species, many species are able to exploit hosts belonging to different families.

14.2.2 Is host specificity a flea species character?

The crucial problem in understanding the evolution of specialization in general, and of host specificity in particular, is to understand the role played by natural selection. This, in turn, leads to another important question of whether a given level of host specificity ('niche breadth') is a species attribute that can be subjected to natural selection, or whether it merely reflects the local restrictions caused by a variety of ecological, morphological, biochemical and/or genetic factors (Fox & Morrow, 1981). The latter possibility is supported by the substantial variation in the degree of host specificity observed among populations within a flea species (e.g. Castleberry et al., 2003).

If the level of specialization expressed as host specificity is a true species character, it should be relatively constant across different populations of the same flea species. Krasnov et al. (2004e) tested this hypothesis using data on fleas from 21 regional surveys, mainly from the Palaearctic, and determined whether host specificity showed some constancy across populations of the same flea species using a repeatability analysis (see details in Arneberg et al., 1997) of 118 flea species that were recorded in at least two of the regions. The repeatability analysis was also carried out across 48 flea genera, to see if host specificity is a generic character. Whether measured as the number of host species used or as the taxonomic distinctness (S_{TD} and $VarS_{TD}$) of these hosts, host specificity estimates from the same flea species were more similar to each other than to values

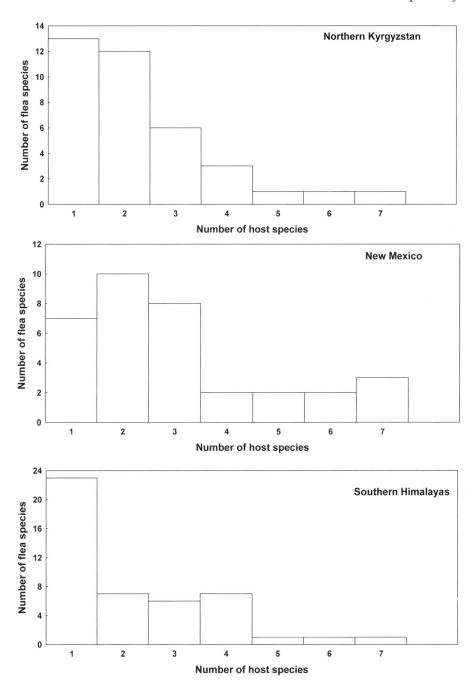

Figure 14.1 Frequency distribution of host specificity measured as the number of host species used among flea species parasitic on small mammals (Rodentia, Soricomorpha and Lagomorpha) from northern Kyrgyzstan, New Mexico and the southern Himalayan Mountains. Data from Shwartz et al. (1958), Morlan (1955) and Guo & Xu (1999), respectively.

288　Ecology and evolution of host specificity

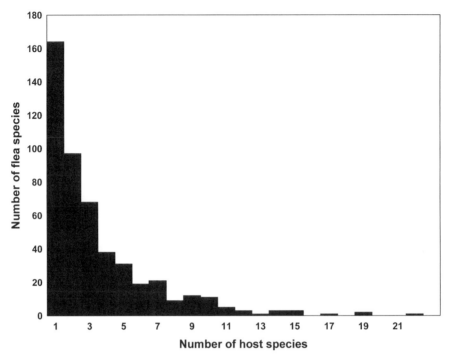

Figure 14.2 Frequency distribution of host specificity measured as the number of host species used among flea species parasitic on small mammals (Rodentia, Soricomorpha and Lagomorpha) from 20 geographic regions. Data from Krasnov et al. (2004f).

from other flea species (Fig. 14.4), with 30.4%, 14.4%, and 17.4%, respectively, of the variation among samples accounted for by differences between flea species. In other words, estimates of host specificity were repeatable within the same flea species. The number of host species exploited and the taxonomic asymmetry (VarS_{TD}) of host assemblages, but not the S_{TD}, were also repeatable across flea genera with 20.8% and 15.4%, respectively, of the variation among samples accounted for by between-genus differences.

The level of host specificity of any parasite species is determined by the range of conditions to which this species is adapted. These conditions are related to ecological, behavioural, physiological and biochemical traits of a particular host taxon (Ward, 1992; Poulin, 1998, 2007b). For fleas, these traits can be the structure of host skin, the physical and chemical properties of host blood, the parameters of the host's immune response and the environmental conditions of the host burrow/nest. If we adopt a Hutchinsonian representation of the ecological niche of a parasite species as an *n*-dimensional hypervolume, the axes of which are host traits, then a parasite species would demonstrate either broad or narrow tolerance along each of these axes. This also corresponds to the more

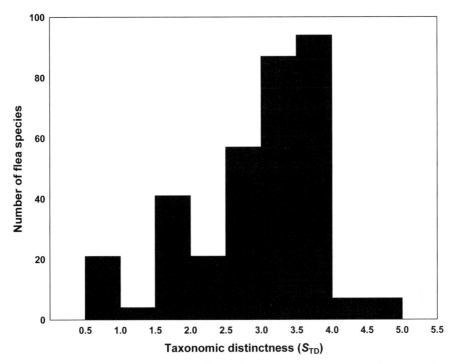

Figure 14.3 Frequency distribution of host specificity measured as the taxonomic distinctness of host species (computed only for flea species with at least two host species) among flea species parasitic on small mammals (Rodentia, Soricomorpha and Lagomorpha) from 20 geographic regions. Data from Krasnov et al. (2004f).

contemporary definition of the functional niche (Tokeshi, 1999; Rosenfeld, 2002). The repeatability of the degree of host specificity among populations of the same flea species suggests that host specificity is a true attribute of a flea species and can be envisaged as the entire set of a flea's responses along all axes representing host traits. Natural selection can, thus, act on this set of responses as a whole. On the other hand, no within-genus repeatability in S_{TD} was found, suggesting that congeneric fleas can vary in their host specificity in terms of the level of taxonomic distinctness of their host assemblage. This supports, albeit indirectly, the hypothesis that specificity, at least for the taxonomic diversity of the hosts, is not an evolutionarily blind alley and that the evolution of specialization has no intrinsic direction (Thompson, 1994; Desdevises et al., 2002b; see below).

14.2.3 Geographical variation of host specificity

Although host specificity appears to be repeatable within flea species, the similarity among host-specificity values in different populations of the same flea species is still subject to wide variations. This suggests some geographical

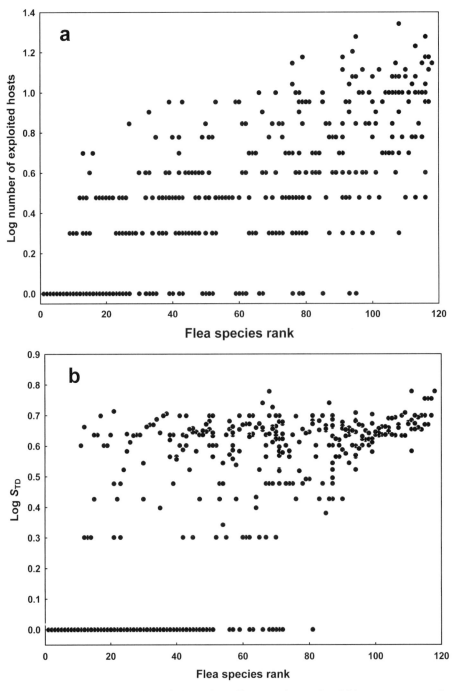

Figure 14.4 Rank plots of (a) number of host species used and (b) average taxonomic distinctness, S_{TD}, of these hosts across 118 flea species ranked from lowest to highest mean host specificity. If geographical variation is small within compared to between flea species, the points are expected to fall in an area of the plot stretching from the lower left to the upper right corner, with few or no points in either the upper left or lower right corner. Redrawn after Krasnov et al. (2004e) (reprinted with permission from Blackwell Publishing).

differences in host availability. Indeed, the degree of resource specialization of a flea species may differ depending on the scale of observation because of two interrelated factors, namely local resource (i.e. host) specialization and the substitutability of those resources (i.e. hosts) across locations (Hughes, 2000). As a result, a species that is considered to be host-specific on a local scale may appear to be host-opportunistic on a regional scale. For example, *Amalaraeus penicilliger* has been recorded in various hosts belonging to different orders across a great part of its geographical range (Holland, 1985), whereas in Mongolia it exploits only two closely related host species (Kiefer et al., 1984). In other words, availability of a resource can profoundly affect the degree of host specificity in a local population. The set of host species used by a flea species in a given region could be either a random draw of locally available host species, or highly dependent on the taxonomy/phylogeny of the hosts.

Using the data set described in the previous section and randomization tests, Krasnov et al. (2004e) tested the null hypothesis that the S_{TD} and $VarS_{TD}$ values for host species used by a flea in a region are no different from those of random subsets of the regional pool of available host species. Among 23 common flea species (those that occurred in at least five regions) from 21 geographical regions, there were 86 cases in which a flea species occurred on at least three host species in a region. In 58 of these 86 cases, the observed S_{TD} value for the flea species did not differ significantly from those of the 10 000 random selections of host species from the regional pool (see Fig. 14.5 for an example with fleas from the Tarbagatai Mountains). However, in 26 of the 28 cases where there was a significant difference, the observed S_{TD} was lower than the values of the random subsets. Lower values than expected were significantly more frequent than higher values (26 versus 2). This means that when host use departed from random, the parasite utilized host species that were more closely related to each other than on average across the regional pool. However, the taxonomic affinities of host species chosen by a flea did not tend to be more or less symmetrical than those of random selections from the regional pool of available host species.

The explanation of these results may have something to do with host compatibility. The appropriateness of a host species for a flea species is determined by (a) the ability of the parasite to acquire the resources provided by this host and (b) the ability of the parasite to use these resources successfully. The successful acquisition of resources is related to host defence mechanisms (e.g. grooming behaviour and immune responses) and the capability of a parasite to cope with these mechanisms, whereas the successful use of resources is related to certain properties of these resources (e.g. physical and chemical parameters of host blood). The tendency for a set of host species used by many fleas to be

292　Ecology and evolution of host specificity

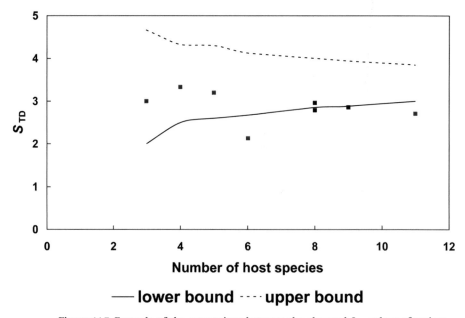

Figure 14.5 Example of the comparison between the observed S_{TD} values of various flea species and randomized selections of host species from the regional host species pool. Shown here are results for eight flea species from the Tarbagatai Mountains (data from Mikulin, 1958). Each point represents the observed value for a flea species. The upper and lower bounds encompass 95% of the simulated random values, forming a funnel that gets narrower as the number of host species exploited by a flea increases. Points above the funnel indicate flea species with higher S_{TD} values than expected from random selections of host species, and points below indicate flea species with lower S_{TD} values than expected. Redrawn after Krasnov et al. (2004e) (reprinted with permission from Blackwell Publishing).

taxonomically clustered may stem from similarities in both the 'defence' and 'resource' parameters of closely related host species. Indeed, blood parameters are likely to be similar in closely related host species. Consequently, the efficiency of feeding on closely related hosts is expected to be similar and to differ from that on distant hosts. For example, *Xenopsylla skrjabini* digested blood of the house mice *Mus musculus* and the great gerbils *Rhombomys opimus* (both belonging to the family Muridae) over 15–18 h, whereas it took this flea 24–30 h to digest blood from guinea pigs (Vashchenok, 1974; see Table 10.1). Behavioural anti-parasite defence mechanisms also tend to be similar in closely related species (Mooring et al., 2004). Although the similarity of the immune responses of closely related hosts to the same parasite species has never been tested specifically, some findings indirectly suggest that this is the case (e.g. Galbe & Oliver, 1992). Consequently, when a flea species adds a host species to its repertoire, this new host species is, as a rule, taxonomically related to one or more host species from the existing host spectrum.

Nevertheless, no effect of the characteristics of a regional host species pool on the characteristics of the host assemblage used by a flea was found. This means that, in general, local host availability (in terms of host number and taxonomic diversity of the host pool) does not influence the flea's host specificity. This finding supports the idea that the ability of a flea species to use a certain set of hosts is genetically constrained. However, the local availability of taxonomically related species in a region can affect to some extent the realized level of host specificity of a given flea species in this region and thus contribute to variability in host specificity across regions.

As we already know, arthropod ectoparasites, such as fleas, are strongly influenced by their off-host environment (e.g. the microclimate of host's nest/burrow: Krasnov et al., 2002b; see Chapters 10 and 11). This dependence on the off-host environmental conditions can mask the true spatial pattern of host specificity in these parasites. This masking is likely to be unidirectional, decreasing rather than increasing the true level of host specificity. For example, some hosts that are exploited by a flea species in one locality may be dropped out of the host spectrum of this flea in another locality due to the unsuitability of the microclimate and/or substrate of the host's burrow in this locality for pre-imaginal flea development. Indeed, *Xenopsylla ramesis* parasitizes several gerbilline species throughout the Middle East (Lewis & Lewis, 1990). However, in some areas Sundevall's jird *Meriones crassus* is left out of the host spectrum of this flea (Krasnov et al., 1997) due to the unsuitability of the microclimatic and substrate conditions in *M. crassus* burrows for successful survival of eggs, larvae and newly emerged imagoes of *X. ramesis* (Krasnov et al., 2001a, 2002b). In many flea species examined by Krasnov et al. (2004e) (especially those that tend not to be very host-specific), at least one measure of host specificity correlated positively or negatively with at least one of the parameters that described the deviation of environmental conditions in a location from their mean value calculated across the entire geographical range of a flea species. This indicated that host specificity was also influenced by local factors.

In addition, the true level of host specificity in a parasite species can be distorted in its peripheral populations because the latter often live under conditions very different from those of core populations. Peripheral areas, on the edges of a flea's geographical range, are often characterized by variable and suboptimal conditions (in terms of both host populations and off-host environment), relative to core areas. Peripheral populations are thus expected to be more variable, since the variable conditions induce fluctuating selection, which itself maintains high genetic diversity (e.g. Volis et al., 1998). Alternatively, due to marginal ecological conditions at the periphery, populations there may be small and isolated and adapted to a narrower range of ecological conditions (Carson, 1959). However, no correlation between any of the host-specificity measures and the relative distance of a location from the centre of the geographic range was found in any of the 23 flea species studied (Krasnov et al., 2004e).

To conclude, though far from being constant and despite some effect of the availability of taxonomically related host species in the region, and some modulation of local environmental conditions, host specificity in fleas can still be considered as a species character.

14.2.4 Distribution of specialization and flea species richness

Studying parasite assemblages of Canadian freshwater fish, Poulin (1997) found host-specific parasites mainly in species-rich assemblages and host-opportunists mainly in species-poor assemblages. This trend was observed in four out of five fish families considered (except for Salmonidae). A reason for this non-random pattern of community composition might be heterogeneity among host species in the rate at which they accumulate parasites (Poulin, 1997). In contrast, Valtonen et al. (2001) found the opposite pattern in parasite assemblages of fish from the northeastern Baltic Sea, namely species-poor assemblages included more host-specific parasites than richer assemblages. Vázquez et al. (2005) argued that both these studies have limitations and approached the question about distribution of parasite specialization as affected by parasite community composition in the context of host–parasite interaction networks. One of the data sets for this study was represented by flea assemblages on small mammals from the Holarctic.

The results of Vázquez et al. (2005) not only supported the findings of Poulin (1997), but represented an important step toward a better understanding the evolution of specialization and the coevolution of host–parasite interactions. It appeared that mammal–flea interactions were asymmetrically specialized. Most flea species were specialists interacting with 'generalist' hosts (i.e. hosts that harboured rich flea assemblages), or were generalists themselves. Similarly, most host species either were exploited by few generalist fleas or were 'generalists' themselves. As a result, hosts should represent strong selective agents for host-specific parasites as they rely on a small number of host species only. However, 'generalist' hosts are parasitized by many parasite species, so the selective importance of these parasites is probably weak. In contrast, selective pressure from each of the many host species of a host-opportunistic parasite is probably weak. But the selective pressure of each host-opportunistic parasite on the host is likely to be high because many of these hosts are 'specialists' (i.e. parasitized by few species). Thus, specialists are strongly selected by generalists but not vice versa.

14.3 Host specificity and evolutionary success

14.3.1 Abundance and fitness achieved in hosts

One of the most pervasive macroecological patterns is an interspecific positive relationship between local abundance and occupancy (e.g. Gaston, 2003). This is a trend for species with larger local populations to be more widespread across many localities, whereas species with smaller local populations tend to occur in fewer localities. One of the mechanisms explaining the positive local abundance–regional distribution relationship is based on interspecific variation in niche breadth (the resource breadth hypothesis: Brown, 1984, 1995). When applied to parasites, this hypothesis predicts that host-opportunists will attain higher local abundances and will have broader distributions that host-specific species. In other words, the same attributes that enable a species to exploit a variety of hosts allow it to attain a high density in these hosts. Studies on some parasites supported the resource breadth hypothesis as an explanation for the positive correlation between distribution and local abundance. For example, Barger & Esch (2002) investigated a community of parasites infecting fish in small streams in North Carolina and found that the number of host species infected by each parasite species was positively related to both the frequency of occurrence among streams and the average local abundance. In sharp contrast with these findings, Poulin (1998) observed a negative relationship between the number of fish host species used by 188 species of metazoan parasites and their average abundance in hosts. He explained this apparent trade-off between the number of host species exploited and the abundance achieved by parasites in these hosts by the presumably high cost of parasite adaptations to multiple host defence mechanisms (the trade-off hypothesis) (Poulin, 1998). Parasites that specialize to exploit a few hosts may attain greater abundance in these hosts than if they were exploiting a broader host spectrum and, consequently, were forced to invest more in a wider range of adaptations against host defence mechanisms (Combes, 1997; Poulin, 1998). Other living conditions, such as microhabitat characteristics and food quality, will also vary among host species, and parasites exploiting many host species may require further adaptations to cope with these variable conditions. Given that the abundance of a consumer in a habitat reflects its efficiency of resource exploitation (Morris, 1987a), the trade-off hypothesis states, in other words, that the broader the host spectrum of a parasite the lower the efficiency of exploitation of any particular host.

However, the trade-off hypothesis is mostly true only if a host assemblage is composed largely of phylogenetically distant host species. In contrast, there is no need to invest more in evasive adaptations if a parasite species exploits several closely related hosts because their behavioural and immunological defence mechanisms are likely to be similar (see above). Consequently, if a host assemblage consists mainly of closely related hosts, there may be no trade-off between the number of host species used and the abundance of parasites in these hosts. The above considerations suggest that the abundance–host specificity relationship will vary with respect to the taxonomic composition of the assemblage of available hosts.

The trade-off hypothesis and the resource breadth hypothesis thus make different predictions about the differences in local abundances between parasites with different host specificity. According to the former, adaptations required by host-opportunistic parasites to overcome host defence systems may occur at the expense of the parasites' ability to attain high abundances in these hosts. A 'jack-of-all-trades' parasite would therefore be a 'master of none' and its abundance even in an optimal host species should be lower than that of a host-specific parasite in the same host, all else being equal. In contrast, the latter hypothesis suggests that a 'jack-of-all-trades' parasite will be a 'master of all' (Brown, 1995) and will attain high abundance in all or the majority of its host species. Consequently, the abundance of host-opportunistic parasites, at least in their optimal hosts, will be higher or equal to that of host-specific parasites in the same hosts, all else being equal.

So, how do 'jack-of-all-trades' fleas behave? This was tested by assessing the interspecific relationship between mean abundance and host specificity using data on fleas parasitic on small mammals in 20 geographical regions (Krasnov et al., 2004f). It appears that in most of the regions studied, the breadth of the host range (inverse level of host specificity) correlated positively with local abundance of fleas (Fig. 14.6). Although these two parameters were not interrelated in the other regions, no trade-off between host range size and local abundance was found in any region.

In other words, host-opportunistic flea species are also the ones that attain higher local densities, thus supporting the resource breadth hypothesis but not the trade-off hypothesis. Furthermore, different measures of host specificity, i.e. the number of host species and S_{TD}, demonstrated the same pattern. This suggests that some features of flea species that allow them to attain high densities in a host also allow these fleas to exploit more host species from a wider range of taxa. However, despite the fact that these features, whatever they are, allow an increase in the taxonomic diversity of the host range, they are not usually associated with a greater taxonomic complexity of this range as was suggested

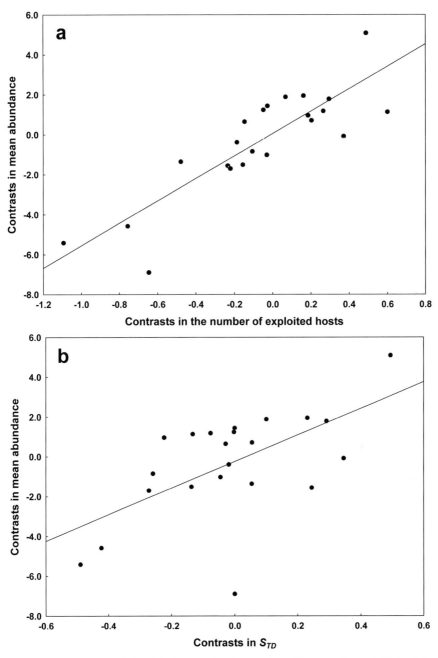

Figure 14.6 Relationship between (a) the number of host species and (b) the index of host specificity (S_{TD}) and mean abundance of fleas per host individual (controlled for host sampling effort and area of host body surface) among flea species from Mongolia using the method of independent contrasts. Redrawn after Krasnov et al. (2004f) (reprinted with permission from the University of Chicago Press).

by the lack of correlation between local abundances and $VarS_{TD}$. Consequently, siphonapteran 'jacks-of-all-trades' appear also to be masters of most, if not all, of them.

There are two main arguments that can explain why resource generalists are expected to have higher local abundances than resource specialists (Hughes, 2000). First, the total amount of resources available to a generalist species may be greater than the total amount of resources available to a specialist. Indeed, for most fleas in each region, host-specific fleas were found to exploit a subset of the host species that are also exploited by host-opportunistic fleas. For example, in the northern Kyrgyzstan, the pygmy woodmouse *Apodemus uralensis* is the only host species of the highly specific *Ctenophthalmus golovi*, but it is also exploited by the host-opportunistic *Amphipsylla rossica*, *Amphipsylla anceps*, *Frontopsylla ornata*, *Neopsylla teratura* and *Neopsylla pleskei* (Shwartz et al., 1958). Second, the ability of generalist species to maintain higher local abundances may be related to lower variability in the total amount of resources available to them (MacArthur, 1955). For example, a highly host-specific flea may suffer from a high risk of population crash if the population of its single host species decreases sharply in a given year; in contrast, if one or a few hosts of a host-opportunistic flea undergo drastic population decrease, even going locally extinct, this flea can easily survive on other host species. Evidence in support of this hypothesis in fleas is extremely scarce. Nevertheless, year-to-year variation in the mean abundance of the host-opportunistic *Megabothris turbidus* and *Ctenophthalmus uncinatus* (17 and 15 host species, respectively) in the Volga–Kama Region of Russia did not depend on density fluctuations of their hosts, whereas the opposite was true for the host-specific *Palaeopsylla soricis* (85% of individuals of this species were recovered from a single host species, the common shrew *Sorex araneus*) (Nazarova, 1981).

Within-region comparisons of maximal abundance of fleas with different degrees of host specificity also supported the resource breadth hypothesis (Fig. 14.7). As mentioned above, the abundance of a species in a habitat reflects the extent to which local conditions in that habitat meet the multiple Hutchinsonian niche requirements of a species (Brown, 1984, 1995) and, therefore, can be considered as a measure of habitat suitability for that species. Although the whole host range of a host-opportunistic parasite can be quite broad, not all host species within this range are equally suitable for this parasite species (e.g. Prasad, 1969; see also Chapters 10 and 11). Consequently, the host in which the abundance of a host-opportunistic flea is the highest can be regarded as an optimal (i.e. principal) host species for this flea (Krasnov et al., 2004c; see also Chapter 9). It appeared that host-opportunistic fleas attained higher abundances than host-specific fleas even when only the optimal hosts were taken

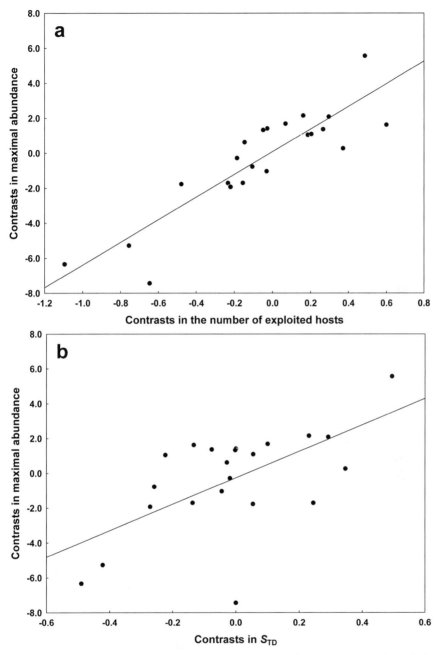

Figure 14.7 Relationship between (a) the number of host species and (b) the index of host specificity (S_{TD}) and maximal abundance of fleas per host individual (controlled for host sampling effort and area of host body surface) among flea species from Mongolia using the method of independent contrasts. Redrawn after Krasnov et al. (2004f) (reprinted with permission from the University of Chicago Press).

into account, again rejecting the idea that a 'jack-of-all-trades' is 'master of none'.

The absence of a negative trend between local flea abundance and their ability to exploit many host species as predicted by the trade-off hypothesis does not, however, mean that host-opportunistic fleas are not faced with the necessity of developing multiple adaptations to evade multiple host defence systems (behavioural or immunological or both). However, in the case of fleas, this trade-off can be counterbalanced by another trade-off related to the non-parasitic lifestyle of flea larvae. Differences among host species in burrow conditions (e.g. the microclimate and organic content of the substrate) are presumably less pronounced, especially within a region, than differences among those hosts in means of defence. The larvae can, therefore, survive similarly in burrows of different host species. In addition, mammals of different species often visit each other's burrows (Kucheruk, 1983; see Chapter 9). Larvae of fleas that exploit multiple hosts can thus achieve a broader spatial distribution and, consequently, the imago upon emergence has a higher probability of successfully attacking a host individual. The cost of adaptations against the defence system of a new host species can, therefore, be compensated for by the higher success of newly emerged imagoes attacking an appropriate host.

Alternatively, the absence of any abundance/specificity trade-off in fleas may be related to cross-resistance of a host against closely related ectoparasites (see Chapter 13). In other words, a host can develop defence mechanisms that are equally effective against multiple ectoparasite species, including those that it has not previously met. A flea species that colonizes a new host species can thus encounter immune responses with which it is already familiar (from its previous hosts). Consequently, it may not decrease its exploitation success and could even attain a relatively high average abundance on this new host.

14.3.2   Geographical range

As we have seen, 'jack-of-all-trades' fleas succeed in attaining a high level of abundance in their hosts. What about geographical range? Do the same features, whatever they are, that allow a flea to attain high densities in a host and exploit more host species from a wider range of taxa also allow it to achieve a broader geographical distribution? This could be expected because the niche of a flea species includes not only a set of host species but also a set of environmental conditions under which these hosts occur. These conditions affect both imagoes (because they periodically leave a host body) and pre-imaginal fleas (because they develop mainly off-host). As a result, a host-opportunistic flea is expected to tolerate a wider range of physical conditions compared with a host-specific flea.

Table 14.1 *Summary of major axis regressions (forced through the origin) of independent contrasts in number of host species (HN) and specificity index (S_{TD}) against independent contrasts in geographical range size for fleas from different regions*

Region	HN	S_{TD}	Reference
Venezuela	0.56	0.34	Tipton & Machado-Allison, 1972
Canada, Alaska and Greenland	0.60	0.45	Holland, 1985
Australia	0.58	0.58	Dunnet & Mardon, 1974
South Africa	0.77	0.47	Segerman, 1995
Morocco	0.21	0.12*	Hastriter & Tipton, 1975
Mongolia	0.26	0.33	Kiefer et al., 1984
Asian Far East	0.61	0.61	Yudin et al., 1976

Note: All values for r, except those marked by *, are significant ($p < 0.007$).
Source: Data from Krasnov et al. (2005a).

Using published data on the geographical distribution and host occurrences of fleas from seven large geographical regions, Krasnov et al. (2005a) searched for a correlation between the degree of host specificity of flea species and its geographical range. In most of the regions studied, this correlation measured via either the number of host species used or the index S_{TD} was strongly negative, although no relationship between the taxonomic asymmetry of host spectrum and geographical range was found (Table 14.1).

Again, the resource breadth hypothesis was strongly supported. Host-specific fleas have more restricted geographical ranges than host-opportunistic fleas. Taking into account that flea species with a broad geographical range not only are capable of exploiting more host species, but also exploit host species from a wider range of taxa, it appears that as the geographical range of fleas expands, not only are additional hosts being used, but these come from increasingly phylogenetically distant taxa.

First, this supports the idea that both host species and physical off-host conditions should be taken into account when an ecological niche of a flea species is considered. As I have already shown in Chapter 10, the cosmopolitan *Xenopsylla cheopis* can successfully digest blood from laboratory hamsters, mice, rats, guinea pigs and pigeons (Vashchenok et al., 1976). The broadly distributed *Xenopsylla brasiliensis* can complete its life cycle at a wide range of relative humidities, i.e. 51–95% (Bahmanyar & Cavanaugh, 1976). The same mechanisms that enable a flea to exploit either few or many host species occurring under either a narrow or a wide range of environmental conditions can be the reason why it has either a small or a large geographical range.

Second, the negative relationship between the degree of host specificity and size of geographical range can also stem from the pattern of frequency distribution of geographical ranges. The geographical range of a host-specific parasite can be either equal to or smaller than the combined range of its hosts. As mentioned above, the distributions of within-taxon geographical range sizes tend to be unimodal with a strong positive skew, i.e. most species have relatively small range sizes, whereas a few have relatively large ranges (Gaston, 2003; see Chapter 3). This is true in particular for different mammalian taxa (e.g. Eeley & Foley, 1999). A host-specific flea is, therefore, expected to have, on average, a small geographical range simply due to the high probability of its mammalian host also having a small geographical range. This mechanistic approach cannot, though, clearly explain the large geographical ranges of host-opportunistic fleas, except if the geographical range of a host-opportunistic flea is simply the summation of the geographical ranges of its multiple hosts (if the degree of overlap among the geographical ranges of these hosts is relatively small).

Third, the negative correlation between the degree of flea specificity and size of geographical range can be related to latitudinal gradients in species richness, niche breadth and geographical ranges. The reason for niche breadth being narrower at low latitudes may be increased interspecific competition due to a higher number of co-occurring species (e.g. Brown, 1975), the relative stability of environmental conditions that allows the persistence of specialized species (e.g. Chesson & Huntly, 1997; but see Vázquez & Stevens, 2004) and/or global processes such as Milankovitch oscillations (climatic changes due to periodical changes in the orbit of the Earth) (Dynesius & Jansson, 2000). The latter cause changes in the size and location of species' geographical distributions called 'orbitally forced species' range dynamics', the magnitude of which is positively correlated with latitude. In the data set used in this study, the average values of the specificity index (S_{TD}) differed significantly between fleas from regions of similar longitude but different latitude, being higher closest to the equator (e.g. 2.80 for fleas from Canada versus 3.42 for fleas from Venezuela). However, this index did not differ between fleas from regions of similar longitude and latitude on different sides of the equator (2.55 for fleas of Morocco versus 2.53 for fleas of South Africa).

Finally, there is a positive correlation between species range size and latitude (Rapoport, 1982; Stevens, 1989) identified as 'Rapoport's rule' by Stevens (1989). Taken together, these three latitudinal gradients could result in the pattern observed in this study, i.e. the increase of geographical range size with decreasing host specificity.

In contrast, the level of taxonomic heterogeneity among a group of host species (estimated as $VarS_{TD}$) did not differ between narrowly and widely distributed flea species and seemed to depend on other still unknown factors.

Relationships between geographical range and $\text{Var}S_{TD}$ were surprisingly very consistent across all regions considered. This suggests that the irregularity of the host taxonomic tree is an invariant parameter, i.e. that when the geographical range increases, new host species add more diversity but not more complexity to the host taxonomic tree.

The pattern of a negative relationship between the degree of host specificity and size of geographical range persists despite some notable exceptions, such as highly host-specific fleas that demonstrate broad geographical distributions due to the broad distribution of their hosts. For example, the principal host of the rather host-specific *Tarsopsylla octodicemdentata* is the Eurasian red squirrel *Sciurus vulgaris* which is distributed across most of Eurasia. In the New World, *T. octodicemdentata* occurs on the closely related North American red squirrel *Tamiasciurus hudsonicus* which is distributed across most of North America. Another type of exception from the observed pattern consists of host-opportunistic fleas with restricted geographical distributions. For example, the South African *Dinopsyllus ellobius* and *Dinopsyllus lypusus* were found on a wide range of rodent hosts (32 and 20, respectively). However, their geographical distributions are restricted mainly to the higher-rainfall areas of the eastern part of the Southern African subcontinent (Segerman, 1995). The reasons for the absence of these fleas from the other parts of the geographical ranges of their numerous hosts might be some as yet unknown abiotic preferences of the pre-imaginal and/or adult insects. An analogous trend was found for avian fleas (Tripet *et al.*, 2002a). Although *Ceratophyllus hirundinis* and *Ceratophyllus farreni*, specialized parasites of the house martin *Delichon urbica*, have large geographical ranges, most specialized avian fleas exhibit a relatively narrow geographical distribution.

High numbers of 'exceptional' species in the flea fauna can cause deviations from the reported trend. Indeed, no correlation between the degree of flea host specificity and host geographical range was found for fleas in Morocco. The composition of the Moroccan flea fauna is characterized by a relatively high percentage of highly host-specific fleas with relatively broad geographical distributions (*Xenopsylla nubica*, *Spilopsyllus cuniculi*, *Leptopsylla taschenbergi*, *Leptopsylla algira*) and some host-opportunistic fleas with restricted geographical distributions (*Xenopsylla blanci*, *Nosopsyllus oranus*, *Nosopsyllus barbarus*, *Ctenophthalmus andorrensis*). These eight flea species represent 35% of the species-poor Moroccan flea fauna.

14.4 Host specificity and host features

From an evolutionary perspective, selection for higher or lower levels of ecological specialization (i.e. host specificity) is affected by a variety of both parasite (i.e. forager)-related and host (i.e. resource)-related factors (Fox & Morrow,

1981; Futuyma & Moreno, 1988; Fry, 1996; Desdevises *et al.*, 2002b). In particular, several host features such as predictability (longevity, level of abundance, spatial distribution) and defensibility may determine whether natural selection will favour an increase or a decrease in a parasite's host specificity.

14.4.1 Predictability: host body mass and longevity

The main resource for a parasitic species is its host, which provides a parasite with food, habitat and mating grounds. The models by Ward (1992) suggested that species tend to specialize on predictable resources, i.e. resources that are relatively stable in both space and time. This minimizes the extinction rate for an optimal forager. Consequently, specialization in parasites is expected to be associated with the level of predictability of its host resources. Persistence of a host individual in time is in turn associated with its size. In general, larger host species live longer and, thus, represent a more predictable resource for a parasite (Peters, 1983). In addition, larger hosts may offer more niches for parasites. If so, they are expected to harbour mainly parasite species with higher host specificity, whereas small-bodied hosts should be exploited mainly by generalist parasites. This hypothesis was found to be supported for monogenean ectoparasites of fish (Sasal *et al.*, 1999; Šimková *et al.*, 2001).

This prediction was also tested for fleas by quantifying the association between the level of host specificity and the mean body mass of their mammalian hosts, using published data from two large, distinct geographical regions (South Africa and North America) (Krasnov *et al.*, 2006b). The approach of the study was twofold. First, for each flea species, the mean body mass of all exploited host species and the above-mentioned measures of host specificity were calculated and correlated across flea species. This approach determines whether host-specific fleas do indeed exploit larger host species than generalist fleas. Second, another way of addressing the same issue is to determine whether large-bodied host species harbour more host-specific fleas than small-bodied ones. Consequently, host specificity measures were averaged across all flea species recorded on a particular host and correlated with host body masses across all host species.

A weak but consistent association between the level of flea host specificity and host body mass was found. However, this association was not always supported by the method of independent contrasts and was somewhat differently expressed in the two geographical regions. The relationships between host body mass and flea host specificity showed a similar pattern when considered from both the host and flea perspectives. In general, host-opportunistic fleas (with a high number and/or low taxonomic 'evenness' of exploited hosts) exploited mainly small-bodied hosts, whereas host-specific fleas tended to use larger hosts (Fig. 14.8).

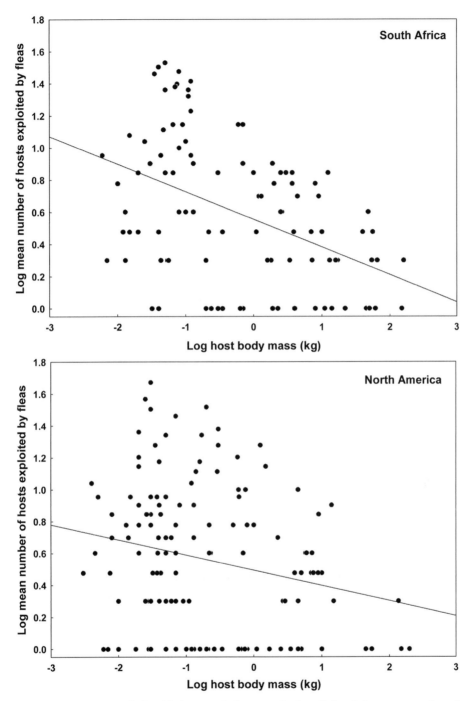

Figure 14.8 Relationship between body mass of a host (kg) and the mean number of hosts exploited by the fleas harboured by this host, across host species from two geographical regions. Redrawn after Krasnov et al. (2006b) (reprinted with permission from Cambridge University Press).

The negative but triangular distribution of points in the relationship between body mass and flea host specificity (Fig. 14.8) implies that smaller hosts can harbour both host-specific and host-opportunistic fleas, whereas larger hosts harbour mainly host-specific fleas. This, as well as the relative weakness of the association between host body size and flea host specificity, suggests that resource predictability alone cannot explain the patterns and pathways of the evolution of host specificity in fleas. Other, still largely unknown, factors must be involved as well. In addition, in most cases when the relationship was found, it was stronger for South African than for North American flea–mammal associations (Krasnov et al., 2006b). This difference may be associated with the differences between these two regions during the Cenozoic, when the main evolutionary development of flea–mammal associations occurred (Medvedev, 2005). Glaciations with repeated advances and retreats of ice-sheets were characteristic for North America but not for South Africa, especially during the Quaternary. Glaciation–interglaciation cycles may have led to the repetitive break-ups and restorations of associations between flea and host species in North America, whereas these associations in South Africa were possibly more stable on a geological timescale. The 'historical instability' of the relationships between flea and host species in North America could have resulted in a relatively weak association between flea and host traits.

14.4.2 Predictability: host abundance, coloniality and spatial distribution

Other parameters that may determine the predictability of host resources for parasites are host abundance, social structure and pattern of spatial distribution. If populations of a host species are large, or a host lives in large colonies or social groups, or the spatial distribution of a host is characterized by clusters of individuals, then natural selection in a parasite that exploits this host may favour high host specificity. On the one hand, availability of a host to exploit is almost guaranteed for such a parasite, whereas on the other hand, it does not need to develop multiple countermeasures to cope with multiple defence mechanisms originating from several different hosts.

Another, not necessarily alternative, effect of host sociality or/and spatial distribution may be related to gene flow among parasite populations. This is supposedly low when a host is characterized by a clumped distribution, promoting high levels of host specificity (Tripet & Richner, 1997b; Tripet et al., 2002a, b). Indeed, parasites exploiting highly social hosts with colonies persistent for long periods sometimes demonstrate extreme host specificity by being adapted to a single colony of a host species (Schönrogge et al., 2006). The effect of host abundance, sociality and/or spatial distribution on parasite host specificity has been shown for several host and parasite taxa (e.g. Ezenwa, 2004; Šimková et al., 2006).

Our knowledge of this effect in fleas is scarce. Nevertheless, as was repeatedly mentioned above, the highly host-specific *Parapulex chephrenis* exploits *Acomys cahirinus* which is characterized by a communal nesting pattern and a considerable level of sociality (Shargal et al., 2000). In contrast, there are no host-specific species among a plethora of fleas parasitic on one of the most social rodents, the great gerbil *R. opimus* (e.g. Zagniborodova, 1968). This may be because colonies of these species are inhabited by a variety of other small mammals (Kucheruk et al., 1972).

Recently, Tripet et al. (2002a) investigated the relationship between bird spatial distribution and flea host specificity (measured as host range) in a comparative analysis of the major group of avian fleas from the family Ceratophyllidae. It appeared that flea species parasitizing colonial birds had narrower host ranges than those infesting territorial nesters or birds with an intermediate level of nest aggregation. It should be noted, however, that this analysis was carried out at the level of host genera and that no phylogenetic corrections for hosts were done. Tripet et al. (2002a) interpreted these results to imply that the effect of host spatial distribution on the metapopulation dynamics and genetics of fleas is via probability of transmission and amount of variation in environmental conditions met by subpopulations of fleas in different host nests. In particular, fleas from colonial nesters (e.g. swallows), have a high chance of re-infesting a host of the same species within the same colony. In such a case, both gene flow and variation in ecological conditions are low. This supposedly favours high host specificity.

When associations between all flea species and all host species in a locality are considered as an interaction network (see Vázquez & Aizen, 2003 for definition), the influence of species abundance patterns on the distribution of specialization in this network can be evaluated with the aid of null models (Vázquez et al., 2005). Using data on fleas and their small mammalian hosts from 25 Holarctic localities, Vázquez et al. (2005) demonstrated that abundant hosts tend to harbour richer parasite faunas, many of which are specialists. The results of this study also suggested that the causal link between abundance of hosts and distribution of specialization in parasite communities is that parasite species interact with host species as they encounter them. In other words, they will encounter abundant hosts more often than rare ones.

14.4.3 Host defensibility

The role of host defences in parasite specialization is largely unknown. Nevertheless, it has been hypothesized that adaptation by a given population of parasites to a given host and its defences should reduce the ability of this parasite population to exploit other hosts (Møller et al., 2005). Møller et al. (2005) used immunocompetence, i.e. intensity of T-cell-mediated immune response measured via the PHA test (see Chapter 13), as an estimator of avian hosts species' defence abilities and investigated the relationship between this parameter and the level of specialization in fleas parasitic on these hosts. When the relationship between host immunocompetence and the level of flea specialization was studied controlling for phylogenetic relationships among birds, a negative relationship was found. As the T-cell-mediated immune response of bird species increased, the number of main and accidental hosts parasitized by their flea species decreased. The distribution of data points was triangular showing that weakly immunocompetent hosts mainly supported both host-specific and host-opportunistic flea species, whereas highly immunocompetent hosts were only exploited by host-specific fleas. When the relationship between host immunocompetence and the level of flea specialization was studied controlling for phylogenetic relationship among fleas, the negative relationship between the two parameters with a triangular distribution of data points held. In other words, the number of species of flea per host increased significantly with an increase in host immune response. Path analysis carried out by Møller et al. (2005) suggested that host coloniality and host immunocompetence independently contributed to the level of flea host specificity, but that the effect of immunocompetence was considerably stronger than the effect of host coloniality.

14.5 Evolution of host specificity: direction, reversibility and conservatism

14.5.1 Is host specificity directional and irreversible?

Comparisons between the phylogeny of a parasite taxon and that of its hosts can help us to understand how parasites and hosts have coevolved since the origin of their association (see Chapter 4) and why some parasites are highly host-specific, whereas others are not (Brooks & McLennan, 1991; Page, 2003; Poulin, 2007a). If the two phylogenies are completely congruent, the following scenario can be suggested (Poulin, 2007a). A speciation event that took place in an ancestral host population harbouring one species of parasite was caused by some barrier to host gene flow between two subpopulations of a host. This same barrier could also prevent gene flow between the two new subpopulations of parasite. Repetitions of this process would result in a particular number of host species and the same number of parasite species, with each parasite being strictly host-specific. This scenario appears to be rather rare in reality and has been supported for a rather limited set of parasite–host associations (e.g. Hafner & Nadler, 1988, 1990; Hafner & Page, 1995; but see Chapter 4).

On the other hand, host specificity can also decrease over time when (a) the original host speciates without concomitant speciation of the parasite; and/or (b) the addition of new host species to a parasite's host spectrum results from host-switching or the colonization of new host species. Both these processes cause incongruence between the topologies of the two phylogenetic trees. Furthermore, it seems that host-switching is a much more common case than cospeciation (Barker, 1991; Krasnov & Shenbrot, 2002; Taylor & Purvis, 2003; see Chapter 4).

As was shown in Chapter 4, the only two studies that have attempted to compare phylogenies of fleas and their hosts (Krasnov & Shenbrot, 2002; Lu & Wu, 2005) found that the history of these associations was not characterized by cospeciation. Krasnov & Shenbrot (2002) concluded that host-switching was the most common case in flea–jerboa associations due to ecological and geographical factors. As a result, fleas parasitic on jerboas are quite host-opportunistic. The lack of cospeciation with ochotonid hosts was also reported for fleas of the genus *Geusibia* (Lu & Wu, 2005). However, many *Geusibia* species are fairly host-specific. Again, ecological and historical–geographical factors have been implicated to explain the relatively high level of host specificity of these fleas despite an apparent lack of cospeciation with their hosts.

Having found a positive relationship between avian host coloniality and host specificity in ceratophyllids (see above), Tripet et al. (2002a) asked whether the further speciation of already specialized fleas was promoted by environmental conditions associated with host aggregation or if host-specific taxa repeatedly originated from host-opportunistic ones. There is no clear answer to this question as yet. On the one hand, Tripet et al. (2002a) argued that in the period of extensive radiation that followed the switch of ceratophyllids from arboreal mammalian hosts to birds, colonial birds and their fragmented habitat offered more speciation opportunities to the early non-specialized fleas than communities of territorial birds. On the other hand, using various ceratophyllids as examples, these authors suggested that new host-specific species evolved both from host-specific and host-opportunistic ancestors, although secondary host opportunism originating from a host-specific ancestor is also possible (see example with *Mioctenopsylla traubi kurilensis* in Tripet et al. (2002a)). This latter example raises another question related to the evolution of host specificity, namely is this evolution directional and irreversible?

Indeed, parasite specialization is generally presumed to be irreversible, leading into evolutionary blind alleys that do not give rise to new lineages. On the one hand, specialist taxa, capable of using only a narrow range of host species, should be less likely to colonize new hosts, and therefore the potential of specialists to give rise to new lineages should be limited (Jaenike, 1990). If this is so, we might expect that generalists could evolve into specialists, but that the likelihood of specialists evolving into generalists would be much lower. Thus, within a clade, the more specialized species should on average be the more derived, i.e. the more recent ones. For example, Jameson (1985) argued that there is a tendency for the more basal flea families (Stephanocircidae, Pygiopsyllidae and Hystrichopsyllidae) to contain a few pleioxenous (i.e. exploiting hosts belonging to the same family) but many polyxenous genera (i.e. exploiting hosts from several families), whereas evolutionary younger families of fleas (Leptopsyllidae and Ceratophyllidae) are characterized by a high percentage of pleioxenous genera. On the other hand, specialist taxa should be more prone to extinction than generalists, because of their strict dependence on a narrow range of host species, and thus we might expect generalist taxa to be favoured and to proliferate over evolutionary time. It is therefore not straightforward to predict in which direction host specificity will evolve in a given taxon, i.e. whether it will tend to increase or decrease over evolutionary time. Recent studies on various animal taxa, including parasites, cast doubt on the paradigm that specialization is both directional and irreversible (Nosil, 2002; Stireman, 2005).

Poulin et al. (2006b) tested for directionality in the evolution of host specificity in fleas parasitic on small mammals. They determined whether host specificity, measured both as the number of host species used and their taxonomic distinctness (S_{TD}), is related to clade rank of the flea species. The latter was evaluated as the number of branching events between an extant species and the root of the phylogenetic tree; it can be used to distinguish flea species that are basal in the phylogenetic tree from those that are highly derived, i.e. those with low and high clade rank, respectively (see details in Poulin et al., 2006b). It was found that there were weak positive relationships between clade rank and the number of host species used (Fig. 14.9). This was true across all flea species in the data set (297 species parasitic on Didelphimorphia, Soricomorpha, Lagomorpha and Rodentia), as well as within the family Hystrichopsyllidae (but not within Ceratophyllidae, Leptopsyllidae and Pulicidae) and within the genus *Neopsylla* (but not within *Amphipsylla*, *Frontopsylla*, *Paradoxopsyllus*, *Ctenophthalmus*, *Rhadinopsylla* and *Xenopsylla*). Positive relationships between clade rank and the index S_{TD} were found within *Xenopsylla* only. These results suggested a slight evolutionary trend of decreasing host specificity, with many flea lineages increasing over evolutionary time the number of host species they can exploit. However, using a more conservative test, these trends could not be distinguished from a non-directional random-walk model, suggesting a lack of directionality in the evolution of host specificity in fleas (Poulin et al., 2006b). This can be seen from the scatter of points

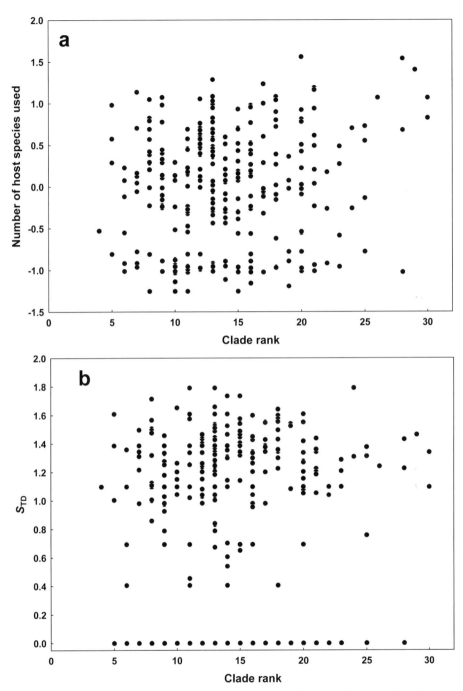

Figure 14.9 Relationship between (a) the number of host species (controlled for host sampling effort) used by a flea species and (b) the taxonomic distinctness of these hosts (S_{TD}) and clade rank, among 297 species of fleas parasitic on small mammals. Redrawn after Poulin et al. (2006b) (reprinted with permission from Elsevier).

in Fig. 14.9. Given the fact that generalist fleas achieve higher abundances on their hosts (see above; Krasnov et al., 2004f), it is not surprising that host specificity shows signs, albeit not strong ones, of having decreased over time. These results suggest that there may be no directional trend and no irreversibility in the evolution of host specificity.

14.5.2 Is host specificity evolutionarily heritable?

Niche conservatism among species over evolutionary timescales has been predicted theoretically (Peterson et al., 1999). This means that closely related species should tend to share common ecological attributes inherited from a common ancestor. This conservatism may be due to active, stabilizing selection or developmental constraints. Furthermore, the mode of speciation may affect the conservatism of ecological traits because sister species sharing morphological traits are perhaps more likely also to share ecological traits when they live in sympatry than when they live in allopatry, as they are evolving in a similar environment. Parasites offer an interesting opportunity to study the conservatism of ecological traits because sympatric speciation may be more common in parasitic than in free-living animals (e.g. Théron & Combes, 1995). Furthermore, host specificity of parasites is a trait that may be used to test the hypothesis of niche conservatism. If conservatism of host specificity occurs, it should be expected mainly for sympatric species which are more susceptible to sharing host species and similar environmental conditions than allopatric species.

Using 68 pairs of flea sister species, Mouillot et al. (2006) tested whether (a) the host specificities of closely related flea species are more similar to one another than expected from randomly associated pairs of species; and (b) the host specificities of parasite species are predictable and, thus, evolutionarily heritable. A significant positive relationship was found between the numbers, but not for the taxonomic distinctness, of host species infested by sister flea species (see Fig. 14.10 for the number of host species). This result was consistent whether sympatric or allopatric flea species were used, suggesting no influence of the mode of speciation on this conservatism of specificity. Furthermore, observed pairs of sister flea species showed a significantly higher coefficient of correlation of their host specificities than randomly associated pairs of species from the pool of fleas (see Mouillot et al., 2006 for details).

The lack of difference in the host specificity conservatism between sympatric and allopatric fleas may be caused by secondary sympatry of the sympatric sister species; they could have become sympatric after an initial allopatric speciation event (Via, 2001). Still, it is likely that sympatric speciation is a major reason

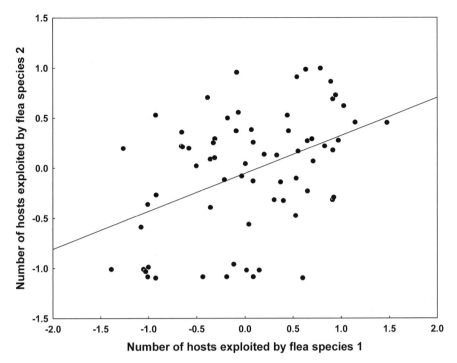

Figure 14.10 Plot of host specificity measured as number of host species (controlled for host sampling effort) used by 68 pairs of flea sister species. Redrawn after Mouillot et al. (2006) (reprinted with permission from Blackwell Publishing).

why fleas now live in sympatry. Nevertheless, the mode of speciation seems to be secondary in its influence on the observed conservatism of host specificity in fleas, the main drivers of host specificity being phylogeny and extrinsic (host- and environment-related) factors (e.g. Krasnov et al., 2004e).

In spite of this, the explained variation in host specificity was low (about 18%: Mouillot et al., 2006). Thus, although host specificity in fleas demonstrates considerable conservatism, this conserved element is obviously not strong enough to allow any accurate prediction of the host specificity of one species based on the specificity of a closely related species. It can only be stated that two closely related parasite species are more likely to have similar levels of host specificity than expected by chance.

14.5.3 Coevolution and host opportunism

One of the most common misconceptions of ecological and evolutionary parasitology is related to the association between host–parasite coevolution and host specificity. According to Fahrenholz's rule, if the phylogenies of host and parasite lineages are fully congruent and if host speciation was repeatedly accompanied by parasite speciation, then the descendant parasites will be strictly host-specific (Poulin, 2007a). Few would deny that this scenario is true. Sometimes, however, this idea is transformed into an opinion that if a strictly specialist parasite is expected to coevolve with a particular host, then a host-opportunistic parasite did not coevolve with any particular host but rather was subject to diffuse coevolution involving several host lineages (Futuyma & Slatkin, 1983; Lapchin & Guillemaud, 2005).

In contrast to the above notion, Tripet & Richner (1997b) argued that the hen flea *Ceratophyllus gallinae*, a flea that is commonly considered as a generalist (e.g. Cyprich *et al.*, 1999, 2002, 2006), coevolved nevertheless with tits (Paridae), one specific lineage among a variety of its hosts. They used data from the literature on the prevalence and intensity of infestation of nests of various birds by this flea and estimated the number of flea individuals produced in the nest of each host species. The results of the comparative analysis demonstrated that the prevalence of *C. gallinae* is highest in hole-nesting birds and that the majority of flea individuals are harboured by hole-nesting Paridae. This suggests that, despite demonstrating a broad host spectrum, *C. gallinae* might have coevolved with Paridae and, thus, host opportunism of this species is somewhat 'secondary'. The evolutionary scenario suggested by Tripet & Richner (1997b) underlined underdispersion of the territorial hole-nesting hosts as an important factor facilitating the exploitation of alternative host species. Underdispersion of nests (the main habitat of *C. gallinae*) results in high among-nest variability in ecological conditions and high potential gene flow between sub- and metapopulations of a flea. This, in turn, may favour broad tolerance to environmental parameters and an ability to exploit additional host species. Under such conditions, a broad host spectrum may be maintained despite the majority of individuals breeding on one particular host species.

14.6 Applicative aspects of host specificity studies

The previous subchapters dealt mainly with theoretical aspects of flea host specificity. The two next sections address more applicative aspects of this property of fleas, although no applicative aspect is possible without a considerable theoretical component. In these sections I discuss how the degree of host specificity of a flea species (a) affects the probability of this species being discovered and (b) determines the ability of this species to transmit the plague pathogen.

14.6.1 Probability of flea species discovery

Discovery and scientific description of new species of living organisms is an ongoing process which, in its modern form, dates back to the pioneering approach of Linnaeus. The rate of discovery of new species not only differs sharply among different taxa for various reasons, but can also greatly fluctuate on a temporal scale within a taxon, with some species being described much later than their close relatives. In other words, the probability of a species being discovered can be profoundly different among species of the same taxon (e.g. Collen et al., 2004) and, although involving an element of chance, is strongly affected by the biological characteristics of a given species (Cabrero-Sanudo & Lobo, 2003).

What does this have to do with host specificity of parasites? The link between these two parameters is quite straightforward. A parasite species exploiting a larger number of host species may be more likely to be encountered and described early than more host-specific parasite species. Beyond the number of host species, the phylogenetic or taxonomic relatedness of host species may also matter. If closely related species share functional or ecological attributes, then parasite species exploiting closely related hosts are less likely to be discovered than parasite species exploiting totally unrelated host species which have different ecological requirements, body size or behaviour. Thus for a given number of exploited hosts a parasite infesting a higher taxonomic diversity of hosts might be expected to be discovered before one infesting only closely related hosts.

Indeed, helminths that exploit more species of freshwater fish hosts, and to a lesser degree those that exploit a broader taxonomic range of host species, tend to be discovered earlier than the more host-specific helminths (Poulin & Mouillot, 2005b). What about fleas? Krasnov et al. (2005f) studied this using data on 297 flea species parasitic on 197 species of small mammals from 34 different regions of the Holarctic and one region from the Neotropics. Separate analyses of the relationships between the date of flea description and measures of host specificity demonstrated that the date of flea description was weakly, but significantly, negatively correlated with either the number of exploited hosts or the taxonomic distinctness of these hosts (see Fig. 14.11 for the number of exploited hosts).

The mechanism behind this is obvious. The probability of a flea being encountered by a collector is higher when this flea occurs on a higher number of host species. The reason for this is that, in field surveys, the primary targets are, usually, hosts rather than fleas. Consequently, a survey of fleas results from a survey of hosts. The probability of the host being found, in turn, is to a large degree determined by its geographical range (Gaston et al., 1995; Allsopp, 1997; Collen

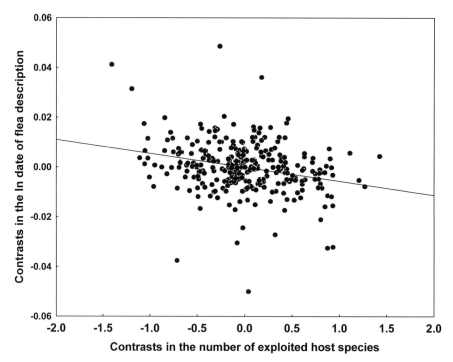

Figure 14.11 Relationship between the date of description and the number of host species among 297 flea species using the method of independent contrasts. Redrawn after Krasnov *et al.* (2005f) (reprinted with permission from Springer Science and Business Media).

et al., 2004). Moreover, both conventional analyses and phylogenetically corrected analyses provided roughly the same results in the study of Krasnov *et al.* (2005f), suggesting the absence of a phylogenetic effect on the relationships between the date of flea discovery and flea host specificity. Indeed, the year in which a species has been described measures, in fact, the efficiency of humans at finding flea species. This efficiency proved not to be influenced by flea phylogeny.

Nevertheless, different measures of flea host specificity were found to have differing predictive power in relation to the flea description date, with the number of exploited host species having the highest predictive power followed by S_{TD}. In other words, the number of host species exploited by a flea species is much more important than the taxonomic composition of these hosts from the perspective of the probability of this flea being found by a collector. Still, the relationship between S_{TD} and the date of flea description estimated by a linear regression was statistically significant. Consequently, flea species exploiting a broad taxonomic range of hosts are more likely to be discovered early than fleas exploiting only closely related hosts. The reason for this could be that the diversity of a host's ecological attributes increases the diversity of ways in which its fleas can be discovered. For instance, mammalian body size is highly conserved across the taxonomic hierarchy (Smith *et al.*, 2004). Flea species infesting hosts with a large range of body size have a higher likelihood of being discovered because this diversity of body sizes increases the number of ways in which a mammal may be sampled: poisoning, trapping or hunting efficiency are definitely linked to particular size ranges. For example, *Ctenophthalmus breviatus*, which exploits hosts from four rodent and one soricomorph families ranging in body size from 8000 g (*Marmota bobac*) to 8 g (*S. araneus*) was described in 1926, whereas *Ctenophthalmus shovi*, which parasitizes hosts from two rodent and one soricomorph families ranging in body size from 40 g (*Apodemus mystacinus*) to 8 g (*Sorex satunini*) was only discovered in 1948.

In spite of a significant correlation between the date of flea discovery and the degree of flea host specificity, the proportion of the total variance explained by these variables was low. This suggests that there are other key determinants of the probability of a flea being found and described. For example, the low proportion of total variance in year of description of a flea species explained by its host specificity may be caused by a 'human factor' in the history of flea taxonomy (see Chapter 1, Fig. 1.2). In addition, one should remember that a new species may be sometimes described after the revision of a taxon or re-identification of previously misidentified specimens. This could lead to the attribution of a relatively late description date to a host-opportunistic flea species. As a consequence, the resulting relationship between the date of description and the level of host specificity could be wrongly underestimated. The possibility of misidentification of some flea species during field surveys should also not be dismissed. The attribution of a taxon to either the species or subspecies level by different taxonomists with different approaches to the species concept may also influence the results of any search for the biological correlates of the probability of a species being discovered. Finally, flea taxonomy is based mainly on morphological characters (Holland, 1964; Medvedev, 2004, 2005). Therefore, in some cases, the existence of cryptic flea species can mask the true patterns of host specificity and, thus, confound the results of studies like Krasnov *et al.*'s (2005f). Despite all these potential sources of background noise, analyses outlined in this section have demonstrated that the number of host species used by a flea (as well as how long its principal host has been known to science and how widely distributed it is: see Krasnov *et al.*, 2005f) influences its probability of being found. By extrapolation, it can be inferred that the flea species not yet discovered are highly host-specific, and that they exploit little-known host species with limited geographical distribution. Indeed, one of the most recently discovered flea species is *Gymnomeropsylla margaretamydis*. It was found only on a murine *Margaretamys parvus* endemic to Sulawesi and described as recently as in 2002 (Durden & Beaucournu, 2002).

14.6.2 Transmission of the plague pathogen

A variety of mammals, mainly rodents and lagomorphs, serve as the main natural reservoirs of the plague pathogen, *Yersinia pestis*, whereas fleas are specific vectors of plague (Pollitzer & Meyer, 1961). Although various modes of plague transmission are possible (pneumonic, septicaemic, exposure to plagued carcasses), the flea-borne mode of transmission is commonly accepted as the main route of plague circulation (Suleimenov, 2004; Gage & Kosoy, 2005).

About 250 flea species are naturally infected by plague (Pollitzer, 1960; Serzhan & Ageyev, 2000). However, not all of them can effectively transmit the disease between hosts (e.g. He *et al*., 1997; Engelthaler *et al*., 2000). The suitability of a flea species to act as a plague vector is determined by a peculiarity of the mechanism of plague transmission described as early as 1914 (Bacot & Martin, 1914). After arriving in a flea during a blood meal, the bacteria multiply in the flea's gut and clog the proventriculus, blocking the passage of blood from the foregut to the midgut (Bacot & Martin, 1914; Pollitzer & Meyer, 1961; Konnov *et al*., 1986; Darby *et al*., 2005). Blocked and starving fleas repeatedly attempt to feed but host blood cannot pass the proventricular block. Eventually the flea regurgitates blood (now infected with bacteria) into the open wound, thereby infecting a new host. Although some fleas that are blocked by the plague bacteria appear unable to transmit the disease, whereas others transmit plague without being blocked, transmission by blocked vectors is still thought to be the most important michanism of plague circulation (Burroughs, 1947; Kartman *et al*., 1958; Bibikova & Klassovsky, 1974; Hinnebusch *et al*., 1998; Bazanova *et al*., 2004; Darby *et al*., 2005; Gage & Kosoy, 2005; Lorange *et al*., 2005; but see Webb *et al*., 2006). Consequently, the blocking rate by the plague pathogen under experimental conditions has been used as an indicator of a flea's transmission ability (Bibikova & Klassovsky, 1974). Although it should be admitted that gut blockage is not strictly synonymous with plague transmission, the blockage rate represents the best quantitative surrogate measure of the ability of a flea species to act as a plague vector. Earlier authors have dismissed the possibility that ecological differences among flea species can also influence whether or not they play roles as vectors (Bibikova & Klassovsky, 1974; Gubareva *et al*., 1976).

The effect of ecological properties of fleas on their suitability to transmit plague has been tested only recently (Krasnov *et al*., 2006c). Data on the blocking rate by *Y. pestis* and data on host specificity in the regions enzootic for plague were obtained for 40 flea species. Whether controlling for phylogenetic

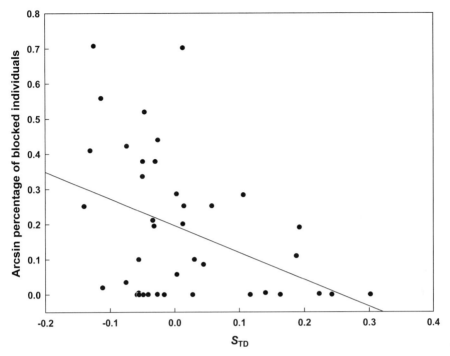

Figure 14.12 Relationships between the percentage of flea individuals blocked by plague bacteria in experiments and the index of taxonomic distinctness, S_{TD}, of a flea's host spectrum in nature across 40 flea species. Redrawn after Krasnov et al. (2006c) (reprinted with permission from Springer Science and Business Media).

influences or not, the percentage of flea individuals blocked by the plague bacteria was not correlated with the number of host species exploited by a flea in nature, but was strongly negatively correlated with taxonomic distinctness of a flea's host spectrum (S_{TD}) (Fig. 14.12). The distribution of data points in Fig. 14.12 is clearly triangular, suggesting that flea species exploiting closely related host species may or may not be efficient plague vectors, whereas a high blocking rate never occurs in flea species exploiting distantly related hosts.

These results demonstrate fundamental ecological differences between fleas that serve as plague vectors and those that do not play an important role in plague circulation. The presence of a suitable reservoir host in a flea's host spectrum is, evidently, another necessary prerequisite for a flea species to be a suitable plague vector. Thus, the use of flea species as vectors by the plague pathogen is not merely a consequence of their physiological attributes: contrary to conventional wisdom, their ecological traits matter as well. This strategic use of the vector species with features that facilitate disease spread highlights the rapid evolutionary fine-tuning of the plague's transmission cycle. Similar non-random use of vector species can be expected to exist in transmission of other vector-borne diseases, a phenomenon that needs to be taken into account when studying their epidemiology and control (Gillespie et al., 2004).

14.7 Concluding remarks

Host specificity determines, among other things, (a) whether a parasite can survive the extinction of a host species; (b) whether a parasite has the potential to colonize new hosts and to invade new habitats; and (c) whether a parasite can become established and spread following colonization of a new geographical area. Siphonaptera is a very convenient taxon to be used as a model for studying large-scale (both spatially and temporally) patterns of host specificity. In addition, studies of the ecology and evolution of host specificity of fleas are not purely theoretical exercises, but can have important implications for nature conservation, epidemiology and prevention of dangerous diseases for which they serve as vectors.

15

Ecology of flea populations

Charles Krebs defined ecology as 'the scientific study of the interactions that determine the distribution and abundance of organisms' (Krebs, 1994: 3). In other words, the main unit of ecological interest is not the individual organism but rather an assemblage of individuals belonging to the same species and coexisting in time and space. Contrary to that of most free-living species, spatial distribution of parasites is not continuous but consists of a set of more or less uniform inhabited 'islands' or patches represented by the host organisms, while the environment between these patches is decidedly unfavourable. In the majority of fleas, a 'habitat patch' also includes the host burrow or nest. This, however, does not negate the fragmented character of spatial distribution of an ensemble of conspecific fleas. This ensemble is fragmented amongst (a) host individuals; (b) host species within a location; and (c) locations. Strict terminology is required in order to distinguish between these different levels of fragmentation.

The scale involving host individuals does not represent a problem. An assemblage of parasites of a particular species inhabiting a particular individual host of a particular species is commonly defined as an *infrapopulation* (Margolis et al., 1982; Sousa, 1994; Combes, 2001; Poulin, 2007a). In contrast, there is no agreement regarding the terminology related to the host species and location (spatial) scales. Margolis et al. (1982) suggested referring to the assemblage of conspecific parasites inhabiting a particular host species in a particular location as a *suprapopulation*, while referring to that inhabiting an assemblage of host species in a particular location as a *metapopulation*. Combes (2001), however, argued that the term 'suprapopulation' in fact relabels what ecologists call 'population' and is thus redundant and confusing. In addition, he also opposed the term 'metapopulation' *sensu* Margolis et al. (1982) because (a) the initial definition of the term involves genetic exchange between fragments of a metapopulation

(Hanski, 1998); and therefore (b) 'metapopulations' of parasites should not differ from those of free-living species; and, thus, (c) it is problematic to use the term 'metapopulation' for a level of fragmentation below that of the population. Instead, Combes (2001) suggested defining (a) an assemblage of conspecific parasites infesting an assemblage of hosts of a particular species in a particular location as a *xenopopulation*; (b) an assemblage of conspecific parasites infesting an assemblage of sympatric host species as a *population*; and (c) an assemblage of all interconnected ensembles of conspecific parasites infesting all host species as a *metapopulation*.

Throughout this chapter, I use Combes's (2001) version of this terminology. I start with basic information on how to measure abundance and distribution of fleas. Then I consider variation in patterns of flea abundance among flea species as well as among host species, between host genders, and among host age cohorts. Next, I discuss the aggregative pattern of flea distribution, its causes and its consequences. Finally, I focus on host- and environment-related factors affecting flea abundance and distribution.

15.1 Measuring abundance and distribution

The fragmented pattern of distribution of a parasite among host individuals prevents us from characterizing the abundance of this parasite by a single value. This pattern stems mainly from the fact that the distribution of a parasite population across a host population is usually aggregated. In other words, most parasite individuals occur in a few host individuals, while most host individuals have only a few, if any, parasites (Anderson & May, 1978; Poulin, 1993; Shaw & Dobson, 1995; Wilson et al., 2001). Furthermore, aggregation of parasites is an almost universal phenomenon (Anderson & May, 1978; May & Anderson, 1978; Anderson & Gordon, 1982; Shaw & Dobson, 1995; Shaw et al., 1998; Poulin, 2007a, b). This ubiquity suggests that similar processes may be involved in generating the same pattern in different host–parasite systems. The aggregated distribution of parasite individuals among hosts is caused by a variety of factors (Poulin, 2007a) and can have important consequences for different aspects of the evolutionary ecology of parasites (e.g. Morand et al., 1993). As a result, the fraction of uninfested hosts should also be taken into account when the abundance of a parasite is considered. In other words, given the aggregated distribution of a parasite across hosts, a parasite's abundance should be considered in conjunction with its distribution.

Common measures of parasite abundance and distribution are mean abundance, intensity of infestation and prevalence. Mean abundance is simply the mean number of parasites per host individual and is calculated by summing both infested and uninfested hosts. Intensity of infestation (sometimes called parasite burden or parasite load) is the mean abundance of parasites per infested host individual, whereas prevalence is the proportion of infested hosts. Obviously, intensity of infestation is a product of mean abundance and prevalence. These measures are straightforward and simple to understand, and may be easily calculated.

Measuring the level of aggregation is a much greater challenge. There are several methods for this (Southwood, 1966; Elliott, 1977; Wilson et al., 2001). The most popular ones are calculation of the variance-to-mean ratio and parameter k of the negative binomial distribution. A variance-to-mean ratio of greater than 1 points to a departure from randomness and a tendency towards aggregation, while an increase in the value of the ratio indicates an increase in the aggregation level. Fitting the negative binomial distribution to an observed distribution is also a common practice to evaluate aggregation. Aggregation increases as k gets smaller until it converges on the logarithmic series with k close to zero. At large k, the distribution approaches the Poisson. The value of k can also be calculated in other ways besides fitting the negative binomial distribution, for example, by using the moment estimate of Elliot (1977), corrected for sample size:

$$k = \left[M^2 - \frac{V(M)}{n} \right] \bigg/ [V(M) - M], \quad (15.1)$$

where M is mean abundance, $V(M)$ is variance of abundance and n is host sample size.

Another method is estimation of k using Taylor's power law (Taylor, 1961). According to this law, mean abundance (M) and variance of abundance [$V(M)$] of an organism's distribution are related as:

$$V(M) = aM^b. \quad (15.2)$$

This pattern of abundance and distribution is astonishingly similar in both free-living and parasitic organisms and is supported by numerous data (Taylor & Taylor, 1977; Taylor & Woiwod, 1980; Anderson & Gordon, 1982; Perry & Taylor, 1986; Shaw & Dobson, 1995; Morand & Guégan, 2000; Šimková et al., 2002). The exponent (parameter b or slope of Taylor's relationship, i.e. slope of linear regression of variance of abundance on mean abundance in the log–log space) of this power function usually varies among species as $1 < b < 2$, but the causes of the variation in this parameter between species (e.g. Kilpatrick & Ives, 2003) as well as within species (at spatial or temporal scale) are poorly understood. For parasites, it has been thought to be an inverse indicator of parasite-induced host mortality (Anderson & Gordon, 1982), as an increase in b suggests that at least some of the hosts are infected with heavy burdens of parasites. In addition, the value of the exponent b has been suggested to be an indicator of a tendency of organisms to be mutually attracted (Perry, 1988). According to Taylor et al. (1979), parameters a and b of Taylor's power law are related to k as:

$$\frac{1}{k} = aM^{b-2} - \frac{1}{M}. \tag{15.3}$$

Two less popular measures of aggregation are Lloyd's (1967) index of mean crowding (m^*) and the index of intraspecific aggregation, J, proposed by Ives (1988a, 1991). Index of mean crowding (m^*) is useful when studying aggregation from the parasite point of view (Wilson et al., 2001). It quantifies the degree of crowding experienced by an average parasite within a host by the following expression:

$$m^* = M + \left(\frac{V(M)}{M} - 1\right), \tag{15.4}$$

where M is the mean and $V(M)$ is the variance of the number of parasites on an average host. It is therefore a measure based on individual counts.

A measure of intraspecific aggregation, J, represents the proportional increase in the number of conspecific competitors experienced by a random individual of species k, relative to a random distribution:

$$J_k = \frac{\sum_{i=1}^{p} \frac{n_{ki}(n_{ki}-1)}{m_k} - m_k}{m_k} = \frac{\frac{V_k}{m_k} - 1}{m_k}, \tag{15.5}$$

where n_{ki} is the number of parasite species k on host individual i, and m_k and V_k are the mean number and the variance in number of parasite species k, respectively. A zero value of J indicates random distribution of individuals, whereas $J = 0.5$ indicates an increase of 50% in the number of conspecific competitors expected in a patch ($=$ host) compared to a random distribution.

15.2 Is abundance a flea species character?

It is commonly accepted that the density (abundance per unit area) of a species in a location results from the interplay between the intrinsic properties of that species and the extrinsic properties of the local habitat, both biotic and abiotic. For example, the density of a species has been shown to depend on intrinsic characters such as body size and associated metabolic rate (Blackburn & Gaston, 2001), fecundity (Hughes et al., 2000) and social structure (López-Sepulcre & Kokko, 2005). On the other hand, the density of a species is undoubtedly determined by characteristics of the habitat it occupies such as the identity and composition of coexisting competitors (Rosenzweig, 1981), the amount of resources available (Newton, 1998) and the pattern of resource acquisition (Morris, 1987a).

Consequently, because it results from interactions among a variety of factors, the predictability of the density level of any given species is often low, causing problems for conservationists and pest managers (e.g. Beissinger & Westphal, 1998; Ludwig, 1999; Fieberg & Ellner, 2000). One of the probable reasons for this low predictability is the fact that the density of a species is determined simultaneously by extrinsic factors generating variation among populations of this species, and by intrinsic factors promoting between-population stability (i.e. repeatability) in density.

High intraspecific variation in the population parameters of parasites, such as their intensity of infestation, abundance and prevalence, is well documented. For example, the abundance of parasites is strongly dependent on the abundance of their host (Anderson & May 1978; see below), which, in turn, is spatially and temporally variable. Moreover, the relationship between parasite and host abundance varies from being positive (e.g. Krasnov et al., 2002e) to being negative (e.g. Stanko et al., 2006) among different species of the same parasite taxa depending on species-specific reproductive rate and seasonality (see also below). In addition, the dependence of survival and, consequently, abundance of parasites on spatially variable abiotic factors (e.g. microclimate) has been reported for both endoparasites (via effects on transmission: Galaktionov, 1996) and ectoparasites (direct effect: Metzger & Rust, 1997). However, in spite of the strong dependence of parasite population parameters on extrinsic factors and, therefore, the expected spatial and temporal variation of these parameters, species-specific features of parasites such as body size and egg production could constrain this variation (Poulin, 1999). Indeed, Arneberg et al. (1997), studying nematodes parasitic on mammals, demonstrated that intensity of infection as well as abundance were repeatable within nematode species, i.e. were less variable within than between species. The conclusion from their study is, therefore, that the levels of intensity of infection and abundance are 'true' attributes of a nematode species. Another study of intraspecific variability versus stability of parasite population parameters was carried out on different taxa of metazoan parasite species of Canadian freshwater fish (Poulin, 2006). Again, prevalence, intensity of infection and abundance values from different populations of the same parasite species were more similar to each other, and more different from those of other species, than expected by chance alone. These results suggest that intensity of infestation and abundance are real characters of parasite species, supporting the view that the biological features of parasite species can potentially override local environmental conditions in driving parasite population dynamics.

Fleas are much more strongly influenced by their off-host environment than either the endoparasites or permanent ectoparasites studied by Arneberg *et al.* (1997) and Poulin (2006). This suggests that the reported patterns may not be valid for them. However, low variation of within-flea species density on a temporal scale (e.g. *Oropsylla silantiewi* and *Rhadinopsylla li* in Central Asia: Berendyaeva & Kudryavtseva, 1969; *Citellophilus tesquorum*: Tchumakova *et al.*, 2002) or a spatial scale (e.g. *Xenopsylla conformis*, *Nosopsyllus laeviceps*, *Stenoponia tripectinata* and *Rhadinopsylla ucrainica* in the Apsheron Peninsula: Kadatskaya & Kadatsky, 1983) has been reported. Launay (1989) noted that temporal density fluctuations of *Xenopsylla cunicularis* in Spain are less expressed than those of its host, the European rabbit *Oryctolagus cuniculus*.

To compare within- and between-species variation in flea abundance, Krasnov *et al.* (2006d) used data on fleas parasitic on small mammals for samples of 548 flea populations, representing 145 flea species and obtained from 48 different geographical regions. First, a strong positive correlation was found between the lowest observed abundances and all other observed abundance values across flea species. In other words, different fleas demonstrated a relatively narrow range of abundances when exploiting the same host species in different regions (Fig. 15.1a). Second, the results of the repeatability analysis (Arneberg *et al.*, 1997) showed that abundances of the same flea species on the same host species but in different regions were more similar to each other than expected by chance, and varied significantly among flea species, with 46% of the variation among samples accounted for by differences between flea species (Fig. 15.1b). Thus, estimates of abundance are repeatable within the same flea species. The same repeatability was also observed, but to a lesser extent, across flea genera, tribes and subfamilies, but not families.

The above analysis demonstrated that patterns found for mammalian endoparasites (Arneberg *et al.*, 1997) and parasites of fish (Poulin, 2006) are also valid for fleas despite their greater sensitivity to external factors. Flea abundance can thus be considered as a true flea species character. Furthermore, abundance can also be considered an attribute characteristic of a flea genus, tribe or subfamily, but not family. This implies that some flea species-specific life-history traits determine the limits of abundance.

Lower limits of flea abundance can be affected by species-specific mating systems and/or the relationship between mating and blood-feeding, whereas upper limits of abundance can be determined by species-specific reproductive outputs, generation times, preferences for blood-sucking on a specific body part of a host and/or the ability of both imagoes and larvae to withstand crowding. For example, site-specific fleas are more prone to crowding and thus may achieve lower abundance than non-site-specific species. Numerous examples of variation in these parameters among flea species can be found in the previous chapters of this book.

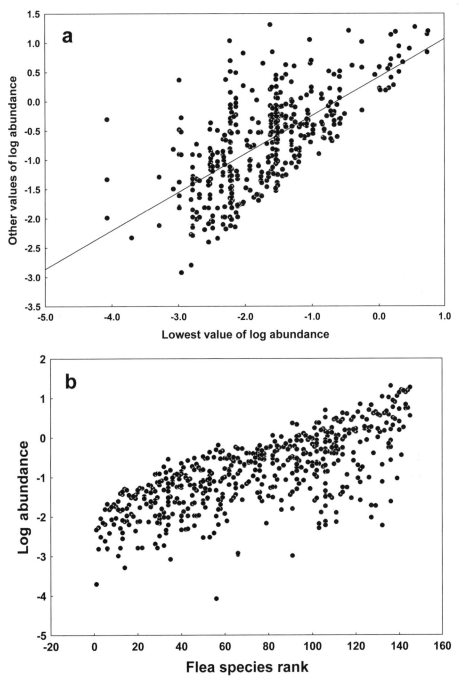

Figure 15.1 (a) Relationship between the lowest abundance and other abundance values on the same host species for 145 flea species recorded in at least two regions; (b) rank plot of flea abundance (see Fig. 14.4 for explanations). Redrawn after Krasnov et al. (2006d) (reprinted with permission from Springer Science and Business Media).

The repeatability of flea abundance within genera, tribes and subfamilies suggests that limits of abundance as well as intrinsic properties of fleas that determine these limits are phylogenetically constrained. In other words, a relatively narrow, species-specific range of interpopulation variation in abundance is the result, at least in part, of natural selection. Consequently, the observed range of abundance for a given flea species has not only evolved for some particular ecological reason, but also seems to be nested within the flea phylogeny, and thus can be inherited by descent. This advocates the necessity of using phylogenetic information in comparative analyses of abundance (e.g. Stanko *et al.*, 2002; Gaston & Blackburn, 2003).

15.3 Aggregation of fleas among host individuals

15.3.1 Are fleas aggregated among hosts?

One of the first reports of flea aggregation among host individuals dates back to the late 1940s, when Cole & Koepke (1946, 1947a, b, c, d) studied the distribution of *Xenopsylla cheopis* on the Norway rat *Rattus norvegicus* in Alabama, Georgia and Hawaii and reported that a large proportion of the flea population was concentrated on a few host individuals. This pattern, in which most host individuals are weakly (if at all) infested whereas a few individuals harbour most of the flea population, has since been found in many flea species such as *Malaraeus telchinus* (Stark & Miles, 1962), *Spilopsyllus cuniculi* (Allan, 1956), *Tunga penetrans* (Chadee, 1994), *C. tesquorum* (Li & Ma, 1999) and *Neopsylla bidentatiformis* (Li & Ma, 1999). Heeb *et al.* (1996) reported that the distribution of *Ceratophyllus gallinae* among nests of the great tit *Parus major* in Switzerland did not differ from the negative binomial distribution. A good fit of a flea distribution among host individuals to the negative binomial distribution was also found by Robbins & Faulkenberry (1982) for *Atyphloceras multidentatus* and *Catallagia charlottensis* parasitic on the grey-tailed vole *Microtus canicaudus* in Oregon. Lundqvist & Brinck-Lindroth (1990) reported that the frequency distribution of ectoparasites, including fleas, in Fennoscandia was negative binomial for the bank vole *Myodes glareolus*, field vole *Microtus agrestis* and root vole *Microtus oeconomus* but not so for the common shrew *Sorex araneus*, red-backed vole *Myodes rutilus* and grey-sided vole *Myodes rufocanus*. However, data on several ectoparasite taxa were pooled in this study for the evaluation of frequency distribution, which could mask species- or taxon-specific patterns.

To confirm aggregated distribution of fleas within populations of host species (i.e. within flea xenopopulations) as well as across whole host communities (i.e. within flea populations), Krasnov *et al.* (2005h, 2006e) investigated the abundance and distribution of fleas parasitic on small mammals in (a) the Negev Desert of Israel and (b) central and eastern Slovakia. In both these studies, aggregation was evaluated either via fitting of the negative binomial distribution or via parameters of Taylor's power law (see above), or both.

In fleas and hosts from the Negev Desert, the negative binomial models provided a statistically satisfactory fit to the observed frequency distribution of each flea species on each host species. Moreover, the negative binomial distribution of fleas across host individuals held at whatever scale of consideration was selected, namely within host within flea species, within host across flea species and within flea species across hosts (see Fig. 15.2a for the illustrative example with Wagner's gerbil *Dipodillus dasyurus*). The slopes of the relationship between log-transformed variance and mean abundance of fleas were significantly greater than unity at a xenopopulation scale (i.e. in all flea–host associations) as well as at a population scale (i.e. within flea species across all host species) and within host species across all flea species (see Fig. 15.2b for illustrative example of *D. dasyurus* and *Xenopsylla dipodilli*). The values for the slopes varied from a low of 1.19 in *Nosopsyllus pumilionis* on the Baluchistan gerbil *Gerbillus nanus* to a high of 1.93 in *Synosternus cleopatrae* on Anderson's gerbil *Gerbillus andersoni*. In fleas and hosts from Slovakia, the slopes of the relationship between log-transformed variance and mean abundance of fleas were also significantly greater than or equal to unity in xenopopulations of most species and varied from a low of 0.83 in *Ctenophthalmus solutus* on *M. glareolus* to a high of 1.59 in *Doratopsylla dasycnema* on *S. araneus* (see Fig. 15.2c for illustrative example of *Ctenophthalmus agyrtes* and the fieldmouse *Apodemus agrarius*). A good fit of the distribution of fleas from Slovakia to the negative binomial distribution had been reported earlier by Miklisova & Stanko (1992) and Stanko (1994).

The results of both these studies confirm the aggregated distribution of fleas across their host individuals. Another study of fleas parasitic on rodents in Slovakia and the Negev Desert used the index of aggregation *J* (Ives, 1988a) and resulted in the same conclusions (Krasnov *et al.*, 2006f). The aggregated distribution may arise as a result of heterogeneities in host populations and/or infection pressure (Anderson & May, 1979; Shaw & Dobson, 1995; Wilson *et al.*, 2001). However, while fleas appear to be aggregated, strong differences exist in the degree of aggregation both within and between fleas and hosts.

15.3.2 Why does flea aggregation vary interspecifically?

Taylor (1961) argued that the value of b of Taylor's power relationship (see above) is constant for a given species and thus represents a true biological character of a species. If so, then the value of b will be repeatable within a

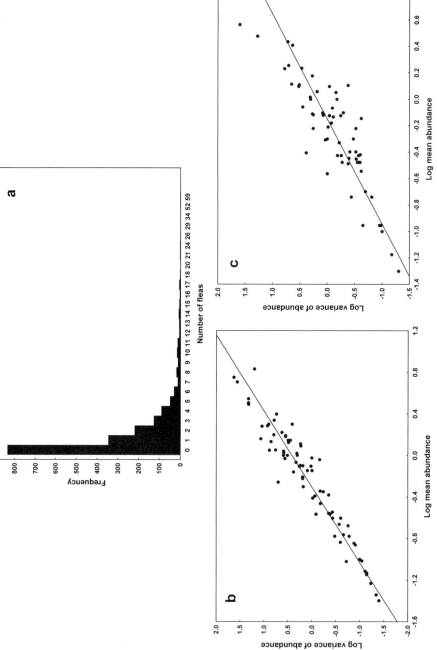

Figure 15.2 (a) Frequency distribution of the number of fleas per individual Wagner's gerbil *Dipodillus dasyurus* in the Negev Desert. Relationship between log variance of abundance and log mean abundance of (b) *Xenopsylla dipodilli* on *Dipodillus dasyurus* in the Negev Desert and (c) *Ctenophthalmus agyrtes* on the fieldmouse *Apodemus agrarius* in Slovakia. (a, b) Redrawn after Krasnov et al. (2006e) (reprinted with permission from Cambridge University Press); (c) redrawn after Krasnov et al. (2005h) (reprinted with permission from Springer Science and Business Media).

flea species, that is, it will be a feature that varies less among populations of the same flea species exploiting different host species than among different flea species. Repeatability of b values within a flea species will thus indicate that intraspecific variation of this character has some limits and is determined by certain biological properties of a species. Krasnov et al. (2006e) carried out a repeatability analysis (see above) for 19 flea species from Slovakia for which at least two flea–host associations were available (among 36 flea–host associations). Estimates of b for the same flea species were closer to each other than expected by chance, and varied significantly among flea species with 40.1% of the variation among samples accounted for by differences between flea species.

The lower variability of abundant populations compared to scarce populations in each flea species implies the increasing intensity and/or efficiency of processes of population regulation with increasing abundance (less increase in variability with increasing abundance). A difference in the slope of Taylor's relationship suggests that the level of abundance at which population regulation aimed at population stability becomes efficient varies among species. Population regulation starts to act efficiently at relatively high or relatively low abundance, depending on some species properties. Parasites face a trade-off between being too aggregated and being too randomly distributed (Anderson & Gordon, 1982; Shaw & Dobson, 1995). A parasite could be lost due to high mortality of overly infested hosts if the level of aggregation is too high, whereas a parasite's mating opportunities could be sharply decreased if its distribution across host individuals is random. Consequently, an optimal level of aggregation and a level of abundance at which regulation processes are manifested seem to be species-specific and depend on demographic factors such as intrinsic birth and death rates, mating behaviour and mobility. This suggests that the pattern of optimal spatial distribution evolves together with the evolution of life-history traits. Indeed, flea species differ drastically in such traits as oviposition rate, clutch size (see Chapters 5 and 11) and dispersal rate (e.g., Tripet et al., 2002a). All these traits, being species-specific, contribute to the precise species-specific relationship between mean abundance and variance.

The relationship between mean abundance and variance is a consequence of the interactions between demographic and environmental stochasticity and depends on the stochastic contributions of various processes such as birth, death and immigration rates (Anderson et al., 1982). If so, values of b could be expected to vary with the level of host specificity of a flea species. For example, it is likely that a highly aggregated distribution could increase the risk of extinction for a highly host-specific parasite. Such parasites should, therefore, decrease the level of their aggregation (resulting in a relatively lower b). On the other hand, the value of b in parasitic species has been suggested to be a reflection of regulation

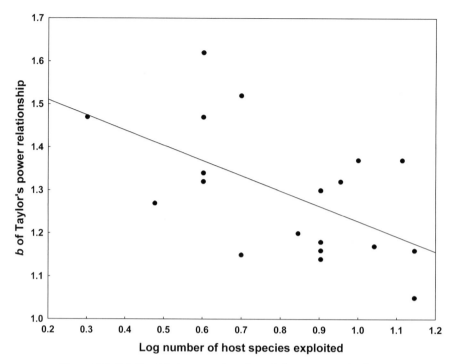

Figure 15.3 Relationship between the value of the parameter b of Taylor's power law and the number of hosts exploited by 19 flea species from central and southeastern Slovakia. Redrawn after Krasnov et al. (2006e) (reprinted with permission from Springer Science and Business Media).

processes in host–parasite systems and, as has been mentioned above, to be inversely indicative of parasite-induced host mortality (Anderson & Gordon, 1982; Madhavi & Anderson, 1985). Highly host-specific parasites are thought to be less virulent than host-opportunistic parasites because a specific host and its descendants are necessary to maintain a parasite population (e.g. Clayton & Tompkins, 1994, 1995; but see Møller et al., 2005 and Poulin, 2007a). As a consequence of this consideration, highly host-specific parasites are expected to be characterized by higher b values.

Krasnov et al. (2005h, 2006e) asked whether flea species with different degrees of host specificity differ in the values of b (the degree of the aggregation). The slope of Taylor's power relationship decreased significantly with an increase in the number of host species exploited by a flea (Fig. 15.3). This relationship held even when the confounding effect of phylogeny was taken into account. When fleas were classified as specialists and generalists with respect to their seasonality, the values of b appeared to be higher in seasonal than in year-round-active species (Krasnov et al., 2005h). Consequently, the species-specific level of flea aggregation can be said to be determined, at least partly, by the level of ecological specialization of a flea. In general, host-specific and/or seasonal fleas demonstrated higher aggregation than host-opportunistic and/or all-year-round fleas.

In particular, these results support the idea of low virulence of highly host-specific parasites. However, the common view that parasite-induced host mortality can decline during a long evolutionary history of the relationship between a particular host species and a particular parasite species has been strongly criticized in the context of the optimal virulence concept (Combes, 2001; Poulin, 2007a). The selection of a strategy of host exploitation by a parasite should depend on a concern for maximization of fitness rather than the welfare of the host. Indeed, one host-specific species, *N. pumilionis*, appeared to be an outlier from the trends found by Krasnov et al. (2005h). The b value for this species was low suggesting a highly negative effect on the host. However, this flea is active during only three to four winter months. Perhaps, it faces a trade-off between (a) the risk of being lost together with over-parasitized hosts and (b) the necessity of finishing the life cycle during a very short time. The contradiction between high mean b in seasonal fleas and low b in *N. pumilionis* (Krasnov et al., 2006h) can be related to this flea being both highly host-specific and strictly seasonal. Possibly, these two traits invoke different constraints on the pattern of aggregation of *N. pumilionis* with the effect of seasonality being stronger. In other words, variation in b values among different flea–host systems can depend upon a variety of factors related to particular life-history traits that can affect the demographic patterns of both parasites and hosts (Downing, 1986).

15.3.3 Why does flea aggregation vary intraspecifically?

Repeatability analyses carried out by Krasnov et al. (2005h, 2006e) and described in the previous section have demonstrated that differences between flea species only partially explain the variation in parameter b among samples, indicating that other factors should be responsible for the rest of the variation. Although each flea species may be evolutionarily constrained in the degree of variation observed for any given abundance (Shaw & Dobson, 1995), the variation in the slope of mean abundance/variance relationship within certain species-specific limits may be caused by extrinsic factors. In particular, species interactions within a community can affect population dynamics of a given species (Kilpatrick & Ives, 2003). Indeed, even sampling of the same reproductive population under different conditions (e.g. stage, phenology or habitat) resulted in different power-law estimates (R. A. J. Taylor et al., 1998). Furthermore, if aggregation of parasites arises due to heterogeneities in host populations, it should heavily depend on external (in relation to a species of interest) factors.

Kilpatrick & Ives (2003), using simulation models, demonstrated that the parameter b can vary on a temporal scale, being dependent on negative interactions among species in a community. These reasons can also be applied to spatial variation of b within species as well as among species. Kilpatrick & Ives (2003) showed that slopes of the Taylor's power relationship average 2 in a lack of competition, and decrease from 2 to 1 with an increase in interspecific competition. Applying this to parasites, a decrease in the b value of each given parasite species could be expected with an increase in size of the community of parasites, provided that the community under consideration is interactive (see Chapter 16). In other words, if parasite species in a community exert negative effects on each other either directly (direct competition) or via the host (apparent competition), these negative interactions would result in an increase of regulation power (less increase in variance of abundance with increasing mean abundance) in species-rich assemblages.

Testing the hypothesis regarding the relationship between the level of aggregation and composition of flea assemblages on fleas from the Negev Desert (Krasnov et al., 2005h) and Slovakia (Krasnov et al., 2006e) provided positive results. The values of b were negatively affected by the number of co-occurring flea species in both regions (Fig. 15.4).

These results indicate that the intensity and/or efficiency of population regulation processes increases with community size, probably due to interspecific competition. This competition can be both direct and indirect (via shared predator or competitor). Although direct interspecific competition in fleas has been suggested (Day & Benton, 1980) and even experimentally demonstrated (Krasnov et al., 2005e), it seems that interactions between flea species are mainly mediated by the host and can be both negative and positive (Krasnov et al., 2005g; see Chapter 16). The host defence system can be a primary mechanism of this mediation. Furthermore, b calculated for a flea species within a host species decreased not only with an increase in the number of co-occurring fleas but also with an increase in their taxonomic distance (Krasnov et al., 2006e). In other words, the relative stability of a flea xenopopulation or population is attained at relatively lower abundance if the number of other co-habitating fleas is relatively high and their taxonomic relatedness is relatively low. This supports an idea that the host can mediate interactions among flea species via its defence system. Indeed, a host subjected to multiple challenges from multiple distantly related parasites can either mount stronger immune responses (e.g. Morand & Poulin, 2000; Møller et al., 2005) and/or increase sharply its grooming effort (e.g. Cotgreave & Clayton, 1994). However, the effectiveness of energy allocation to immune defence decreases as the diversity of attack types increases (Jokela et al., 2000) due to

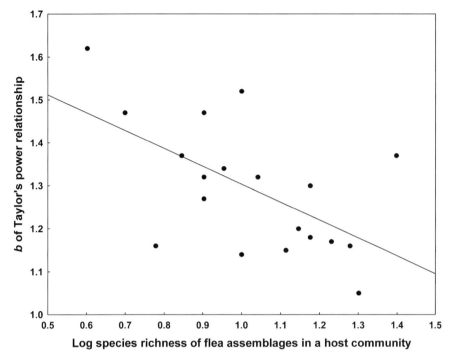

Figure 15.4 Relationship between the value of the parameter b of Taylor's power law and flea species richness of host communities across 19 flea species parasitizing 24 mammalian species in central and southeastern Slovakia. Redrawn after Krasnov et al. (2006e) (reprinted with permission from Springer Science and Business Media).

the cost of maintaining several different means of defence (L. H. Taylor et al., 1998). Consequently, a host subjected to attacks from multiple parasite species is forced to give up its defence and to surrender (Jokela et al., 2000). As a result of any of these processes, a flea will tend to be eliminated from a heavily infested host due to either its strong defence response (multiple challenges) or its death (inefficiency of defence). As a consequence, the level of aggregation of each species from species-rich and diverse parasite assemblages is likely to decrease.

15.3.4 Dynamics of flea aggregation: does an uninfested host stay uninfested for ever?

One of the most common explanations of the main cause for aggregation is that a particular host individual offers the parasites a higher-quality habitat than other host individuals (Poulin, 2007a). This higher-quality habitat could be a host that the parasite encounters easily (e.g. its behaviour favours encounter) and/or that is compatible with a parasite (e.g. its resources may be exploited and its immune defences avoided) (Combes, 2001). Analogously, an

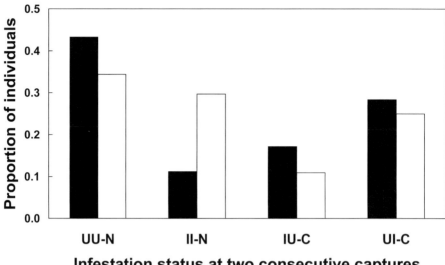

Figure 15.5 Status of infestation by fleas in Wagner's gerbil *Dipodillus dasyurus* (black columns) and Egyptian spiny mouse *Acomys cahirinus* (white columns) at two consecutive captures. UU-N — individuals uninfested at both captures; II-N — individuals infested at both captures; IU-C — individuals infested at the first capture but flea-free at the second; UI-C — individuals flea-free at the first capture but infested at the second. Data from Krasnov *et al.* (2006g).

uninfested host represents an unfavourable habitat for parasites for reasons such as successful avoidance of parasites or effective anti-parasitic defence.

Accounts of both infested and uninfested hosts are equally important in parasitological studies (e.g. when prevalence of infestation is calculated). However, the reasons why a particular host individual is either infested or uninfested at a particular time are often not clear. Moreover, it is unclear whether the infestation status of a host is temporal or permanent. Indeed, the uninfested host individuals could be unexploited by parasites either because they are permanently unsuitable as hosts or because they are not colonized at a given time by chance, although in principle they are suitable hosts (Gotelli & Rohde, 2002). Most studies of endoparasites are not able to answer this question because hosts are usually sacrificed to be examined for parasites. However, ectoparasitological studies, in particular those that concern fleas, provide the possibility of monitoring the infestation status of a host individual over time because fleas can be counted on a live animal which can then be recaptured and re-examined.

Krasnov *et al.* (2006g) studied temporal variation in the presence or absence of fleas on individual *D. dasyurus* and *Acomys cahirinus* and demonstrated that the infestation status of a host individual is not permanent but is temporally variable

(Fig. 15.5). In general, the probability of an individual changing its infestation status from infested to uninfested and vice versa was high. In other words, an individual host found to be infested by fleas during a short-term survey can lose these parasites and be flea-free in subsequent surveys, whereas initially uninfested individuals could be colonized by fleas and be flea-infested in subsequent surveys. Consequently, a short-term survey for parasites represents a snapshot of the parasite distribution among host individuals, which may vary temporally without replacement of host individuals.

These results do not refute the notion that infested hosts represent better habitats for parasites than uninfested hosts, but rather suggest that the suitability of a host individual for a parasite can vary temporally. It is clear that this pattern of change in the infestation status of a host individual can vary with parasite and/or host abundance (e.g. Šimková et al., 2002; Krasnov et al., 2002e) and seasonality (e.g. Krasnov et al., 2002c), whereas the suitability of a host for parasites can vary with host age (e.g. Sorci, 1996). Less obvious mechanisms that may determine the temporal variation in an infestation status of an individual host are related to both encounters with parasites (e.g. temporal variation in behaviour) and compatibility for parasites (e.g. energy deprivation). For example, resident individuals possessing burrows may be preferred by fleas over dispersing individuals that do not possess permanent burrows. However, the status of disperser or resident of a rodent individual (in terms of possessing the individual home range and burrow) can be inconclusive. For many rodent species, dispersing individuals often become residents, whereas resident individuals often abandon their burrows and disperse (Nikitina, 1961; Smith, 1980; Brandt, 1992; Khokhlova et al., 2001). Therefore, a disperser that usually lacks fleas could be colonized by them when it becomes a resident, whereas a resident could be abandoned by fleas when it starts to disperse.

Another example of behavioural changes in an individual that can lead to changes in infestation status is related to variation in the social status that can influence the encounter with parasites in either a favourable or an unfavourable way. For example, dominant individuals may be less (e.g. Gray et al., 1989) or more (e.g. Halvorsen, 1985) parasitized than individuals of lower social status. In many rodents, the social status of an individual is not permanent and can change during its lifetime (e.g. Bartolomucci et al., 2001).

Parasite-compatible hosts are those whose resources can be exploited and whose anti-parasitic defences can be avoided. Energy-deprived hosts seem to be less resistant, and thus represent better patches for fleas (Krasnov et al., 2005d; see Chapter 13). The energy intake of an individual host varies temporally for a variety of biotic and abiotic reasons. This can affect its defensive abilities and, consequently, its infestation status.

Another reason for temporal change in the infestation status of an individual host could involve reproductive outbreaks of fleas, leading to the infestation of hosts that were not infested previously (Krasnov et al., 2002a). In addition, a host individual could lose or acquire fleas merely by chance.

Initial infestation status of an individual affected its subsequent infestation status in *D. dasyurus* but not in *A. cahirinus* (Krasnov et al., 2006g). This effect was manifested in initially uninfested but not in initially infested individuals. Therefore, it is less probable for an uninfested *D. dasyurus* to acquire fleas than for an infested *D. dasyurus* to lose fleas. In contrast, *A. cahirinus* has a similar probability of both acquiring and losing fleas. The difference between species can be related to different social patterns in these two host species, solitariness in *D. dasyurus* and communal nesting in *A. cahirinus*. As a result, an exchange of fleas between different individuals is much more probable in *A. cahirinus* than in *D. dasyurus*. This suggests that, despite the infestation status of a host being temporally variable, the pattern of this variation can be influenced by host species-specific biological parameters.

15.4 Biases in flea infestation

15.4.1 Host species

Any field zoologist knows that some host species are characterized by higher flea abundance than other species. Consequently, a level of flea abundance can be determined not only by flea, but also by host identity. In other words, one may ask whether flea abundance can also be seen as a property of the host. To test this, Krasnov et al. (2006d) analysed the repeatability of flea abundance within host species for fleas infesting 48 host species in at least two regions. The analysis demonstrated significant repeatability of flea abundance independent of flea species among host species, although the proportion of the variance accounted for by differences among host species as opposed to within a host species was lower than that accounted for by differences among flea species, being only 24.1%.

Arneberg et al. (1997) reported the abundance of nematodes to be repeatable within mammalian species. In other words, independently of nematode species, some mammal species have many nematodes per individual, whereas other mammals have only a few nematodes per individual. However, the repeatability of nematode abundance among host species was weaker than that among nematode species (18% of the variation in abundance associated with differences among host species versus 36% associated with differences among nematode species: Arneberg et al., 1997). The results of Krasnov et al.'s (2006d) study provided a very similar ratio between percentages of the variation of flea abundance associated with differences among host species and among flea species. Consequently, the repeatability of flea abundance within host species suggests that some host properties also constrain to some extent the number of fleas harboured by an individual. These constraints can be related to processes on the host body that affect imago fleas and/or to processes within a host burrow or nest that affect pre-imaginal fleas. The abundance of imago fleas can be determined by host body features, host immune defence and host anti-parasitic behaviour, whereas the abundance of larvae is determined by the structure of a host's burrow or nest.

Repeatability of flea abundance within host species and variation in flea abundance among host species suggest that fleas may be aggregated not only at the scale of host individuals (e.g. at the level of a xenopopulation), but also at the scale of host species (e.g. at the level of a population). In other words, some host species represent better habitats for fleas than other host species. Indeed, Krasnov et al. (2006f) found a trend of intraspecific aggregation of fleas (measured as the index J) to be correlated positively with the composite variables representing variation in (a) host body mass and burrow complexity (Fig. 15.6) and (b) host mass-independent basal metabolic rate (BMR). This means that the occurrence of heavily infested individuals is more characteristic of larger host species that possess more complicated burrows and/or of host species with a higher mass-independent BMR.

Host body size can determine the number of parasites per host, as more space is available for multiple parasite individuals in larger hosts (e.g. Morand & Guégan 2000). However, this appears not to be the case for fleas parasitic on small mammals (Stanko et al., 2002; Krasnov et al., 2005g; but see Marshall, 1981a). Host species also differ in the level of their immune and behavioural defence against parasites. For example, species-specific differences in the ability to mount both humoral and cell-mediated immune responses have been reported even for closely related host species (Klein & Nelson, 1998b). The same is true for behavioural defence (Mooring et al., 2000). In addition, larger species may be less defensive against haematophages (see Chapter 13). Consequently, a host species with lower defence abilities can be exploited by a higher number of fleas than a host species with higher defence abilities, all else being equal. Species-specific differences in sheltering behaviour and burrow/nest structure (depth, length of tunnels, nest composition, defecation in or outside burrow etc.) can determine the amount of space in the burrow/nest for larvae, the amount of the organic matter in substrate that larvae use as a food and burrow microclimate (Kucheruk, 1983). Both the amount of organic matter and the microclimatic parameters affect flea abundance (Krasnov et al., 2001a, 2005e) which is likely to

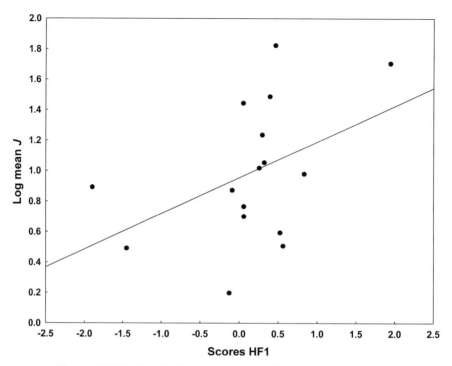

Figure 15.6 Relationship between the index of aggregation J (averaged for all flea species within a host species) and scores of the composite variable representing variation in host body mass and burrow complexity (positive correlation, cumulative factor loading 0.89) across 17 small mammalian hosts from Slovakia and the Negev Desert. Data from Krasnov et al. (2006f).

be reflected in between-host differences in the abundance of imago fleas. Hosts with higher BMR may increase their exposure to parasites because a high BMR suggests an increased activity (Gregory et al., 1996; but see Morand & Harvey 2000). Hosts with higher BMR may also invest more energy to compensate for the negative impact of parasites. Finally, the fecundity of fleas depends on which host species they exploit (see Chapter 11) which might, in turn, determine host-specific levels of flea abundance.

15.4.2 Host gender

Gender biases in parasite infestation are known for various host–parasite systems (see review in Zuk & McKean, 1996), involving both endoparasites (e.g. Poulin, 1996a) and ectoparasites, including fleas (Smit, 1962b; Stark & Miles, 1962; Ulmanen & Myllymäki, 1971; Botelho & Linardi, 1996; Bursten et al., 1997; Soliman et al., 2001; Anderson & Kok, 2003; Morand et al., 2004; Kirillova *et al.*, 2006; Lareschi, 2006). In most cases, males of higher vertebrates (birds and mammals) are infested by more parasites than females. This can be manifested in an intraspecific gender difference in one or more parasitological parameters such as mean abundance of parasites (e.g. Rossin & Malizia 2002) and prevalence of infestation (e.g. Soliman *et al.*, 2001). Male-biased parasitism is often related to gender differences in mobility and immunocompetence. Indeed, males in higher vertebrates are usually more mobile than females, and this difference in mobility increases the chances of males of being exposed to a larger variety and number of parasites (Davis, 1951; Stark & Miles, 1962; Brinck-Lindroth, 1968; Lang, 1996). Differences in the immunocompetence between males and females may occur because of the immunosuppressive effect of androgens (see Chapter 13). As a result, feeding performance and, consequently, reproduction may be more successful on male than on female hosts (e.g. *Xenopsylla vexabilis* in Hawaii: Haas, 1965; see Chapters 10 and 11).

Looking at the male-biased pattern of flea parasitism from the flea perspective, it was suggested that if fleas increased their fitness through dispersal, then they would benefit from a tighter association with male hosts (Lundqvist, 1988; Bursten *et al.*, 1997; Smith *et al.*, 2005). Indeed, Bursten *et al.* (1997) found that juvenile males of the California ground squirrel *Spermophilus beecheyi* were infested with more fleas *Oropsylla montana* than juvenile females, whereas adult rodents (a) had fewer fleas and (b) did not demonstrate gender differences in flea infestation. This was despite a clear lack of behavioural difference between male and female juvenile rodents except for the further natal dispersal distances of males. Interestingly, the disproportionate infestation of juvenile males was mainly due to male fleas. Bursten *et al.* (1997) suggested that male-biased parasitism characteristic for the juvenile cohort of a host may be an adaptation of fleas to decrease the chances of inbreeding. Greater numbers of male fleas on farther-dispersing host individuals presumably allows these males to increase their chances of mating with unrelated females. On the other hand, female fleas prefer to stay on shorter-dispersing hosts to ensure successful development of their offspring in hosts' burrows with guaranteed resources. In other words, emigration is a risky strategy. However, an emigrating male flea jeopardizes its own life only, whereas an emigrating female flea jeopardizes not only its own life but also the life of its offspring.

Examples of female-biased parasitism also exist. For example, female wood ducks *Aix sponsa* in the USA had a higher prevalence of plathelminth infections than males (Drobney *et al.*, 1983). Female-biased parasitism by haematozoans was reported for European kestrels *Falco tinnunculus* in Finland (Korpimäki *et al.*, 1995). Female little brown bats *Myotis lucifugus* in Pennsylvania and Leisler's bats *Nyctalus leisleri* in Slovakia hosted higher densities of ectoparasites, including fleas (e.g. *Myodopsylla insignis*) than did males (Dick et al., 2003; Kaňuch et al., 2005). Among European vespertilionid bats studied by Zahn & Rupp (2004) in Germany and Portugal, the greater mouse-eared bat *Myotis myotis*, whiskered bat *Myotis mystacinus*, noctule bat *Nyctalus noctula* and common pipistrelle *Pipistrellus pipistrellus* demonstrated female bias in their infestation by ischnopsyllid fleas, whereas male bias was found in Natterer's bat *Myotis nattereri* only. In Chile, prevalence of *Sternopsylla distincta* was higher in female than in males of the Mexican free-tailed bat *Tadarida brasiliensis* (Muñoz et al., 2003). Medvedev et al. (1984) reported strong female bias in the infestation of seven bat species by seven species of ischnopsyllid fleas in Central Asia. In Sudetes, female bank voles *M. glareolus* were found to be more heavily infested than males by fleas in early spring and, sometimes, in early autumn, although male bias was observed in other seasons (Haitlinger, 1973, 1975).

Gender-biased parasitism has sometimes been related to sexual size dimorphism. For example, Mooring et al. (2004) argued that higher ectoparasite infestation in male ungulates can be related to sexually dimorphic grooming in which breeding males groom less than females. This has coevolved with sexual body size dimorphism, suggesting that intrasexual selection has favoured reduced grooming that enhances vigilance of males for oestrous females and rival males. Another explanation of gender-biased ectoparasitism that invokes a differential between-gender defensiveness is that individuals of a smaller gender should be more defensive due to their larger body surface:body mass ratio (Gallivan & Horak 1997; see Chapter 13). A positive relationship between male-biased parasitism and sexual size dimorphism was demonstrated by Moore & Wilson (2002). They also demonstrated a positive correlation, across various mammalian taxa, between male-biased mortality and male-biased parasitism, thus supporting the hypothesis that parasites contribute to the association between sexual size dimorphism and male-biased mortality.

Morand et al. (2004) carried out a comparative analysis of flea abundance for 10 species of rodents and soricomorphs from Slovakia and found that the abundance of fleas was higher in males than in females. After taking into account potential confounding effects such as host phylogenetic relationships and host density, they also found that an increase in sexual size dimorphism was not related to an increase in male infestation by fleas. The lack of relationship between host sexual size dimorphism and flea abundance and prevalence was also reported for the Negev Desert rodents (Krasnov et al., 2005c). Lareschi (2006) explained male-biased flea parasitism in the water rat *Scapteromys aquaticus* in Argentina by male-biased sexual size dimorphism. However, Morand et al. (2004) and Lareschi (2006) did not look at gender differences in parasite-induced host mortality, perhaps because of certain limitations of the census data. Nevertheless, for the census data, these differences can be inferred from the gender differences in the pattern of parasite aggregation.

Gender differences in traits that are responsible for the levels of defensibility and/or exposure to parasitism can vary seasonally. For example, in small mammals, sexual difference in mobility and home range size is well expressed during the breeding season, but practically disappears in other seasons (e.g. Randolph, 1977). The level of androgens is known to fluctuate seasonally, being lowest during low reproductive activity (e.g. Lee et al., 2001). Consequently, gender differences in the infestation parameters are also expected to vary seasonally.

Krasnov et al. (2005c) investigated seasonal manifestation of gender differences in the patterns of flea infestation in nine rodent species from the Negev Desert. Specifically, they tested whether sex-biased parasitism by fleas in terms of mean abundance, prevalence and the level of aggregation (a) varies among host species and (b) is seasonally stable within host species.

In summer, male rodents were significantly more infested with fleas than females in *G. andersoni* (Fig. 15.7a) only, whereas in winter this was true also for Wagner's gerbil *D. dasyurus*, the pygmy gerbil *Gerbillus henleyi*, Baluchistan gerbil *G. nanus* and Sundevall's jird *Meriones crassus* (Fig. 15.7b). Female-biased parasitism has never been recorded in summer, whereas it occurred in winter in the golden spiny mouse *Acomys russatus* (Fig. 15.7b). Gender differences in flea prevalence were found in winter populations of *D. dasyurus* and *G. henleyi* only, with a higher percentage of infested males than females.

Aggregation level measured as parameter b of Taylor's power relationship differed significantly between males and females in three species only (b for males was significantly higher than that for females in *A. russatus* in both seasons and in the greater Egyptian gerbil *Gerbillus pyramidum* in summer, whereas the opposite was true for *G. henleyi* in summer). No relationship was found between the degree of sexual size dimorphism and the value of b in any season, but when the data were pooled for both seasons, the gender difference in parameter b of Taylor's relationship was found to be positively correlated with the degree of sexual size dimorphism (Fig. 15.8). Correction of the data for the confounding effect of phylogeny did not change the results.

These results demonstrated that gender differences in rodent hosts in the pattern of flea parasitism vary among species. This can be related to interspecific differences in male and female mobility and spatial behaviour of the hosts. Indeed, gender differences in spatial behaviour are known for at least four rodents for which gender-biased parasitism was found in this study. Furthermore, gender-biased parasitism was manifested mainly in winter when most rodents in the study region showed peak reproduction (Shenbrot et al., 1994, 1997, 1999b; Krasnov et al., 1996a). The reasons for this can be sharper

344 Ecology of flea populations

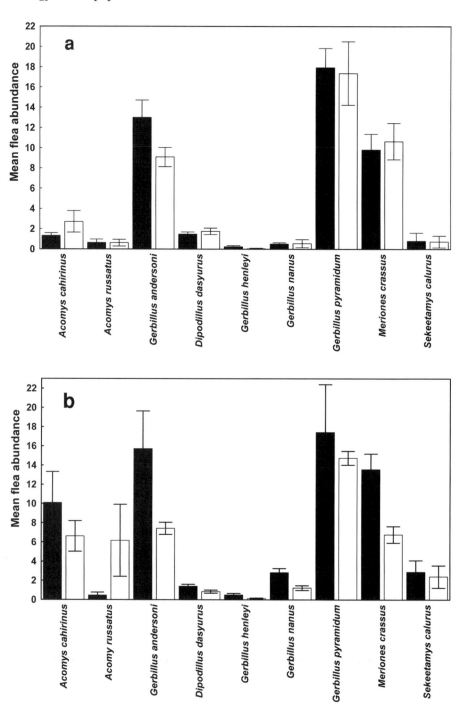

Figure 15.7 Mean (±S.E.) flea abundance in males (black columns) and females (white columns) of nine rodent species from the Negev Desert in (a) summer and (b) winter. Redrawn after Krasnov et al. (2005c) (reprinted with permission from Springer Science and Business Media).

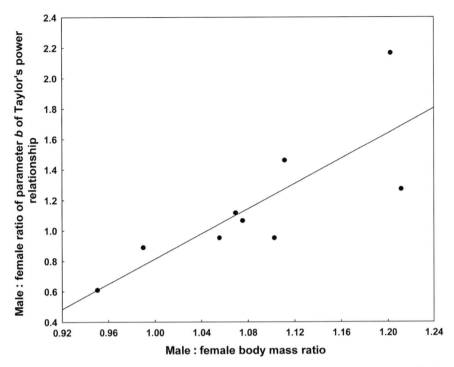

Figure 15.8 Relationship between male : female body mass ratio and male : female ratio of the parameter b of Taylor's power relationship between mean flea abundance and its variance in nine rodent species from the Negev desert. Data from Krasnov et al. (2005c).

between-gender differences in territoriality (Lott, 1991) and/or immunocompetence (see Chapter 13) in the reproductive period.

A positive relationship between the degree of gender difference in flea aggregation and the degree of sexual size dimorphism suggests that gender-biased mortality cannot be rejected as an explanation of gender-biased flea infestation, at least in some hosts. A degree of male bias in adult mortality in mammals has been shown to be positively correlated with the degree of sexual size dimorphism (Promislow, 1992). Moore & Wilson (2002) found that parasites can mediate, or at least contribute to, this male-biased mortality. Furthermore, gender-biased parasitism can be associated not only with male-biased but also with female-biased mortality. Nevertheless, in some studies, extrinsic factors often explained better intraspecific variation in infestation patterns than intrinsic factors such as gender (Moura et al., 2003; Behnke et al., 2004). This suggests that the gender difference in the immune abilities cannot be the primary factor that explains gender-biased parasitism.

15.4.3 Host age

Parasite abundance and patterns of distribution often vary between younger and older hosts (e.g. Goater & Ward, 1992). Data from a number of host–parasite systems demonstrated that relationships between host age and parasite abundance and distribution may be either linearly positive, asymptotic or convex (see review in Hudson & Dobson, 1995). Each of these patterns can be generated by various mechanisms.

Anderson & Gordon (1982) concluded that the convex pattern of variation in parasite distribution among hosts of different ages (relatively low proportion of heavily infested individuals among young or old hosts) provides clues to mortality patterns. This idea was supported by Rousset et al. (1996). They demonstrated that, in case of parasite-load-dependent mortality, the occurrence of parasite-induced host mortality and variability of infection rate for hosts of different ages led to a decrease in mean parasite abundance or aggregation level (variance-to-mean ratio) in younger and older host classes (less resistant or more exposed categories of hosts). In addition, the loss of the infected individuals from population due to parasite-induced host mortality would generate either convex or asymptotic prevalence–age relationships (Grenfell et al., 1995). Differential attractiveness of hosts of different age for parasites can also be a reason for convex age-intensity or age-prevalence patterns (e.g. Christe et al., 2003).

Another process that affects the shape of the host age–parasite abundance and parasite aggregation curves is age-dependent development of host defence mechanisms (Pacala & Dobson, 1988; Gregory et al., 1992; Grenfell et al., 1995; Hudson & Dobson, 1995). Convex or asymptotic relationships between host age and prevalence can be expected if the level of acquired resistance against parasites is low either in younger (by definition) and older (because of immunosenescence: Møller & de Lope, 1999) or in younger-only host cohorts, respectively. Age-dependent behavioural defence can also generate asymptotic age–intensity curves, especially in the case of haematophagous ectoparasites (e.g. Schofield & Torr, 2002). If, however, the host does not acquire either immunity or defensive behaviour with age and the rate of acquisition of new parasites therefore exceeds the rate of parasite mortality due to host defence, parasite abundance, aggregation and prevalence would increase with the age of a host (Hudson & Dobson, 1995).

Age bias in flea parasitism has been reported in various flea–host associations, but no common trend can be gleaned from these reports (see also Marshall, 1981a). For example, in Kazakhstan, prevalence and abundance of *Echidnophaga oschanini*, *N. laeviceps* and *Citellophilus trispinus* were higher on adult than on young great gerbils *Rhombomys opimus* (Morozov, 1974). A similar pattern was observed for the scrub hare *Lepus saxatilis* in South Africa (Louw et al., 1995) and the edible dormouse *Glis glis* in the Middle Volga Region (Kirillova et al., 2006). However, in the Sudetes and Pieniny Mountains, fleas were more abundant on adult than on young *M. glareolus* in autumn, but not in summer (Haitlinger, 1974, 1975). In the Terskey–Alatau Ridge, young male and female grey marmots *Marmota baibacina* were less infested by *O. silantiewi* than adult males and adult reproductive females, but the level of infestation of adult non-reproductive females was also low (Bibikova, 1956). In California, young *S. beecheyi* were more highly infested by *O. montana* and *Hoplopsyllus anomalus* than adult conspecifics (Bursten et al., 1997). In Egypt, the prevalence and/or intensity of infestation by *X. cheopis*, *Leptopsylla segnis* and *Echidnophaga gallinacea* increased with an increase of age of their rat (*Rattus rattus* and *R. norvegicus*) hosts (Soliman et al., 2001).

Krasnov et al. (2006a) studied age-dependent patterns of flea infestation in seven species of rodents from Slovakia. The expectations were that the mean abundance and level of aggregation of fleas would be lowest in hosts from the smallest (youngest) and largest (oldest) size classes and highest in hosts of medium-sized classes, whereas the pattern of variation of prevalence with host age would be either convex or asymptotic. In general, mean abundance of fleas increased with an increase in host age, although the pressure of flea parasitism in terms of number of fleas per unit host body surface decreased with host age. Two clear patterns of the change in flea aggregation and prevalence with host age were found (Fig. 15.9). The first pattern demonstrated a peak of flea aggregation and a trough of flea prevalence in animals of middle age classes (*Apodemus* species and *M. glareolus*). The second pattern was an increase of both flea aggregation and flea prevalence with host age (*Microtus* species).

Consequently, there was no unequivocal evidence for the main role of either parasite-induced host mortality or acquired resistance in host age-dependent patterns of flea parasitism. Furthermore, these results suggested that this pattern can be generated by various processes and strongly affected by natural-history parameters of a host species such as dispersal pattern, spatial distribution and structure of shelters. Indeed, a relatively low level of flea abundance in young individuals can be explained by their relatively low level of exposure to fleas due to their post-weaning dispersal during which they do not possess burrows. Consequently, age-related variation in the level of exposure to fleas can at least partly explain low levels of flea abundance and lack of flea aggregation in young *Microtus*, but cannot explain high prevalence in young *Apodemus* and *M. glareolus*. The reason for this difference may be dissimilarity in the age at juvenile dispersal between these groups of species (Viitala et al., 1994). Young common voles *Microtus arvalis* start to disperse at the age of 21–25 days (Bashenina, 1962, 1977), whereas the youngest age of dispersal by *M. glareolus* was estimated at

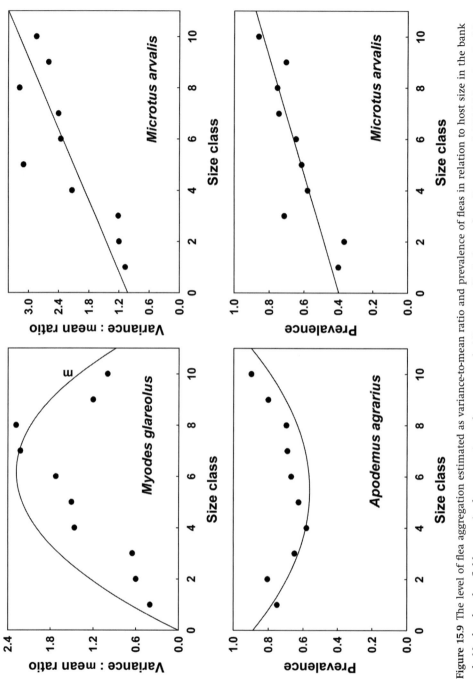

Figure 15.9 The level of flea aggregation estimated as variance-to-mean ratio and prevalence of fleas in relation to host size in the bank vole *Myodes glareolus*, fieldmouse *Apodemus agrarius* and common vole *Microtus arvalis*. Redrawn from Krasnov et al. (2006a) (reprinted with permission from the American Society of Parasitologists).

40–50 days (Bashenina, 1981). Another factor possibly contributing to the relatively low abundance and low levels or lack of aggregation of fleas in young age classes is direct or indirect parasite-induced mortality related to the strong negative effect of fleas on juvenile individuals (see Chapter 13).

Decrease in flea aggregation in large size classes, i.e. the oldest age cohort, accompanied by a high flea abundance (but lower size-specific intensity of flea infestation) and high prevalence in *Apodemus* and *M. glareolus* may be indicative of either high mortality of the individuals with a high level of infestation (because of deterioration of immune function in older individuals: see review in Miller, 1996), or low attractiveness of older (and, thus, with poorer body conditions) individuals (but see above). It may also be merely an artefact of the low relative abundance of large individuals. However, all these explanations should also be valid for *M. arvalis*, which demonstrated an increase in the level of flea aggregation in large individuals. This suggests that some features of host natural history may be responsible for the difference in age-dependent patterns of flea aggregation between *M. arvalis* and *Apodemus/Myodes*. The sharpest difference between these groups of rodents is in the structure of their shelters. *Apodemus* and *M. glareolus* shelter mainly in above-ground nests, tree hollows or cavities under tree roots, whereas *Microtus* construct complicated deep burrows. Flea accumulation is higher in deep and relatively stable burrows than in above-ground nests (Kucheruk, 1983; Němec, 1993). As a result, the probability of finding highly infested older *M. arvalis* is greater than the probability of finding highly infested older *Apodemus* or *M. glareolus*. Moreover, the level of flea aggregation in *M. arvalis* of the middle age cohorts was as high as the highest level of flea aggregation in *Apodemus* (see Fig. 15.9). Given that the number of fleas on a host body has been shown to be positively correlated with the number of fleas in the host nest or burrow (Krasnov *et al.*, 2004d), this suggests the continuous flea accumulation in *M. arvalis* burrows (Němec, 1993). This is supported by the continuous increase of flea prevalence with an increase in host size in *Microtus*. By contrast, accumulation of fleas in adult woodmice may be interrupted because these rodents frequently abandon their home ranges, and consequently their nests, and relocate (Nikitina, 1961).

Surprisingly, flea prevalence decreased in individuals of middle size classes in *Apodemus* and *M. glareolus*. This means that a relatively large proportion of individuals were completely free of fleas. An individual may be free of fleas either because of a high ability to resist them or due to an extremely low exposure to fleas, or both. The support for the latter explanation comes from numerous observations that the increase in density in populations of woodmice and bank voles often results in a surplus of adult individuals that have no individual home ranges and burrows (Gliwicz, 1992). This dispersal is represented mainly by males, in contrast with juvenile dispersal which is linked to both genders. Flea populations can be established and maintained on those host individuals that have burrows, whereas 'homeless' hosts can take part in flea transmission, but are not able to sustain fleas. This heterogeneity of hosts in relation to burrows as habitats for successful flea breeding can lead to a decrease in prevalence. Microtus too can have either peaked resistance at adulthood or a high number of transient adults, or both, but flea accumulation in its burrows can negate whatever strong defence responses may occur in individual animals and mask the true pattern of flea prevalence in the middle-aged individuals. In addition, mobility of 'non-settled' individuals can be a more important factor affecting patterns of flea distribution in Apodemus and M. glareolus than in Microtus. This is because the latter are colonial and, thus, a period of 'transience' for a given Microtus individual is likely to be shorter than that for Apodemus or M. glareolus (Bashenina, 1962, 1981).

Hawlena et al. (2005) tested two alternative hypotheses about host preference by S. cleopatrae in relation to the age of G. andersoni. The first hypothesis suggested that fleas select adult over juvenile rodents because the latter represent a better nutritional resource, whereas the second hypothesis suggested that fleas prefer the weaker and less resistant juveniles because they are easier to colonize and exploit. Flea preference for juvenile hosts was estimated as the ratio between mean flea density on juvenile hosts and total flea density on both adult and juvenile hosts within a sampling plot (Abramsky et al., 1990). Preference values greater than 0.5 indicated that a majority of fleas occurred on juvenile hosts, while values of less than 0.5 indicated that a majority of fleas occurred on the adult hosts. Surprisingly, flea distribution changed as a function of flea density, namely from juvenile-biased flea parasitism at low densities to adult-biased flea parasitism at high densities (Fig. 15.10).

This effect of intraspecific density on flea preference may be explained by at least three, not entirely mutually exclusive, hypotheses. First, the switch in preference towards adult hosts may be the result of a decrease in juvenile host quality at high flea densities. If the quality of juveniles is reduced at high flea densities, it may thus be beneficial for some flea individuals to relocate to the higher-quality, though riskier, adult host until average flea fitness on both hosts equalizes. Second, the ability to change host age-related distribution in a density-dependent fashion may have an evolutionary component. If fleas benefit more from exploiting juveniles compared to adults at low densities but benefit more from exploiting adults at high densities, then an ability to choose juvenile hosts at low densities and adults at high densities may be favoured by natural selection. Finally, the switch in preference towards adult hosts may also be a result of differences in the effect of flea density on anti-parasitic defensiveness of juvenile

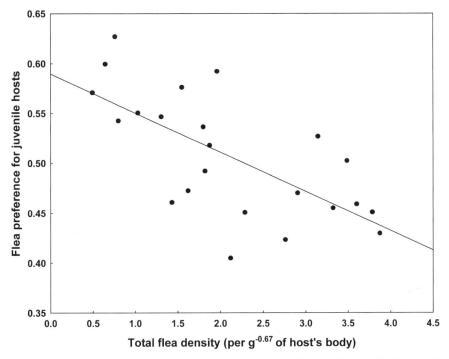

Figure 15.10 Relationship between the index of preference for juvenile Anderson's gerbil *Gerbillus andersoni* and total density (number of fleas per $g^{-0.67}$ of host's body) of *Synosternus cleopatrae*. Data from Hawlena et al. (2005).

versus adult hosts. For example, while the grooming rate of both adult and juvenile hosts may increase with flea density, the incremental increase in grooming rate in juveniles may be higher, turning heavily infested juveniles into riskier habitats (see Chapter 13).

Other factors influencing flea preference were soil temperature and the presence of ixodid ticks. Flea preference for juveniles decreased with an increase in tick density and soil temperature (though the effect of soil temperature on flea distribution seemed to be indirect via fluctuations of total flea densities). However, ticks might reduce the preference of fleas for juvenile hosts. These results suggested that host selection is not an explicit alternative choice between adults and juveniles but rather a continuum where the distribution between adults and juveniles depends on host-, parasite- and environment-related factors. Moreover, density-dependent host selection (not only between host species — see Chapter 9 — but also within host species) may promote coexistence between parasites and their hosts. At high parasite densities, at which the effects of parasites on hosts may be detrimental to hosts, it is beneficial for the parasites to shift their preference toward the less vulnerable hosts. This may have an adaptive value for the parasites because density-dependent distribution may decrease the intensity of a host's countermeasures against parasites and keep the host in a better condition. Consequently, density dependence of parasite distribution among different cohorts of a host may decelerate the arms race between the two evolutionary players.

An additional cause of the higher level of infestation of young hosts may be avoidance of inbreeding because in most cases young individuals represent a dispersing cohort. Described above the results of observations of *O. montana* and *S. beecheyi* by Bursten *et al.* (1997) suggest that this can be the case in at least some flea–host associations.

15.5 Relationship between flea abundance and prevalence

As mentioned above, a positive relationship between local abundance and occupancy is a well-known pattern (Gaston, 2003; but see Boeken & Shachak, 1998). In the application of this relationship to parasite populations, a positive correlation between the mean abundance of parasites and their prevalence (i.e. host occupancy) has been supported in many studies (Mohr, 1958; Shaw & Dobson, 1995; Morand & Guégan, 2000; Krasnov *et al.*, 2002e; Šimková *et al.*, 2002). The positive abundance/occupancy relationship has been explained by a variety of mechanisms. In fact, Gaston *et al.* (1997) and Gaston (2003) listed nine different hypotheses that might explain it. Morand & Guégan (2000) tested several of these hypotheses using nematodes parasitic on mammals. In particular, they found that the prevalence of nematodes could be successfully predicted using an epidemiological model with a minimal number of parameters such as mean abundance of a parasite and its variance, and an indicator of aggregation. The latter parameter, in turn, can be calculated from the positive relationship between mean abundance and its variance (Taylor's power law; see above). Consequently, Morand & Guégan (2000) concluded that the abundance/distribution relationship in parasites could be explained by demographic and stochastic mechanisms revealed by simple epidemiological models without invoking more complex explanations.

To confirm a positive relationship between abundance and distribution of fleas within and across host species and to test whether the prevalence of fleas can be reliably predicted from a simple epidemiological model that takes into account mean flea abundance and its variance, Krasnov *et al.* (2005h, i) used data on the abundance and distribution of fleas parasitic on small mammals in Slovakia and the Negev Desert. In both regions, prevalence of a flea species increased with an increase in its mean abundance both in xenopopulations and populations (i.e. within and across host species, respectively; see example in

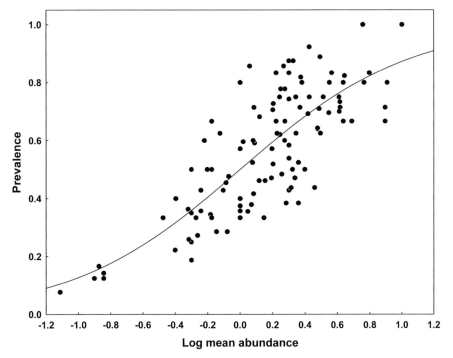

Figure 15.11 Logistic function fit for prevalence of fleas parasitizing Wagner's gerbil *Dipodillus dasyurus* plotted against log mean flea abundance. Redrawn after Krasnov et al. (2005h) (reprinted with permission from Cambridge University Press).

Fig. 15.11). Fleas thus do not differ from the majority of plant and animal taxa for which a positive relationship between abundance and occupancy has been reported (Gaston, 2003).

Epidemiological models (Anderson & May, 1985) predict that the probability distribution of parasite numbers per host individual, being a negative binomial, determines the relationship between the prevalence of infection P_t at any given time t as:

$$P_t = 1 - \left(1 + \frac{M_t}{k}\right)^{-k}, \tag{15.6}$$

where M_t is the mean number of parasites per host individual at time t and k is the parameter of the negative binomial distribution. Based on (15.3) and (15.6), the predicted prevalences (P_{p1}) for each flea–host association and each flea across all hosts were calculated. Another family of predicted prevalences (P_{p2}) was also calculated for the Negev Desert data. For these, k were replaced with the moment estimates from (15.1). Predicted prevalences were then compared with the observed prevalences.

For the Slovakian data, observed prevalences for each flea–host association or for each flea species across all hosts did not differ significantly from prevalences predicted by the epidemiological model using parameters a and b of Taylor's power relationship. Regressions of predicted prevalences on observed prevalences produced slope values that did not differ significantly from unity and were independent of scale (xenopopulation or population) (Fig. 15.12).

For the Negev data, regressions of flea prevalences expected from epidemiological models with k values calculated from Taylor's power law with observed flea prevalences produced slopes significantly less than 1. However, if moment estimates of k corrected for host number were used instead, expected prevalences did not differ significantly from observed prevalences in most flea–host associations (Fig. 15.12). Furthermore, the model with k calculated using Taylor's power law tended to overestimate relatively low prevalences and underestimate relatively high prevalences (Fig. 15.12).

Thus, a simple epidemiological model can, in general, successfully predict the occurrence of fleas in populations of their hosts. Parameters of the epidemiological model used in these two studies appear to be, generally, the most parsimonious set of factors that explain much of the variance in flea prevalence without involving other mechanisms such as the degree of flea host specificity or the level of host resistance against flea parasitism. Consequently, these studies support the demographic hypothesis of parasite abundance and distribution (Anderson et al., 1982), which suggests that the observed distributions of parasites across host individuals are generated by two opposing forces, namely those leading to overdispersion (aggregation) and those leading to underdispersion (regularity). Stochastic variability in demographic parameters may generate both overdispersion (pure birth process) and underdispersion (pure death process), whereas stochasticity in environmental processes creates overdispersion. Environmental processes include those acting in host individuals, being both (a) purely dependent on hosts and hosts' environment (e.g. variability in the susceptibility to parasitism or the level of resistance in dependence on genetic background, age, nutritional state) and (b) dependent on host–parasite relationships (e.g. the level of parasite-induced host mortality, past experience with a parasite, innate and acquired resistance against parasite). In addition, Hanski et al. (1993) argued that the positive correlation between abundance and occupancy can arise from a simple random process. However, significant departure of flea distributions among host individuals from randomness (as indicated by $b > 1$; see above) suggests that this explanation of positive abundance/occupancy relationship is not satisfactory (see also Brown, 1995; Morand & Guégan, 2000).

Nevertheless, prevalences in some xenopopulations were higher (e.g. *Parapulex chephrenis* on *A. cahirinus* in the Negev Desert and *Megabothris turbidus* on

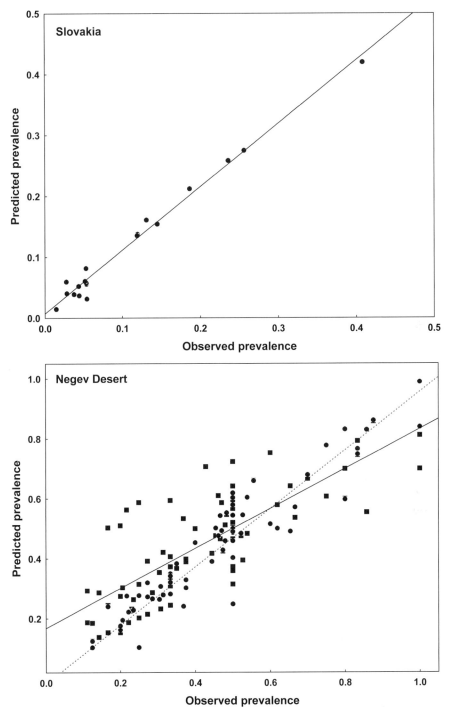

Figure 15.12 Relationship between observed and predicted (from the epidemiological models and Taylor's power law) prevalences of 19 flea species across host species for the data from Slovakia and of *Nosopsyllus iranus* across host species for the Negev Desert. For *N. iranus*, P_{p1} — squares, solid line; P_{p2} — circles, dashed line (see text for explanations). Redrawn after Krasnov *et al.* (2005h) and Krasnov *et al.* (i) (reprinted with permissions from Cambridge University Press and Elsevier, respectively).

M. glareolus in Slovakia) or lower (*D. dasycnema* on the European water shrew *Neomys fodiens* in Slovakia) than those predicted from the epidemiological model. In other words, there were either more or fewer infested animals than expected from the model. Flea prevalence overestimated by predictions can be explained by either a relatively low negative effect of flea parasitism on a host and/or a strong resistance of a host to flea parasitism, whereas underestimations of prevalence can be explained either by high flea-induced host mortality or low preference for a particular host by a particular flea.

For the data from the Negev Desert, the epidemiological model predicted accurately the observed prevalence only when it was corrected for host sample size. This confirms the important effect of host density on flea distribution (e.g. Krasnov et al., 2002c; Stanko et al., 2002; see below) and suggests that calculation of k from Taylor's relationship should be used cautiously, although it has been used successfully in other studies (Morand & Guégan, 2000; Šimková et al., 2002). In addition, the effect of the off-host environment should not be forgotten. Environmental factors such as ambient temperature and relative humidity can strongly affect fleas' survival and birth and death rates (see Chapters 10 and 11). Therefore, a confounding effect of the off-host environment on the relationship between flea abundance and distribution can be expected. Nonetheless, this appeared not to be the case in Slovakia. This means that purely environmental factors played a minor role in flea distribution among hosts in this region. However, in temperate regions, stochastic (as opposed to seasonal) environmental fluctuations are less sharp and the environment is more predictable than in regions such as deserts. Consequently, the abundance/distribution relationship in fleas from the arid environment appeared to be more complicated than in those from the temperate region.

15.6 Factors affecting flea abundance and distribution

15.6.1 Host density

It is obvious that some kind of relationship exists between abundance of a consumer and abundance of a resource. A host is an ultimate resource for a parasite and, consequently, its abundance must influence the abundance of a parasite (Dobson, 1990).

Classic models describing the dynamics of host (H) and parasite (P) populations were presented by Anderson & May (1978), May & Anderson (1978), Grenfell (1992), Arneberg et al. (1998) and Wilson et al. (2001). The main assumptions of these models are that (a) host population growth is density dependent, given that there is no parasite-induced reduction of host reproduction; (b) hosts are

Table 15.1 *Parameters used in the basic host–parasite dynamics model*

Parameter	Description
a	Instantaneous host birth rate (per host per unit time)
b	Instantaneous host death rate due to causes other than parasite (per host per unit time)
α	Instantaneous host death rate due to parasite influence (per host per unit time)
β	Instantaneous rate of reduction in host reproduction due to parasite influence (per host per unit time)
ϖ	Severity of density dependence in host population growth
λ	Instantaneous birth rate of parasite transmission stages (imago in case of fleas) (per parasite per unit time)
μ	Instantaneous death rate of parasites due to either natural or host-induced (e.g. autogrooming) causes
H_0	Transmission efficiency constant, varying inversely with the proportion of parasite transmission stages that infest individuals of the host population (see Anderson & May, 1978 for further explanations)
k	Parameter of the negative binomial distribution

Source: Adapted from Anderson & May (1978), Dobson (1990) and Arneberg *et al.* (1998).

long-lived, relative to their parasites; and (c) the frequency of parasites within hosts follows the negative binomial distribution. The model assumes also that a parasite induces mortality of a host (Anderson & May, 1978; May & Anderson, 1978). However, Arneberg *et al.* (1998) argued that the positive relationship between host density and parasite abundance should also be expected when parasites have no pathogenic effects.

The basic model utilizes the following equations:

$$\frac{dH}{dt} = (a - b - \varpi H)H - (\alpha + \beta)P \tag{15.7}$$

$$\frac{dP}{dt} = \frac{\lambda P H}{H_0 + H} - (b + \varpi H + \alpha + \mu)P - (\alpha + \mu)\frac{(k+1)P^2}{kH}. \tag{15.8}$$

A description of the parameters of these equations is presented in Table 15.1. This model predicts that the abundance of a parasite should increase in a curvilinear fashion to a plateau with increasing host density because the greater the host density, the greater the probability that each parasite individual or respective transmission stage will contact a host (e.g. Haukisalmi & Henttonen, 1990). These relationships between parasite and host abundances have been reported for fleas. For example, this pattern was described for *C. tesquorum* and *Neopsylla setosa* parasitic on the pygmy ground squirrel *Spermophilus pygmaeus* in the southern part of European Russia (Gerasimova *et al.*, 1977; Pushnitsa *et al.*, 1978);

X. cunicularis parasitic on *O. cuniculus* in Spain (Launay, 1989); *Xenopsylla gerbilli*, *Xenopsylla hirtipes*, *X. conformis*, *Xenopsylla skrjabini* and *Coptopsylla lamellifer* parasitic on the great gerbil *R. opimus* and the midday jird *Meriones meridianus* in various regions of Central Asia (Maslennikova *et al.*, 1967; Rudenchik *et al.*, 1967; Kunitsky *et al.*, 1971b; Shevchenko *et al.*, 1971; Sukhanova *et al.*, 1978; Rzhevskaya *et al.*, 1991; Rapoport *et al.*, 2007); *N. laeviceps*, *X. conformis* and *Rhadinopsylla dives* parasitic on the Mongolian gerbil *Meriones unguiculatus* in China (Z.-L. Li *et al.*, 1995, 2002); and *Xenopsylla astia* parasitic on *R. norvegicus* in Qatar (Abu-Madi *et al.*, 2005). An exponential increase in flea burden with density growth of the Daurian ground squirrel *Spermophilus dauricus* was also found for *C. tesquorum*, *Frontopsylla luculenta*, *Neopsylla abagaitui* and *N. bidentatiformis* in Inner Mongolia (China) (Li & Zhang, 1997, 1998). In Michigan, Hoogland & Sherman (1976) showed that both mean abundance of *Ceratophyllus styx* per burrow and proportion of infested burrows of the sand martin *Riparia riparia* increased linearly with an increase in active burrows in a colony. Ma (1988) studied density dynamics of 10 flea species in relation to density fluctuations of four rodent and lagomorph hosts in the Jilin, Gansu and Qinghai Provinces of China and found similar trends in density fluctuations of hosts and their fleas. In particular, an experimental density decrease of *S. dauricus* has led to a concomitant decrease of abundance and prevalence of *C. tesquorum* and *N. abagaitui*. Abundance and prevalence of *O. silantiewi* and *Callopsylla dolabris* were higher in areas of higher density of the Himalayan marmot *Marmota himalayana*, while in areas with lower density of this host abundance and prevalence of these fleas were significantly lower. The same was true for *Amphipsylla vinogradovi*, *Ophthalmopsylla kukuschkini* and *N. bidentatiformis* parasitic on the Chinese striped hamster *Cricetulus barabensis* and for *Amphalius clarus*, *Frontopsylla aspiniformis*, *Rhadinopsylla dahurica* and *Ochotonobius hirticrus* parasitic on the Plateau (Daurian) pika *Ochotona daurica*. Yensen *et al.* (1996) compared parasite abundance and distribution (including those of fleas) between two subspecies of the Idaho ground squirrel *Spermophilus brunneus*. The proportion of parasitized individuals as well as intensity of infestation were significantly lower in *S. b. brunneus*, which lives in small, isolated populations, than in *S. b. endemicus*, which has larger, less fragmented populations, suggesting a relationship between host population structure and parasite abundance.

Krasnov *et al.* (2002e) studied how the abundance of *D. dasyurus* in the Negev Desert affected the abundance and distribution of two fleas exploiting this host. Indeed, the abundance of *X. dipodilli*, a flea reproducing during the reproductive period of its host, increased with an increase in *D. dasyurus* density to a plateau (Krasnov *et al.*, 2002e) (Fig. 15.13). How can the occurrence of this plateau be explained? Perhaps this can be related the limited carrying capacity of host individuals that can harbour only a limited number of parasites, and/or to

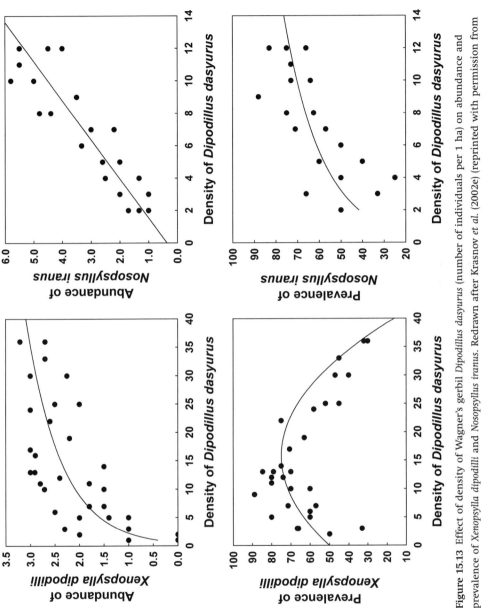

Figure 15.13 Effect of density of Wagner's gerbil *Dipodillus dasyurus* (number of individuals per 1 ha) on abundance and prevalence of *Xenopsylla dipodilli* and *Nosopsyllus iranus*. Redrawn after Krasnov et al. (2002e) (reprinted with permission from the Ecological Society of America).

flea-induced host mortality. However, this explanation seems to be unsatisfactory because, in contrast to *X. dipodilli*, the relationship between abundance of the seasonal *N. iranus*, which reproduces in the periods when its host does not reproduce, and density of *D. dasyurus* appeared to be linear (Fig. 15.13).

An alternative explanation implies the heterogeneity of hosts in relation to burrows as habitats for successful flea breeding, as well as the carrying capacity of the host habitat. It has already been mentioned that an increase in density of small mammal populations, particularly in solitary species, often results in a surplus of individuals ('transients' or 'dispersers') that have no individual home ranges (Brandt, 1992; Gliwicz, 1992 and references therein; see also below). In theory, these 'homeless' individuals should not be putative hosts for fleas, because they do not possess burrows that are necessary for flea reproduction and development of pre-imaginal stages, although they can take part in flea transmission (Janion, 1968). In turn, the density of resident hosts is determined by the carrying capacity of the given habitat. For example, this carrying capacity can be dependent on the number of available burrows or places for burrowing *ceteris paribus*. As overall host density increases, the number of residents attains a particular level determined by the carrying capacity of host habitat. Further increases in host density result in an increase in the number of transients, while the number of residents remains stable. Because transient hosts do not participate in the siphonapteran life cycle, the overall increase in host density after saturation of a habitat by residents has no effect on flea abundance. This may also explain the absence of a plateau in *N. iranus* under a high density of gerbils. In contrast to *X. dipodilli*, this flea reproduces in periods when most host individuals in a given habitat are residents, whereas transient individuals are absent (Shenbrot et al., 1997; Khokhlova et al., 2001).

What of the relationship between flea prevalence and host density? Parasites infesting different host individuals are analogous to free-living organisms inhabiting discrete patches. According to metapopulation theory, the percentage of occupation of the patches increases with the decrease in patch isolation (Thomas & Hanski, 1997), and thus prevalence could be expected to increase with increasing host density. Another argument for this is the same as that for the abundance, namely the increased probability of the transmission stage of being transmitted under high host density. However, the relationship between host density and parasite prevalence is expected to attain a plateau under high host densities, given that there are no parasite extinction events across host individuals. The explanation is that once host density becomes high enough for all parasites to find a host, a further increase in host density would be inconsequential. However, field data on the relationship between flea prevalence and host density are contradictory. For example, the increase and stabilization of flea prevalence at around 100% with an increase in host density has been reported for *Oropsylla bruneri* and *Oropsylla rupestris* parasitizing Richardson's ground squirrel *Spermophilus richardsoni* in Canada (Lindsay & Galloway, 1997), whereas the prevalence of *Xenopsylla bantorum* increased with a decrease in density of its host, the Nile grass rat *Arvicanthis niloticus*, in Kenya (Schwan, 1986). Conversely, prevalence cannot be lower than a specific eradication threshold (*sensu* Nee *et al.*, 1997), because a minimum number of hosts is required for the parasite population to persist, analogous to a minimum number of patches for free-living animals (Nee, 1994; Kareiva & Wennergren, 1995; Hanski *et al.*, 1996).

The data from the Negev Desert demonstrated that with an increase in overall density of *D. dasyurus*, prevalence of *X. dipodilli* changed unimodally, whereas that of *N. iranus* increased logarithmically (Krasnov *et al.*, 2002e) (Fig. 15.13). A unimodal response of prevalence of *X. dipodilli* to host density can result from host heterogeneity in spatial behaviour described above (i.e. resident versus transient individuals). If only resident host individuals support the flea population, the prevalence increases with an increase in host density until all resident individuals are infested and the habitat is saturated with residents (all available burrows are occupied or individual home ranges reduced to minimal possible size). Under higher host densities, the prevalence of flea infestation decreases, because the resident and infested hosts compose an ever-decreasing fraction of the overall population. In the case of *N. iranus*, the number of habitat patches (hosts) at each given location does not increase. As a result, the decreasing part of a unimodal curve of prevalence plotted against host density is absent.

The pattern of the relationship between host and parasite abundances may differ depending on a particular type of association between a particular parasite and a particular host. For example, Darskaya *et al.* (1970) showed that in the central part of European Russia, the effect of density of *M. arvalis* on flea abundance varied among flea species, being strong in *Amphipsylla rossica* and weak (if present at all) in *Ctenophthalmus assimilis* and *Palaeopsylla soricis*. Zhovty & Kopylova (1957) studied variation in the abundance of fleas parasitic on *O. daurica* in the Trans-Baikalia. They found that infrapopulations of *F. luculenta* increased, whereas infrapopulations of *N. bidentatiformis* decreased with the host population growth.

As we have already seen, even a highly host-opportunistic parasite varies in its abundance among different host species. If the difference in the abundance of a parasite in various hosts stems from differential fitness rewards, then different hosts obviously play different roles in the long-term persistence of a parasite population. In such cases the parasite population would thus depend mainly on one or a few key host species. Consequently, a tight link between host abundance and the abundance and distribution of a particular parasite should be expected for only some hosts from the entire host spectrum of this parasite.

To test this hypothesis, Stanko et al. (2006) studied the effect of host abundance on flea abundance and distribution using data on 57 flea–host associations from Slovakia. Surprisingly, relationships between flea abundance or prevalence and host abundance were found to be either negative or absent (23 and 34 flea–host associations, respectively) (Fig. 15.14). Negative relationships between flea and host abundance were always accompanied by negative relationships between flea prevalence and host density. However, in eight flea–host associations, negative relationships between host abundance and flea prevalence, but not abundance, were found.

The link between host and flea abundances thus differed among host species exploited by the same flea. A strong link between host and parasite population dynamics suggests close interactions between host and parasite demographic parameters (Anderson & May, 1982). The lack of this link in a parasite–host system implies that the demographies of a given parasite and a given host are unrelated. In other words, a host-opportunistic parasite is not equally dependent on all its host species, but rather the parasite's population dynamics are determined by populations of some but not other hosts. This consideration provides firmer grounds for a classification that distinguishes between principal and auxiliary hosts among the entire host spectrum of a parasite. As we have seen in Chapter 9, assigning a host species to one of these categories is usually done according to the abundance attained by a parasite in this host species. However, a host in which a parasite attains the highest abundance and a host whose population dynamics are linked with that of a parasite are not always the same species. For example, *Amalaraeus penicilliger* in the study of Stanko et al. (2006) exploited mainly three vole species. Nevertheless, its abundance and prevalence were heavily dependent on the abundance of the European pine vole *Microtus subterraneus* only, whereas it attained the highest abundance on *M. glareolus*.

In some cases, though, the relationships between the density of a flea and the density of even an undoubtedly principal host are not clear. Bekenov et al. (2000) analysed data on densities of *X. skrjabini* and its principal host *R. opimus* collected during 28 years in Kazakhstan. Although at the spatial scale the densities of the flea and the host were positively correlated, no relationship between their density dynamics has been found at the temporal scale.

Furthermore, Stanko et al. (2006) found no increase in flea abundance or prevalence with host abundance growth in any flea–host association, as predicted by epidemiological models. Patterns of relationships between parasite distribution and host abundance contradicting the predictions of epidemiological models have been reported in other studies (e.g. Sorci et al., 1997). For example, the

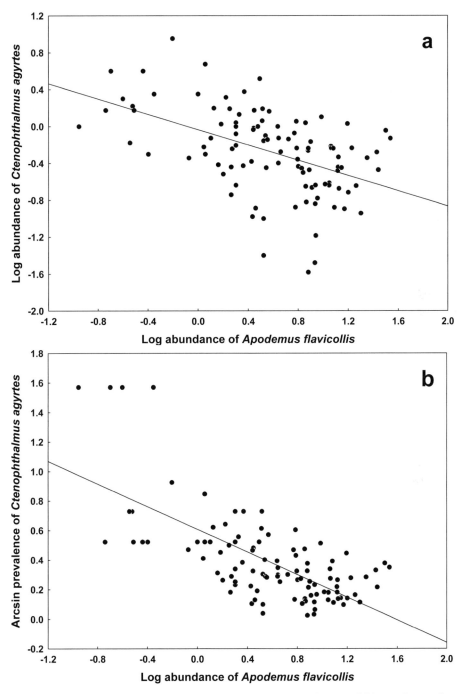

Figure 15.14 Relationships between (a) the mean abundance and (b) prevalence of *Ctenophthalmus agyrtes* and abundance of the yellow-necked mouse *Apodemus flavicollis* (number of individuals captured per 100 trap–nights). Redrawn after Stanko et al. (2006) (reprinted with permission from Blackwell Publishing).

negative relationship between flea and host abundance has been mentioned by Haitlinger (1977, 1981) and Krylov (1986) for the same flea species from the Middle Sudetes and Lower Silesia (Poland) and the vicinity of Moscow (Russia), respectively, as those used in the study by Stanko *et al.* (2006), although no formal analysis has been carried out. A negative relationship or a lack of relationship between parasite abundance or prevalence with an increase in host abundance can arise from a number of causes. One of these may be the lower rate of flea reproduction and transmission compared to the rate of reproduction and dispersal of hosts. In other words, the rate of establishment of new patches (newly born or dispersing young mammals) is faster than the rate of their infestation. Consequently, under high host density, a fraction of host individuals may remain 'underused' by the fleas, merely because they cannot keep pace with host reproduction and dispersal. The 'delay' in fleas' reproductive response to changes in host density is indirectly supported by observations by Kotti & Kovalevsky (1995) that the increase in abundance of *Megabothris advenarius* and *Megabothris asio* in the area between the Amur and Bureya Rivers followed the increase in the abundance of their small mammalian hosts with a year-long lag. In other words, a parasite can be 'diluted' among hosts when the latter attain high densities. Indeed, the development time of many fleas (see Vashchenok, 1988) is longer than the time of pregnancy and postnatal development (until dispersal) of many small mammals (see Bashenina, 1977).

The contradictions between theoretical models and real data demonstrate that nature is much more complicated that any human-constructed model. The diversity of patterns of the relationship between host abundance and flea abundance and distribution mentioned above may be a result of different regulating mechanisms governing different flea–host associations. Furthermore, different regulating mechanisms may act simultaneously within the same flea–host association. Can we infer these mechanisms from field observations?

15.6.2 Inferring regulating mechanisms

Theoretically, the plateau attained by parasite abundance with an increase or a decrease in host abundance is expected due to regulatory mechanisms such as parasite-induced host mortality, host-induced parasite mortality and density-dependent reductions in parasite fecundity and survival (Grenfell & Dobson, 1995). These mechanisms are difficult to demonstrate in the field because dead hosts are rarely found and if they are, the cause of death can rarely be unequivocally attributed to the parasite (McCallum & Dobson, 1995). However, parasite-induced host mortality or host-induced parasite mortality can be inferred from the pattern of parasite distribution and aggregation (Anderson

& Gordon, 1982; Rousset et al., 1996). An attempt to do this for fleas has been undertaken by Stanko et al. (2006).

In their study of Slovakian fleas and small mammals, Stanko et al. (2006) assumed that if host mortality is induced by parasite accumulation, then flea-induced host mortality may be inferred from the relationship between flea aggregation and flea abundance, whereas host-induced flea mortality may be inferred from the relationship between flea abundance and aggregation and host abundance. If flea-induced host mortality plays a regulating role in host populations and imposes an upper limit on the number of fleas that a host is able to endure without dying, then a degree of aggregation will approach an asymptote with an increase in flea abundance (because of loss of heavily infested hosts at high flea abundances). If host-induced flea mortality plays the main regulating role and imposes an upper limit on the number of fleas that a host is able to tolerate without mounting defence mechanisms, then (a) flea abundance will decrease or, at least, will not increase with an increase of host abundance; and (b) a degree of flea aggregation will approach an asymptote with an increase in host abundance (because of the lack of heavily infested hosts despite an increase in transmission rate).

They found that the link between flea abundance or prevalence and host abundance, evaluated as the coefficient of determination of the respective regressions, decreased with an increase in parameter b of Taylor's power law across flea–host associations. In addition, mean crowding of fleas (m^*) in eight associations decreased with an increase of host abundance and in 49 associations increased with an increase of flea abundance (Fig. 15.15). The absolute value of the slope in four of eight and in 15 of 49 associations was <1, suggesting that flea crowding decreased or increased, respectively, with an increase in host or flea abundance, respectively, to an asymptote.

Parasites may cause the death of their hosts for various reasons, both direct and indirect (see Chapter 12). Examples of indirect causes are increased susceptibility to predation (Kavaliers & Colwell, 1994) and modification of the outcome of competitive interactions (Hudson & Greenman, 1998). A correlation between the coefficient of determination of the regressions of flea abundance or prevalence against host abundance and the slope of Taylor's b suggests some role of parasite-induced host mortality in a relationship between flea and host population dynamics. Anderson & Gordon (1982) showed that when the rate of parasite-induced host mortality was high, the slope of the relationship between the logs of the variances and means was low, while, conversely, when the rate of mortality was low, the slope was high. The strength of the link between flea and host population dynamics decreasing with an increase in b, thus suggested that parasite-induced host mortality is expected to be

366 Ecology of flea populations

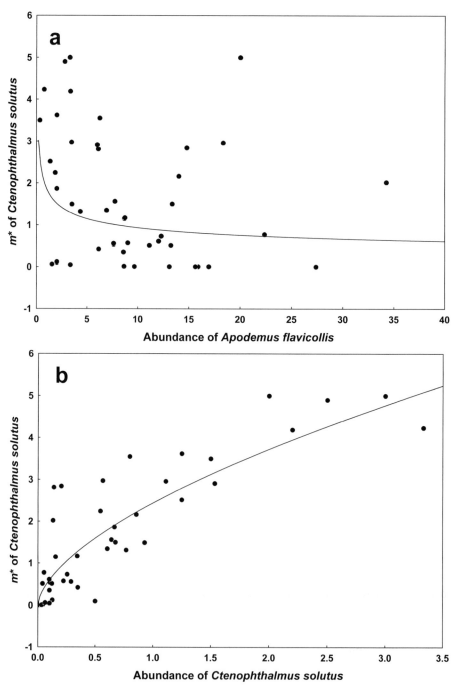

Figure 15.15 Relationships between the mean crowding of *Ctenophthalmus solutus* and (a) abundance of the yellow-necked mouse *Apodemus flavicollis* (number of individuals captured per 100 trap–nights) and (b) its own mean abundance. Redrawn after Stanko et al. (2006) (reprinted with permission from Blackwell Publishing).

high in flea–host associations where flea abundance is affected strongly by host abundance.

Parasite-induced host mortality can also be inferred by comparing the degree of flea aggregation in host populations that are characterized by different levels of flea abundance. Indeed, in some flea–host associations, flea aggregation increased initially with an increase of flea abundance until a certain level and then did not change at high flea abundances (Fig. 15.15). In other words, flea population growth beyond a certain level did not lead to extreme infestation of some host individuals. The most parsimonious explanation of the absence of heavily infested hosts at high flea abundances implying regulation is flea-induced (direct or indirect) host mortality. However, an asymptote in the relationship between mean flea crowding and flea abundance was found in only a few flea–host associations. This suggests that flea-induced host mortality might be important in shaping the pattern of the flea abundance/host density relationship in some cases but not in others. Alternatively, the reason for this may simply be that some fleas in this study area did not attain a level of abundance high enough for flea-induced host mortality to be detected. It should be noted, however, that parasite-induced host mortality is not the only mechanism that can produce an asymptote of flea aggregation level with an increase in flea abundance; other mechanisms such as acquired immunity and spatial and/or temporal variation in body condition and behaviour among host individuals may operate as well (Anderson & Gordon, 1982; Wilson et al., 2001).

Another, not necessarily alternative, mechanism of the negative relationship between flea abundance or prevalence and host abundance could be host-induced parasite mortality. For example, an increase in grooming (Mineur et al., 2003) or a decrease in the production of immunosuppressive steroid hormones (Rogovin et al., 2003) under social stress, which, in turn, increases under high density (Krebs, 1996), can be a mechanism of increased host-induced flea mortality at high host abundances. Host-induced flea mortality is also suggested by a negative relationship between the degree of flea aggregation and host abundance. In other words, the number of fleas on 'heavily infested' hosts could differ between periods of host abundance. 'Heavily infested' hosts at periods of high host abundance harbour fewer fleas than 'heavily infested' hosts at periods of low host abundances. The regulating role of hosts in flea mortality is strengthened by an asymptote in the relationship between mean flea crowding and host abundance. This means that the rate of decrease in the degree of flea aggregation decreases with an increase in host abundance until a certain level which seems to be tolerable for a host, beyond which the host ceases to defend itself.

Abundance of flea species can be affected by competitive interactions with other flea species. Indeed, interspecific competition that leads to competitive exclusion does occur among larval fleas (Krasnov *et al.*, 2005e; see Chapter 16). However, the occurrence of such interspecific competition among imagoes is dubious, although it has been suggested (Day & Benton, 1980). For instance, Clark & McNeil (1981, 1991) demonstrated a positive correlation between the density of *Ceratophyllus hirundinis* with those of *Ceratophyllus farreni* and *Ceratophyllus rusticus* in the nests of the house martin *Delichon urbica* in Great Britain.

Density-dependent intraspecific processes in fleas can also mediate relationships between flea and host abundance (Hudson & Dobson, 1997). For example, intraspecific interactions can keep flea density at a certain level (see Chapter 11), and thus be responsible for the lack of a relationship between flea and host abundance in some flea–host associations. It should be noted, however, that in different flea species an increase in density can affect the growth of infra- or xenopopulations in either a negative or a positive way. Negative effects of density on the reproductive success and, consequently, on the growth rate of their populations were found in *Ceratophyllus gallinae* in Switzerland (Tripet & Richner, 1999a; Tripet *et al.*, 2002c; see Chapter 11). In the nests of *P. major* experimentally infested with *C. gallinae*, two discrete flea age cohorts were distinguished (Heeb *et al.*, 1996). It was suggested that competition and/or cannibalism between cohorts negatively affected flea population size. In contrast, an increase in density of *X. cheopis* and *C. tesquorum* led to an increase in frequency of the multiple matings which, in turn, increased egg production and the number of fleas of the new generation (Tchumakova *et al.*, 1978).

15.6.3 Host spatial behaviour

Contradictions between real data and theoretical predictions may be explained by factors that are not usually taken into account in simple epidemiological models. One of these factors is density-dependent changes in the host spatial behaviour.

I have already mentioned 'homeless' host individuals that are characteristic for any population of small mammals at high density and that cannot support flea populations due to lack of shelters necessary for flea reproduction and pre-imaginal development (see also Janion, 1962). Let us assume that flea abundance increases in a curvilinear fashion to a plateau with increasing host density as is the case with *X. dipodilli* and *D. dasyurus* (see above). We should expect that the breakpoint of this curve will be determined by the density of resident hosts rather than by the overall density of the hosts. Consequently, flea abundance will decrease with the proportion of resident individuals under the same overall abundance of the host and the same overall abundance of parasites. The plateau of the flea abundance curve should be lower than is expected from the basic model.

Krasnov et al. (2002e) corrected the pre-existing models based on those of Anderson & May (1978), substituting the overall host density in (15.7) and (15.8) by the density of resident host individuals. Although both basic and 'corrected' models describing the relationships between flea burden and host density fitted the observational data well, simulations of the fraction of resident hosts demonstrated that this parameter influenced the relationship between host density and flea abundance only when residents comprised no more than 50% of all host individuals (Fig. 15.16a). The breakpoint of the curve changed logarithmically with a different fraction of residents reaching a plateau when this fraction achieved 50–60% (Fig. 15.16b).

It appears that when the percentage of non-residents is relatively low, they do not contribute heavily toward flea transmission, so their influence on flea dynamics and distribution is negligible (Khokhlova & Knyazeva, 1983). Thus, the introduction of a parameter for non-resident density into the model did not produce any significant shift in comparison with the basic model. However, when the fraction in the host population is relatively high, their contribution to flea transmission becomes significant. In such a case, a model that takes into account this non-resident component should describe the observational data better than that which considers the overall host density only. The role of resident hosts in flea dynamics is also supported by the fact that flea abundance was correlated negatively with the percentage of resident gerbils, showing that all flea individuals were distributed almost exclusively among resident hosts and did not occur on transients (Krasnov et al., 2002e).

In addition, the simulation demonstrated that the threshold of establishment (the minimal size of host population necessary to sustain a parasite population) was higher when there was a lower (5–30%) fraction of residents in a host population (Fig. 15.16b). The lower the percentage of resident gerbils, the greater the number needed to maintain flea reproduction. Again, this is consistent with the assumption that only residents support flea populations and that heterogeneity of hosts in terms of spatial behaviour is an important component of host–flea relationships.

Spatial aggregation of hosts may also lead to an increase of flea density on clumping host individuals. This was reported for *Orchopeas howardi* in the nests of the southern flying squirrel *Glaucomys volans* in Virginia (Lauer & Sonenshine, 1978). Abundance of fleas was higher in nests occupied by several squirrels than in nests with a single squirrel. A similar pattern was observed by Verzhutsky et al. (1990) for *C. tesquorum* parasitic on the long-tailed ground squirrel *Spermophilus undulatus* in Tuva (southern Siberia). In this host, related females

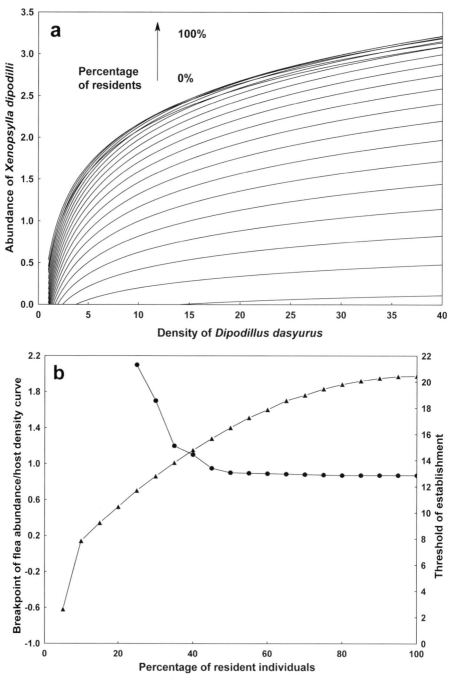

Figure 15.16 (a) Relationships between abundance of *Xenopsylla dipodilli* and density of Wagner's gerbil *Dipodillus dasyurus* (number of individuals per 1 ha) produced by manipulation of the fraction of host residents from 5% to 100% whilst holding all other parameters constant. (b) Values of breakpoint (triangles) and the threshold of establishment (circles), calculated for simulation curves of *X. dipodilli* abundance/*D. dasyurus* density curves plotted against the simulated percentage of resident host individuals. Redrawn after Krasnov *et al.* (2002e) (reprinted with permissions from the Ecological Society of America).

(mothers, daughters, sisters) tend to aggregate together (Popov & Verzhutsky, 1988) which results not only in an increase in flea abundance but also in a sharply female-biased flea distribution.

15.6.4 Host community structure

Changes of host community structure can strongly affect populations of a parasite that exploits all members of a community (Norman et al., 1999; Ostfeld & Keesing, 2000; Holt et al., 2003). For example, a resident parasite may respond to the introduction of a new host species into a community in two different ways. Its abundance and/or prevalence could either increase because of an increase in the amount of available resources (i.e. total number of hosts) or decrease in a resident host because a parasite will be 'diluted' among resident and introduced hosts. The latter hypothesis was supported by Telfer et al. (2005). They reported the effect of the introduced to Ireland M. glareolus on the relationships between prevalence of fleas (A. penicilliger, Ctenophthalmus nobilis and Hystrichopsylla talpae) and abundance of the native woodmouse Apodemus sylvaticus. In areas where M. glareolus were absent, an increase of A. sylvaticus density has led to an increase in flea prevalence. However, in areas invaded by M. glareolus, flea prevalence in A. sylvaticus decreased with an increase in density of this host.

15.6.5 Environmental factors

To investigate the role of local environmental conditions in flea abundance, Krasnov et al. (2006d) tested the repeatability of this parameter not only within flea and host species, but also within regions (see details of the analysis above). The proportion of the variance accounted for by differences among regions, as opposed to within-region differences, was rather low (12.7%), but significant, suggesting that some locations are characterized by higher flea abundance than other locations.

To understand the possible causes of geographical variation in flea abundance within a flea–host association, Krasnov et al. (2006d) calculated the parameters that characterized the abiotic environment for each region and for each of 24 flea species with at least six different regional records on the same host species. Environmental parameters were latitude, mean annual surface temperature, mean January surface air temperature, mean July surface air temperature, mean annual precipitation and mean altitude. Because most of the variables were strongly correlated with each other, they were substituted with the scores calculated from the principal-component analysis of all these variables. No correlation was found between abundance and either of the two environmental composite factors in 19 of the 24 flea species investigated. However, in the remaining five species there was a correlation between abundance and one of the environmental composites. The abundance of *A. penicilliger* increased with an increase in the mean altitude and a decrease in the mean annual precipitation of a region. The abundance of the remaining four species correlated with air temperature and latitude, either decreasing (*H. talpae*, *Amphipsylla schelkovnikovi*, *Peromyscopsylla silvatica*) or increasing (*M. turbidus*) with an increase in these variables (Fig. 15.17).

The lack of a clear relationship between flea abundance on a given host species and regional climatic parameters found for most species in this study suggests that abiotic characteristics of a location play only a minor role in the observed pattern of within-region repeatability of flea abundance. Nevertheless, the strong effect of environmental factors such as air temperature and humidity on feeding, reproduction and mortality of fleas (see Chapters 5, 7, 10 and 11) suggests that their density dynamics must be influenced by the environment. For example, humidity was found to be the main factor affecting the abundance of *Pulex irritans* in Kazakhstan (Yakunin et al., 1971). It is also possible that it is microclimate that matters rather than macroclimate as considered in the study of Krasnov et al. (2006d). Indeed, Heeb et al. (1996) did not find any effects of macroclimatic and seasonal factors on the density dynamics of *C. gallinae* in Switzerland. Clark & McNeil (1993) demonstrated low winter mortality in *C. hirundinis*, *C. farreni* and *C. rusticus* in Great Britain and concluded that abiotic weather conditions are not major determinants of density in these species. In contrast, the abundance of *C. tesquorum* on *S. undulatus* in the Altai Mountains was strongly affected by climatic factors, being lower at higher altitudes *ceteris paribus* (Letova et al., 1969), whereas the opposite was reported by Flux (1972) for *Ctenocephalides felis* on the Cape hare *Lepus capensis* in Kenya. Li et al. (1995) argued that the abundance of *N. laeviceps*, *X. conformis* and *R. dives* in Inner Mongolia (China), although influenced by the density of their host *M. unguiculatus*, is nevertheless affected much more strongly by meteorological factors. Similar conclusions have been reached for *Monopsyllus anisus* and *Nosopsyllus fasciatus* parasitic on *R. norvegicus* in the southern Russian Far East (Zhovty & Leonov, 1958) and *X. gerbilli*, *X. hirtipes*, *X. conformis* and *Xenopsylla nuttalli* on *R. opimus* in Turkmenistan (Novokreshchenova et al., 1975) and Kazakhstan (Dubyansky et al., 1975).

Finally, it should be remembered that the effect of environmental factors on flea density dynamics may differ among flea species. For example, Harper et al. (1992) found that air temperature was important in determining population size in *C. gallinae*, but not in *Dasypsyllus gallinulae* from the same area (Herefordshire, Great Britain). Abundance responses to air temperature and relative humidity

Flea abundance and distribution 373

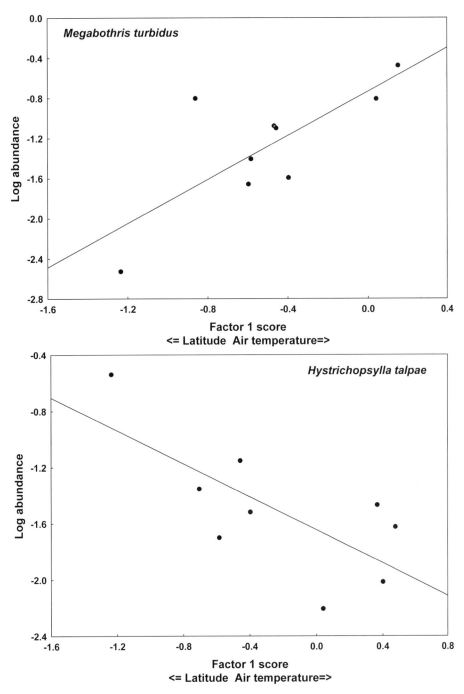

Figure 15.17 Relationship between local abundance of *Megabothris turbidus* (on the bank vole *Myodes glareolus*) and *Hystrichopsylla talpae* (on the common vole *Microtus arvalis*), and scores of the composite environmental factor across different regions. Data from Krasnov *et al.* (2006d).

differed substantially between *O. montana* and *H. anomalus* parasitic on *S. beecheyi* in California (Lang, 1996).

15.7 Concluding remarks

Abundance and distribution of a flea species are true species characters. Nevertheless, within species-specific limits, infrapopulations, xenopopulations and populations of fleas are to some extent determined by host species identity, density and spatial behaviour as well as affected by local biotic and abiotic conditions. On the one hand, this supports the ideas of Arneberg *et al.* (1997) and Poulin (2006) that the biological attributes of parasite species are primary determinants of parasite dynamics compared with characteristics of hosts and of the local environment. On the other hand, materialization of flea dynamics depends on a variety of factors. Although the distribution of fleas among infrapopulations or xenopopulations may be described by relatively simple models with a limited set of interconnected parameters (i.e. mean abundance and number of hosts sampled), the relationships between these parameters are rather complicated. Patterns of these relationships are variable, so they are manifested differently in relation to flea species, host species, season, geographical location and interactions between all or some of these variables.

16

Ecology of flea communities

As is the case with almost all free-living organisms, parasites belonging to a particular species rarely occur alone, although this can happen at the scale of host individual. In most cases, parasites belonging to different species co-occur forming a community. Analogously with assemblages of conspecific parasites, spatial distribution of parasite communities is fragmented among host individuals, among host species within a location and among locations. It is logical thus that the hierarchical terminology aimed at distinguishing between parasite communities at different scales will be parallel to that adopted for parasite populations. Consequently, Combes (2001) suggested that we should distinguish *infracommunity* (an assemblage of parasites of all species infesting an individual host), *xenocommunity* (= component community; an assemblage of parasites of all species infesting a host population) and *community* (= compound community; an assemblage of parasites of all species infesting a host community). I use this terminology throughout this chapter, albeit with some modifications.

The scale of consideration is extremely important when parasite communities are studied. In particular, this stems from the fact that there are at least two principal differences between infracommunities, on the one hand, and parasite assemblages at higher scales, on the other hand. First, this is related to temporal persistence of an assemblage. Infracommunities are short-living by definition, whereas xenocommunities and communities persist much longer. Second, this is related to possible interactions among parasite species. Occurrence of the interactions among heterospecific parasites is not necessarily the rule. Depending on the presence or absence of interspecific interactions, both isolationist and interactive parasite communities can be distinguished (Holmes & Price, 1986; Bush et al., 1997). In particular, a parasite community is considered as interactive if parasite species in this community exert selective pressures on each other, which then induces the selection of traits that limit competition by separating niches (Holmes & Price, 1986; Combes, 2001). In interactive communities, interspecific interactions are expected to be expressed mainly within infracommunities, whereas these interactions within xenocommunities and communities are likely to be weak (if they occur at all). Furthermore, much previous research on parasite communities has been conducted at the level of parasite infracommunities (e.g. Bush & Holmes, 1986; Haukisalmi & Henttonen, 1993; Forbes et al., 1999). In contrast, xenocommunities and communities of parasites (*sensu* Combes, 2001) have received much less attention. One reason for the scarcity of studies of the relationships among parasite species at higher scales as well as the expectation of the weakness of interspecific interactions at these scales may be the difficulty of assessing interactions between parasite species occurring in different host individuals and/or different host species. This is undoubtedly true for endoparasites such as intestinal helminths. However, ectoparasitic species, especially those such as fleas, spend a significant amount of time off-host and easily switch between individual hosts both within and between host species (Rödl, 1979; Krasnov & Khokhlova, 2001; Kuznetsov & Matrosov, 2003; see above). Consequently, interspecific interactions among fleas may occur not only on the host body but also in the host's burrow or nest. Moreover, many larval fleas from different host species co-occur in multi-species host and parasite assemblages in burrow colonies constructed by some rodents (e.g. the great gerbil *Rhombomys opimus*: Kucheruk, 1983), making interactions at the level of xenocommunities and communities possible. Consequently, species interactivity in the communities of fleas cannot be refuted a priori. Interspecific interactions, in turn, may shape the structure of a community, although they represent only one of many factors affecting a community.

In addition, interactions within communities can be quite strong in an odd sort of way. Consider two host species that compete, each with its own parasite. The more strongly a parasite affects its host, the poorer will be the host's competitive ability. This may lead to an increase in the other host and its parasite. In this way, the interaction between the parasites would be indirect and mutualistic. This has never been studied in fleas, but this type of interaction has been observed in other parasite taxa (see Combes, 2001).

This chapter deals mainly with structure of flea communities. The main aim is to demonstrate that flea infracommunities, xenocommunities and communities represent predictable sets of species rather than stochastic assemblages. After addressing patterns of species composition in these communities, I discuss the possible forces shaping their structure. Finally, I consider geographical patterns in flea communities.

16.1 Are flea communities structured?

16.1.1 Patterns of species co-occurrences

A plethora of patterns of community non-randomness has been suggested, including the assembly rules of Diamond (Diamond, 1975), species nestedness (Patterson & Atmar, 1986), core–satellite organization (Hanski, 1982) and favoured and unfavoured species combinations (Fox & Brown, 1993) (although the latter can be considered as part of assembly rules). In spite of this variety of assembly principles, the underlying question, when the pattern of the community organization is considered, is as follows: does frequency of co-occurrence of different species in a real community differ from that expected in a community with a random species assemblage? If species co-occur more often than expected by chance, then the assemblage is aggregatively structured, whereas if species co-occur less frequently than expected by chance, then the assemblage is competitively structured. Until recently, the only mention of co-occurrence of flea species was that of Lundqvist & Brinck-Lindroth (1990) who reported that *Amalaraeus penicilliger* and *Megabothris rectangulatus* co-occurred on the field vole *Microtus agrestis* and the root vole *Microtus oeconomus* in Fennoscandia more frequently than expected by chance.

In recent years, the number of studies on parasite community structure has increased drastically (Guégan & Hugueny, 1994; Sousa, 1994; Rohde *et al.*, 1995, 1998; Poulin, 1996b; Worthen, 1996; Worthen & Rohde, 1996; Hugueny & Guégan, 1997; Rohde, 1998; Carney & Dick, 2000; Matejusová *et al.*, 2000; Poulin & Guégan, 2000; Dezfuli *et al.*, 2001; Poulin & Valtonen, 2001, 2002; Gotelli & Rohde, 2002; Goüy de Bellocq *et al.*, 2003; Timi & Poulin, 2003; Vidal-Martinez & Poulin, 2003; González & Poulin, 2005; Krasnov *et al.*, 2005j). Results of these studies were equivocal and revealed great variability in the expression of structure among parasite communities. Some parasite communities appeared to be structured (González & Poulin, 2005), whereas others appeared to be randomly assembled (Gotelli & Rohde, 2002). On the other hand, some communities demonstrated aggregative structure (Rohde *et al.*, 1995), whereas others seemed to be produced by competitive interspecific interactions (Dezfuli *et al.*, 2001). In addition, the structure of many communities appeared not to be repeatable in space (when parasite communities in different populations of the same host species are sampled) or stable in time (when parasite communities of the same host population are sampled in different years) (Poulin & Valtonen, 2001, 2002; Timi & Poulin, 2003; Vidal-Martinez & Poulin, 2003). This suggests that either the manifestation of community structure is dependent on some temporally and/or spatially variable factors or we are actually unable to measure the truly important facets of structure.

Gotelli & Rohde (2002) investigated ectoparasite communities of marine fish and concluded that they provide little evidence for the non-random species co-occurrence patterns. Furthermore, in comparing the expression of non-randomness in these communities with that in communities of other organisms, they suggested that non-randomness is characteristic of communities of large-bodied taxa with high vagility and/or large populations (birds, mammals), whereas communities of small-bodied taxa with low vagility and/or small populations (marine ectoparasites, reptiles) are mostly random. The relationship between population size and the detection of non-randomness in these communities suggests that community structure can vary temporally due to temporally variable population size.

The community structure of fleas and its temporal variation were studied by Krasnov *et al.* (2006h) for temporal samples of fleas parasitic on small mammals in eastern Slovakia (*Apodemus agrarius*, *Apodemus uralensis*, *Apodemus flavicollis*, *Myodes glareolus* and *Microtus arvalis*) using null models (Gotelli & Graves, 1996; Gotelli, 2000; Gotelli & Entsminger, 2001; Gotelli & Rohde, 2002). The null model analysis compares frequencies of co-occurrences of species (fleas) across sites (host individuals or host populations) with those expected by chance, i.e. derived from randomly assembled species × site matrices. These frequencies were evaluated by four metrics, namely the *C*-score, the number of chequerboard species pairs, the number of species combinations and the variance ratio (*V*-ratio) (Gotelli, 2000; Gotelli & Rohde, 2002). In brief, the *C*-score is the average number of chequerboard units that are found for each pair of species. The number of chequerboard species pairs is the number of species pairs that never co-occur in any site (Diamond, 1975), whereas the number of species combinations represent the number of co-occurring species. Finally, the *V*-ratio (Robson, 1972; Schluter, 1984; Gotelli & Rohde, 2002) is the ratio between the variance in species richness and the sum of the variance in species occurrence. It equals 1 if the species are distributed independently, whereas negative or positive covariance between species pairs lead to the value of the index smaller or greater than 1, respectively (Gotelli, 2000).

When species are aggregated, the *C*-score and the number of chequerboard species pairs are smaller than expected by chance (observed (O) < expected (E)), while the number of species combinations and the *V*-ratio are larger than expected by chance ($O > E$). When species are segregated, the *C*-score and the number of chequerboard species pairs are larger than expected by chance ($O > E$), while the number of species combinations and the *V*-ratio are smaller than expected by chance ($O < E$) (Haukisalmi & Henttonen, 1993; Gotelli, 2000; Gotelli & Arnett, 2000; Gotelli & McCabe, 2002; Gotelli & Rohde, 2002).

An important technical detail in the null model analysis is related to the algorithm used for the assembling of simulated matrices. Two main algorithms, namely fixed–fixed (FF) and fixed–equiprobable (FE), have been suggested (Gotelli, 2000; Gotelli & Entsminger, 2006). The FF algorithm maintains the differences among host individuals or species in the number of flea species they harbour. This means that the FF model considers uninfested host individuals as permanently unsuitable hosts, and thus uninfested hosts harbour no parasites in both observed and simulated communities. In contrast, the FE algorithm does not constrain the number of flea species that can be harboured by a host individual or species. In other words, this model suggests no differences in the probability of supporting a particular number of flea species among host individuals or species. From this point of view, uninfested hosts are not colonized by chance, although in principal, they may harbour parasites. Consequently, in the FE model, the uninfested hosts in the real data could become infested in simulated data. Which algorithm should be used for fleas from the point of view of biological reality? Obviously, the answer is the FE algorithm (at least for the scale of host individuals) as we have already seen that an individual host recorded as uninfested at its initial capture could be found to be infested if captured after a week to several months, whereas an individual host infested at one capture could be found to be uninfested in future captures (Krasnov et al., 2006g; see Chapter 15).

It appeared that in many, albeit not all, cases the observed indices differed significantly from the null expectations (see Fig. 16.1 for illustrative examples with the fieldmouse *A. agrarius*). Furthermore, in all these cases the C-score and the number of chequerboard species pairs were smaller, whereas the number of species combinations and the V-ratio were larger than expected by chance. These results demonstrate two important patterns, namely (a) flea assemblages on small mammalian hosts are structured at some times, whereas they appear to be randomly assembled at other times; and (b) whenever non-randomness of flea co-occurrences is detected, it suggests aggregation but never segregation of flea species in infra- and xenocommunities.

Aggregative patterns of flea co-occurrences suggest apparent facilitation among flea species, which seems to be mediated via the host. In general, many sources of heterogeneity among host individuals could cause these aggregative patterns, such as differences in grooming abilities, immunocompetence and age (see Chapters 9, 12, 13 and 15). For example, host-mediated facilitation could arise from immunodepression in a host subjected to multiple challenges from a variety of parasites (Bush & Holmes, 1986; Cox, 2001; Chapter 13). As a result, the effectiveness of energy allocation to immune defence decreases with an

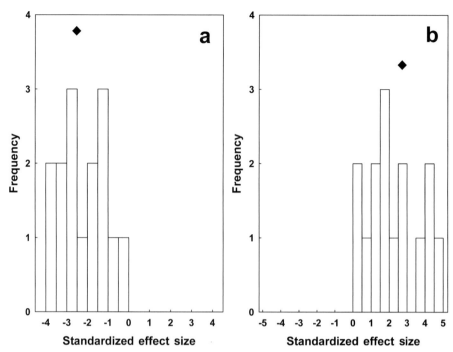

Figure 16.1 Histograms of the standardized effect size (SES) of (a) the C-score and (b) the V-ratio of flea co-occurrences on the fieldmouse *Apodemus agrarius*. Each observation is a presence/absence matrix of flea occurrences in the individual hosts constructed for each trapping session. The diamond denotes the tail of the distribution for which species co-occurrence is higher than expected by chance. Redrawn after Krasnov et al. (2006h) (reprinted with permission from Blackwell Publishing).

increase in the diversity of parasite attacks (Jokela et al., 2000). However, this explanation is weakened if cross-resistance against closely related ectoparasites is taken into account (see Chapter 13). Nevertheless, cross-resistance has been reported mostly for congeneric rather than heterogeneric ectoparasites (McTier et al., 1981). In addition, the same host species have different cross-resistant abilities against different ectoparasite species (Kumar & Kumar, 1996 versus Rechav et al., 1989). Thus, in spite of cross-resistance, immunodepression resulting from multiple attacks can, at least partly, explain the aggregation of fleas on host individuals or species. An additional explanation of the observed aggregative pattern in flea species co-occurrences in infracommunities is heterogeneity of spatial behaviour of host individuals (residents versus transients; see Chapter 15). As we already know, the 'homeless' host individuals should not be suitable hosts for any flea species, whereas the resident individuals possessing burrows may be preferred hosts for all or most flea species.

Why were flea communities structured at some times and randomly assembled at other times? To understand this temporal variation, Krasnov et al. (2006h) calculated the standardized effect size (SES) for each matrix (i.e. for each host species and for each sampling period or for the entire host assemblage and for each sampling period). The SES is the number of standard deviations that the observed index is above or below the mean index of simulated matrices (Gotelli & McCabe, 2002). Then they analysed the relationships between the values of the SES of the C-score and V-ratio and the parameters of flea and host populations and assemblages that vary temporally (host density, mean abundance of fleas and prevalence of flea infestation). These parameters were represented by a composite variable that correlated positively with host density and negatively with flea abundance and prevalence. The SES of the C-scores and the V-ratio of temporal samples of flea co-occurrences correlated significantly with this composite variable for four (except for the pygmy woodmouse A. uralensis) or three (except for A. uralensis and the bank vole M. glareolus) of five host species, respectively. Furthermore, significant correlations were negative for the C-score and positive for the V-ratio. In other words, the expression of structure in flea assemblages depended on the level of both flea and host abundances.

This suggests that a snapshot of a pattern in the structure of a parasite community observed at one time and one location is not representative of the patterns of parasite communities of the same host species in different populations and/or at different times. Instead, it seems that some temporally or spatially variable factors are responsible for the expression of community structure. As we have already seen, in many cases, parasite abundance and prevalence increase with increasing host abundance (Anderson & May, 1978; Krasnov et al., 2002e; see Chapter 15). Poulin & Valtonen (2002) demonstrated that as the prevalence or mean intensity of parasites increased in a fish population, the likelihood that the structure of parasite infracommunities significantly deviated from random also increased. However, in the flea–mammal system of Slovakia analysed by Krasnov et al. (2006h), the relationship between host density and flea abundance has been shown to be the opposite to that predicted by epidemiological models, and flea abundance and prevalence decreased with an increase in host density (Stanko et al., 2006; see Chapter 15). The likelihood of the non-randomness in flea communities increased with an increasing host density and concomitant decreasing flea abundance and prevalence in the majority of host species. The deceptive contradiction between these results and those of Poulin & Valtonen (2002) suggests a strong host influence on the expression of structure in parasite communities. Partly, this may due to an increase in the proportion of the 'unsuitable-for-all-fleas' hosts—transients in the growing host population. However, the lack of a relationship between host and flea abundance and the detection of non-randomness of flea communities on *A. uralensis* suggest that some other, still unknown, factors could affect the manifestation of structure of a flea community.

16.1.2 Nested pattern

Another possible departure from random community assembly is the nested pattern. This is a pattern in which species comprising depauperate assemblages constitute non-random subsets of the species occurring in successively richer assemblages (Patterson & Atmar, 1986; Brown, 1995; Hecnar & M'Closkey, 1997; Wright *et al.*, 1998). Nested patterns have been found in communities of different taxa of both plants and animals (e.g. Patterson & Brown, 1991; Hecnar & M'Closkey, 1997; Honnay *et al.*, 1999; Matthews, 2004). The rationale for the possible existence of nestedness in parasite communities is that nested species subsets are a common pattern in many types of communities found in insular or fragmented habitats (Patterson & Atmar, 1986; Bolger *et al.*, 1991), whereas hosts can be viewed as biological islands for parasites (Kuris *et al.*, 1980). Indeed, in recent years, the search for nested patterns has been applied to parasite communities (Guégan & Hugueny, 1994; Poulin, 1996b; Worthen, 1996; Worthen & Rohde, 1996; Hugueny & Guégan, 1997; Poulin & Valtonen, 2001, 2002; Timi & Poulin, 2003; Vidal-Martinez & Poulin, 2003).

However, studies of nestedness in parasite communities have provided contradictory results, and there is currently no consensus on the occurrence of this pattern in parasites. In part, these contradictions stem from the type of statistical technique used (e.g. Guégan & Hugueny, 1994 versus Worthen & Rohde, 1996). Another reason for the conflicting results might be the complicated relationships between extrinsic and intrinsic factors that affect species composition of parasite communities. For example, Poulin & Valtonen (2001) reported that host size and dietary specialization can affect the occurrence and manifestation of nested patterns.

In addition, earlier nestedness analyses may have focused on the wrong spatial scale. Indeed, the majority of studies have focused on infracommunities, whereas xenocommunities have received less attention (but see Guégan & Kennedy, 1996; Calvete *et al.*, 2004), although this higher hierarchical level is more relevant to nestedness analyses. Infracommunities are ephemeral assemblages greatly influenced by epidemiological processes, whereas xenocommunities are structured by the kind of biogeographical processes relevant to the original idea of the nested subsets pattern developed by Patterson & Atmar (1986).

Krasnov *et al.* (2005j) searched for the occurrence of nested patterns in flea assemblages among host populations of the same species across the species' geographical range, using data on 25 Holarctic small-mammal species. A flea community was considered to be structured if it departed from random assembly in either direction, i.e. towards either nestedness or anti-nestedness. As defined above, nestedness occurs if flea assemblages can be arranged in such a way that depauperate assemblages consist of proper subsets of progressively richer ones. Alternatively, if flea species are always absent from assemblages richer than the most depauperate one in which they occur, then the assemblages are considered as anti-nested (see Poulin & Guégan, 2000). Anti-nested patterns, although rarely reported in the literature, indicate non-random assembly of species, via the action of unknown structuring factors (Poulin & Guégan, 2000). Krasnov et al. (2005j) assesssed the frequency of non-random (nested and anti-nested) flea assemblages across hosts and evaluated the nestedness using the metric matrix 'temperature' (T) proposed by Atmar & Patterson (1993) which is independent of matrix size; T provides a standardized measure of matrix disorder by quantifying the deviation of an observed matrix from one of the same size and fill that is perfectly nested and ranges from 0 (perfectly nested matrix) to 100 (completely disordered matrix). In this approach, nestedness is clearly an estimate of the degree of non-random pattern in species distribution. It should be noted, however, that the use of Atmar & Patterson's (1993) algorithm has recently been criticized and new methods to calculate nestedness temperature have been suggested (Rodríguez-Gironés & Santamaría, 2006; Almeida-Neto et al., 2007).

Flea assemblages among 25 host species were significantly nested in six hosts (the small five-toed jerboa *Allactaga elater*, European water vole *Arvicola amphibius*, midday jird *Meriones meridianus*, common vole *M. arvalis*, house mouse *Mus musculus* and common shrew *Sorex araneus*), significantly anti-nested in two hosts (the migratory hamster *Cricetulus migratorius* and tamarisk jird *Meriones tamariscinus*) and non-significantly structured in the remaining 17 hosts. Using the looser criterion of tending toward nestedness or anti-nestedness (see Krasnov et al., 2005j), there were eight assemblages that tended toward nestedness (with addition of the grey-sided vole *Myodes rufocanus* and forest dormouse *Dryomys nitedula* to the former list) and four assemblages that tended toward anti-nestedness (with addition of the narrow-skulled vole *Microtus gregalis* and great gerbil *R. opimus* to the former list). Thus, the organization of flea assemblages across host populations within host species forms a continuum among host species from true nestedness to true anti-nestedness.

A continuum of community organization from nestedness to anti-nestedness was reported in several studies of infracommunities of ecto- and endoparasites of fish (Poulin & Guégan, 2000; Poulin & Valtonen, 2001), birds (Šimková et al., 2003; Calvete et al., 2004) and mammals (Fellis et al., 2003; Goüy de Bellocq et al., 2003). Furthermore, the proportion of nested and anti-nested flea assemblages is similar to that found for fish parasites. This suggests that nested patterns in parasite xenocommunities may be as uncommon as they are in parasite infracommunities (Poulin & Guégan, 2000; Poulin & Valtonen, 2001) and hints at the possibility that community organization of parasites at both hierarchical levels, namely host individuals and host populations, is governed by similar rules.

In free-living organisms (mainly mammals and birds), nested patterns observed across insular or fragmented habitats are considered to be the result of differential colonization or extinction probabilities among the available species (Patterson & Atmar, 1986; Patterson, 1990; Bolger et al., 1991). Nested patterns in parasite communities have also been explained by colonization–extinction dynamics (Guégan & Hugueny, 1994). Moreover, Poulin & Guégan (2000) argued that interspecific differences in colonization rate may also be responsible for the anti-nested pattern, if rare and highly host-specific species are characterized by poor colonization ability compared with locally abundant host-opportunistic species. They demonstrated that there is a positive relationship between parasite prevalence and parasite local mean intensity, and that as the prevalence or mean intensity of parasites increases in a fish population, the likelihood that the parasite infracommunities are nested also increases. These findings confirm a link between the positive prevalence–intensity relationship and the nestedness–anti-nestedness continuum. The prevalence–intensity relationship is the parasitological equivalent of the more general spatial distribution–local abundance relationship (see above). Let's recall that, in general, fleas exploiting many host species, or those exploiting taxonomically unrelated hosts, achieved higher abundance than host-specific fleas (Krasnov et al., 2004f; Chapters 14–15). Therefore, a linkage between the positive prevalence–intensity relationship and the nestedness–anti-nestedness continuum occurs not only at the level of parasite infracommunities but also at the higher hierarchical level.

To test whether host biology affects the organization of flea assemblages, Krasnov et al. (2005j) explored the effects of extrinsic factors (the descriptors of the host geographical range, environment and the taxonomic composition of the host's community) on the occurrence of nested patterns and found that across host species, the tendency for flea assemblages to approach nestedness increased with increasing host geographical range size and with decreasing latitude of the host's geographical range (Fig. 16.2). This tendency also decreased with an increase in a composite variable combining data on mean January and July air temperature, whereas the taxonomic composition of the host's community had no influence on whether or not the structure of flea assemblages among its populations departed from randomness.

Thus nested flea assemblages are more prevalent in southern and warmer (in the data set used, those inhabiting desert regions) compared to northern and

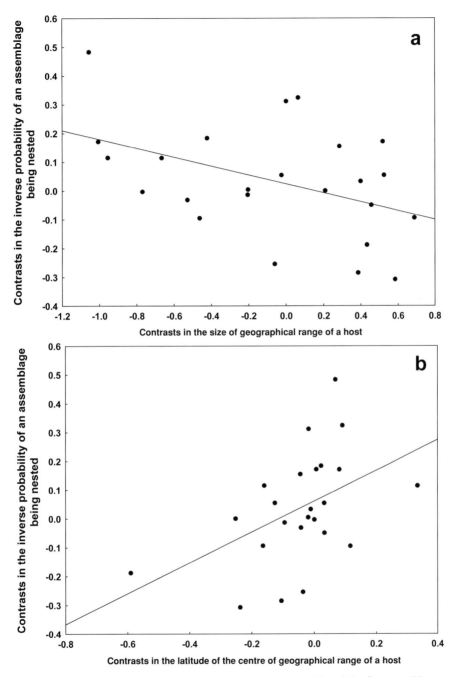

Figure 16.2 Relationships between the inverse probability of the flea assemblage being nested and (a) the size of the host geographical range and (b) the latitude of the centre of the host geographical range using the method of independent contrasts. Redrawn after Krasnov et al. (2005j) (reprinted with permission from Blackwell Publishing).

cooler (in the data set used, those inhabiting steppe and forest regions) host species. Perhaps this is due to the higher density of small mammals in steppe and forest regions compared to that in the desert regions. This could increase host-switching, and therefore lead to more scattered presences of flea species across the populations of a host species and thus to a lower likelihood of nestedness. However, Calvete *et al.* (2004) reported a positive relationship between the probability of nestedness and host density for the helminth communities of the red-legged partridge *Alectoris rufa*. This apparent contradiction between flea and helminth communities might arise from differences in the way a host acquires the parasites, with helminths being actively obtained by oral ingestion and fleas being passively picked up. Consequently, flea habitat heterogeneity does not seem to play an important role in promoting nestedness as is the case for helminths (variation in dietary specialization and/or feeding rate of individual hosts: see Calvete *et al.* (2004) for details). On the other hand, the greater likelihood of the desert flea assemblages being nested may stem from the fact that among flea assemblages in different populations of a desert host, flea extinction processes might prevail over flea colonization processes. The reasons for this could be the lack of sources for colonizing flea species (because of relatively low host density) and/or the extremely dry climate (because low relative humidity in burrows has a strong negative effect on the survival of pre-imaginal fleas; see Chapter 5). In conclusion, the structuring of flea assemblages within host species across host populations strongly depends on host biology, although this does not mean that ecological processes acting among flea species are not important.

16.2 Local versus regional processes governing flea communities

Patterns in local communities are governed not only by local processes (such as competition, predation and habitat heterogeneity) but also by regional and historical processes (such as long-distance migration and speciation) (Cornell & Lawton, 1992; Gaston & Blackburn, 2000). For example, in order to persist in a locality, a species has to (a) originate there and maintain its population in the local environment; or (b) to arrive there by dispersal once and then maintain its population; or else (c) to arrive there repeatedly by dispersal. The relative importance of local and regional processes in governing local species composition can be inferred from examination of the relationship between local and regional species richness (Srivastava, 1999; but see Rosenzweig & Ziv, 1999). If, for example, regional processes strongly control local communities by, for example, dispersal limiting local species richness, then the relationship between local and regional species richness will be linear (Cornell & Lawton, 1992). Local communities are thus unsaturated (species are often absent from suitable habitats) and exhibit 'proportional sampling' of the available regional species pool. If, however, local processes (e.g. competition) play the main role in structuring local communities and impose upper limits on the number of species that are able to coexist, then local species richness will approach an asymptote with an increase in regional richness (Terborgh & Faaborg, 1980; Cornell & Lawton, 1992). At higher regional species richness, local richness becomes independent of regional richness. Local communities demonstrating a curvilinear relationship of local versus regional species richness are considered to be saturated with species (Guégan et al., 2005; but see Rohde, 1998).

Testing the relationship between local and regional species richness is, at first glance, rather straightforward and can be carried out using regression analysis (e.g. Oberdorff et al., 1998). However, some methodological problems arise (Cresswell et al., 1995; Caley & Schluter, 1997; Griffiths, 1999; Srivastava, 1999; Fox et al., 2000; Loreau, 2000; Shurin et al., 2000; Hillebrand, 2005), and thus the use of local/regional richness plots to test for saturation of diversity has been strongly criticized. The criticism has been mainly related to statistical issues (Srivastava, 1999), the definition of two spatial scales (Loreau, 2000; Shurin et al., 2000; Hillebrand & Blenckner, 2002) and the effects of different types of interactions (Shurin & Allen, 2001). Nevertheless, the use of regional to local diversity regressions remains widespread (Valone & Hoffman, 2002; Heino et al., 2003; Calvete et al., 2004; Karlson et al., 2004). One of the most important methodological issues is a precise definition of borders for local and regional communities. It is sometimes self-evident for freshwater organisms (e.g. a pond: see Shurin et al., 2000), but it is much more difficult for terrestrial or marine organisms. However, for parasites the definition of a community at the lowest hierarchical scale is relatively easy. This is the infracommunity. The measure of local parasite species richness is, thus, mean (e.g. Morand et al., 1999) or maximum (e.g. Calvete et al., 2004) infracommunity parasite species richness. Obviously, the next hierarchical level is the xenocommunity. Finally, all xenocommunities' parasite communities within a given host species represent either a regional parasite community or a parasite fauna (Poulin, 2007a). Although Srivastava (1999) argued that in the case of parasites, an equivalent of 'regional' species richness is the parasite fauna, I believe that xenocommunity richness can be considered as 'regional' in relation to infracommunity richness. This is because the species pool of a xenocommunity contains all species than can colonize an infracommunity assuming the absence of competitive exclusion. Dispersal of species within a xenocommunity may be slow but, nevertheless, it is much more frequent than host-switching (equivalent to dispersal between regions; see Srivastava, 1999).

Most studies of the relationship between species richness of communities of free-living organisms at different spatial scales have demonstrated that unsaturated communities are the norm (see Srivastava, 1999 for review). However, analyses of local versus regional species richness in helminth parasites have revealed that saturated and unsaturated communities are equally common (e.g. Poulin, 1996b; Morand et al., 1999; Calvete et al., 2004). Getting back to fleas, Krasnov et al. (2006i) investigated this relationship in communities of fleas on small mammalian hosts from Slovakia, the Negev Desert and California, at two different spatial scales: between the richness of flea infracommunities and that of flea xenocommunities, and between the richness of xenocommunities and that of the entire regional species pool. At both spatial scales, consistent curvilinear relationships (a slope <1 in a regression using log-transformed data) between species richness of the more 'local' communities and richness of the more 'regional' communities were found (Fig. 16.3).

The curvilinear relationship between infracommunity and xenocommunity flea species richness suggests that the number of species in species-rich flea infracommunities is independent of the species richness of the xenocommunity of which they are part. This appeared to be true for both comparisons across host species and for comparisons across populations of the same host species. Thus, at first glance, the flea infracommunities are 'saturated' (at about five to six flea species per infracommunity) and vacant niches seem to be generally unavailable in these communities. The observed pattern may arise because some species can be eliminated or not allowed to invade local communities due to some ecological constraints such as negative interactions among species in an infracommunity (Srivastava, 1999; Calvete et al., 2004).

However, if negative competitive interactions among flea species in an infracommunity are indeed important, one would expect density compensation in species-poor infracommunities (Cornell, 1993). Consequently, demonstrating the existence or absence of saturation in parasite assemblages requires the additional investigation of interspecific interactions (Guégan et al., 2005). To test for this, Krasnov et al. (2006i) assessed the relationship between mean flea abundance per host individual and richness of the 'local' flea community. There was no strong evidence for density compensation at the infracommunity level (significant linear relationship between mean flea abundance and infracommunity species richness) (Fig. 16.4), although its existence at the xenocommunity level appeared likely (no relationship between mean flea abundance and xenocommunity species richness). This approach is, of course, a weak substitute for removal experiments but the latter for fleas are hardly possible.

Consequently, negative interspecific interactions appear not to occur for flea infracommunities suggesting that the curvilinear relationship between infracommunity and xenocommunity species richness may occur for reasons other than 'saturation' due to competitive interspecific interactions. Indeed, Rohde

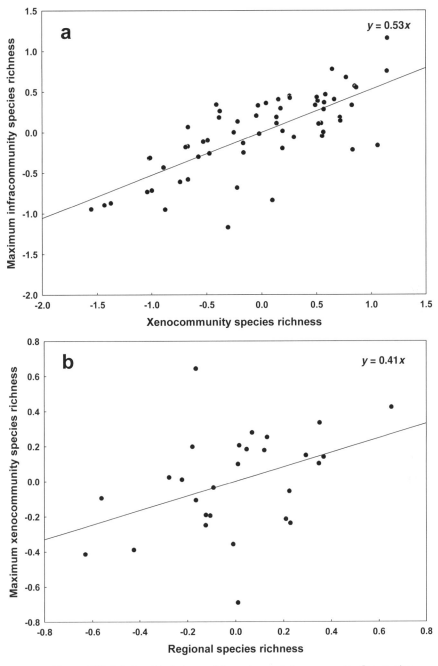

Figure 16.3 Relationship between (a) maximum infracommunity flea species richness and xenocommunity flea species richness for small mammal hosts from Slovakia, the Negev Desert and California and (b) maximum xenocommunity flea species richness and regional flea species richness for small mammal host species from Slovakia and the Negev Desert. Data were controlled for host sampling effort and/or size of host geographical range. Data from Krasnov et al. (2006i).

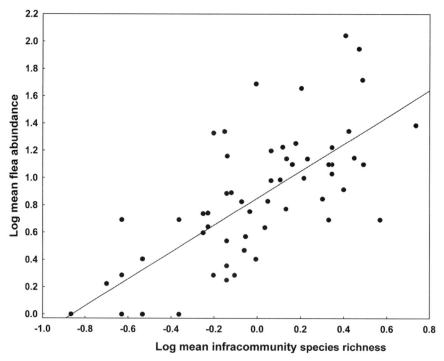

Figure 16.4 Relationships between mean flea abundance and mean infracommunity flea species richness on the bank vole *Myodes glareolus*. Data from Krasnov et al. (2006i).

(1998) demonstrated that curvilinearity in the local versus regional species richness relationship may be caused by processes other than species interactions within a local community. In particular, this curvilinearity may be a consequence of the differential likelihood of parasite species of occurring in an infracommunity because of different transmission rates and lifespans (Rohde, 1998). In the case of fleas, these reasons can also be related to differential abiotic preferences of either imagoes or larval fleas of different species which contribute to the elimination of some flea species from some infracommunities (Krasnov et al., 2001, 2002b; see Chapter 11). All the above indicate that flea infracommunities are governed by processes acting at higher than 'local' levels, and that further species could possibly be added over evolutionary time (Rohde, 1998).

The relationship between local and regional flea species richness appeared to be the same at the larger scale as at the smaller scale; in other words, the relationship between the richness of xenocommunities and that of the regional flea pool seems to be similar to that found for infracommunities versus xenocommunities. However, the absence of a relationship between mean flea abundance and xenocommunity species richness suggests the existence of density compensation.

Therefore, xenocommunities appeared to be saturated. The causes of this saturation are likely to be some intrinsic limiting factors that may play an important role in shaping flea xenocommunities. As we know, one of the common factors responsible for community saturation is negative interspecific interactions (Cornell, 1993). Although direct interspecific competition between flea imagoes within a host population has not been explicitly proved (see below), such competition can undoubtedly occur among larval fleas (Krasnov et al., 2005e; see below). To conclude, similar patterns in the relationships between 'local' and 'regional' species richness in the same host–parasite system but at different spatial scales may arise because of different mechanisms. This could be one explanation for the contrasting relationships reported between local and regional species richness in earlier studies of different host–parasite systems.

16.3 Inferring patterns of interspecific interactions

16.3.1 *Patterns of interspecific interactions inferred from patterns of abundance*

The structure of a community is the net result of the interactions among species composing this community. These interactions can be both negative and positive. Domination by either negative or positive interactions among species indicates either a competitive or a facilitating type of community structure (Rosenzweig, 1981; Schall, 1990). In general, strong competitive interactions in a community may lead to species exclusion (e.g. Levine & Rees, 2002), whereas the operation of positive processes promotes species coexistence (e.g. Moeller, 2004). However, multiple pair-wise competitive interactions within a community can lead to net positive effects by, for example, incorporating interactions with a shared competitor (Levine, 1999). In other words, species coexist in a community even when the dominant interaction type is competitive.

An understanding of patterns in community organization and mechanisms that produce and support them requires information on how the abundance of an individual species in a community is affected by the abundance and diversity of the coexisting species. In general, negative or positive relationships between the abundances of species co-occurring in a community can indicate competitive or facilitating relationships between these species, respectively. Negative interactions can also signal asymmetric competition leading to exclusion (Levine & Rees, 2002) or antagonistic relationships between species (Lombardero et al., 2003).

Demonstrating species interactions in the field is logistically difficult. Although manipulative studies of competition, i.e. by field removal experiments, have produced a number of examples of pair-wise interspecific competitive interactions under natural conditions (see Gurevitch et al., 1992 for review), this approach is limited by the practical impossibility of performing experimental studies on all pairs of species in a community. Therefore, alternative methods have been devised to measure species interactions from census data (e.g. MacArthur & Levins, 1967; Schoener, 1974; Crowell & Pimm, 1976; Fox & Luo, 1996; Shenbrot & Krasnov, 2002). Although some of these methods have been criticized (Rosenzweig et al., 1985; Abramsky et al., 1986), census data can nevertheless provide hints about the type of species interactions that might prevail in a community, and thus can serve as a basis for further manipulative experiments and/or more sophisticated analysis. In addition, experimental removals are extremely difficult (if at all possible) in parasite communities.

Utilizing this approach, Stepanova & Mitropolsky (1971, 1977) and Zolotova & Iskhanova (1979) studied the abundance of *Xenopsylla hirtipes* and *Xenopsylla gerbilli* in burrows of the great gerbil *R. opimus* in Central Asia and found that the abundances of these species were positively correlated (although no formal analysis has been carried out). Faulkenberry & Robbins (1980) used the o- and Q-statistics based on relative odds to measure the degree of association between *Atyphloceras multidentatus*, *Catallagia charlottensis* and six other flea species (pooled together) parasitic on the grey-tailed vole *Microtus canicaudus* in Oregon. The results of their analyses showed that fleas belonging to different species co-occurred more frequently than expected by chance and/or if they were distributed independently of each other. Brinkerhoff et al. (2006) found a non-random, positive association between *Oropsylla hirsuta* and *Oropsylla tuberculata* on the black-tailed prairie dog *Cynomys ludovicianus* in Colorado: hosts highly infested with one flea species were also highly infested with the other species.

Broadening the scale of consideration, Krasnov et al. (2005g) used a data set that involved 230 flea species and 92 host species across 27 geographical regions, totalling in 1798 flea species/host species/region combinations, and examined how the overall abundance and diversity of flea communities affect the abundance of individual flea species in these communities. This approach allows investigation of the potential role of diffuse competition in communities because fleas use the same resource. This type of competition occurs when a species competes with a constellation of other species in various combinations and densities (MacArthur, 1972). For example, the decrease in the abundance of a flea with an increase in the abundance of all other co-occurring fleas would suggest the occurrence of diffuse competition (Bock et al., 1992). The latter can also be revealed by negative relationships between the abundance of a given flea species and the species richness or any other measure of diversity of the entire flea assemblage, given that a higher number of species leads to more intense competition (MacArthur, 1972).

The above is true if the species in a community interact directly. Indeed, the original model of MacArthur (1972) does not incorporate indirect interactions. Later models of diffuse competition that account for indirect interactions have concluded that a high number of species could reduce the intensity of interactions or even lead to facilitation (Davidson, 1980; Vandermeer, 1990; Stone & Roberts, 1991). As we have seen earlier, suppression of host defence systems resulting from the high abundance of one or more parasite species or/and high parasite diversity could lead to facilitation among parasite species. As a result, the abundance of a given species should be positively correlated with either the abundance of other co-occurring species or their diversity or both. Facilitation could also occur if only one of the two parameters of the entire community (overall abundance and diversity) is positively correlated with the abundance of a given species, whereas the other is not. In any case, studying the relationships between the abundance of a parasite species and the descriptors of the entire parasite community can give insights into processes that govern parasite communities.

As a measure of flea diversity, Krasnov et al. (2005g) took not only the mere number of flea species in assemblages, but also indices of taxonomic distinctness and taxonomic asymmetry (Δ^+ and Λ^+) (Clarke & Warwick, 1998, 1999, 2001; Warwick & Clarke, 2001) similar to the S_{TD} and $VarS_{TD}$ indices of host specificity of Poulin & Mouillot (2003) (see Chapter 14). The index of taxonomic distinctness incorporates basic phylogenetic information on the co-occurring species and places the emphasis on the taxonomic distance between species rather than on their number, providing a different perspective on diversity, namely a measure of the composition, and not the size, of an assemblage (see further details in Chapter 17).

At all scales of analysis, i.e. whether the comparison was done for the same flea species on different host species, or different flea species, two consistent results emerged. First, the abundance of a given flea species correlates positively with the total abundance of all other co-occurring flea species in the community (Fig. 16.5a), providing thus absolutely no support for diffuse competition in the traditional sense because the more individuals of other species are present on a host population, the more individuals of the focal species are there as well. This confirms that some host species or populations, but not others, represent better habitats for multiple flea species (see Chapters 9–11). As we already know, the superiority of a host species or population compared to other host species or populations from the 'viewpoint' of a flea can be related to quantitative (the amount of resources; e.g. the amount of organic matter in the host's burrow or nest available for flea larvae) or qualitative (the pattern of resource acquisition; e.g. the ease of blood-sucking and/or flea movement, efficiency of blood

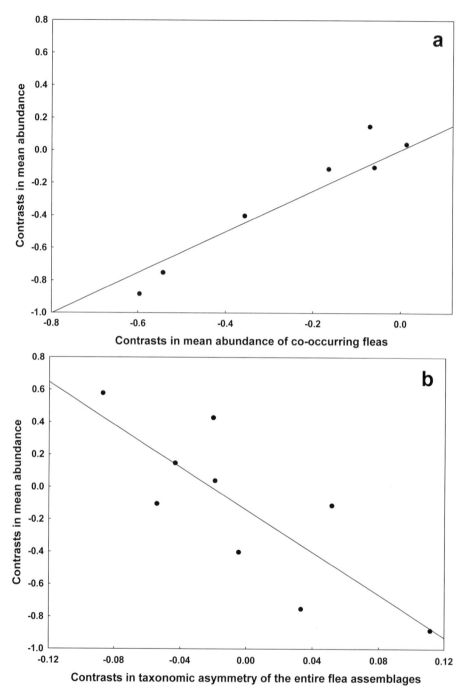

Figure 16.5 Relationships between the mean abundance of *Frontopsylla hetera* and (a) the abundance of all other co-occurring flea species and (b) taxonomic asymmetry of flea assemblages across different host species using the method of independent contrasts. Redrawn after Krasnov et al. (2005g) (reprinted with permission from Blackwell Publishing).

digestion, variation in the immunocompetence and/or behavioural defence abilities) or both. In other words, the results of this study suggested important indirect (host-mediated) facilitation among flea species within xenocommunities and communities living on small mammals. The fitness of a host is determined by various factors (availability of spatial and energetic resources, predation, intra- and interspecific competition). The fitness of a parasitized host is also limited by competition with the parasite, which may hijack part of the available resources. In this sense, for the host a parasite is only a competitor, while for the parasite the host is both a competitor and the resource (Combes, 2001). Therefore, the main pattern of interactions among flea species in component communities can be referred to as apparent facilitation that in terms of parasite-induced immunodepression has been repeatedly reported for parasites, although mainly at the infracommunity scale (Bush & Holmes, 1986; Cox, 2001). Nevertheless, apparent facilitation is thought to be more likely to arise in assemblages where the different pairs of competitors compete for different resources, or use different mechanisms to acquire resources (Davidson, 1985). Few would deny that flea species (both imagoes and larvae) may compete for the same resource (for blood or organic matter, respectively). However, the acquisition of the resource (e.g. the pattern and location of a blood meal) can be strikingly different in different flea species (e.g. Hsu *et al.*, 2002; see Chapters 9–10).

Second, the abundance of any given flea species correlates negatively with either the species richness or taxonomic diversity of the flea community (Fig. 16.5b). In general, the abundance of a given flea species is highest in assemblages consisting of few species of limited taxonomic diversity. While this supports the existence of some form of negative interactions among species, such that the abundance of a given flea species is lower when many other species are also present, it also supports the occurrence of facilitation mediated via the host, since the abundance of a given flea species is higher when co-occurring flea species are closely related (taxonomically) with it. The latter can be linked to the higher likelihood of immunosuppression if the immunogens of the parasites involved are similar which, in turn, is more likely if the parasites are phylogenetically close (a phenomenon opposite to cross-resistance). In addition, flea species showing a significant negative association between their abundance and the taxonomic diversity (Δ^+) of the flea assemblage, did not show a similar trend with the taxonomic asymmetry (Λ^+), and vice versa. This finding suggests that the various processes affecting abundance of flea species can operate separately. Thus, an alternative hypothesis could be related to the niche-filtering process where environmental conditions (both biotic (related to the host) and abiotic) act as a filter to restrict co-occurring species to a certain functional, phylogenetic or taxonomic subset (Tofts & Silvertown, 2000; Statzner *et al.*, 2004).

Strong niche-filtering, possibly mediated by host immune responses, would lead to only closely related flea species occurring together on hosts where conditions are favourable for them to achieve high abundance. Weaker niche-filtering would allow the co-occurrence of several unrelated flea species, some of which may achieve low abundance because local (host) conditions are not optimal for their requirements.

The results of Krasnov et al.'s (2005g) study demonstrate that flea communities may be a convenient model for community ecology because they are governed by the same rules as communities of other organisms. In particular, these results support the idea that both facilitation and competition operate among the same species either simultaneously or with the strength of each process varying in time or space (Callaway & King, 1996; Callaway & Walker, 1997; Levine, 1999). Consequently, a community has to be considered from a more synthetic perspective, where species interactions should be viewed as complex combinations of negative and positive components (Callaway & Walker, 1997; Levine 1999). As mentioned above, it is also possible that multiple pair-wise competitive interactions within a community can amount to net positive effects by incorporating interactions with a shared competitor (host) (Levine, 1999). In other words, both direct and indirect effects should be taken into account when considering interactions in the context of entire communities (Stone & Roberts, 1991).

16.3.2 Patterns of interspecific interactions inferred from patterns of aggregation

The results of Krasnov et al.'s (2005g) study described in the previous section unequivocally showed that coexistence of flea species is facilitated and, consequently, there are some mechanisms producing this facilitation. An additional mechanism by which multiple species in a community may coexist is reducing the overall intensity of competition via aggregated utilization of resources (Shorrocks & Rosewell, 1986; Jaenike & James, 1991; Hartley & Shorrocks, 2002). This model, known as the aggregation model of coexistence, states that if competing species are distributed such that interspecific aggregation is reduced relative to intraspecific aggregation then species coexistence is facilitated. The aggregation model has attracted much attention and has been studied in many ways, using various criteria for coexistence based on different assumptions (see Hartley & Shorrocks, 2002 and references therein). The main assumptions of this model are: (a) fragmented nature of resources; (b) aggregated distribution of species among resource patches; and (c) independent distribution of species (Jaenike & James, 1991). However, the third assumption is not necessary as it has been shown that aggregation can promote coexistence among species even if their abundances are correlated among patches (Ives, 1988b). These assumptions suggest that communities of parasites, in general, and fleas, in particular, have numerous advantages for the investigation of the aggregation model of coexistence compared with communities of free-living organisms.

The only attempt to test the aggregation model of coexistence in relation to flea communities has been undertaken by Krasnov *et al.* (2006f) again using data on fleas parasitic on small mammals in Slovakia and the Negev Desert. They calculated three measures of aggregation. A measure of intraspecific aggregation, J, has been described in Chapter 15 (equation 15.5). A measure of interspecific aggregation, C, reflects the proportional increase in the number of heterospecific competitors relative to a random association (Shorrocks, 1996) as:

$$C_{kl} = \frac{\sum_{i=1}^{p} \frac{n_{ki} n_{li}}{m_k P} - m_l}{m_l} = \frac{\mathrm{Cov}_{kl}}{m_k m_l}, \qquad (16.1)$$

where Cov is a covariance between a pair of species k and l, n_{ki} and n_{li} are the numbers of parasite species k and l on host individual i, m_k and m_l are mean numbers of parasite species k and l, and P is number of host species. A value of $C>0$ indicates a positive association between two species, whereas $C<0$ indicates a negative association between two species. Finally, the relative strength of intraspecific aggregation versus interspecific aggregation in two species was assessed by a measure A as:

$$A_{kl} = \frac{(J_k + 1)(J_l + 1)}{(C_{kl} + 1)^2}. \qquad (16.2)$$

This measure represents a product of inverse indicators of the relative mutual effects of aggregation of two competitors on each other (Sevenster, 1996). A value of $A_{kl} > 1$ (for species k and l) indicates that intraspecific aggregation is stronger than interspecific aggregation, and is a necessary, albeit not a sufficient, criterion for the persistence of both species in the presence of each other (see Sevenster, 1996; Hartley & Shorrocks, 2002 for further discussion). Both C and A were calculated for each pair of flea species within a host species, and then averaged for each xenocommunity of fleas.

Among 17 host species, fleas were negatively associated in seven communities (mean $C < 0$) and positively associated in 10 communities (mean $C > 0$) (Fig. 16.6a). Intraspecific aggregation was stronger than interspecific aggregation in all xenocommunities of fleas (log mean $A > 0$) (Fig 16.6b). In other words, flea assemblages were not dominated by negative interspecific associations, and the level of interspecific aggregation in flea assemblages was reduced in relation to the level of intraspecific aggregation, thus facilitating flea coexistence. Consequently, aggregation of conspecific flea individuals may limit their own population growth to

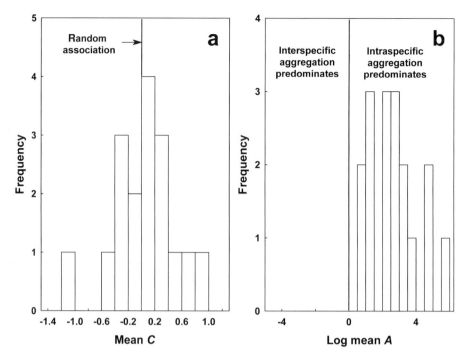

Figure 16.6 Frequency distributions of (a) mean interspecific aggregation C and (b) the mean relative strength of intraspecific aggregation versus interspecific aggregation A (log) across 17 flea xenocommunities from Slovakia and the Negev Desert. Redrawn after Krasnov et al. (2006f) (reprinted with permission from Blackwell Publishing).

such an extent that the remaining resources appear to be sufficient to support other flea species (Hartley & Shorrocks, 2002).

As has been repeated many times in this book, some host individuals or species represent better habitats for a parasite species or for multiple parasite species than other host individuals or species. This heterogeneity among host individuals or species can be related to such host parameters as body size, mobility, and depth and complexity of host burrows (Rózsa et al., 1996; Lo et al., 1998). Consequently, some host characters may affect coexistence of parasite species via their effect on intra- and/or interspecific parasite aggregation. For example, a higher degree of facilitation of coexistence of fleas is expected to be found (a) in larger rather than in smaller hosts; and (b) in hosts possessing deep and complicated burrows rather than in those that use simple and shallow shelters. Larger hosts and/or hosts possessing deep and complicated burrows should provide more space and a greater variety of niches for adult and/or pre-imaginal fleas, respectively (in addition, larger hosts have a larger number of body areas that need different defensive abilities against parasites). Alternatively, other host characteristics (e.g. energy expenditure for the immune response; see Chapter 13) may be affected by the size of a parasite assemblage, which, in turn, is affected by the relationship between intra- and interspecific parasite aggregation that may either facilitate or impede parasite coexistence. Consequently, the reduction of the level of interspecific aggregation in relation to the level of intraspecific aggregation is expected to be positively correlated with flea abundance and species richness which, in turn, should occur in hosts with higher rather than with lower basal metabolic rate (BMR) because hosts exposed to diverse challenges (from multiple flea species) should invest in a high BMR in order to provide for a costly immune response (Morand & Harvey, 2000; but see Chapter 17).

To test these hypotheses, Krasnov *et al.* (2006f) analysed interspecific aggregation and relative strength of intraspecific aggregation versus interspecific aggregation in dependence of parameters characterizing host species and flea communities. Parameters of host species were evaluated using two composite variables, one of which positively correlated with body mass and complexity of burrow/nest, whereas the other positively correlated with mass-independent BMR. The original values of the parameters of flea communities were also substituted with two composite variables, namely that (a) positively correlated with mean and maximum infracommunity species richness and xenocommunity species richness and (b) positively correlated with mean abundance of fleas and taxonomic distinctness (Δ^+) of flea component communities.

It appeared that interspecific aggregation tended to correlate positively with (a) infracommunity and xenocommunity species richness and (b) body mass and burrow complexity of the host (Fig. 16.7a, b). The former correlation again indicated apparent facilitation between different flea species (likely via suppression of a host's immune system due to multiple challenges), whereas the latter correlation supported the ideas about (a) higher number of niches for fleas on larger hosts with deeper and more complicated burrows and (b) larger number of host body areas that need different defensive abilities. A trade-off in a host's defensive abilities and in a flea's evasive abilities among different host body areas may lead to segregation of parasite site-specificities (Reiczigel & Rózsa, 1998; see Chapter 9).

The level of intraspecific aggregation versus interspecific aggregation (A) was also positively correlated with species richness of flea infra- and xenocommunities, but negatively correlated with host body mass and burrow complexity (Fig. 16.7c, d). An increase in the intraspecific aggregation compared to interspecific aggregation with an increase in infra- and xenocommunity species richness supports analogous observations on ectoparasites of marine fish by Morand *et al.* (1999). In addition, the lack of relationship between the ratio of the effect of

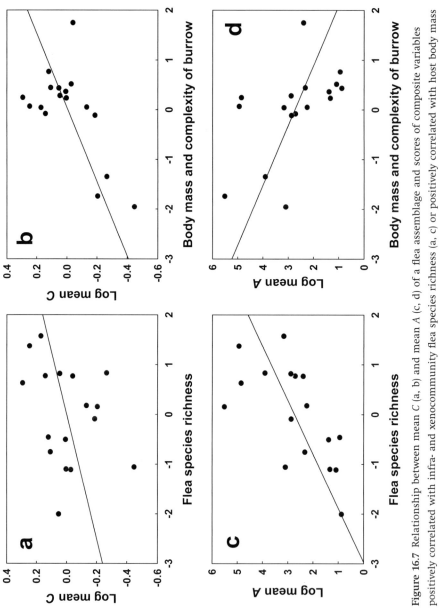

Figure 16.7 Relationship between mean C (a, b) and mean A (c, d) of a flea assemblage and scores of composite variables positively correlated with infra- and xenocommunity flea species richness (a, c) or positively correlated with host body mass and burrow complexity (b, d). Redrawn after Krasnov et al. (2006f) (reprinted with permission from Blackwell Publishing).

intra- to interspecific aggregation and taxonomic distinctness of xenocommunity suggests that aggregation may be important for coexistence of both closely and distantly related species. However, the negative relationship between A and a composite variable reflecting among-host variation in body mass and burrow complexity was opposite to the prediction. This may be related to the reduction of the size of resource patches (and, consequently, resource amount) that may promote coexistence by increasing intraspecific and decreasing interspecific interactions in some circumstances (Atkinson & Shorrocks, 1981; Kouki & Hanski, 1995). If competitively inferior species show a preference for small patches (i.e. smaller host species and/or hosts with simple and/or shallow burrows) that have a reduced probability of encounters, they may experience lower levels of interspecific interaction, thus increasing their chance of coexistence.

Concluding, the pattern of flea coexistence appeared to be related to both the structure of flea communities and affinities of host species. The effect of host factors on the relationship between intraspecific and interspecific aggregation supports the idea that interaction between parasites and their hosts is more important in determining parasite community structure than direct interaction between parasite species (Dobson, 1985; Dobson & Roberts, 1994). It should be also noted that aggregation may be only one of various mechanisms responsible for coexistence of species in a community. In reality, aggregation will affect coexistence together with other processes such as temporal segregation between flea species (e.g. Ageyev et al., 1984), niche segregation by specialization (e.g. Šimková et al., 2000) and niche-filtering (Tofts & Silvertown, 2000). Furthermore, the relative importance of aggregation compared to other mechanisms of coexistence depends on the phylogenetic and guild diversity of the community (Takahashi et al., 2005).

16.4 Negative interspecific interactions

The results of the studies described above suggest that interspecific flea interactions are generally positive and are mediated by a host. However, this conclusion comes mainly from census data that considered the entire assemblages of imago fleas, while the net positive effects can result from negative pair-wise interactions. In addition, interactions between imago fleas and interactions between larval fleas can differ, as the mediating effect of a host is probably much weaker in the latter than in the former. Therefore, experimental studies are needed to establish the exact type of interactions between each pair of species and, consequently, the exact mechanism of community structuring.

However, there is also an alternative opinion that interspecific interactions among parasites are much weaker in comparison with intraspecific interactions

(Overal, 1980; Rohde, 1994; Rohde & Heap, 1998; Morand *et al.*, 1999). Nevertheless, the occurrence of interspecific interactions, in particular competition, between parasite species has been supported by studies of helminths (e.g. Holland, 1984; Patrick, 1991), although only a few experimental studies of ectoparasites have been undertaken. For example, Dawson *et al.* (2000) studied experimentally the attachment rates of two parasitic copepods *Lepeophtheirus thompsoni* and *Lepeophtheirus europaensis* which are naturally isolated on their sympatric fish hosts, turbot *Psetta maxima* and brill *Scophthalmus rhombus*, respectively. The results suggested (a) a greater sensitivity to competition for the generalist species *L. europaensis* than for the specialist *L. thompsoni* and (b) the role of interspecific competition in the maintenance of *L. thompsoni* and *L. europaensis* on their respective natural hosts, preventing turbot invasion by *L. europaensis*.

To the best of my knowledge, interspecific interactions between imago fleas have never been studied experimentally, although the possibility of the occurrence of the interspecific competition between imagoes has often been implied for the explanation of distributional patterns (Evans & Freeman, 1950; Barnes, 1965; Arzamasov, 1969; Barnes *et al.*, 1977; Stepanova & Mitropolsky, 1977; Day & Benton, 1980; Lindsay & Galloway, 1997). For example, Burdelova & Burdelov (1983) found that a decrease of abundance of *X. gerbilli* and *X. hirtipes* in the Akdala Valley of Kazakhstan was accompanied by an increase of abundance of *Xenopsylla conformis*, *Coptopsylla lamellifer*, *Nosopsyllus laeviceps*, *Nosopsyllus turkmenicus*, *Nosopsyllus tersus* and *Ctenophthalmus dolichus*. Evans & Freeman (1950) reported that *Ctenophthalmus nobilis* and *A. penicilliger* in Berkshire (Great Britain) co-occurred on the bank vole *M. glareolus*, but not on the woodmouse *A. sylvaticus*. It was suggested that the longer pelage of voles allows the fleas to avoid competition (at least, interferential) which occurs in the shorter pelage of a mouse. Barnes *et al.* (1977) observed that two sympatric species of the North American genus *Anomiopsyllus* were never found together in the same nest of their rodent host, the dusky-footed wood rat *Neotoma fuscipes*, and explained this by competitive exclusion. It should be noted, that attempts to explain patterns of spatial (e.g. among host body areas: Allan, 1956, or among host individuals: Layene, 1963) or temporal (e.g. seasonal: Verts, 1961) separation of fleas by interspecific competition were mainly made during the time when competition held a central place in ecological and evolutionary theory and was a dominant paradigm in ecology.

Larval fleas received a bit more attention in this relation. At least one study specifically and experimentally tested interspecific interactions between flea larvae (Krasnov *et al.*, 2005e). This study considered fleas *X. conformis* and *Xenopsylla ramesis* that exploit Sundevall's jird *Meriones crassus* and focused on the performance of larvae of the two species in terms of their developmental success in mixed-species and single-species treatments under different air temperatures, relative humidities, substrate textures and food abundance. It was found that the developmental success (evaluated as the number of individuals that survived until emergence) of *X. conformis* depended on the presence of competing species, being, in general, lower in mixed-species compared with single-species treatments (Fig. 16.8a). The decrease in developmental success of *X. conformis* in mixed-species treatments was found mainly during food shortage. In contrast, presence of the competitor did not affect developmental success of *X. ramesis* (Fig. 16.8b).

These results clearly demonstrated that the larvae of the two fleas competed for food. Furthermore, when food resources were limited, *X. ramesis* outcompeted *X. conformis*. However, when the amount of food was extremely scarce (25% of requirements), the results of competition were masked by high mortality of larvae in both species. Potential proximate mechanisms of interspecific competition between insect larvae have been shown to be both interferential (e.g. cannibalism: Fox, 1975 and references therein) and exploitative (e.g. Braks *et al.*, 2004) or else mediated by the interaction of a parasite (Park, 1948). These mechanisms are not mutually exclusive, although the results of Krasnov *et al.*'s (2005e) study suggest that interspecific cannibalism (interference competition) seem to be the most likely mechanism for outcompeting of *X. conformis* by *X. ramesis*. Indeed, manifestation of competition in treatments with food deficiency as well as higher developmental success in *X. ramesis* in some mixed-species treatments when food amount was limited suggests the occurrence of cannibalism. Furthermore, obvious asymmetrical competition (Lawton & Hassell, 1981; Rosenzweig, 1991) between the two flea species suggests that *X. ramesis* possesses some traits that allow it to outcompete *X. conformis* when food is limited. These traits can be related to higher specialization of *X. ramesis* in terms of environmental preferences as well as the degree of host specificity. Indeed, *X. conformis* and *X. ramesis* not only differ in their environmental preferences (Krasnov *et al.*, 2001a, b, 2002b, d), but *X. ramesis* appears to be less tolerant of microclimatic fluctuations than *X. conformis* (Krasnov *et al.*, 2001a). The geographical range of *X. ramesis* is narrower than that of *X. conformis* (Theodor & Costa, 1967) and is more climatically homogeneous. In addition, the host spectrum of *X. conformis*, at least in the Negev Desert, is broader than that of *X. ramesis* (seven versus four host species: Krasnov *et al.*, 1999). Consequently, these results are in accordance with a model of asymmetrical competition between specialist and generalist species, where the generalist is expected to be more sensitive to competition than the specialist (Futuyma & Moreno, 1988; Rausher, 1993; but see Iwao & Ohsaki, 1996).

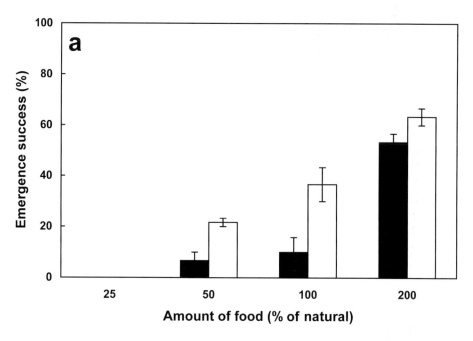
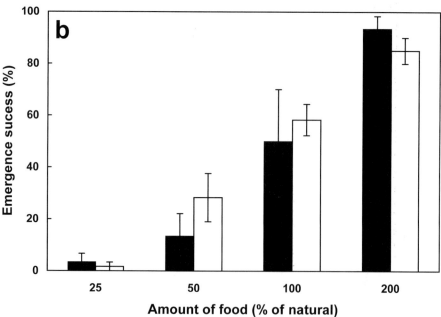

Figure 16.8 The effect of competing species and food availability on the mean (± S.E.) proportion of individuals of (a) *Xenopsylla conformis* and (b) *Xenopsylla ramesis* that survived until emergence at 28 °C and 75% relative humidity, on loess substrate. White columns – single-species treatments, black columns – mixed-species treatments. Data from Krasnov et al. (2005e).

Table 16.1 *Percentage of flea larvae belonging to different genera and found in different parts of a burrow of the great gerbil* Rhombomys opimus

	Total larvae found	Tunnels	Food storage chambers	Winter nest	Delivery and nursing nest
Echidnophaga	26	–	69.2	23.1	7.7
Xenopsylla	1380	15.2	82.9	0.3	2.3
Coptopsylla	213	10.0	6.6	27.2	56.2
Nosopsyllus	358	1.5	0.9	83.1	14.5
Paradoxopsyllus	41	7.3	70.6	2.5	19.6
Ctenophthalmus	6907	1.4	–	98.5	0.1
Rhadinopsylla	215	–	1.9	79.4	18.7
Stenoponia	519	0.4	–	99.6	–

Source: Data from Kunitskaya *et al.* (1979).

Competitive superiority of more specialized species has been repeatedly reported for both parasites (e.g. Dawson *et al.*, 2000; Perlman & Jaenike, 2001; Poulin, 2007a) and free-living organisms (e.g. Chase, 1996; Bohn & Amundsen, 2001). Poulin (2007a) provided numerous examples of asymmetric interspecific competition between parasite species of various taxa and concluded that it is extremely common in parasites.

In general, the results of the study of larval interactions between the two fleas indicate that interspecific competition, at least among pre-imagoes, may play a certain role in structuring flea assemblages and that some species coevolved life-history strategies as the result of this competition. In particular, differential preferences of larval fleas of different species or genera to different parts of rodent burrows can be seen as a 'ghost of competition past' (Rosenzweig, 1991), although mediated by microclimatic and seasonal preferences (Kunitsky, 1970). For example, Kunitskaya *et al.* (1979) reported that in the burrows of *R. opimus* in Kazakhstan, the majority of larvae of *Echidnophaga*, *Xenopsylla* and *Paradoxopsyllus* occurred in chambers with stored food, whereas larval *Rhadinopsylla*, *Ctenophthalmus*, *Stenoponia* and *Nosopsyllus* preferred the nest chambers used by gerbils during winter (Table 16.1).

16.5 Similarity in flea communities: geographical distance or similarity in host composition?

Abiotic and biotic factors affecting individuals of the same host species but occupying different geographical locations are obviously different. If so, species composition of the parasite assemblages on conspecific hosts should vary geographically. Indeed, this has been repeatedly supported by studies on helminths (e.g. Kisielewska, 1970; Kennedy & Bush, 1994; Poulin, 2003). Flea communities have been studied in this relation rarely. Nevertheless, Bogdanov et al. (2001) reported values of similarity in flea species composition among different populations of small mammals occurring in different habitats of the Omsk Region of Siberia. The highest similarity between flea xenocommunities was found in shrews of the genus *Sorex* (up to 86.9%), whereas rodents of the genera *Myodes*, *Microtus* and *Apodemus* demonstrated lower values of similarity (up to 75.6%, 64.1% and 64.1%, respectively).

One of the most common geographical patterns observed in biological communities is distance decay of similarity (Nekola & White, 1999; Poulin, 2003). Community composition follows this pattern if the proportion of species shared by two communities decreases with increasing distance between them. The decrease of biological similarity with distance can arise because of various mechanisms, one of them being a decrease in environmental similarity with increasing distances (Nekola & White, 1999). Poulin (2003) found that the similarity in the composition of helminth communities in vertebrate hosts decreased with increasing geographical distance between host populations, although this was only true for some host species and not others.

However, environmental similarity for parasites involves not only the physical environment but also an environment represented by the species composition of the host community (host 'faunal' environment). For example, host species in communities of similar composition but under different environmental conditions can support similar parasite assemblages and vice versa. Furthermore, given that fleas are strongly influenced by the characteristics of their off-host environment, similarity in species composition of flea xenocommunities should demonstrate a stronger geographical effect than that of helminths.

To examine how the species composition of flea assemblages on a host species varies with geographical or 'faunal' distance among host populations, Krasnov et al. (2005k) used data on flea assemblages of 11 host species (*Neomys fodiens*, *A. agrarius*, *A. uralensis*, *M. musculus*, *A. amphibius*, *M. glareolus*, *Myodes rutilus*, *M. arvalis*, *M. gregalis*, *Microtus oeconomus* and *C. migratorius*). For each host species, similarity in flea species composition (Jaccard and Morisita–Horn similarity indices), as well as geographical and 'faunal' distances (reciprocal of the Jaccard similarity), were computed for all possible pairs of host populations. No relationship between either similarity measure and either distance measure was found in the water shrew *N. fodiens*, whereas in the other 10 species, at least one of the similarity measures was negatively correlated with at least one of the distance measures. Similarity in species composition of flea xenocommunities decreased with an increase in both distance measures in two species (*A. uralensis*, *M. glareolus*) (Fig. 16.9). 'Faunal', but not geographical, distance determined the

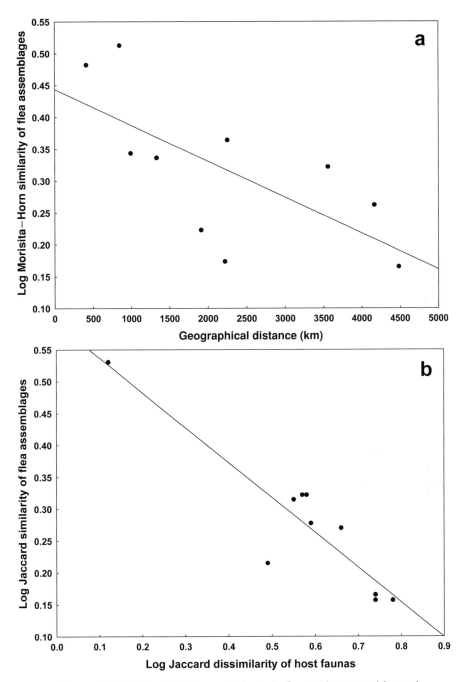

Figure 16.9 Relationship between similarity in flea species composition and (a) geographical (km) and (b) 'faunal' distance between two populations of the bank vole *Myodes glareolus*. Data from Krasnov et al. (2005k).

dissimilarity in species composition of flea xenocommunities in *A. agrarius*, *M. musculus*, *M. arvalis* and *C. migratorius*, whereas similarity in species composition of flea xenocommunities decreased with increasing geographical distance but was not affected by 'faunal' distance in *A. amphibius*, *M. rutilus*, *M. gregalis* and *M. oeconomus*.

These results showed that the pattern of distance decay of biological similarity found in other organisms is universal. Furthermore, a negative exponential function provided the best fit to the distance decay of similarity in flea assemblages, as was the case for plant communities (Nekola & White, 1999) and helminth assemblages in fish and mammals (Poulin, 2003), meaning that similarity in flea assemblages declines constantly and proportionally per unit distance. However, the slope values for geographical distance were lower than those for the 'faunal' distance, suggesting that, perhaps, difference in the surrounding milieu between host populations is a more important determinant of the composition of flea assemblages than mere physical distance. Nevertheless, the difference between host species in the effect of either the 'faunal' or geographical distance on the similarity of their flea assemblages requires some explanation. This difference can be related to either differences in the structure of geographical ranges of different species or in patterns of sampling across the geographical range of a species, or both. Indeed, all species for which the relationship between the similarity in flea assemblages and the 'faunal', but not geographical, distance was found are characterized either by continuous geographical ranges (*M. musculus*, *M. arvalis*, *C. migratorius*) or they were sampled in the continuous part of their geographical range (*A. agrarius*) that is situated in the same biome or group of biomes. For these species, environmental variability across the geographical range is likely to be lower than the 'faunal' variability. In contrast, species for which the relationship between the similarity in flea assemblages and the geographical but not the 'faunal' distance was found are either characterized by a fragmented geographical range (*M. gregalis*) or were sampled in those parts of their geographic ranges where they are distributed patchily in intrazonal habitats (*A. amphibius*, *M. rutilus* and *M. oeconomus*). For these species, environmental variability can be higher than 'faunal' variability (*M. gregalis*, *M. oeconomus* and *M. rutilus*), and thus geographical distance plays the major role in determining similarity between different populations.

16.6 Concluding remarks

There are two important messages from the studies described in this chapter. First, biological differences between larval and imago fleas produce a striking difference in mechanisms prevailing in their communities.

Communities of imago fleas are characterized mainly by facilitation mediated via the host, whereas competition seems to be the main type of interactions in communities of larval fleas. This means that, on the one hand, a flea during its individual life is subjected to different forces and should switch its competitive capabilities from struggle with individuals belonging to other species to struggle with a host. On the other hand, facilitation-governed communities of imago fleas are the source of competition-governed communities of larval fleas and vice versa. This suggests that a flea community in its entirety, i.e. comprising both imagoes and larvae, is governed by complicated rules that still remained to be revealed.

Second, although studies of structure of flea communities are rather scarce and the community ecology of fleas (as well as other parasites) is only starting its first steps, it is already clear that the rules governing these communities are similar to the rules governing communities of free-living organisms. Concepts developed for communities of free-living animals can be successfully applied when studying communities of parasites (see also Guégan *et al.*, 2005). This means that both parasitologists and 'mainstream' community ecologists should adopt broader research perspectives. The former may largely gain from the use of the theoretical achievements of mainstream ecology, whereas the latter may better understand the complicated mechanisms of nature by including parasites, such as fleas, into their field of vision.

17

Patterns of flea diversity

The search for patterns of biodiversity across locations and through time and the explanation of these patterns is one of the most popular themes in ecology. Understanding of biodiversity patterns in application to parasites is especially important because parasites play important roles in the regulation of populations and communities of their hosts (e.g. Combes, 2001; Poulin, 2007a) and because this understanding is crucial for successful control of diseases that hit humans as well as wild and domestic animals. In addition, parasites are living organisms and are de facto part of biodiversity. Moreover, parasitism is possibly more common than any other feeding strategy (Sukhdeo & Bansemir, 1996). Consequently, studying biodiversity patterns of parasitic organisms is also important from the point of view of conservation efforts because parasites deserve to be protected as much as any other living species (Windsor, 1990, 1995; Whiteman & Parker, 2005; Christe et al., 2006).

When a biodiversity study concerns parasites, at least one extremely important difference between parasites and free-living organisms should be taken into account. Any parasitic species is characterized by 'dual location'. On the one hand, a 'location' for a parasite is the host species it exploits, whereas, on the other hand, it is the geographical location where a host (and, consequently, a parasite) occurs. It is well known that different host species harbour different number of parasite species (e.g. Caro et al., 1997). It is highly improbable that parasite species are distributed randomly among their hosts but rather parasite diversity probably results from multiple interwoven factors. In fact, Combes (2001) listed as many as 16 different hypotheses related to correlates of parasite diversity. However, most of these hypotheses have never been tested. Also, the biodiversity of fleas has been studied relatively less in comparison with other parasitic taxa, especially various helminths.

In this chapter, I review the observed patterns of flea species richness and diversity associated with parameters of individuals, populations and communities of their hosts as well as some of the major spatial patterns of flea diversity. As considered here, flea diversity is evaluated using one or both of the two main measures. One measure is a mere species richness which is self-explanatory, whereas another measure represents the taxonomic distinctness of a flea assemblage (Δ^+) (see Chapter 16). Using the taxonomic classification of Hopkins and Rothschild (1953, 1956, 1962, 1966, 1971), Traub et al. (1983) and Medvedev (1998), flea species were fitted into a taxonomic structure with eight hierarchical levels above species, i.e. subgenus (or species group), genus, tribe, subfamily, family, superfamily, infraorder and order (Siphonaptera). The use of taxonomic levels was restricted to these basic ones because they are the only ones available for all flea taxa. The maximum value that the index Δ^+ can take is thus 8, and its lowest value is 1 (see Chapter 14). However, since the index cannot be computed for an assemblage composed by a single flea species, a Δ^+ value of 0 was assigned to such assemblages, to reflect their extreme species poorness. The variance in Δ^+, Λ^+, provides information on asymmetries in the taxonomic distribution of flea species in assemblages. It is analogous to the index $VarS_{TD}$ (see Chapter 14), and can only be computed when a flea assemblage comprises no fewer than three species.

Estimates of parasite species richness may be biased if hosts in some regions are studied more intensively than in others (see Morand & Poulin, 1998). Consequently, unequal between-study sampling effort may result in confounding variation in estimates of both host and flea species richness. Therefore, whenever the species richness of flea assemblages is mentioned in this chapter, the confounding effect of sampling effort has been controlled for. Analogously with indices S_{TD} and $VarS_{TD}$, Δ^+ and Λ^+ may be sensitive to species richness (see Chapter 14). If this was the case, then the values of Δ^+ and Λ^+ were corrected for unequal number of species.

17.1 Flea diversity and host body

The host body is the ultimate habitat for the majority of parasites, including fleas. Consequently, variation in host body characteristics (body size, metabolic rate, lifespan) has often been considered as a primary factor determining among-host variation in parasite diversity (Feliu et al., 1997; Morand & Poulin, 1998; Morand & Harvey, 2000; Arneberg, 2002; Krasnov et al., 2004g). The reasons why a correlation between parasite (e.g. flea) diversity and host body mass is expected are rather straightforward. Larger hosts are likely to sustain richer flea assemblages because they provide more space and a greater variety of niches, and thus can provide different parasite species with an opportunity for spatial niche diversification. Indeed, on relatively large hosts, different fleas prefer different body areas (Hsu et al., 2002; see Chapter 9).

Basal metabolic rate (BMR) is expected to correlate positively with flea diversity because hosts exposed to diverse infections should invest in a high BMR in order to compensate for a costly immune response (Morand & Harvey, 2000; see Chapters 13 and 15), although others have argued that immune response cost is energy cost above that of BMR (Degen, 1997). Nevertheless, BMR in mammals has been shown to correlate positively with helminth species richness without compromising host longevity (Morand & Harvey, 2000). If flea species richness is expected to be positively correlated with BMR, correlation of parasite richness with average daily metabolic rate (ADMR) is expected to be more pronounced. This is because ADMR is considered as a more appropriate measurement than BMR for evaluating the energy requirements and efficiency of energy utilization of an animal (Degen, 1997). However, ADMR has been largely ignored in studies of energy expenditure, partly due to the lack of standard methods of measurement.

These hypotheses were tested correlating species richness of flea assemblages with body mass, BMR (both for 92 species) and ADMR (for 39 species) of rodent hosts (Krasnov et al., 2004g). It appeared that none of the above-mentioned parameters of the host body correlated with species richness of flea assemblages either when the original data were analysed or when the confounding effect of phylogeny was controlled for. These results suggest that the conclusions of Poulin (1995b) and Morand & Poulin (1998) about the lack of relationship between body size and species richness of mammalian endoparasites are also valid for ectoparasites. It may differ in fish, however, as a correlation between host body size and parasite richness was reported for fish ectoparasites when the effect of phylogeny was removed (Guégan & Morand, 1996). Nevertheless, Arneberg (2002) found a positive relationship between strongylid nematode richness and mammalian body mass, but to see this the effect of host population densities had to be controlled for. However, in rodents, host density can vary greatly on a temporal scale, with 10-fold fluctuations often observed. Consequently, consideration of the mean density of rodent populations in the present context is not feasible. The main reason for the lack of correlation between host body size and flea species richness may be that a principal habitat for many fleas is not only the body of a host per se but also its burrow or nest. Consequently, flea species richness might be related to the size and the degree of complexity of the host burrow rather than to its body size, although this has never been tested. That this may be the case finds support in the case of the great gerbil *Rhombomys opimus*. This rodent constructs highly complicated and deep burrows (Kucheruk, 1983) and has the richest flea assemblages (15 species in the database used) among gerbillines. The same is true for the northern grasshopper mouse *Onychomys leucogaster*. This species constructs deep and intricate burrows (Ruffer, 1965) and is parasitized by a very high number of flea species (Thomas, 1988). Nevertheless, the relationship between host body mass and flea diversity can be envisaged when comparisons among species are carried out at a smaller scale (e.g. within a region). For example, Bossard (2006) noted that among mammals of the Great Basin Desert, larger species had fewer flea species. In the coastal area of the Aral Sea, larger mammal species had richer flea assemblages (Burdelov et al., 1989). However, it is difficult to compare the two latter studies as only small mammals (rodents and soricomorphs) were taken into account by Burdelov et al. (1989), whereas Bossard (2006) considered a broader spectrum of mammalian species including carnivores.

The lack of correlation between either BMR or ADMR and flea species richness may suggest that either flea parasitism does not negatively affect a rodent host or does not trigger an immune response, or else that the immune response to a particular flea species is equally effective against other flea species (cross-resistance; see Chapter 13). However, we have already seen that flea parasitism has an energy cost for at least some hosts (see Chapter 12) and that stimulation of an immune response by derived molecules from the salivary glands of the fleas is a general case (see Chapter 13). Thus, the occurrence of cross-reactions of the immune response to different fleas (see Chapter 13) can be responsible for the absence of correlation between host metabolic parameters and flea species richness. In addition, an already described study of energy requirements for maintenance in Wagner's gerbil *Dipodillus dasyurus* under flea parasitism (Khokhlova et al., 2002; see Chapter 12) demonstrated that these requirements increased in parasitized individuals in spite of a relatively small blood loss, indicating thus that the energy cost of an immune response was above the ADMR of the rodent. This suggests that a more relevant parameter relating to flea species richness on a host might be the ability of the host to increase its metabolic rate above requirements and not the ADMR itself.

Host longevity can also be an important factor determining the diversity of parasites as a consequence of the continued accumulation of parasites in long-lived species (Bell & Burt, 1991; Morand, 2000). Considering fleas, such a relationship can be expected in sedentary host species with particularly complex and developed burrow system. However, longevity of small mammalian hosts and flea species richness were found not to be correlated (Stanko et al., 2002). At present, there are no other studies relating this host trait and flea diversity. Nevertheless, the longevity of an individual host (= host age) seems to be linked to variation in flea diversity among host individuals within host species. Energy-costly

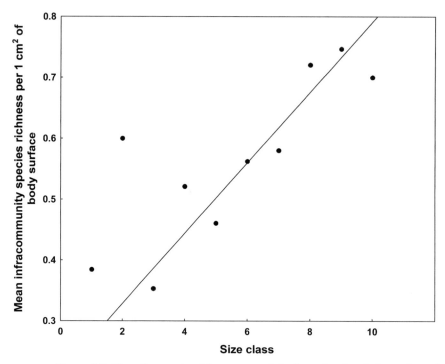

Figure 17.1 Mean flea species richness per unit body surface in relation to host size in the fieldmouse *Apodemus agrarius*. Data from Krasnov *et al.* (2006a).

immunity against parasites is expected to decrease in senescent individuals (see Chapters 13 and 15). This, in turn, can facilitate the co-occurrence of multiple parasite species in long-lived hosts. In some rodent species (the fieldmouse *Apodemus agrarius*, yellow-necked mouse *Apodemus flavicollis*, pygmy woodmouse *Apodemus uralensis*, bank vole *Myodes glareolus* and common vole *Microtus arvalis*) from Slovakia, senescence was accompanied by an increase in mean infracommunity species richness (Krasnov *et al.*, 2006a; see Chapter 15) (Fig. 17.1). The mechanisms responsible for this pattern can differ in different species, not being, however, mutually exclusive. Analogous to what we have already seen for conspecific fleas (see Chapter 15), the main reason for the increase in species richness of flea assemblages in older *M. arvalis* could be due to the accumulation of different fleas in permanent burrows (let's recall that this species constructs complicated deep burrows where flea accumulation is high: Kucheruk, 1983; Němec, 1993). An increase of flea species richness in old *Apodemus* spp. and *M. glareolus* can be due their high mobility (Nikitina, 1961) and, thus, an increase in the probability of lateral transfer of fleas.

17.2 Flea diversity and host gender

As we already know, gender differences in mobility and defence competence can lead to gender biases in the infestation not only by one particular flea species but by other flea species as well (see Chapter 15). As a consequence, gender bias in parasite diversity should have the same direction as gender bias in the infestation parameters due to the same reasons. In other words, if, for example, males are more mobile than females, then males would have higher chances of being exposed not only to a higher number but also to a larger variety of parasites. If increased levels of androgens decrease the ability of males to withstand infection or their larger size allows them to support a higher number of parasites per unit body size without compromising their health and/or fitness, then male bias not only in the size but also in the diversity of parasite infracommunities may be expected.

Indeed, in small mammals from Slovakia, the composition of flea infracommunities of males was richer than those of females, except for the woodmouse *Apodemus sylvaticus* (female-biased flea species richness) and pygmy shrew *Sorex minutus* (no gender differences in flea species richness) (Morand *et al.*, 2004) (Table 17.1). Krasnov *et al.* (2005c) found gender differences in species richness of flea assemblages in five of nine rodent species from the Negev Desert (Table 17.1). It is interesting that these gender differences were (a) observed in winter only (when most rodents showed peak reproduction) and (b) were male-biased in four hosts and female-biased in one host. Lareschi (2006) reported the results of the study of ectoparasite infestation in the water rat *Scapteromys aquaticus* in Argentina. Both males and females were predominantly infested by *Polygenis atopus*. Other four species of *Polygenis* exploited this host accidentally. However, only one accidental flea species was recorded on female rats, whereas three accidental flea species were found on males. The lack of the link between sexual size dimorphism and gender bias in species richness of flea infracommunity in the among-host comparisons (Morand *et al.*, 2004; Krasnov *et al.*, 2005c) rules out the hypothesis of differential behavioural defensibility but rather hints at the validity of a mobility- and immunocompetence-related explanation.

17.3 Flea diversity and host population

Density is the main characteristic of any population of living organisms. It affects a variety of individual and population parameters, including the spread and distribution of parasites among host individuals (e.g. Anderson & May, 1978). This is because the rate at which host individuals acquire the

Table 17.1 *Mean species richness of flea infracommunities on small mammalian hosts from Slovakia (all seasons; data from Morand et al., 2004) and the Negev Desert (winter only; data from Krasnov et al., 2005c)*

Region	Host	Mean species richness of flea infracommunities	
		Males	Females
Slovakia	Apodemus agrarius	3.24	2.67
	Apodemus flavicollis	3.36	2.83
	Apodemus uralensis	1.88	1.57
	Apodemus sylvaticus	1.10	1.50
	Myodes glareolus	3.11	2.67
	Microtus arvalis	1.93	1.75
	Microtus subterraneus	2.15	1.45
	Neomys fodiens	2.67	2.00
	Sorex araneus	1.26	0.86
	Sorex minutus	0.16	0.17
Negev	Acomys cahirinus	0.82	0.98
	Acomys russatus	0.09	0.67
	Dipodillus dasyurus	0.73	0.72
	Gerbillus andersoni	1.51	1.24
	Gerbillus henleyi	0.15	0.07
	Gerbillus nanus	0.94	0.89
	Gerbillus pyramidum	1.33	1.50
	Meriones crassus	1.62	1.32
	Sekeetamys calurus	1.00	0.78

parasite species may be determined by how many individuals are available for parasite colonization (Morand & Poulin, 1998). In addition, high host density can facilitate a process of horizontal parasite transmission both within and among host species and, thus, increase the mean number of parasite species per host individual. Indeed, a positive relationship was found between helminth species richness and host density of both small and large mammals (Morand & Poulin, 1998). In contrast, Stanko *et al.* (2002) did not find any relationship between rodent and soricomorph density and flea species richness in among-host species comparison. Instead, they found the effect of host density on flea species richness among host populations within host species. This lack of a relationship between host density and flea species richness in among-host comparisons is not especially surprising because of the huge temporal fluctuations in the density of small mammals for which the idea of 'typical' density hardly makes sense (see above). However, the existence of this relationship among host populations within host species is easily explained by epidemiological theory (Anderson & May, 1978; May & Anderson, 1979).

Host social structure is another factor that can be linked with parasite diversity as has been shown by Côté & Poulin (1995) and Tella (2002) for various parasite taxa (but not for fleas). In contrast, both non-phylogenetic and phylogenetically corrected analyses of data on fleas parasitizing rodent hosts demonstrated that social and solitary species did not differ in either species richness or taxonomic distinctness of their flea assemblages (Krasnov et al., 2006j). A possible reason for the contradicting results for fleas and mammals and other parasite–host associations is that host sociality can affect differently parasites with different transmission strategies (Altizer et al., 2004). In addition, it is sometimes difficult to distinguish the role of host social structure from that of host density in their effect on parasite diversity (Morand & Poulin, 1998). For example, some rodent species live solitarily at low density while becoming social at high density (e.g. Schradin & Pillay, 2005).

17.4 Flea diversity and host community

17.4.1 Flea diversity among host species and the effect of host community structure

Factors associated with the structure of host communities can have a profound influence on the parasite diversity of a host species (but see Hallas & Bang, 1976). One of the most important factors related to host community structure is the species composition of this community, for example the number of sympatric closely related host species. An increase in the richness of the taxonomic milieu of a host increases the probability of lateral transfer of parasites, and thus can increase the parasite species richness of any given host species in that community (Combes, 2001). Indeed, in rodents, the number of coexisting species belonging to the same subfamily (both across the entire geographical range and locally) as a host of interest appeared to be positively correlated with species richness of flea assemblages on this host (Krasnov et al., 2004g) (Fig. 17.2).

This also means that the ability of a flea species to exploit successfully several host species strongly depends on the phylogenetic and/or taxonomic relatedness of these hosts. The reasons for this can be similarities among host species in ecological, physiological and/or immunological characters (see Chapter 9). For example, fleas assemblages on majority of dipodid species (jerboas) inhabiting eastern Mongolia (e.g. the hairy-footed jerboa *Dipus sagitta*, Andrews's three-toed jerboa *Stylodipus andrewsi* and Mongolian five-toed jerboa *Allactaga sibirica*)

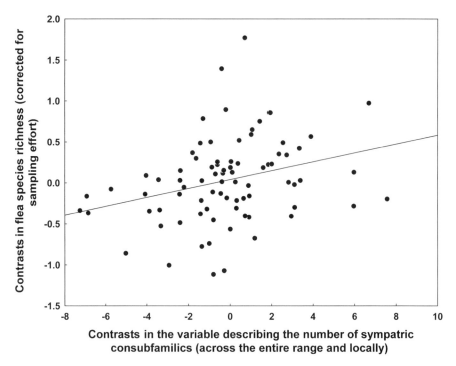

Figure 17.2 Relationship between flea species richness (controlled for host sampling effort) and the principal component of the number of sympatric members of the same subfamily across the entire geographical range and in the location of flea richness study across 82 rodent species using the method of independent contrasts. Redrawn after Krasnov *et al.* (2004g) (reprinted with permission from Blackwell Publishing).

have the same species composition (*Frontopsylla wagneri*, *Ophthalmopsylla praefecta*, *Ophthalmopsylla kiritschenkovi* and *Mesopsylla hebes*) (Scalon, 1981).

Furthermore, the positive link between the number of co-habitating host species and flea diversity of a given host species was supported at another scale (Krasnov *et al.*, 2006k). In this study, 14 species of small mammals were sampled and fleas collected in 18 locations representing nine habitat types across central Slovakia. Species richness of flea assemblages of a host species expressed as flea fauna (overall number of flea species), mean infra- or xenocommunity richness correlated positively with the mean number of co-habitating host species independently of whether co-habitants were phylogenetically related to a species of interest (Fig. 17.3), although no significant relationship was found between any of the measurements of flea species richness and the number of habitats occupied by a host species. This supports the idea that a high number of co-habitating hosts can facilitate flea exchange between hosts (Ryckman, 1971; Rödl, 1979; Krasnov & Khokhlova, 2001).

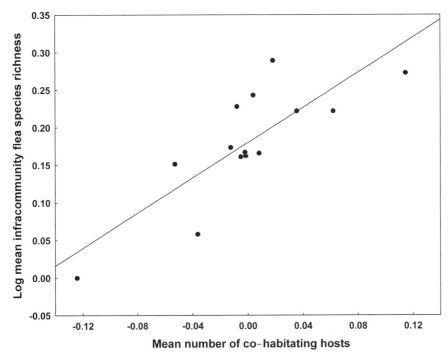

Figure 17.3 Relationship between mean infracommunity flea species richness and the number of co-habitating hosts (controlled for host sampling effort) across 14 host species from central Slovakia. Redrawn after Krasnov et al. (2006k) (reprinted with permission from Springer Science and Business Media).

17.4.2 Flea diversity among host communities

A positive correlation between species diversity and habitat variety is a well-known phenomenon (Rosenzweig, 1995 and references therein). However, when diversity is considered in this context, the question of what is a 'habitat' and, consequently, 'habitat diversity' arises. Is a habitat predefined and related to an area of a particular relief, vegetation and soil structure? Or is it a patch with a set of environmental conditions and resources promoting occupancy, survival and reproduction by individuals of a given species and, thus, represents a species-specific response to a set of conditions (Morrison et al., 1992; Rosenzweig, 1992)? Parasites offer a conceptual advantage over free-living animals in this respect (see Chapter 9). Considering a host species as a habitat for a parasite species, in particular for a flea species, avoids the disagreement between the two above-mentioned concepts of habitat. First, a host species is a clearly predefined entity. Second, fleas clearly distinguish among different host species both in terms of host choice behaviour and fitness reward (see Chapters 9–11). Given a positive correlation between species diversity and habitat

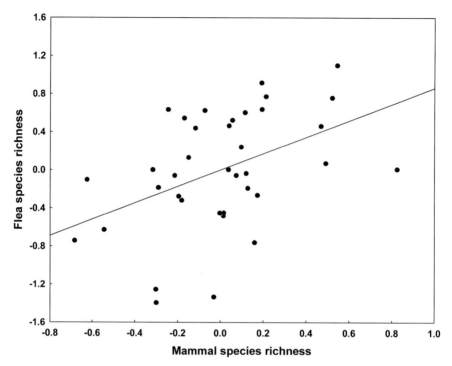

Figure 17.4 Relationship between host species richness and flea species richness (both controlled for area and host sampling effort) using cross-region comparisons. Redrawn after Krasnov et al. (2004h) (reprinted with permission from Blackwell Publishing).

diversity in free-living organisms, a positive correlation between host diversity and flea diversity can also be expected. Indeed, the notion of a positive link between these two parameters can sometimes be found in faunistic reports (e.g. Sapegina et al., 1980b, 1990; Sapegina, 1988). In other cases, no relationship between flea diversity and host diversity has been found (Zhang et al., 2002).

Krasnov et al. (2004h) studied the relationship between host species richness and flea species richness using simultaneously collected data on small mammals (Soricomorpha, Rodentia and Lagomorpha) and fleas in 37 different regions. The data were controlled for the area sampled and sampling effort, and then this relationship was tested using both cross-region conventional analysis and the independent-contrasts method (to control for the effects of biogeographical historical relationships among different regions). Both analyses showed a positive correlation between host species richness and flea species richness (Fig. 17.4).

An ecological reason for this pattern can be the enhancement of consumer diversity by resource diversity (higher diversity of resources may allow a larger number of consumer species to coexist: Pimm, 1979). An evolutionary reason for the pattern could be the diversification of fleas as a response to the diversification of hosts, similar to the process of specialization of free-living species to a range of habitat properties (Rosenzweig, 1992). Diversification of hosts can facilitate an increase in the number of their fleas either by a higher probability of flea co-diversification (if host diversification stems from host speciation) (Combes, 2001; Clayton et al. 2003; but see Chapter 4) or by the introduction of new flea species (if host diversification stems from host immigration) or both. In any case, the evolutionary reason for the positive host diversity–flea diversity pattern can be a process of specialization of fleas on different host species, exactly as the specialization of free-living species to a limited range of habitat properties is the reason behind the positive species diversity–habitat diversity pattern (Rosenzweig, 1992). This is because 'fine habitat subdivision is a coevolved property of the species in a biome' (Rosenzweig, 1992: 715). The conceptual difference in comparisons between species versus habitat diversity and parasite versus host diversity is mainly in our inability to recognize different habitats in the same manner as animals and plants do, whereas it is much easier to recognize different host species. The absence of a negative relationship between flea species richness and mammal species richness suggests that the relationships between flea parasites and their mammalian hosts reach an equilibrium when neither does host defence cause parasite extinction nor does parasite pressure lead to host extinction.

On the other hand, the relationship between host and parasite species richness can be considered as a bottom–up effect which occurs when the diversity at lower trophic levels controls the diversity at higher trophic levels (Siemann, 1998; Brändle et al., 2001). It is known that the manifestation of bottom–up effects varies among communities. The reasons for this variability include a variety of factors, such as heterogeneity within a trophic level (Hunter & Price, 1992) and length of the food chain (Duffy et al., 2005). Manifestation of bottom–up effects can also vary among similar communities from different locations (Pennings & Silliman, 2005). If so, the pooling of the data from several different regions as was done in Krasnov et al. (2004h) could mask the true region-specific relationship between parasite and host diversity, which can vary due to differences in the history of the relationships between the various communities (Fleming et al., 1987) or differences in abiotic conditions (Pennings & Silliman, 2005). The history of the relationships may, in turn, be affected by region-specific evolution driven by local climate and/or landscape.

Indeed, when the relationships between small mammal and flea species richness were considered for 26 Palaearctic and 19 Nearctic regions separately, the results indicated geographical variation in the occurrence of bottom–up effects in the relationship between flea and host species richness (Krasnov et al., 2007c).

The bottom–up pattern was strongly expressed in the Palaearctic but not in the Nearctic (Fig. 17.5).

From an ecological perspective as defined above, this suggests that host diversity controls flea diversity in the Palaearctic, but no control of flea diversity by host diversity occurs in the Nearctic. In addition, analyses of non-transformed data indicated that it requires fewer host species to support one flea species in the Palaearctic compared with the Nearctic (slopes 1.16 versus 0.85, respectively). This suggests that flea–host interactions in the Palaearctic are relatively specialized compared with those in the Nearctic (because each flea species interacts with relatively fewer host species in the Palaearctic than in the Nearctic).

A potential explanation for the difference between the patterns observed in the Palaearctic and Nearctic might be that the quality of host resources differs between the two biogeographical realms. For instance, the mammalian taxa of Eurasia are not the same as those of North America: some families (e.g. Heteromyidae) are endemic to one land mass and absent from the other. If mammalian hosts of the Nearctic were consistently less immunocompetent than those of the Palaearctic, then it might be that host resources do not limit flea diversity in the Nearctic as they do in the Palaearctic. Less immunocompetent hosts may be able to support a richer flea fauna than more immunocompetent hosts. Thus, the relationship between flea species richness and host species richness would not be expected to hold in the Nearctic if host taxa there were less immunocompetent than in the Palaearctic. However, there are no data that would support this possibility.

The geographical variation in the host versus flea diversity pattern could also be explained by between-realm historical differences in flea–host associations. As mentioned above, positive correlation can either stem from host speciation promoting parasite diversification by an increase of the probability of parasite codiversification, or be the result of host immigration and ensuing parasite diversification by the introduction of new parasite species. Although co-diversification of parasites and their hosts is rarely congruent and is often complicated by a number of evolutionary events (which is characteristic for fleas; see Chapter 4), the predictability of the relationship between the number of parasites and the number of their hosts would probably be higher if the main reason for parasite diversification was a response to host speciation rather than a response to host immigration. Indeed, the relationship between the number of Palaearctic fleas and their hosts appeared to be more predictable than that in the Nearctic, and Palaearctic fleas are, on average, more host-specific than Nearctic fleas. Although no strong evidence of co-diversification between some flea taxa and their mammalian hosts is available (Krasnov & Shenbrot, 2002; Lu & Wu, 2005; see Chapter 4), fleas, nevertheless, are thought to exploit the group of hosts with

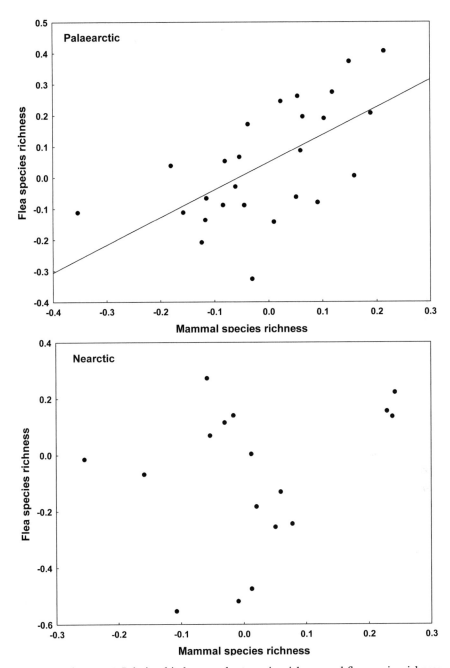

Figure 17.5 Relationship between host species richness and flea species richness (both controlled for area and host sampling effort) across different regions in the Palaearctic and Nearctic. Redrawn after Krasnov et al. (2007c) (reprinted with permission from Blackwell Publishing).

which they coevolved or hosts that evolved later, but not more ancestral mammalian lineages (Traub, 1980). The hosts that support the majority of flea species are representatives of several families and subfamilies of rodents (such as Arvicolinae, Murinae, Gerbillinae, Cricetinae) and soricomorphs (such Soricidae) that originated in the Old World (see Traub, 1980 and references therein). Furthermore, the only flea family that radiated mainly in the New World (Ceratophyllidae) is also the evolutionarily youngest family (Traub, 1980; Medvedev, 2005; see Chapter 4). These different lines of evidence suggest a longer history of flea–host associations in the Palaearctic than in the Nearctic, and thus can explain the stronger link and higher predictability of flea–mammal relationships in the former. Another, albeit indirect, line of evidence supporting earlier Palaearctic compared with Nearctic associations between fleas and their hosts is that the number of Palaearctic flea species exceeds by almost three times the number of Nearctic fleas (890 species versus 299 species, respectively: Medvedev, 2005). However, Medvedev (1996, 2005) suggested a non-historical explanation for the higher diversity of the flea fauna in the Palaearctic compared with the other biogeographical realms, argueing that this could be associated with the high landscape diversity of the Palaearctic as well as with the dynamics of glacial transgressions and regressions during the Pleistocene.

Additional explanation for the occurrence of the bottom–up pattern in the Palaearctic but not in the Nearctic is that this pattern is a consequence of a relatively high level of consumer specialization. The higher level of specialization of Palaearctic fleas can be the evolutionary result of a higher number of flea species in the Palaearctic than in the Nearctic regions that exploit a similar number of host species (22.3 ± 1.8 versus 14.3 ± 2.1, respectively). Competition among fleas could lead to their specialization on different host species (but see Chapter 16).

Concluding, it appears that macroecological trends can vary among biogeographical realms. A strong trend occurring in one realm may mask a weaker trend or even the absence of a trend in another realm. Consequently, one should be cautious when pooling data from regions with different biogeographical histories into a single data set for a macroecological analysis.

17.5 Flea diversity and host geographical range

17.5.1 Patterns within a host species

Diversity of parasites can vary among different populations of the same host species across its geographical range (see Chapter 16). However, when spatial variation in the diversity of fleas on 69 species of small mammals from

24 different regions of the Holarctic was examined, it appeared that flea species richness varied less within than among host species, and was thus a repeatable host species character (Krasnov et al., 2005k) (Fig. 17.6). This suggests the existence of some threshold of defence against fleas in a host species that limits the host's ability to cope with multiple flea species (e.g. because of costly defence systems) while maintaining their pressure (expressed as a number of parasite species) at a 'tolerable' level. In contrast with species richness, the taxonomic distinctness of flea assemblages and its variance were not repeatable among populations within a host species. This means that whenever a new exploiter is added to a host's parasite community, this exploiter is a random addition from the regional pool of exploiter species that manages somehow to adapt itself to the new host species.

However, in spite of flea species richness being a true host character, this character varied across the geographical range in many hosts, indicating that diversity of flea assemblages is also influenced by local factors. In most host species, the diversity of flea assemblages correlated with one or more environmental (climatic) variables, in particular mean winter temperature (Fig. 17.7). This demonstrates that the diversity of flea assemblages on small mammalian hosts is to an important extent mediated by local climatic conditions, highlighting thus how ecological processes interact with coevolutionary history to determine local parasite biodiversity.

17.5.2 Patterns among host species

Hosts that differ in the size and position of their geographical range are expected to differ also in the diversity of their parasite assemblages. Hosts with larger geographical ranges would presumably encounter more parasite species, and thus a positive correlation between parasite species richness and the size of the host geographic range is expected. Indeed, this pattern was found in flea assemblages of rodents (Krasnov et al., 2004g; Bossard, 2006) (Fig. 17.8). Combes (2001) suggested that this relationship be interpreted in the framework of the theory of island biogeography (species richness on islands correlates positively with the size of the island: MacArthur & Wilson, 1967).

17.6 Flea diversity and the off-host environment

As mentioned above, the diversity of flea assemblages within a host species is to an important extent mediated by local environmental conditions. Consequently, it can be expected that factors other than host species richness (e.g. climate and landscape) might also explain well variation in flea species

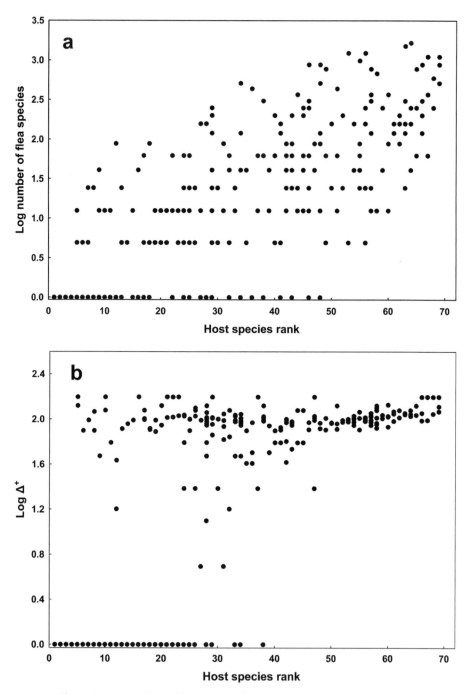

Figure 17.6 Rank plots of (a) number of flea species and (b) average taxonomic distinctness (Δ^+) between fleas in flea assemblages on 69 small mammals (see Fig. 14.4 for explanations). Redrawn after Krasnov et al. (2005k) (reprinted with permission from Blackwell Publishing).

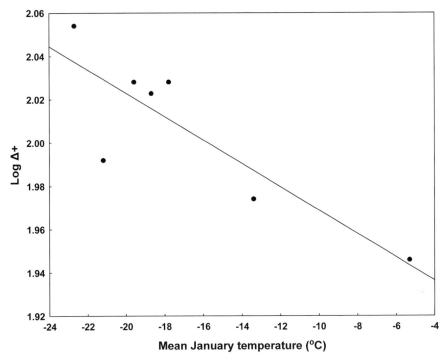

Figure 17.7 Relationship between taxonomic distinctness (Δ^+) of flea assemblages and mean January temperature in different populations of the water vole *Arvicola amphibius*. Data from Krasnov et al. (2005k).

richness among habitats or geographical locations (i.e. at the level of host communities).

17.6.1 Flea diversity and productivity

Productivity (the rate of energy flow through an ecosystem) is considered to be an important factor influencing distribution of many taxa of animals and plants (Rosenzweig, 1992, 1995). The only study that specifically aimed to reveal the relationship between parasite species diversity and productivity was carried out by Poulin et al. (2003). In this study, a linear relationship between productivity and endoparasite species richness was found both across 131 vertebrate hosts of various taxa as well as for each of the five vertebrate groups (fish, amphibians, reptiles, mammals and birds). This study dealt with productivity components that are intrinsic (i.e. host-related) to parasite communities. The effect of productivity components related to extrinsic (i.e. environment-related) factors on parasite diversity can be quite different. As fleas are strongly affected by the off-host environment, some relationship between productivity of the

428　Patterns of flea diversity

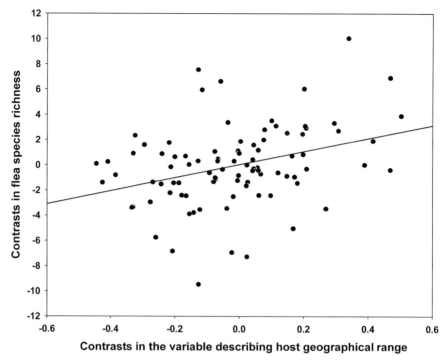

Figure 17.8 Relationship between flea species richness (corrected for host sampling effort) and the first principal component of size and north–south and west–east lengths of host geographical range using the method of independent contrasts. Redrawn after Krasnov et al. (2004g) (reprinted with permission from Blackwell Publishing).

off-host environment and flea diversity can be expected. However, only indirect evidence is available. Although Krasnov et al.'s (1997, 1998) studies on the structure of flea assemblages in *Dipodillus dasyurus* and *M. crassus* in the Negev Desert were not specifically designed to test the relationship between flea diversity and habitat productivity, the highest species richness of fleas on these hosts was found in habitats with the highest abundance of annual vegetation. The latter is a good estimator of primary production for desert environments. In the Great Basin Desert, mammalian hosts from habitats of moderately low productivity (sage and grass) had the highest flea species richness (Bossard, 2006). On the other hand, no relationship between either species richness or taxonomic distinctness of flea assemblages and annual precipitation (another good estimator of productivity in dryland ecosystems) was found among populations of the steppe- and desert-dwelling migratory hamster *Cricetulus migratorius* (Krasnov et al., 2005g).

17.6.2 Flea diversity and elevation

Observations of flea species richness at different altitudes have been reported for several geographical areas. For example, Tipton & Méndez (1966) found that in Panama the number of flea species was greater at altitudes higher than 1500 m than at lower altitudes, whereas in Mexico this altitudinal threshold was as high as 3050 m (Tipton & Méndez, 1968). Among 34 flea species studied by Méndez (1977) in Colombia, seven species never occurred in mountain habitats, nine species occupied a variety of habitats, including mountains, whereas as many as 18 species occurred only in habitats higher than 1500 m. Nevertheless, very high altitudes (e.g. above 5000 m) are characterized by low flea richness (see Barrera (1968) for Mexico). In contrast, the number of flea species decreased with an increase of altitude in the Tatras (Bartkowska, 1973), the Beskids (Haitlinger, 1989) and the Mongolian Altai (Kiefer et al., 1984). Small mammals in the subalpine and alpine regions of the Tyrol Mountains harboured less diverse flea assemblages than in lower regions (Manhert, 1972), whereas the opposite is true in the Sudetes (Haitlinger, 1975). In the Yunnan Province of China, flea diversity was higher in the mountain area than in lowlands (Guo et al., 2000; Gong et al., 2004b). Moreover, the relationships between flea species richness and elevation appeared to be unimodal and peaked at 3000 above sea level in the southern part of the province (Gong et al., 2000) and at 2500–3800 m above sea level in its western part (Gong et al., 2005, 2007). A peak of flea species richness at 2100–2500 m above sea level was reported for the Trans-Ili Alatau Ridge in the Tien Shan Mountains (Busalaeva & Fedosenko, 1964).

The contradictions in these results could stem from a potential difference in the evolutionary history of a flea assemblage in a region. For example, if the majority of flea species in a location have boreal origins they would probably inhabit the higher elevations (e.g. Durden & Kollars, 1997). Another reason for the contradictory results on the effect of elevation on flea species richness may be that potentially confounding factors such as host species richness have not been taken into account in these studies. Recently, an analysis of the effect of elevation on flea species richness taking into account host species richness as well as other confounding factors such as air temperature and precipitation has been carried out (Krasnov et al., 2007c). It was found that in the Palaearctic, the number of flea species in a region increased with an increase in the region's mean elevation. This, however, does not suggest that flea species richness is higher at higher elevations. Rather, high mean altitude of a region reflects the presence of mountains, which presumably increases the environmental variation within the region, resulting in a high number of flea species. However, no relationship between the mean altitude of a region and flea species richness was found for the Nearctic regions (Krasnov et al., 2007b). This between-realm difference can be the result of the higher variety of mountain systems, where fleas and hosts were sampled, in the Palaearctic than in the Nearctic. Indeed, the mountain Palaearctic data sets in Krasnov et al.'s (2007c) study included those from the Tatra Mountains, Scandinavian Mountains, the Caucasus, the Ural Mountains, the Tien Shan Mountains, the Khangay Mountains, Dzhungarian Alatau, the Tarbagatai Mountains, the Sayan Range, the Koryak Mountains and the Atlas Mountains, whereas the mountain Nearctic data sets came from the Rocky Mountains, the Sierra Nevada and the Cascade Mountains only.

17.6.3 Flea diversity and latitude

Another pattern of parasite richness that is expected to vary geographically is the latitude. It is predicted to correlate negatively with parasite richness according to the well-known pattern of the latitudinal gradient of species richness. The latitudinal gradient is that, in general, the inventory of species declines as one moves further from the equator, either north or south (Rosenzweig, 1995), as has repeatedly been shown for free-living animals (e.g. Rohde, 1992; Rosenzweig, 1992). Krasnov et al. (2004g) found a clear correlation between the latitude of the host geographical range and species richness of flea assemblages, but this trend was the exact opposite of the main latitudinal gradient rule, namely flea species richness increased with the latitude of the centre of the geographical range (see also Gong et al., 2001) (Fig. 17.9).

One of the reasons for the unusual pattern that has been found for fleas may be that only a few flea assemblages of both tropical and arctic rodents have been studied. In the data set of Krasnov et al. (2004g), the centre of the geographical range was situated at latitudes lower than 20° in only two species and at latitudes higher than 60° in only three species. Rohde (1996, 1999) questioned the generality of the latitudinal gradient rule and suggested that this rule is a 'local' phenomenon that is restricted to the Holarctic above latitudes of 40°–50° N. However, when the data set of Krasnov et al. (2004g) was limited to 62 Holarctic species with geographical range centres above 40° N, the positive correlation between latitude and flea richness remained.

As I have already noted, a flea species is characterized by 'dual location'. At one level of consideration, a 'location' for a flea is a host species it exploits, whereas, at another level of consideration, it is the geographical location where it occurs together with its hosts. Consequently, the search for a latitudinal gradient of flea diversity can be carried out not only among different host species but also among different geographical regions.

When flea diversity was compared among the Palaearctic and Nearctic regions located at different latitudes, it appeared that in the Palaearctic, the number

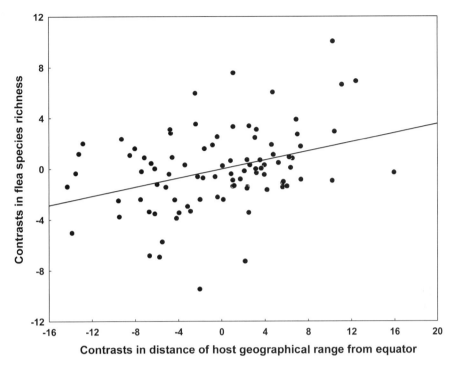

Figure 17.9 Relationship between flea species richness (controlled for host sampling effort) and distance of the centre of host geographic range from equator using the method of independent contrasts. Redrawn after Krasnov et al. (2004g) (reprinted with permission from Blackwell Publishing).

of flea species was not affected by latitude, whereas in the Nearctic, flea species richness tended to decrease at higher latitudes (Krasnov et al., 2007c). In these analyses, the factor of host species richness was controlled for. Thus, the standard latitudinal trend in flea species richness was revealed when the analysis was performed at the level of host communities rather than of host species. Moreover, the standard latitudinal trend for flea diversity appeared to be true only for one of the two biogeographical realms studied. However, this trend was relatively weak and was manifested only when no other factor affected flea richness. This means that in the Palaearctic, the latitudinal trend in the number of flea species could be merely obscured by other, stronger determinants of flea richness.

17.7 Flea diversity and parasites of other taxa

Different parasite taxa exploit different host resources and are often unlikely to interact directly (e.g. an intestinal helminth and an ectoparasitic mite). I have already mentioned the concept of isolationist and interactive parasite communities (Chapter 16). It is commonly accepted that interactive

Table 17.2 *Pearson's correlation coefficients (r) for all pair-wise associations of species richness and taxonomic distinctness (Δ^+) between fleas and assemblages of other three ectoparasite taxa found on the same host species. Both parameters were controlled for confounding variables (sampling effort in the case of species richness, and number of species in the case of Δ^+)*

Data points	Taxon	Species richness	Δ^+
Original data	Lice	0.08	−0.27
	Mesostigmatid mites	0.31*	0.10
	Ticks	0.47*	0.07
	Endoparasitic helminths	0.15	−0.04
Independent contrasts	Lice	0.03	−0.29
	Mesostigmatid mites	0.49*	−0.07
	Ticks	0.58*	0.08
	Endoparasitic helminths	0.49*	−0.06

Note: * − $p < 0.05$.
Source: Data from Krasnov et al. (2005l).

communities are those that comprise parasite species belonging to the same guild, e.g. sharing the same trophic level, whereas parasite species in isolationist communities, though sharing a host, do not exploit the same resources (Poulin, 2007a). Nevertheless, interactions, although rather indirect than direct, between parasite species belonging to different guilds are also possible. Some components of host immune defences may operate simultaneously against all kinds of parasites, whereas investment by the host in specific defences against one type of parasites may come at the expense of defence against other parasites. Consequently, both negative and positive relationships among species diversity of parasites belonging to different taxa can be expected. Investigation of the relationships between the flea diversity and diversity of the assemblages of three other taxa of ectoparasites (sucking lice, mesostigmatid mites and ixodid ticks), and between the species richness of fleas and endoparasitic helminths, across different species of rodent hosts demonstrated positive pair-wise correlations between the species richness of fleas, mites and ticks (Table 17.2, Fig. 17.10a) as well as endoparasite assemblages across host species (Fig. 17.10b) (Krasnov et al., 2005l), although the latter was true only when the confounding effect of phylogeny was removed.

These results, combined with an earlier demonstration that the species richnesses of different groups of endoparasitic helminths covary positively among their vertebrate hosts (Poulin & Morand, 2004), provide strong evidence of

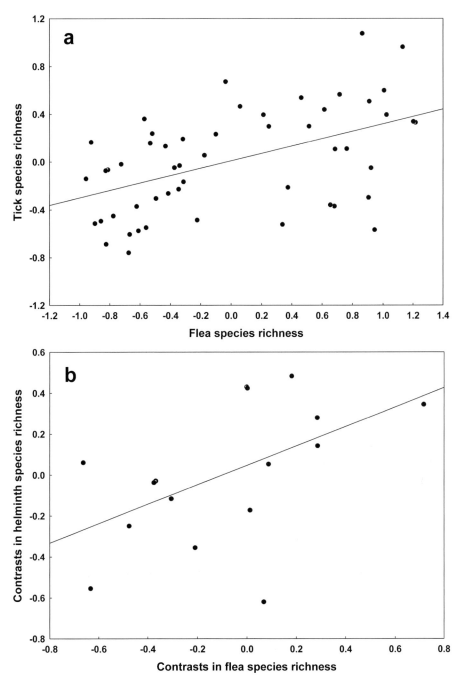

Figure 17.10 Relationships between (a) the number of tick species and the number of flea species on the same host species among 52 rodent species and (b) the number of helminth species and the number of flea species on the same host species among 17 rodent species (controlled for host sampling effort) using the method of independent contrasts. Redrawn after Krasnov et al. (2005l) (reprinted with permission from Cambridge University Press).

apparent facilitation (see Chapter 16) that, therefore, exists not only among related but also among unrelated parasite taxa in the organization of parasite communities. The existence of relationships between the species diversities of different parasite taxa (even those from different guilds) suggests that the host represents an important force shaping parasite communities. The positive relationships among species diversities of the assemblages of different ectoparasites as well as between fleas and endoparasites could arise from the already mentioned immunodepression in a host subjected to multiple immune challenges from a variety of parasite species exactly as is the case with multiple flea species (see Chapters 13 and 16).

When the diversity of unrelated taxa of free-living organisms covaries positively across localities, the general explanation usually invokes intrinsic differences in rates of colonization and extinction among localities (Gaston, 1996). It is thus possible that intrinsic properties of the various host species could lead to some hosts accumulating parasites of all taxa at a high rate. This can be because of some biochemical, physiological or ecological properties of the host. On the other hand, a host species that is able to resist attacks from many species of one parasite taxon appears to be able also to resist attacks from many species of other parasite taxa. This suggests that some level of cross-resistance may occur against distantly related parasites (Ribeiro, 1996; see also Chapter 13). Another explanation for the observed patterns is that host species can differ in their intrinsic ability to defend themselves against parasites using their immune system. Different, sometimes even closely related, rodent species have been shown to have different abilities to mount both humoral and cell-mediated immune responses (Klein & Nelson, 1998a, b; Goüy de Bellocq *et al.*, 2006b). As a result, a rodent with lower intrinsic immunocompetence can be exploited by a higher number of parasite species, including fleas, compared with a more immunocompetent species.

Nevertheless, negative associations between fleas and parasites belonging to other taxa have also been reported, although these studies considered mainly relations between abundances of fleas and parasites of other taxon such as mesostigmatid mites (Sosnina, 1967a, b; Vysotskaya, 1967; Ryba *et al.*, 1987; Saxena, 1987, 1999; Rodríguez *et al.*, 1999) and lice (Wegner, 1970; Bartkowska, 1973; Haitlinger, 1977). For example, Ryba *et al.* (1987) carried out a field experiment on the effect of mesostigmatids on fleas *Neopsylla setosa*, *Citellophilus simplex* and *Ctenophthalmus orientalis* in the nests of the ground squirrel *Spermophilus citellus* in the Czech Republic and found that the presence of a large number of mites in the nests decreased the number of fleas, but increased their feeding and reproductive performance.

17.8 Concluding remarks

The diversity of flea assemblages on their hosts is affected by a variety of factors. These factors include those related to the host body and those related to the off-host biotic and abiotic environment, as well as those related to flea community structure (see Chapter 16). However, an absolute majority of the studies have been carried out on small mammalian hosts in the Palaeoarctic and Nearctic. Consequently, two main directions for future studies can be envisaged. First, we still need further investigations of various hosts (birds and large mammals) in clearly understudied regions (e.g. Afrotropics, Neotropics, Oriental). Second, the application of modern ecological theories and models that were initially developed for free-living organisms to flea assemblages has proved to be a promising approach.

18

Fleas, hosts, habitats

Composition of parasite communities can vary across host individuals, populations, species and communities (Carney & Dick, 2000; Poulin & Valtonen, 2002; Calvete et al., 2004; see Chapter 16). The source of this variation is the diversity of the hosts' biotic and abiotic environment. For example, a richer community of co-habitating hosts increases the probability of lateral transfer of parasites, and thus affects the species richness and composition of the xenocommunity (Caro et al., 1997). The abiotic environment external to a host, such as air temperature or substrate, can also affect parasite species composition (Galaktionov, 1996). Therefore, some part of a parasite community encountered in a host is due to its specific location, another part due to host identity, and yet another part due to the host's environment (Kennedy & Bush, 1994). However, the relative importance of spatially variable factors in variation of community composition is poorly known for most parasite and host taxa. Moreover, most studies of spatial variation in parasite communities have been done on helminth communities (e.g. Bush & Holmes, 1986; Carney & Dick, 2000; Calvete et al., 2004). Therefore, it is not surprising that the hypothesis that host identity is a major determinant of parasite community structure has been supported (Bell & Burt, 1991; Buchman, 1991; Guégan et al., 1992). One of the reasons for this can be the relative stability of the internal environment of a host organism (Sukhdeo, 1997). However, the habitat of an ectoparasite should not be just a particular host, but a particular host in a particular habitat because of their sensitivity to the factors in the off-host environment.

In this chapter, I address the questions of (a) how variable is the composition of flea assemblages among different populations of the same host occurring in different habitats; and (b) whether the composition of flea assemblages in a habitat is affected either by species composition of hosts or by the environment of the habitat. I consider these questions based on two case studies by myself and my colleagues in two regions, namely the hyper-arid Middle East (the Negev Desert: Krasnov *et al.*, 1997, 1998) and the forests, meadows and brook valleys of central Europe (central and eastern Slovakia: Krasnov *et al.*, 2006k). I focus on the relative importance of 'environmental' and 'host' parameters of a habitat as factors affecting composition of flea assemblages.

18.1 The Middle East

18.1.1 Description of habitats, hosts and fleas

This study was carried out in the Ramon erosion cirque, an area of about 200 km^2 situated at the southern boundary of the Negev Highlands. The cirque is incised into the crest of a northeast–southwest-tending asymmetrical anticline and is a valley surrounded by steep walls of hard rocks (limestone and dolomite) at the top and friable rocks (sandstone) at the bottom. The altitude of the north rim of the cirque ranges between 900 and 1200 m above sea level, while the south rim is about 510 m above sea level. The level of the cirque bottom decreases gradually from the southwest (900 m above sea level) to the northeast and in the deepest part reaches 420 m above sea level.

The climate is characterized by hot, dry summers (mean daily air temperature of July is 34 °C) and relatively cool winters (mean daily temperature of January is 12.5 °C). There is a sharp decrease in annual rainfall from 100 mm on the north rim to 56 mm in the bottom of the cirque. Rainfall also decreases from the southwest of the cirque bottom to the northeast.

Based on 13 environmental variables including soil, vegetation and relief parameters, habitats were classified into six main types, as follows: (1) sand dunes with cover of *Calligonum comosum* or *Echiochilon fruticosum*; (2) dry river beds (*wadis*) of the eastern and central parts of the cirque with sandy-gravel soils and with cover of *Retama raetam*, *Moricandia nitens*, *Tamarix nilotica* and *Artemisia monosperma* among gravel plains (which will be referred to as east *wadis*); (3) flat gravel plains with sparse vegetation of *Hammada salicornica*, *Anabasis articulata* and *Gymnocarpos decandrum*; (4) limestone cliffs and rocks with sparse cover of *Zygophyllum dumosum*, *Helianthemum kahiricum* and *Reaumuria hirtella*; (5) a complex of narrow *wadis* and hills of the western part of the cirque covered with thin loess cover and vegetation of *Anabasis articulata*, *Atriplex halimus* and *Artemisia herba-alba* (which will be referred to as west *wadis*); and (6) a complex of loess hills and wide *wadis* with vegetation of *Anabasis articulata*, *Atriplex halimus*, *Artemisia herba-alba*, *Salsola schweinfurthii* and *Noaea mucronata*.

The region is inhabited by 11 common rodent species (*Psammomys obesus*, *Meriones crassus*, *Dipodillus dasyurus*, *Gerbillus henleyi*, *Gerbillus gerbillus*, *Sekeetamys calurus*, *Acomys cahirinus*, *Acomys russatus*, *Mus musculus*, *Jaculus jaculus* and *Eliomys melanurus*) parasitized by nine common flea species (*Xenopsylla conformis*, *Xenopsylla dipodilli*, *Xenopsylla ramesis*, *Coptopsylla africana*, *Nosopsyllus iranus*, *Stenoponia tripectinata*, *Rhadinopsylla masculana*, *Parapulex chephrenis* and *Myoxopsylla laverani*). However, the house mouse *M. musculus* and the lesser Egyptian jerboa *J. jaculus* have extremely low levels of flea infestation.

18.1.2 Habitat variation in flea species composition

To visualize habitat variation in flea species composition, Krasnov *et al.* (1997) ordinated flea assemblages from each individual host using principal-component analysis and then plotted these assemblages according to their habitat occurrences. This procedure allowed the construction of axes of ordination space that were linear combinations of abundances of each flea species, and thus presented the main directions of change in flea species composition among individual hosts. Importantly, these axes were obtained independently of the affinity of either host species or habitat type.

The resulted four ordination axes explained 86% of total variance with each of them reflecting a change in flea species composition (Fig. 18.1). The first axis represented the change from *X. dipodilli*, *N. iranus*, *S. tripectinata* and *R. masculana* on gerbils (negative zone of the first axis) to *P. chephrenis* on *Acomys*. This change is illustrated by segregation of Wagner's gerbil *G. dasyurus*, bushy-tailed jird *S. calurus* and partly pygmy gerbil *G. henleyi* from *Acomys* species along the first axis. The second axis responded mainly to fleas on Sundevall's jird *M. crassus* (*X. conformis* and *C. africana*). So, *M. crassus* was distanced from all other host species to the positive zone of the second axis (Fig. 18.1). The third and fourth axes were correlated with the relative abundances of *X. ramesis* and *M. laverani*, respectively, and thus they determined the positions of the fat sand rat *P. obesus* and the Asian garden dormouse *E. melanurus*, respectively.

When flea assemblages were plotted in the resulted ordination space according to their habitat occurrences, five among- and within-habitat directions of shift in flea species composition could be envisaged (Fig. 18.2). Among-habitat shifts were (a) from rocky habitats to east *wadis* and sand dunes along the first axis; and (b) from sand dunes and east *wadis* to west *wadis* and loess hills along the second axis. These shifts can be interpreted as responses of flea assemblages to a gradient of rodent burrow availability and a gradient of soil humidity and

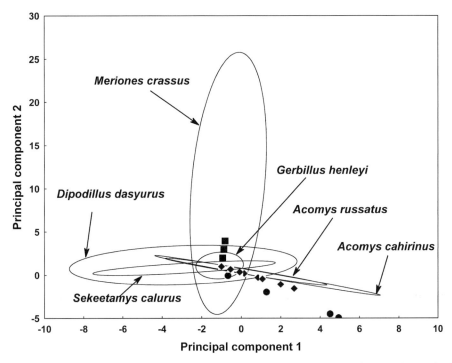

Figure 18.1 Plot of 75% confidence ellipses for flea assemblages from each examined animal in the Negev Desert in the space of two first principal-component axes according to the host species. Fat sand rat *Psammomys obesus*, squares; lesser Egyptian gerbil *Gerbillus gerbillus*, circles; Asian garden dormouse *Eliomys melanurus*, diamonds. Data from Krasnov *et al*. (1997).

temperature, respectively. Within-habitat shifts were (a) within east *wadis* along the first axis; (b) within rocky habitats along the first axis; and (c) within loess habitats along the third axis. Obviously, these shifts in flea species composition were related to among-host differences as follows: (a) between *G. dasyurus* and *M. crassus*; (b) between gerbillines (*D. dasyurus* and *S. calurus*) and *Acomys* species; and (c) between gerbillines (*G. dasyurus* and *M. crassus*) and *Acomys* or *E. melanurus*, respectively. The composition of a flea community is therefore determined by two directions of change in flea species composition, namely among hosts within a habitat and among habitats within a host.

The main conclusion from this study is that host–habitat relations affect species composition of flea assemblages and, therefore, species composition of flea infra- and xenocommunities is determined not only by host–flea relations, but also by host–habitat relations. In turn, flea species composition in a habitat is determined not only by host species composition there (see Chapter 17), but also by the environmental parameters of that habitat. These parameters

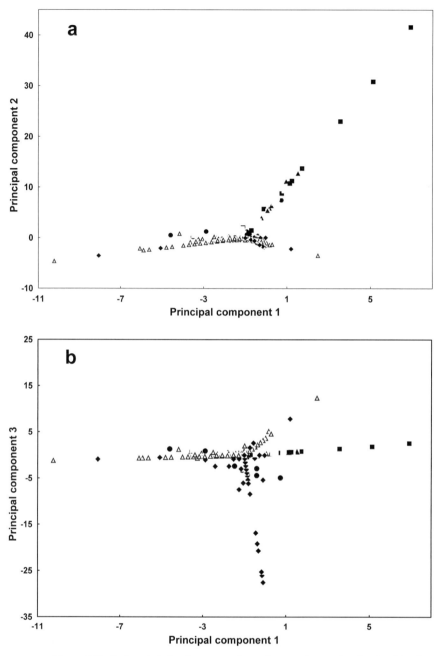

Figure 18.2 Scatterplots for flea assemblages from each examined animal in the Negev Desert in the space of (a) the first and the second and (b) the first and the third principal-component axes according to habitat type. Habitat types are sand dunes (closed squares), east *wadis* (open squares), gravel plains (closed triangles), rocks (open triangles), west *wadis* (circles) and loess hills (diamonds). Recalculated and redrawn after Krasnov et al. (1997) (reprinted with permission from Cambridge University Press).

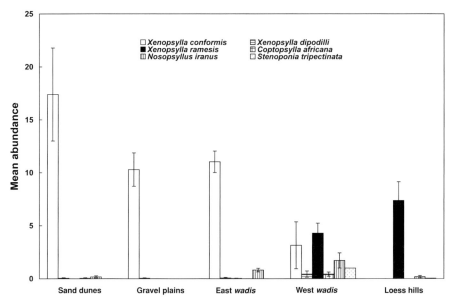

Figure 18.3 Mean (±S.E.) infestation of Sundevall's jird *Meriones crassus* by different flea species in different habitats of the Negev Desert. Data from Krasnov et al. (1997, 1998).

determine the conditions of the burrow or the nest of the host (temperature, humidity, substrate, nest material) and thus affect the flea assemblage.

Among fleas of the region, host-dependent and host-habitat-dependent species can be clearly distinguished. Furthermore, the distribution of even host-dependent fleas suggests that the relatively strong host specificity is also restricted environmentally. For example, *P. chephrenis* is obviously host dependent and associated with the Egyptian spiny mouse *A. cahirinus*. Nevertheless, this flea does not occur on *A. cahirinus* throughout its entire distribution even within Israel; it occurs only in the southern region and in the Jordan Valley north to Lake Tiberias, where climatic conditions are similar to those of the Negev Desert (Theodor & Costa, 1967). *Acomys cahirinus* does not build deep burrows but lives in rock crevices and between stones. This results in the strong dependence of a flea on an individual host, on the one hand, and its relatively high susceptibility to climatic conditions, on the other hand.

Xenopsylla conformis and *X. ramesis* on *M. crassus* exemplify host-habitat-dependent fleas. The former parasitized *M. crassus* in sand dunes, east *wadis* and gravel plains, whereas it was replaced by the latter in loess hills. Both species occurred in habitats of intermediate environment (west *wadis*), although their abundances were relatively low (Fig. 18.3). It seems that the distribution of these fleas conforms to the distinct-preference model of community organization

(Pimm & Rosenzweig, 1981). The pattern of between habitat distribution of these two fleas on *D. dasyurus* was the same, except that, in contrast to *M. crassus*, *X. conformis* and *X. ramesis* were subdominants whereas *X. dipodilli* was dominant.

In addition, there were also fleas in which host dependence or host-habitat dependence was weakly expressed. For example, *N. iranus* occurred on seven host species in all habitat types.

18.1.3 Mechanisms of distribution of the two Xenopsylla species

The most intriguing pattern of habitat dependence in species composition of flea assemblages in the Negev Desert was the clear replacement of *X. conformis* with *X. ramesis* on *M. crassus* and *D. dasyurus* between two groups of habitat types (see above). These habitats are situated at opposite sides of a steep precipitation gradient, with the maximum distance between less than 40 km (Krasnov et al., 1997, 1998, 1999). I will further refer to these habitat groups as 'xeric' (relatively high air temperature, relatively low relative humidity (RH), sandy or sandy-gravel substrate; predominance of *X. conformis*) and 'mesic' (relatively low air temperature, relatively high RH, loess substrate; predominance of *X. ramesis*) habitats. These differences determined the respective differences in microclimate, substrate texture and organic content of the substrate of burrows of *M. crassus* (Shenbrot et al., 2002). The architecture of these burrows also differed between xeric and mesic habitats. As mentioned above, burrows in the mesic habitats were deeper, more complicated but more ventilated than those in the xeric habitats (see Chapter 9). These differences could be a mechanism behind the paratopic pattern of distribution of *X. conformis* and *X. ramesis* via differential environmental preferences of the two species. To test this, a series of studies was carried out (Krasnov et al., 2001a, b, 2002b, d).

Examination of survival and rate of development of pre-imaginal *X. conformis* and *X. ramesis* at two air temperatures (25 and 28 °C) and four levels of RH (40%, 55%, 75% and 92%) (Krasnov et al., 2001a, b) showed that survival of *X. conformis* eggs did not depend on either air temperature or RH, whereas significantly fewer eggs of *X. ramesis* survived at 40% RH than at higher levels of RH. Fluctuations of RH during early stages of the life cycle (from egg to larva) increased the maximal survival time in *X. conformis* larvae but decreased it in *X. ramesis* larvae. In general, pupal survival of both species was higher at higher RH, independent of temperature, but survival of *X. conformis* pupae was lower than that of *X. ramesis* pupae when RH was low. In addition, *X. conformis* pupae developed longer than those of *X. ramesis* at relatively low air temperatures. Fluctuations of RH at later stages of pre-imaginal development (from larva to cocoon) decreased pupal survival and the percentage of emerged females in *X. conformis* and had no effect on *X. ramesis* pupal survival. Survival of *X. conformis* larvae was significantly higher in sand than in loess, whereas survival of *X. ramesis* larvae did not depend on substrate. Moreover, maximal survival time of *X. conformis* larvae that died before pupation did not depend on substrate, whereas *X. ramesis* larvae survived significantly longer in loess than in sand.

Responses to microclimatic conditions differed not only between pre-imagoes of the two species but also between imagoes (Krasnov et al., 2002d). Survival time under starvation of *X. conformis* depended on air temperature but not on RH, whereas in *X. ramesis* it was affected by both air temperature and RH. Furthermore, *X. conformis* generally survived for less time than *X. ramesis*. The only regime at which *X. conformis* survived longer than *X. ramesis* was that with relatively high air temperature and relatively low RH.

Summarizing the results of these experiments, it can be concluded that between-habitat distribution of the two flea species can be explained, in part, by an interaction of three factors (air temperature, RH and substrate texture). The bottlenecks for the occurrence of *X. ramesis* in xeric habitats can be (a) sensitivity of eggs, larvae and newly emerged imagoes to low RH, and (b) faster mortality of larvae in sand substrate; whereas bottlenecks for the occurrence of *X. conformis* in mesic habitats can be (a) sensitivity of pupae to low air temperature, and (b) lower survival of larvae in loess substrate.

However, the paratopic pattern of distribution in the two *Xenopsylla* species may also be caused by other mechanisms such as interspecific larval competition. As we have already seen in Chapter 16, larvae of the two species competed for food and, when food resources were limited, *X. ramesis* was more successful than *X. conformis* (Krasnov et al., 2005e). Importantly, the competition intensity was, to some degree, mediated by microclimatic conditions. This was manifested in a sharper decrease in the developmental success of *X. conformis* in mixed-species treatments under lower air temperatures and RH (Fig. 18.4). Thus, both factors, namely differential environmental preferences and interspecific competition, may lead to among-habitat, within-host variation in flea species composition.

18.2 Central Europe

18.2.1 Description of habitats, hosts and fleas

The study area in Slovakia is characterized by a typical temperate climate with a mean air temperature of 29.3 °C in July and −3.8 °C in January. Two habitat groups (lowland and mountain) with nine habitat types, based on

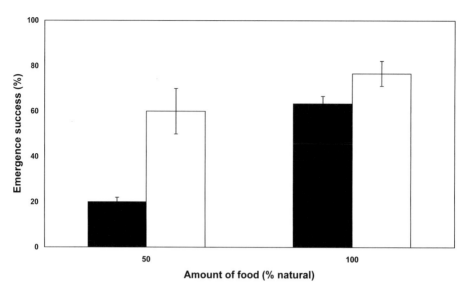

Figure 18.4 Mean (±S.E.) proportion of larval *Xenopsylla conformis* that survived until emergence in presence of larval *Xenopsylla ramesis* at 25 °C, 75% relative humidity (black columns) and 28 °C, 95% relative humidity (white columns) in the sand substrate and 50% or 100% food availability. Data from Krasnov et al. (2005e).

physiognomy, were distinguished. Lowland habitats were those situated at elevations between 100 and 200 m above sea level and included (1) lowland river valleys with willow–poplar and ash–alder floodplain forests dominated by *Salix alba*, *Salix fragilis* and *Populus alba*; (2) woodland belts represented by three to eight rows of *Populus canadensis* and various shrubs (*Prunus* sp., *Rosa* sp., *Sambuccus nigra*) with herbal floor composed mainly of *Urtica dioica*; (3) agricultural fields of wheat, maize and stubble; (4) floodplain lowland forests dominated by *Fraxinus angustifolia*, *Quercus robur*, *Carpinus betulus*, *Salix alba* and *Salix fragilis*; and (5) shrubbery dominated by *Prunus spinosus*, *Rosa canina* and *Crataegus* sp. with sporadic occurrence of poplar and willow trees. Mountain habitats were situated at elevations from 300 to 1100 m above sea level and included (1) narrow submontane and montane brook valleys (later referred to as mountain river valleys) with the main vegetation represented by *Alnus glutinosa*, *Alnus incana*, *Fagus sylvaticus* and *Carpinus betulus*; (2) submontane and montane forests dominated by *Fagus sylvaticus*, *Carpinus betulus*, *Quercus robur* and *Acer platanoides*; and (3) shrubbery patches on pastures dominated by *Prunus spinosus*, *Coryllus avellana* and *Rosa canina*. Finally, urban habitats were represented by gardens and orchards in public green spaces within cities (at elevation 650–750 m above sea level).

In total, a survey in this area resulted in 24 small mammal species (rodents and soricomorphs) and 30 flea species (see complete species lists in Stanko (1987,

1988, 1994) and Stanko et al. (2002)). Of these, 14 common host species and 25 common flea species were used in the analyses to answer the question of whether the composition of flea assemblages in a habitat is affected by species composition of hosts or environment of the habitat.

18.2.2 Habitat variation in flea species composition

The first step in the test for matching the host species composition and flea species composition of a set of habitats was evaluation of between-habitat similarity in either flea or host species composition based on abundances of either fleas or hosts, respectively. This resulted in two between-habitat similarity matrices. Among-habitat similarity in host species composition averaged 52.6%, being minimal between fields and mountain shrubbery and maximal between lowland river valleys and belts. Similarity in flea species composition was slightly higher (average 61.5%), being maximal also between lowland river valleys and belts, but minimal between lowland and mountain shrubbery. Ordinations of habitats based on similarity in either flea or host species composition distinguished groups of habitats that were more similar to each other than to other habitats in both flea and host species composition (Fig. 18.5). These were three mountain habitats (river valleys, forests and shrubbery) and three lowland habitats (river valleys, forests and woodland belts). Field and urban habitats differed sharply from other habitat types in their flea and host composition. However, lowland shrubbery was similar in its host species composition to other lowland habitats, but flea species composition in this habitat differed substantially from any other habitat.

Then, the questions that should be answered were (a) whether and (b) how closely the two sets (hosts' similarity and fleas' similarity) were related. This was done using the ρ statistics that correlates the elements of two similarity matrices and thus indicates significant disagreement or agreement between these matrices ranging from a minimum of 0 (no relation between two matrices) to a maximum of 1 (perfect match between two matrices). The resulting value of ρ equalled 0.36 demonstrating that among-habitat similarity in flea species composition was reflected by among-habitats similarity in flea species composition. Furthermore, between-habitat flea species composition similarity was positively correlated with between-habitat host composition similarity (Fig. 18.6).

When flea assemblages from each host individual were ordinated using principal-component analysis as was done for the Negev data (see above), the first five principal components explained 64% of the total variance in flea species composition. When flea assemblages were plotted according to their hosts, the segregation in flea species composition between rodents and shrews can be

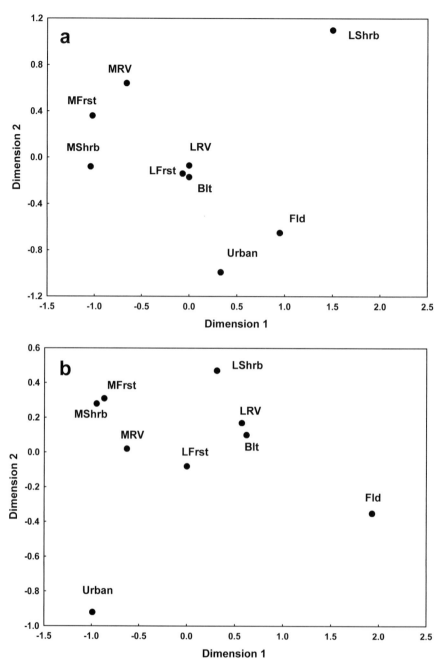

Figure 18.5 Multidimensional scaling distribution of habitats based on Bray–Curtis similarity (Bray & Curtis, 1957) in (a) flea and (b) host composition. Abbreviations for habitat names are: LRV, lowland river valleys; Blt, lowland woodland belts; Fld, lowland agricultural fields; LFrst, lowland forests; LShrb, lowland shrubbery; MRV, mountain river valleys; MFrst, mountain forests; MShrb, mountain shrubbery; Urban, urban habitats. Redrawn after Krasnov et al. (2006k) (reprinted with permission from Springer Science and Business Media).

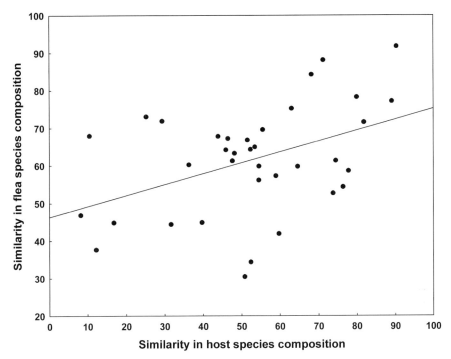

Figure 18.6 Relationship between Bray–Curtis similarity in flea species composition and similarity in host species composition across pairs of habitats in Slovakia. Redrawn after Krasnov *et al.* (2006k) (reprinted with permission from Springer Science and Business Media).

discerned easily (Fig. 18.7a). However, when flea assemblages were plotted according to their habitat affinities, the distinction of habitats based on variation in flea composition was not as clear (Fig. 18.7b). Moreover, the variation in each principal component (analysed using analysis of variance (ANOVAs)) was explained better by the factor of host species than by the factor of habitat type.

Nevertheless, habitat type, in general, affected flea species composition within a host species, although this effect was manifested differently in different hosts (Fig. 18.8). For example, flea assemblages on the fieldmouse *Apodemus agrarius* and pygmy woodmouse *Apodemus uralensis* were similar across most habitats except for lowland shrubbery and mountain river valleys, respectively. Flea assemblages on *A. agrarius* in lowland shrubbery differed from all other habitats due to the high abundance of *Ctenophthalmus solutus*, whereas *Megabothris turbidus* was mainly responsible for the difference in flea composition on *A. uralensis* between mountain river valleys and the rest of habitats. Flea assemblages on the common vole *Microtus arvalis* and common shrew *Sorex araneus* within lowland habitats and within mountain habitats tended to cluster along the first dimension of

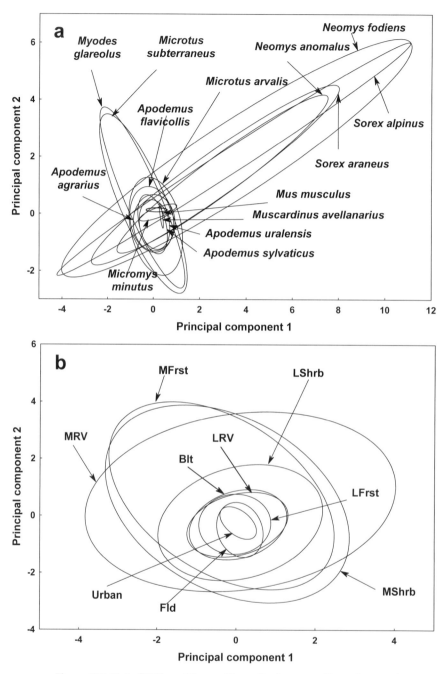

Figure 18.7 Plot of 95% confidence ellipses for flea assemblages from each individual small mammal from Slovakia in the space of two first principal-component axes according to (a) the host species and (b) habitat type. See Fig. 18.5 for the abbreviations of habitat names. Redrawn after Krasnov *et al.* (2006k) (reprinted with permission from Springer Science and Business Media).

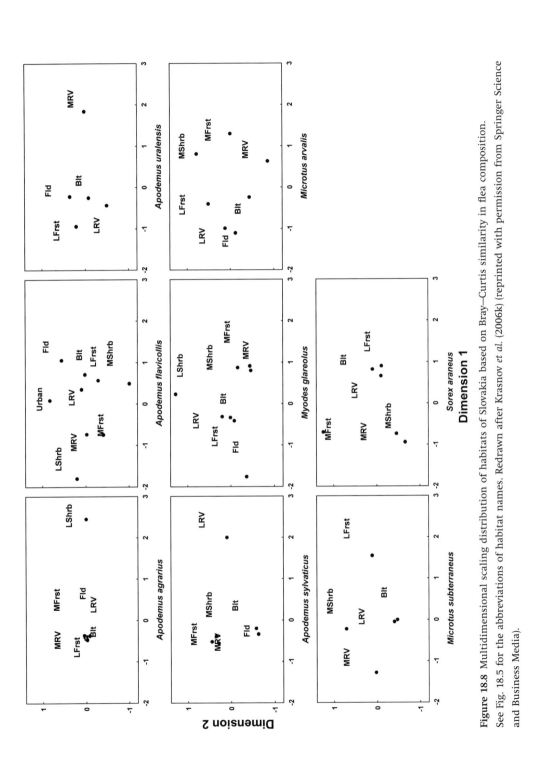

Figure 18.8 Multidimensional scaling distribution of habitats of Slovakia based on Bray–Curtis similarity in flea composition. See Fig. 18.5 for the abbreviations of habitat names. Redrawn after Krasnov et al. (2006k) (reprinted with permission from Springer Science and Business Media).

the ordination space, but differed sharply between these two habitat groups. The reason for these differences in *M. arvalis* was the relatively high abundance of *Ctenophthalmus assimilis* in lowland habitats, whereas lowland and mountain flea assemblages on *S. araneus* differed due to relatively high abundance of *Palaeopsylla soricis* and relatively low abundance of *Palaeopsylla similis* in lowland compared to mountain habitats. Apparent clusters of habitats (= similar flea assemblages) were also evident for the woodmouse *Apodemus sylvaticus*, bank vole *Myodes glareolus* and European pine vole *Microtus subterraneus*. In most cases, these clusters were represented by habitats belonging to either lowland or mountain areas. *Nosopsyllus fasciatus* was the main reason for among-habitat differences in *A. sylvaticus*, whereas flea assemblages of both voles differed among habitats mainly due to *M. turbidus* (28–50% of contribution), *A. penicilliger* (24–42% of contribution) and *C. assimilis* (25–43% of contribution).

These results suggested that species composition of flea infra- and xenocommunities in a given habitat of Slovakia was determined mainly by host identity and, to a lesser extent, by habitat identity. To confirm this, repeatability analysis (Arneberg *et al.*, 1997) based either on host or on habitat identity was carried out. It appeared that flea species composition varied less (a) among populations of the same host species than among host species and (b) among habitats of the same type than among different habitat types (Krasnov *et al.*, 2006k). However, the proportion of the total variance in flea species composition originating from differences among host species, as opposed to within species, was 18.8%, whereas the proportion of the total variance originating from differences among habitat types, as opposed to within habitat type, was only 8.7%.

Nevertheless, the between-habitat similarity in host species composition explained only about 20% of variance of the between-habitat similarity in flea species composition. This means that some habitat pairs were characterized by similar flea assemblages but different host composition, and other habitat pairs by similar host communities harbouring different flea assemblages. For example, the difference in dominating host species between mountain shrubbery (*A. flavicollis* and *M. glareolus*) and agricultural fields (*A. agrarius*, *A. uralensis* and *M. arvalis*) led to a coefficient of similarity as low as 8%. However, flea assemblages on these hosts in both habitats were dominated by *Ctenophthalmus agyrtes*, *C. assimilis*, *Ctenophthalmus solutus* and *M. turbidus*, leading to a coefficient of similarity as high as 47%. On the other hand, mountain and lowland shrubbery were occupied by host communities with 51% similarity, although the presence of *A. penicilliger*, *Amphipsylla rossica*, *Ctenophthalmus bisoctodentatus*, *Doratopsylla dasycnema*, *Hystrichopsylla talpae*, *Monopsyllus sciurorum*, *Peromyscopsylla bidentata*, *P. similis* and two *Rhadinopsylla* species in mountains and their absence in lowland resulted in only 30% similarity in flea assemblages. This difference can be due to the difference in elevation and, consequently, in air temperature between mountain and lowland area.

In other words, although the effect of habitat on flea species composition was lower than that of host species composition, it cannot be ruled out. This effect can be associated with both biotic and abiotic components of a habitat. Among-habitat variation in the biotic component can be related to the number of co-habiting hosts that facilitate between-host flea transfer (see Chapters 16–17). High probability of this transfer might, therefore, mask the effect of host identity on the within-host, among-habitat composition of flea assemblages. Another biotic component of a habitat is represented by the fleas. Different combinations of flea species composition might be a result of competition between some flea species, at least, in their larval stages (Chapter 16).

Comparison of among-habitat variation in composition of flea assemblages within host species (Fig. 18.8) supports also, albeit indirectly, the important role of abiotic components of a habitat in determining flea community structure. Apparent clusters of flea assemblages corresponding to lowland and mountain habitats can be distinguished in six out of eight host species. Environmental variation between these two areas is probably more pronounced than that among habitats within each of these areas.

18.3 Other examples

To the best of my knowledge, the two case studies described above are the only ones that have specifically analysed habitat variation in the species composition of flea infra- and xenocommunities. However, earlier sources have offered numerous narrative examples that host habitat is an important determinant of flea species composition. Below, I present the results of some studies from various geographical regions which, although not specifically aimed at investigating the link between flea infra- and xenocommunity structure and habitat characteristics of a host, have shown a strong habitat effect on species composition of flea assemblages.

In Kazakhstan, Mikulin (1956) reported that some flea species (e.g. *Xenopsylla gerbilli* and *Xenopsylla skrjabini*) on the great gerbil *Rhombomys opimus* were abundant in all habitat types, whereas others were either absent from some habitats (e.g. *Ctenophthalmus dolichus*) or strictly habitat-specific (e.g. *Stenoponia vlasovi*). Further studies revealed that a dissimilar spatial distribution of some flea species was associated with differential microclimatic preferences (Burdelov *et al.*, 1999). On the island of Honshu, Sakaguti & Jameson (1962) observed that *Stenoponia tokudai* occurred on mice *Apodemus speciosus* and *A. sylvaticus* at elevations of about 2000 m above sea level, but not in the lowlands. Similarly, *Dinopsyllus apistus* in Africa (Kenya, Uganda and the Democratic Republic of Congo) was found on the harsh-furred rat *Lophuromys* only at elevations higher than 1700 m above sea level (Haeselbarth *et al.*, 1966). In Poland, *Amalaraeus arvicolae* infested *M. subterraneus* in the alpine but not lowland habitats of the Sudetes (Haitlinger, 1970, 1975).

Mashtakov (1969) compared flea assemblages on jirds *Meriones tamariscinus* and *Meriones meridianus* from sand massifs of the right and left banks of the Ural River. He found that, in general, rodents from the right bank were characterized by richer flea assemblages. However, this was not caused by the absence of some fleas from the left bank habitat. Instead, the proportion of flea species shared by conspecifics from both habitats attained only 0.47 for *M. tamariscinus* and only 0.31 for *M. meridianus* (Table 18.1).

In Kahawa (Kenya), the striped grass mouse *Lemniscomys striatus* in acacia–themeda grasslands harboured mainly *Leptopsylla*, whereas this flea was absent from individuals inhabiting maize fields, being replaced by *Orchopeas* (Oguge *et al.*, 1997). Note that there could be misidentification as *Orchopeas* has North American distribution. In South Australia, *Spilopsyllus cuniculi* did not occur in rabbit burrows where the humidity of burrow air is unsuitable for the survival of larvae or imago fleas (Cooke, 1990a). Among populations of the grass mouse *Akodon montensis* broadly distributed in southeastern Brazil, *Polygenis tripus* was found only on rodents that inhabited areas of Atlantic forest that were close to *cerrado*, a savanna-like habitat, and were absent from hosts in typical forest habitats (de Moraes *et al.*, 2003). In Canada, the short-tailed shrews *Blarina brevicauda* were parasitized by different flea species in different habitats, namely *Doratopsylla blarinae* and *Nearctopsylla genalis* in mixed woods, and *Ctenophthalmus pseudagyrtes* in old-field habitat (Berseth & Zubac, 1987). The eastern grey squirrels *Sciurus carolinensis* inhabiting woodlands in southeastern Georgia had a higher infestation prevalence and intensity of infestation of *Orchopeas howardi* than squirrels in parklands (Durden *et al.*, 2004). In Spain, wild rabbits *Oryctolagus cuniculus* harboured *Echidnophaga iberica* in relatively dry habitats but not in relatively wet habitats (Osacar-Jimenez *et al.*, 2001). In Slovakia, *Citellophilus martinoi* and *Citellophilus simplex* replaced each other on the European ground squirrel *Spermophilus citellus* occurring in different habitats (Cyprich *et al.*, 2003).

Jurík (1983a, b) studied flea assemblages on the European mole *Talpa europaea* in flooded and non-flooded habitats of eastern Slovakia. Moles were parasitized by 10 flea species in the former and by only five species in the latter. Moreover, between-habitat difference was manifested not only in species richness but also in the composition of flea xenocommunities. In non-flooded habitats, 90.3% of all fleas collected were represented by specific parasites of soricomorphs (*P. similis*, *Ctenophthalmus bisoctodentatus* and *Hystrichopsylla orientalis*). In contrast, moles in

Table 18.1 *Flea species composition on* Meriones tamariscinus *and* Meriones meridianus *inhabiting the right and left banks of the Ural River*

	Meriones tamariscinus		Meriones meridianus	
Flea	Left bank	Right bank	Left bank	Right bank
Amphipsylla prima	−	+	−	+
Amphipsylla rossica	+	+	+	+
Amphipsylla schelkovnikovi	−	+	−	−
Citellophilus tesquorum	+	+	+	+
Coptopsylla lamellifer	−	+	−	−
Ctenophthalmus breviatus	+	+	+	−
Ctenophthalmus pollex	−	+	−	+
Frontopsylla frontalis	−	+	−	+
Frontopsylla semura	+	+	−	+
Leptopsylla segnis	−	+	−	+
Leptopsylla taschenbergi	+	−	−	−
Mesopsylla hebes	+	+	+	−
Mesopsylla lenis	+	+	−	−
Mesopsylla tuschkan	+	−	+	−
Neopsylla setosa	+	+	−	+
Nosopsyllus laeviceps	+	+	+	+
Nosopsyllus mokrzeckyi	+	+	−	−
Ophthalmopsylla kasakiensis	−	+	−	−
Ophthalmopsylla volgensis	+	−	+	−
Oropsylla ilovaiskii	+	+	+	+
Pulex irritans	+	−	−	+
Rhadinopsylla cedestis	−	+	−	−
Xenopsylla conformis	+	+	+	+

Source: Data from Mashtakov (1969).

flooded habitats were parasitized mainly by fleas characteristic for mice and voles (*C. agyrtes*, *C. solutus* and *C. assimilis*). The explanation of this pattern was that the environmental conditions in flooded habitats were not favourable for the pre-imaginal development of flea parasites of soricomorphs, so they were substituted by other flea species.

In addition, some studies analysed flea assemblages according to physiognomy or environmental parameters of trapping sites within a geographical region. In some cases, these studies described distinct flea assemblage characteristics for a given landscape unit within a large area such as, for example, the Pri-Irtyshie Region of the West Siberian Plain (Sapegina, 1976), Mongolia (Kiefer et al., 1982), China (Gong et al., 1996, 1999, 2000) and Kazakhstan (Serzhan, 2002).

For instance, Sapegina (1976) argued that flea species parasitic on small mammals in the Pri-Irtyshie Region can be classified into four categories according to their habitat distribution as (a) species of forest landscape; (b) species of forest–meadow landscape; (c) species of bogs and marshes; and (d) species of human settlements. However, parallel analysis of host similarity among these landscape units was not done. Consequently, the relative roles of the environment-related and host-related components in the composition of flea assemblages remained unknown.

18.4 Concluding remarks

The case studies described in this chapter support the idea that a habitat for a flea is not a particular host or a group of hosts but rather a particular host or a group of hosts in a particular habitat. Nevertheless, some geographical variation may be envisaged from the examples presented. Among-habitat differences in flea assemblages within a host species in Slovakia were less pronounced than those in the Negev Desert. Flea assemblages of the same host in different habitats in Slovakia differed mainly in the relative abundances of fleas, whereas those in the Negev Desert differed mainly in their species assortment. This difference between temperate and arid environments remains to be explained more fully. It may be related to the sheltering pattern of host species and, consequently, to the difference between the two geographical regions in the degree of expression of among-habitat variation in environmental conditions of burrows and nests. Another explanation may be associated with the frequency of intra- and interspecific visits of each other's burrows in small mammals, which presumably is higher in temperate regions due to the higher density of small mammals than in deserts. This could lead to more scattered occurrences of flea species across populations of a host species occupying different habitats.

19

What further efforts are needed?

It is not accidental that the title of this book emphasizes functional and evolutionary aspects of flea ecology. It is also not accidental that Part I which focuses on the descriptive ecology of fleas is much shorter than either Part II or Part III. This is because we know much about the taxonomy, geographical distribution and biology of fleas. The level of our knowledge about fleas can be easily envisaged by looking at bibliographic databases. For example, Robert E. Lewis of Indiana State University compiled lists of papers and books that mentioned fleas in one way or another (although it not always is obvious from the titles) and presented them in the biannual newsletter *Flea News* at the website http://www.ent.iastate.edu/fleanews/ (there were 60 issues in total, the last one in July 2000). The list of publications on fleas from 1993 to 1997 contains 974 citations. In other words, the average number of publications on fleas per year from 1993 to 1997 was about 184. This is not much lower than the 230 publications per year in the period from 1969 to 1978 listed by F. G. A. M. Smit (Marshall, 1981a). The number of publications on fleas to the end of the 1970s was estimated at 11 000 (Marshall 1981a). Consequently, if we assume an average of 200 publications per year, there should be 16 000 publications at present! Moreover, this number is obviously an underestimation as it does not takes into account numerous papers in Russian and Chinese published in various collective monographs and proceedings of research institutes, plague-control stations and nature reserves. In this concluding chapter, I try to evaluate where we are now in relation to flea ecology and what we lack, as well as to present examples of how modern theoretical approaches to flea ecology may be applied for practical aims.

456   What further efforts are needed?

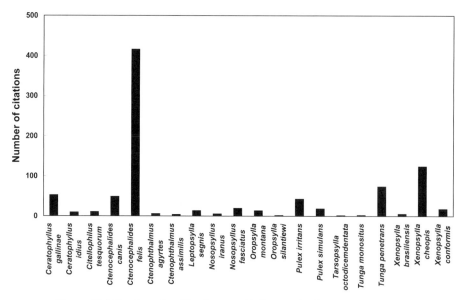

Figure 19.1 Number of publications on 20 flea species from 1965 to 2007.

19.1   Where are we now and what do we have?

There are several strong biases in our knowledge about fleas. The sharpest of them is taxonomic or, more precisely, 'species' bias. Physiology, behaviour and ecology are well known for only a few species, mainly those having medical and/or veterinary importance. Indeed, the number of publications dedicated to a particular flea species varies strongly among species, with the cat flea *Ctenocephalides felis* being the most studied (Fig. 19.1). During the last 40 years more than 400 papers and one monograph have been published about this species. Interestingly, the closely related dog flea *Ctenocephalides canis* received much less attention although it is of not much less medical and veterinary importance than the cat flea. Analogously, *Xenopsylla cheopis* is much more studied than any of its congenerics. This is obviously related to its importance as a vector of the plague.

Another bias in our knowledge about flea ecology is related to geography, although this is true mainly for fleas parasitic on wild hosts. Fleas from regions where they have high medical importance (e.g. in natural foci of plague) have been studied much more intensively than fleas from other regions. The role of historical and economical factors in research efforts also cannot be ruled out. As a result, the best-studied region in relation to flea ecology is Central Asia. Data on the ecology of species inhabiting North America, Europe, Caucasus, China and Siberia are also numerous. In contrast, the ecology of fleas from

Table 19.1 *Publications on fleas in some ecological and parasitological journals from 1961 to 2007*

	Journals							
	English language						Russian language	
Years	JAE	E	O	JME	PS	JP	PZ	ZZ
1961–1970	2	2	0	16	8	8	–	–
1971–1980	1	0	0	68	0	14	67	6
1981–1990	0	1	0	40	2	3	46	4
1991–2000	3	3	2	98	3	14	22	10
2001–2007	11	5	2	31	9	14	20	2

Note: JAE – *Journal of Animal Ecology*, E – *Ecology*, O – *Oikos*, JME – *Journal of Medical Entomology*, PS – *Parasitology*, JP – *Journal of Parasitology*, PZ – *Parazitologiya*, ZZ – *Zoologicheskyi Zhurnal*.

South America, Africa and Australia has been much less studied, although several brilliant monographs describing flea fauna from these regions have been published (e.g. Haeselbarth *et al.* 1966; Wenzel & Tipton, 1966; Dunnet & Mardon, 1974; Méndez, 1977; Segerman, 1995; Linardi & Guimarães, 2000). Furthermore, except for Argentina and Brazil, quantitative data on fleas from these regions are scarce. Finally, the ecology of fleas from Wallacea and some of the South Asian regions is practically unknown.

Most papers on fleas have been authored either by entomologists who were interested mainly in flea taxonomy and faunistics or by medical and/or veterinary parasitologists who were interested in the role of fleas as causes of medical and veterinary disorders and/or vectors of infectious diseases. It is not surprising, therefore, that the outlets for most of these publications were parasitological rather than ecological journals (Table 19.1).

However, it can also be seen from Table 19.1 that publications on fleas in ecological journals increased in the 1990s and 2000s. This happened not only in the English-language journals, but also in the Russian-language periodicals, although the tendency in Russia was weaker. Concomitantly, interest in fleas from medical and/or veterinary studies did not decline. This hints at an increasing trend to use fleas as a model taxon for studies based on modern ecological and evolutionary theories. Thus, on the one hand, 'mainstream' ecologists who ignored fleas in earlier times have started to recognize the advantages and opportunities presented by these animals, whereas, on the other hand, 'mainstream' parasitologists have begun to adopt the ideas, methods and approaches developed in studies of free-living organisms.

The huge amount of data collected on fleas by many scientists in many regions still remains underused. Often, some earlier publications presented mainly information on flea species collected on host species in a particular location. In more detailed cases, the number of individuals of a particular flea species found on the particular host species was specified. These data are extremely important for comparative analyses as well as for macroecological and biogeographic studies. For example, many studies by the author of this book and his colleagues (e.g. Krasnov et al., 2004c, d, e, f, 2005a, b, f, g, 2006b, c, d, 2007c) were based solely on such information. The database used for these studies comprises data on fleas collected from almost 1 500 000 individual small mammals in 68 geographic locations. It is difficult to imagine that somebody nowadays could carry out sampling of that scale.

19.2 What do we lack?

In the concluding parts of most chapters of this book, I have directed the reader's attention to limitations of our knowledge on functional and evolutionary aspects of flea ecology. In my opinion, the main limitation is the general lack of integrity of the approaches.

19.2.1 Multifaceted vision

First, the lack of integrity in flea studies is related to a general shortage in the angles of vision. Many studies have considered flea ecology from a particular aspect and have not given insights into causes and consequences of the patterns that were found. For example, many studies were made on the effect of host species and/or environmental conditions on flea feeding. But these studies rarely tested how the feeding responses of fleas are translated into reproduction which, in turn, may affect the abundance and spatial distribution of fleas on their hosts. Consequences of the latter may be changes in the host's defence efforts which may affect its reproductive success and abundance. These may affect flea abundance, causing changes in flea population and community structure.

Furthermore, there is a huge discrepancy in the knowledge of different facets of flea ecology, both functional and evolutionary. For example, the effects of environmental factors on parameters of flea feeding and reproduction are well studied, although physiological mechanisms of these effects are largely unknown. The effect of host species on these parameters is to some extent known, whereas information about the effect of host gender, age and body condition is scarce and fragmentary.

The evolutionary ecology approach made its first steps in flea studies in the late 1990s, i.e. later than in studies with other parasitic taxa (e.g. Rohde, 1979; Poulin, 1995a; Morand, 1996). It is not surprising, therefore, that answers for many evolutionary ecology questions such as the relationship between size and fecundity, evolutionary reasons for commonness and rarity, and the role of interspecific competition in community organization are either substantially incomplete or do not exist. This book has attempted to fill some of the gaps, but much further effort is needed. I hope that flea ecology will switch from a purely phenomenological to a functional and evolutionary approach, as has happened with studies on some other parasite taxa.

19.2.2 Flea–host associations as integral entities

Another component of integrity usually lacking from flea studies is related to the association between a flea and a host. In many cases, flea studies do not consider interactions with their hosts. Hosts are often seen as a mere resource for fleas and their responses are rarely considered concomitantly with those of fleas. In many other cases, the responses of a host alone are measured and fleas are seen merely as one of the factors that negatively influences an animal of interest. However, as mentioned earlier, a host and a parasite are involved in an intimate and durable interaction (Combes, 2001). The intimacy of this interaction stems from the facts that (a) fitness of a parasite without a host is zero, and (b) each individual parasite meets only a limited set of conditions during its life; whereas durability is related to the fact that associations between a parasite and a host last for extended periods (Combes, 2001, 2005). This idea led Combes (2001) to consider the interaction between a parasite and a host as a 'superorganism' that possesses a 'supergenome'. Within this 'supergenome', the component belonging to a parasite is expressed via the phenotype of a host, whereas the component belonging to a host is expressed via the phenotype of a parasite. These ideas of Claude Combes form the theoretical basis of ecological and evolutionary parasitology and, in my opinion, should be borne in mind in any study of a parasite. Consequently, parasitological studies within ecological and evolutionary frameworks should be on parasite–host associations rather than on parasites alone. However, at present, studies that consider fleas and their host(s) simultaneously and as two partners of the same interaction are scarce. We have already seen that except for studies of the associations (a) of the hen flea and parid birds and (b) a few rodent fleas and gerbils and spiny mice, most flea–host studies have considered either flea responses or host responses but not the responses of both.

19.2.3 External validity

Any scientific study, including those on flea ecology, reveals some patterns and processes. However, a question that remains is how general are these patterns and processes and to what extent they apply to objects, settings or times other than those that were the subject of a study? In other words, the findings of a particular study should be validated in other geographical locations or on other taxa.

In several chapters of this book, I have described two series of similar studies carried out in the Negev Desert and Slovakia on fleas and their small mammalian hosts. Some patterns were quite similar between the two localities. For example, in both cases, the prevalence of flea infestation could be successfully predicted from mean abundance using a simple epidemiological model (Krasnov et al., 2005h, i). It can be concluded, thus, that the relationship between flea abundance and distribution is governed by the same or similar mechanisms, independent of location. Moreover, similar results for other parasite taxa and host taxa, such as nematodes (Morand & Guégan, 2000) and ixodid ticks (Stanko et al., 2007) parasitic on mammals and monogeneans parasitic on fish (Šimková et al., 2002), broaden our understanding and prove that this pattern reflects some profoundly deep demographic processes characteristic for a variety of animals and independent of the environment they occupy and their life-history style.

On the other hand, some patterns differed between two localities. The relationship between flea and host abundance in the Negev Desert was positive (Krasnov et al., 2002e), whereas that in Slovakia was negative (Stanko et al., 2006). At first glance, the main difference between the two localities that may lead to this difference is environment: the Negev is arid and Slovakia is temperate. The results of these two studies cast some doubts on the applicability of the commonly accepted epidemiological theories and models. To improve the models, environmental components should be added or the models should be replaced with new, more realistic ones. Further validations of the host-abundance–parasite-abundance pattern should be done. This can be achieved, for example, via field experiments. Additional ways to understand the relative roles of environment-, parasite- and host-related components in the relationships between host and parasite abundances is to study (a) the same parasite–host (e.g. flea–mammal) associations in different geographical locations; (b) populations of the same parasite taxa exploiting different host taxa (e.g. fleas parasitic on rodents versus bats); and (c) populations of the same host taxa being exploited by different parasites (e.g. rodents parasitized by fleas versus mesostigmatid mites).

The two previous paragraphs dealt with patterns that are formed during ecological time. The generality of patterns formed during evolutionary time should also be validated. For example Krasnov *et al.* (2004g) did not find a correlation between host body mass or basal metabolic rate and diversity of flea assemblages, whereas flea diversity increased with an increase in the size of the host geographical range. However, when Korallo *et al.* (2007) studied the relationships between the same parameters of roughly the same set of host species from about the same geographical regions and diversity of mesostigmatid mites, the results were opposite to those found for fleas. In particular, mite diversity correlated with host body parameters, but not with the host geographical range. The reasons for this difference undoubtedly lie in differences in life history and the type of association between these two arthropod taxa and their hosts. An important conclusion coming from the comparison of the results of these two studies is that relationships between host features and parasite diversity appeared to be specific for each parasite–host association. This example demonstrates the importance of external validation. There appears to be no universal determinant of parasite diversity (see also Poulin & Morand, 2004), and associations between host features and parasite diversity have probably evolved independently in different host–parasite systems.

Some patterns, however, are strikingly similar across immensely different host–parasite systems. For example, Mouillot *et al.* (2006) reported that similarity in the degree of host specificity between closely related species is characteristic for both fleas and trematodes. Vázquez *et al.* (2005) evaluated the distribution of specialization in species interaction networks and found that the tendencies of specialist parasites to be found in species-rich xenocommunities and of generalist parasites in species-poor xenocommunities holds not only for fleas parasitic on small mammals but also metazoans parasitic on freshwater fish (nematodes, acanthocephalans, cestodes, trematodes, monogeneans, leeches, copepods and branchiurans). Moreover, a similar pattern was reported for mutualistic (plant–pollinator) associations (Vázquez & Aizen, 2003). This evidence demonstrates that asymmetric specialization is a general feature of species interactions, independent of their phylogeny, foraging mode and environment. However, this generalization would be impossible without external validation of a pattern initially found in a particular biological system.

19.2.4 *Laboratory, field and comparative approaches*

Functional approach to flea ecology is realized mainly in laboratory experiments using laboratory cultures of fleas. Yet this has been done mainly on a few species, such as *C. felis*, *X. cheopis* and *Ceratophyllus gallinae*. Maintenance of laboratory colonies of fleas parasitic on wild hosts is rarer, although the number of species maintained in laboratories all over the world is growing (e.g. Rassokhina *et al.*, 1985; Cooke, 1990b; Ma, 1990, 1994a, b; Liu *et al*, 1993; Larsen, 1995; Hu *et al.*, 1998, 2001; Ratovonjato *et al.*, 2000; Vashchenok, 2001; Grazhdanov *et al.*, 2002; Krasnov *et al.*, 2002a; Feng *et al.*, 2003; Li *et al.*, 2004; Burdelov *et al.*, 2007). In some cases, however, the colonies were maintained on laboratory animals such as mice and rats rather than on the natural hosts. Data from laboratory studies should be interpreted cautiously because patterns of many flea responses depend heavily on the identity of a host. Teratogenic changes that can occur in laboratory colonies of fleas should also be taken into account (Hu *et al.*, 1996; Ma, 1997). Further fine-tuning and generalization of our theories may be achieved by increasing the range of both flea and host species maintained in laboratories.

Modern population and community ecology was greatly stimulated in the 1960s by the implication of the experimental field approach. In this approach, natural population and communities are manipulated allowing the testing of specific hypotheses and predictions. However, in its application to parasites in general and to fleas in particular, this approach is rather difficult and labour- and time-consuming. This is one of the reasons why field experiments on populations and communities of fleas and on their effect on naturally living hosts are rare. However, in cases when this approach has been applied, it has proved to be highly profitable. Several examples of field experiments have been described in this book, such as studies of *C. gallinae* and several tit species (e.g. Richner *et al.*, 1993; Oppliger *et al.*, 1994; Heeb *et al.*, 1996, 1998, 1999, 2000; Tripet & Richner, 1997a, b, 1999a, b; Fitze *et al.*, 2004a, b; Tschirren *et al.*, 2004, 2005, 2007a, b) and *Synosternus cleopatrae* and Anderson's gerbil *Gerbillus andersoni* (Hawlena *et al.*, 2005, 2006c).

Elucidation of evolutionary trends that cannot be established experimentally can be done using a comparative approach. In this method, a comparison of a trait is made across different species, taking into account the phylogenetic relationships among them (Felsenstein, 1985; Harvey & Pagel, 1991; Garland *et al.*, 1992, 1993). The rationale of using this approach in studies of parasite ecology and evolution has been presented recently by Morand & Poulin (2003), Poulin & Morand (2004) and Poulin (2007a). In particular, in its application to parasites (e.g. fleas), this approach allows answers to general questions about evolution in response to selection pressure from both host- and environment-related factors. However, the comparative approach with control for the confounding effect of phylogeny has not been applied often to flea data. After the pioneering attempt of Kirk (1991) to correlate the body length of fleas with that of their hosts using the phylogenetic regression of Grafen (1989), there was a gap of 10 years until this approach was used again in a flea study (Tripet et al., 2002a). However, two important caveats should be noted in relation to the comparative approach. First, comparative analysis can point out a correlation between traits, but cannot establish causal relationships between these traits (Poulin, 2007a). Consequently, a causal trend suggested by the results of a comparative analysis should be validated, when possible, via experimental studies. Second, there is an urgent need for highly resolved and robust phylogenies of both fleas and their hosts. Molecular studies provided those for birds and mammals, but flea phylogeny needs much improvement. Most comparative studies on fleas have used phylogenies created from scratch and often based on morphological taxonomy (e.g. Tripet et al., 2002a; Krasnov et al., 2006b). Of course, non-robust and weakly resolved phylogeny is better than no phylogeny at all. Nonetheless, there is hope that a molecular phylogeny of Siphonaptera will soon be available (Lu & Wu, 2001, 2002, 2005; Whiting, 2002a, b; Dittmar de la Cruz & Whiting, 2003; Whiting et al., 2003; Luchetti et al., 2005).

19.3 Not only a pure science . . .

A famous definition of science in the professional folklore states that it 'is satisfaction of personal curiosity on taxpayers' money'. Of course, this statement is only a joke, but I must admit that satisfaction of curiosity is an integral component of any scientific research. It may seem that functional and evolutionary ecology studies of fleas are done solely for our satisfaction of curiosity and enhancement of our knowledge and understanding of nature. However, this is not exactly so. Health, veterinary and conservation issues can benefit greatly from the application of these approaches to fleas and the diseases that they transmit. An example of application of the comparative method to understand the evolution of fleas as vectors of plague has been described in Chapter 14. There are also other illustrations showing the value of ecological parasitology studies for society (Poulin, 2007a).

For example, several attempts to understand the dynamics of flea-borne diseases via modelling disease–vector (flea)–reservoir systems have been made recently. Sometimes, these studies have focused on diseases non-pathogenic to humans but they were a convenient model for the investigation of the dynamics of a flea-borne infection in a natural host population. Smith et al. (2005) studied the relationship between the prevalence of a flea-borne protozoan (Trypanosoma microti) in the population of the field vole Microtus agrestis. The results suggested that although male hosts harboured higher flea burdens than female hosts, the disease spread was not driven mainly by male hosts as is the case in some helminth infections (e.g. the nematode *Heligmosomoides polygyrus* in the yellow-necked mouse *Apodemus flavicollis*: Ferrari *et al.*, 2004, 2007). Instead, *Trypanosoma* prevalence depended on the interplay among several factors including host age, season and past flea infestation.

In other cases, modelling was applied to systems relevant to human health. Keeling & Gilligan (2000a, b) used stochastic metapopulation models and showed that the ability of the plague to be transmitted by fleas persists when the population size of a rodent host is small, thus explaining persistence of the plague despite long disease-free periods and its reoccurrence in areas with tight quarantine control. Frigessi *et al.* (2005) demonstrated the usefulness of hierarchical Bayesian models in extracting ecological information from data on *Xenopsylla* fleas and the great gerbil *Rhombomys opimus*. Clear evidence was found of density-dependent over-summer net growth of the flea population, which depended on the flea-to-gerbil ratio at the beginning of the reproductive period. As a continuation of this study, Stenseth *et al.* (2006) generated a model to analyse the field data on the abundances of the great gerbils and *Xenopsylla* fleas, prevalence of the plague pathogen and climatic variables from a natural plague focus in Kazakhstan. They demonstrated that changes in spring temperature appeared to be the most important environmental factor determining plague prevalence. According to their model, warmer conditions in spring may lead to an elevated flea-to-gerbil ratio, which, in turn, may lead to a higher plague prevalence level in the gerbil population. Moreover, an increase in flea survival and reproduction has an important influence on this effect, suggesting that the climate effect on plague prevalence is mediated via flea activity. These results may, thus, explain the causes of the emergence of the Black Death and the Third Pandemic of plague, which could have been triggered by favourable climatic conditions in Central Asia and created a cascading effect via fleas and the great gerbils on the occurrence and prevalence of plague.

In contrast, the modelling approach for the plague epizootics among the prairie dogs *Cynomys ludovicianus* in North America (Webb *et al.* 2006) demonstrated that the transmission of plague via blocked fleas (see Chapter 14 for details on plague blockage) could not drive these epizootics. Instead, the model required a short-term reservoir for epizootic dynamics. Nevertheless, a potential role of fleas that may transmit the plague without blockage of the proventriculus was admitted. There can be various reasons for these results ruling out the common paradigm of the role of blocked fleas in the circulation of the plague pathogen. These reasons may include the complicated relationship between the number of flea attacks and the abundance of fleas during epizootics. The frequency of attacks may increase without a concomitant increase in flea abundance due to an increase in the frequency of feeding attempts by blocked fleas and despite their high mortality. In addition, the plague pathogen invaded North America only recently (about 100 years ago: e.g. Moll & O'Leary, 1945; but see Meyer, 1947), so the structure and functioning of the American plague foci may differ from those in the Old World. Anyway, although the role of fleas in plague epizootics and epidemics has not been doubted in the past, modern studies force us to look at this role from a new angle. Therefore there is a need for further studies combining ecology and epidemiology that will undoubtedly improve our ability to predict, prevent and control dangerous infectious diseases such as the plague (see also Collinge et al., 2005).

Developments in geographical ecology of fleas are also important for the prevention and control of diseases. For example, Adjemian et al. (2006) used the ecological niche modelling system Genetic Algorithm for Rule-Set Production (GARP) to predict the spatial respective distribution of 13 flea species that are potential plague vectors in California. It was found that GARP effectively modelled the distributions of these species. Moreover, all of these modelled ranges were robust, with a sample size of at least six fleas not significantly impacting the percentage of the in-state area where the flea was predicted to be found.

Another direction of studies of flea–host relationships that has a strong applicative aspect and should be pursued is related to the immunological aspects of flea parasitism. Khokhlova et al. (2004a, b) have demonstrated that rodent hosts mounted strong immune responses not only to repeated infestations of fleas (which simulates the natural situation), but also to the injection of whole-body extract of a flea. The latter introduces the host to many antigens not present in flea saliva, which indicates host sensitivity to both exposed and concealed antigens. Exposed antigens are those injected into the host during ectoparasite feeding and evolved under the pressure of host immunity (Tellam et al., 1992), whereas concealed antigens are those parasite antigens that are not inoculated into the host (Willadsen, 1987; Key & Kemp, 1994). Such antigens not 'seen' by the host immune system might be implemented successfully as an anti-parasite vaccine (Willadsen et al., 1995; Willadsen, 2001, 2006), following what has been done for other groups of ectoparasites, such as ticks (Galun, 1975; Mulenga et al., 2000; Andreotti et al., 2002).

Ecologists and parasitologists who are interested in revealing general patterns and processes related to consumer interactions (*sensu* Lafferty & Kuris, 2000, 2002) still have much to do. The search for these patterns and processes can be facilitated by using a convenient model system. I hope that I have succeeded in showing that fleas and their hosts offer such a system. My other hope is that this book will serve as a template for the integrated study of parasitism on all levels, from the physiology of individuals to the evolution of communities.

References

Abouheif, E. & Fairbairn, D. J. (1997). A comparative analysis of allometry for sexual size dimorphism: assessing Rensch's rule. *American Naturalist*, **149**, 540–562.

Abrahams, M. (2004). Mark but this flea. *Guardian*, 30 November 2004.

Abramsky, Z., Bowers, M. A. & Rosenzweig, M. L. (1986). Detecting interspecific competition in the field: testing the regression method. *Oikos*, **47**, 199–204.

Abramsky, Z., Rosenzweig, M. L., Pinshow, B., *et al.* (1990). Habitat selection: an experimental field-test with two gerbil species. *Ecology*, **71**, 2358–2369.

Abu-Madi, M. A., Behnke, J. M., Mikhail, M., Lewis, J. W. & Al-Kaabi, M. L. (2005). Parasite populations in the brown rat *Rattus norvegicus* from Doha, Qatar between years: the effect of host age, sex and density. *Journal of Helminthology*, **79**, 105–111.

Acosta, R. (2005). Relación huésped-parásito en pulgas (Insecta: Siphonaptera) y roedores (Mammalia: Rodentia) del estado de Querétaro, México. *Folia Entomologica Mexicana*, **44**, 37–47.

Acosta, R. & Morrone, J. J. (2005). A new species of *Hystrichopsylla* Taschenberg (Siphonaptera: Hystrichopsyllidae) from the Mexican transition zone. *Zootaxa*, **1027**, 213–238.

Adjemian, J. C. Z., Girvetz, E., Beckett, L. & Foley, J. E. (2006). Analysis of genetic algorithm for rule-set production (GARP) modeling approach for predicting distributions of fleas implicated as vectors of plague, *Yersinia pestis*, in California. *Journal of Medical Entomology*, **43**, 93–103.

Ageyev, V. S. & Sludsky, A. A. (1985). Materials on fleas of small mammals in the eastern Pamir and perspective of the epizootologic monitoring of this region. In *Important Questions of Epidemiological Monitoring in the Natural Foci of Plague: Natural Focality of Plague in High Mountains*, ed. I. F. Taran. Stavropol, USSR: Scientific Anti-Plague Institute of Caucasus and Trans-Caucasus, pp. 10–12 (in Russian).

Ageyev, V. S., Arzhannikova, A. S., Tlegenov, T. T., Samarin, E. G. & Serzhanov, O. S. (1983). Interspecific interactions among five species of fleas co-occurring on the midday jirds. In *Prophylaxis of Diseases in the Natural Foci*, ed. I. F. Taran. Stavropol, USSR: Scientific Anti-Plague Institute of Caucasus and Trans-Caucasus, pp. 209–210 (in Russian).

Ageyev, V. S., Serzhanov, O. S., Arzhannikova, A. S. & Tlegenov, T. T. (1984). Interspecific interactions among imagoes of five species of fleas (Siphonaptera) parasitic on gerbils. *Parazitologiya*, **18**, 185–190 (in Russian).

Akin, D. M. (1984). Relationship between feeding and reproduction in the cat flea, *Ctenocephalides felis* (Bouché). Unpublished M.Sc. thesis, University of Florida, Gainesville, FL.

Alania, I. I., Rostigaev, B. A., Shiranovich, P. I. & Dzneladze, M. T. (1964). Data on the flea fauna of Adzharia. *Proceedings of the Armenian Anti-Plague Station*, **3**, 407–435 (in Russian).

Alekseev, A. N. (1961). On the bionomics of fleas *Ceratophyllus* (*Nosopsyllus*) *consimilis* Wagn., 1898 (Ceratophyllidae, Aphaniptera). *Zoologicheskyi Zhurnal*, **40**, 1840–1847 (in Russian).

Alekseev, A. N., Grebenyuk, R. V., Tchirov, P. A. & Kadysheva, A. M. (1971). On the relationships between the listeriosis pathogen (*Listeria monocytogenes*) and fleas. *Parazitologiya*, **5**, 113–118 (in Russian).

Alexander, J. O. (1986). The physiology of itch. *Parasitology Today*, **2**, 345–351.

Allan, R. M. (1956). A study of the populations of the rabbit flea *Spilopsyllus cuniculi* (Dale) on the wild rabbit, *Oryctolagus cuniculus*, in north-east Scotland. *Proceedings of the Royal Entomological Society of London A*, **31**, 145–152.

Allander, K. (1998). The effects of an ectoparasite on reproductive success in the great tit: a 3-year experimental study. *Canadian Journal of Zoology*, **76**, 19–25.

Allred, D. M. (1968). Fleas of the National Reactor Testing Station. *Great Basin Naturalist*, **28**, 73–87.

Allsopp, P. G. (1997). Probability of describing an Australian scarab beetle: influence of body size and distribution. *Journal of Biogeography*, **24**, 717–724.

Almeida-Neto, M., Guimarães, P. R. & Lewinsohn, T. M. (2007). On nestedness analyses: rethinking matrix temperature and anti-nestedness. *Oikos*, **116**, 716–722.

Altizer, S., Nunn, C. L., Thrall, P. H., *et al.* (2004). Social organization and parasite risk in mammals: integrating theory and empirical studies. *Annual Review of Ecology and Systematics*, **34**, 517–547.

Amin, O. M. (1974). Comb variation in the rabbit flea *Cediopsylla simplex* (Baker). *Journal of Medical Entomology*, **11**, 227–230.

Amin, O. M. (1976). Host associations and seasonal occurrence of fleas from southeastern Wisconsin mammals, with observations on morphologic variations. *Journal of Medical Entomology*, **13**, 179–192.

Amin, O. M. (1982). The significance of pronotal comb patterns in flea–host lodging adaptations. *Wiadomosci Parazytologiczne*, **28**, 93–94.

Amin, O. M. & Sewell, R. G. (1977). Comb variations in the squirrel and chipmunk fleas, *Orchopeas h. howardii* (Baker) and *Megabothris acerbus* (Jordan) (Siphonaptera), with notes on the significance of pronotal comb patterns. *American Midland Naturalist*, **98**, 207–212.

Amin, O. M. & Wagner, M. E. (1983). Further notes on the function of pronotal combs in fleas (Siphonaptera). *Annals of the Entomological Society of America*, **76**, 232–234.

Amin, O. M., Wells, T. R. & Gately, H. L. (1974). Comb variation in the cat flea, *Ctenocephalides f. felis* (Bouché). *Annals of the Entomological Society of America*, **67**, 831–834.

Amin, O. M., Liu, J., Li, S.-J., Zhang, Y.-M. & Sun, L.-Z. (1993). Development and longevity of *Nosopsyllus laeviceps kuzenkovi* (Siphonaptera) from Inner Mongolia under laboratory conditions. *Journal of Parasitology*, **79**, 193–197.

Amrine, J. W. & Lewis, R. E. (1978). The topography of the exoskeleton of *Cediopsylla simplex* (Baker 1895) (Siphonaptera: Pulicidae). I. The head and its appendages. *Journal of Parasitology*, **64**, 343–358.

Anderson, P. C. & Kok, O. B. (2003). Ectoparasites of springhares in the Northern Cape Province, South Africa. *South African Journal of Wildlife Research*, **33**, 23–32.

Anderson, R. M. & Gordon, D. M. (1982). Processes influencing the distribution of parasite numbers within host populations with special emphasis on parasite-induced host mortality. *Parasitology*, **85**, 373–398.

Anderson, R. M. & May, R. M. (1978). Regulation and stability of host–parasite population interactions. I. Regulatory processes. *Journal of Animal Ecology*, **47**, 219–247.

Anderson, R. M., Gordon, D. M., Crawley, M. J. & Hassell, M. P. (1982). Variability in the abundance of animal and plant species. *Nature*, **296**, 245–248.

Andreotti, R., Gomes, A., Malavazi-Piza, K. C., *et al.* (2002). BmTI antigens induce a bovine protective immune response against *Boophilus microplus* tick. *International Immunopharmacology*, **2**, 557–563.

Anholt, B. R. & Werner, E. E. (1998). Predictable changes in predation mortality as a consequence of changes in food availability and predation risk. *Evolutionary Ecology*, **12**, 729–738.

Anonymous. (2004). Human plague in 2002 and 2003. *Weekly Epidemiological Record*, **79**, 301–306.

Apanius, V. (1998). Stress and immune defence. In *Stress and Behavior, Advances in the Study of Behavior*, vol. 27, ed. A. P. Møller, M. Milinski & P. J. B. Slater. New York: Academic Press, pp. 133–153.

Araújo, F. R., Silva, M. P., Lopes, A. A., *et al.* (1988). Severe cat flea infestation of dairy calves in Brazil. *Veterinary Parasitology*, **80**, 83–86.

Arneberg, P. (2002). Host population density and body mass as determinants of species richness in parasite communities: comparative analyses of directly transmitted nematodes of mammals. *Ecography*, **25**, 88–94.

Arneberg, P., Skorping, A. & Read, A. F. (1997). Is population density a species character? Comparative analyses of the nematode parasites of mammals. *Oikos*, **80**, 289–300.

Arneberg, P., Skorping, A., Grenfell, B. & Read, A. F. (1998). Host densities as determinants of abundance in parasite communities. *Proceedings of the Royal Society of London B*, **265**, 1283–1289.

Arzamasov, I. T. (1969). Ectoparasite assemblages of insectivores in Belorussia. In *Fauna and Ecology of Animals in Belorussia*, ed. Anonymous. Minsk, USSR: Academy of Sciences of the Belorussian SSR, pp. 212–221 (in Russian).

Atkinson, W. D. & Shorrocks, B. (1981). Competition on a divided and ephemeral recourse: a simulation model. *Journal of Animal Ecology*, **50**, 461–471.

Atmar, W. & Patterson, B. D. (1993). The measure of order and disorder in the distribution of species in fragmented habitat. *Oecologia*, **96**, 373–382.

Audy, J. R., Radovsky, F. J. & Vercammen-Grandjean, P. H. (1972). Neosomy: radical intrastadial metamorphosis associated with arthropod symbioses. *Journal of Medical Entomology*, **9**, 487–494.

Autino, G. A. & Lareschi, M. (1998). Siphonaptera. In *Biodiversidad de Artópodos Argentinos*, ed. J. J. Morrone & S. Ciscaron. La Plata, Argentina: Ediciones SUR, pp. 279–290.

Bacot, A. W. (1914). A study of bionomics of the common rat fleas and other species associated with human habitations, with special reference to the influence of temperature and humidity at various periods of the life history of the insect. *Journal of Hygiene*, **13**, 447–654.

Bacot, A. W. & Martin, C. J. (1914). Observations on the mechanism of the transmission of plague by fleas. *Journal of Hygiene*, **13**, 423–439.

Bahmanyar, M. & Cavanaugh, D. C. (1976). *Plague Manual*. Geneva, Switzerland: World Health Organization.

Baker, K. P. & Elharam, S. (1992). The biology of *Ctenocephalides canis* in Ireland. *Veterinary Parasitology*, **45**, 141–146.

Balashov, Y. S., Bibikova, V. A., Murzakhmetova, K. & Polunina, O. A. (1961). A flea as an environment of the plague pathogen. I. Feeding and digestion in uninfected fleas. In *Proceedings of the Interdisciplinary Conference Dedicated to the 40th Anniversary of the Kazakh Soviet Socialist Republic*, ed. Anonymous. Alma-Ata, USSR: The Middle Asian Scientific Anti-Plague Institute, pp. 27–30 (in Russian).

Balashov, Y. S., Bibikova, V. A., Murzakhmetova, K. & Polunina, O. A. (1965). Feeding and failure of the foregut valve function in fleas. *Medical Parasitology and Parasitic Diseases* [*Meditsinskaya Parazitologiya i Parazitarnye Bolezni*], **35**, 471–476 (in Russian).

Ball, S. L. & Baker, R. L. (1996). Predator-induced life history changes: antipredator behavior costs or facultative life history shifts? *Ecology*, **77**, 1116–1124.

Ballabeni, P. & Ward, P. I. (1993). Local adaptation of the trematode *Diplostomum phoxini* to the European minnow *Phoxinus phoxinus*, its second intermediate host. *Functional Ecology*, **7**, 84–90.

Banbura, J., Blondel, J., Wilde-Lambrechts, H. & Perret, P. (1995). Why do female blue tits (*Parus caeruleus*) bring fresh plants to their nests? *Journal of Ornithology*, **136**, 217–221.

Banet, M. (1986). Fever in mammals: is it beneficial? *Yale Journal of Biology and Medicine*, **59**, 117–124.

Banfield, A. W. F. (1974). *The Mammals of Canada*. Toronto, ON: University of Toronto Press.

Bansemir, A. D. & Sukhdeo, M. V. K. (1996). Habitat selection of a gastrointestinal parasite: proximal cues involved in decision making. *Parasitology*, **113**, 311–316.

Barger, M. A. & Esch, G. W. (2002). Host specificity and the distribution–abundance relationship in a community of parasites infecting fishes in streams of North Carolina. *Journal of Parasitology*, **88**, 446–453.

Barker, S. C. (1991). Evolution of host–parasite associations among species of lice and rock-wallabies: coevolution? *International Journal for Parasitology*, **21**, 497–501.

Barnard, C. J., Behnke, J. M., Gage, A. R, Brown, H. & Smithurst, P. R. (1998). The role of parasite-induced immunodepression, rank and social environment in the modulation of behaviour and hormone concentration in male laboratory mice (*Mus musculus*). *Proceedings of the Royal Society of London B*, **265**, 693–701.

Barnes, A. M. (1965). Three new species of the genus *Anomiopsyllus*. *Pan-Pacific Entomologist*, **41**, 272–280.

Barnes, A. M. & Radovsky, F. J. (1969). A new *Tunga* (Siphonaptera) from the Nearctic region with description of all stages. *Journal of Medical Entomology*, **6**, 19–36.

Barnes, A. M., Tipton, V. J. & Wildie, J. A. (1977). The subfamily Anomiopsyllinae (Hystrichopsyllidae: Siphonaptera). I. A revision of the genus *Anomiopsyllus* Baker. *Great Basin Naturalist*, **37**, 138–206.

Barrera, A. (1968). The altitudinal distribution of the Siphonaptera of Mount Popocatepetl (Mexico) and a biogeographical interpretation of it. *Anales del Instituto de Biologia, Universidad Nacional Autónoma de Mexico, Serie Zoologia*, **39**, 35–100.

Barriere, P., Beaucournu, J. C., Menier, K. & Colyn, M. (2002). A new flea species of the genus *Allopsylla* Beaucournu et Fain, 1982 (Siphonaptera: Ischnopsyllidae) from Central African Republic, on a poorly known molossid bat. *Parasite*, **9**, 233–237.

Bartholomew, G. A. & Casey, T. M. (1978). Oxygen consumption of moths during rest, pre-flight warm-up, and flight in relation to body size and wing morphology. *Journal of Experimental Biology*, **76**, 11–25.

Bartkowska, K. (1973). Siphonaptera Tatr Polskich. *Fragmenta Faunistica (Warszawa)*, **19**, 227–281.

Bartolomucci, A., Palanza, P., Gaspani, L., *et al.* (2001). Social status in mice: behavioral, endocrine and immune changes are context dependent. *Physiology and Behavior*, **73**, 401–410.

Bar-Zeev, M. & Sternberg, S. (1962). Factors affecting the feeding of fleas (*Xenopsylla cheopis* Rothsch.) through a membrane. *Entomologia Experimentalis et Applicata*, **5**, 60–68.

Bashenina, N. V. (1962). *The Ecology of the Common Vole*. Moscow, USSR: Moscow University Press (in Russian).

Bashenina, N. V. (1977). *Pathways of Adaptations in the Myomorph Rodents*. Moscow, USSR: Nauka (in Russian).

Bashenina, N. V. (ed.) (1981). *The Bank Vole*. Moscow, USSR: Nauka (in Russian).

Bates, J. K. (1962). Field studies on the behaviour of bird fleas. I. Behaviour of the adults of three species of bird fleas in the field. *Parasitology*, **52**, 113–132.

Baumgartner, D. L. & Kane, A. (1986). Wood ducks as accidental hosts of the squirrel flea, *Orchopeas howardi* (Siphonaptera, Ceratophyllidae). *Great Lakes Entomologist*, **19**, 249–250.

Bayreuther, K. & Brauning, S. (1971). Die Cytogenetik der Flohe (Aphaniptera). II. *Xenopsylla cheopis* Rothschild, 1093, und *Leptopsylla segnis* Schonherr, 1811. *Chromosoma*, **33**, 19–29.

Bazanova, L. P. & Khabarov, A. V. (2000). Block formation in fleas *Citellophilus tesquorum altaicus* Ioff, 1936 in relation to the gender of an insect. *Medical Parasitology and Parasitic Diseases* [*Meditsinskaya Parazitologiya i Parazitarnye Bolezni*], **69**, 42–44 (in Russian).

Bazanova, L. P. & Mayevsky, M. P. (1996). The duration of persistence of the plaque microbe in the organism of a flea *Citellophilus tesquorum altaicus*. *Medical Parasitology and Parasitic Diseases* [*Meditsinskaya Parazitologiya i Parazitarnye Bolezni*], **65**, 45–48 (in Russian).

Bazanova, L. P., Voronova, G. A. & Tokmakova, E. G. (2000). Differences in the proventriculus blockage between males and females of *Xenopsylla cheopis* (Siphonaptera: Pulicidae). *Parazitologiya*, **34**, 56–59 (in Russian).

Bazanova, L. P., Nikitin, A. Y. & Mayevsky, M. P. (2004). Seasonal changes in the aggregated state of the causative agent of plague in the organism of *Citellophilus tesquorum altaicus* (Siphonaptera). *Medical Parasitology and Parasitic Diseases* [*Meditsinskaya Parazitologiya i Parazitarnye Bolezni*], **73**, 3–6 (in Russian).

Beaucournu, J. C. & Menier, K. (1998). Le genre *Ctenocephalides* Stiles et Collins, 1930 (Siphonaptera, Pulicidae). *Parasite*, **5**, 3–16.

Beaucournu, J. C. & Pascal, M. (1998). Origine biogéographique de *Nosopsyllus fasciatus* (Bosc, 1800) (Siphonaptera – Ceratophyllidae) et observations sur son hôte primitif. *Biogeographica*, **74**, 135–132.

Beaucournu, J. C. & Wells, K. (2004). Trois espèces nouvelles du genre *Medwayella* Traub, 1972 (Insecta: Siphonaptera: Pygiopsyllidae) de Sabah (Malaisie Orientale, Ile de Bornêo). *Parasite*, **11**, 373–377.

Beaucournu, J. C., le Piver, M. & Guiguen, C. (1993). Actualité de la conquête de l'Afrique intertropical par *Pulex irritans* Linné, 1758. *Bulletin de la Société de Pathologie Exotique*, **86**, 290–294.

Beaucournu, J. C., Kock, D. & Menier, K. (1997). La souris *Mus musculus* L., 1758 est-elle l'hôte primitif de la puce *Leptopsylla segnis* (Schonherr, 1811) (Insecta: Siphonaptera)? *Biogeographica*, **73**, 1–12.

Beaucournu, J. C., Degeilh, B. & Guiguen, C. (2005). Les puces (Insecta: Siphonaptera) parasites d'oiseaux: diversité taxonomique et dispersion biogéographique. *Parasite*, **12**, 111–121.

Beaucournu, J. C., Vergara, P., Balboa, L. & Gonzalez-Acuna, D. A. (2006). Description d'une nouvelle puce d'oiseau provenant du Chili (Siphonaptera: Ceratophyllidae). *Parasite*, **13**, 227–230.

Beck, W. (1999). Landwirtschaftliche Nutztiere als Vektoren von parasitären Epizoonoseerregern und zoophilen Dermatophyten. *Hautarzt*, **50**, 621–628.

Beck, W. & Clark, H. H. (1997). Differentialdiagnose medizinisch relevanter Flohspezies und ihre Bedeutung in der Dermatologie. *Hautarzt*, **48**, 714–719.

Behnke, J. M., Harris, P. D., Bajer, A., *et al.* (2004). Variation in the helminth community structure in spiny mice (*Acomys dimidiatus*) from four montane wadis in the St Katherine region of the Sinai Peninsula in Egypt. *Parasitology*, **129**, 379–398.

Beissinger, S. R. & Westphal, M. I. (1998). On the use of demographic models of population viability in endangered species management. *Journal of Wildlife Management*, **62**, 821–841.

Bekenov, Z. E., Serzhan, O. S., Alashbayev, M. A., et al. (2000). Fluctuations in X. skrjabini population inhabiting the Northern Cis-Aral autonomous plague focus and their causes. *Quarantinable and Zoonotic Infections in Kazakhstan*, **2**, 52–56 (in Russian).

Bell, G. & Burt, A. (1991). The comparative biology of parasite species diversity: intestinal helminths of freshwater fishes. *Journal of Animal Ecology*, **60**, 1046–1063.

Bell, P. J., Burton, H. R., & Vanfraneker, J. A. (1988). Aspects of the biology of *Glaciopsyllus antarcticus* (Siphonaptera, Ceratophyllidae) during the breeding-season of a host (*Fulmarus glacialoides*). *Polar Biology*, **8**, 403–410.

Belokopytova, A. M., Tchumakova, I. V. & Kozlov, M. P. (1983). Feeding of *Xenopsylla conformis*, *Nosopsyllus laeviceps* and *Citellophilus tesquorum* on reptiles. In *Prophylaxis of Diseases in the Natural Foci*, ed. I. F. Taran. Stavropol, USSR: Scientific Anti-Plague Institute of Caucasus and Trans-Caucasus, pp. 214–215 (in Russian).

Belyavtseva, L. I. (2002). Phenology of fleas *Citellophilus tesquorum* inhabiting different areas of North Caucasus. *Quarantinable and Zoonotic Infections in Kazakhstan*, **6**, 30–34 (in Russian).

Bengtson, S. A., Brinck-Lindroth, G., Lundqvist, L., Nilsson, A. & Rundgren, S. (1986). Ectoparasites on small mammals in Iceland: origin and population characteristics of a species-poor insular community. *Holarctic Ecology*, **9**, 143–148.

Benjamini, E., Feingold, B. F., Young, J. D., Kartman, L. & Shimizu, M. (1963). Allergy to flea bites. IV. *In vitro* collection and antigenic properties of the oral secretion of the cat flea. *Experimental Parasitology*, **13**, 143–154.

Bennet-Clark, H. C. & Lucey, E. C. (1967). The jump of the flea: a study of the energetics and a model of the mechanism. *Journal of Experimental Biology*, **47**, 59–67.

Bennett, G. F., Caines, J. R. & Bishop, M. A. (1988). Influence of blood parasites on the body mass of passeriform birds. *Journal of Wildlife Diseases*, **24**, 339–343.

Benton, A. H., Cerwonka, R. & Hill, J. (1959). Observations on host perception in fleas. *Journal of Parasitology*, **45**, 614.

Benton, A. H., Surman, M. & Krinsky, W. L. (1979). Observations on the feeding habits of some larval fleas (Siphonaptera). *Journal of Parasitology*, **65**, 671–672.

Berendyaeva, E. L. & Kudryavtseva, K. F. (1969). Density of fleas on *Marmota sibirica* in the Upper-Arym autonomous plague focus. In *Proceedings of the 6th Scientific Conference of the Anti-Plague Establishments of the Middle Asia and Kazakhstan*, vol. 2, ed. M. A. Aikimbaev. Alma-Ata, USSR: The Middle Asian Scientific Anti-Plague Institute, pp. 55–58 (in Russian).

Bernotat-Danielowski, S. & Knülle, W. (1986). Ultrastructure of the rectal sac, the site of water vapor uptake from the atmosphere in larvae of the oriental rat flea *Xenopsylla cheopis*. *Tissue and Cell*, **18**, 437–455.

Berseth, W. C. & Zubac, P. (1987). Habitat associations of fleas, Siphonaptera, parasitizing the short-tailed shrew, *Blarina brevicauda*. *Canadian Field Naturalist*, **101**, 594–596.

Betz, O. & Fuhrmann, S. (2001). Life history traits in different life forms of predaceous *Stenus* beetles (Coleoptera, Staphylinidae), living in waterside environments. *Netherlands Journal of Zoology*, **51**, 371–393.

Beutel, R. G. & Gorb, S. N. (2001). Ultrastructure of attachment specializations of hexapods (Arthropoda): evolutionary patterns inferred from a revised ordinal phylogeny. *Journal of Zoological Systematics and Evolutionary Research*, **39**, 177–207.

Beutel, R. G. & Pohl, H. (2006). Endopterygote systematics: where do we stand and what is the goal (Hexapoda, Arthropoda)? *Systematic Entomology*, **31**, 202–219.

Beveridge, I. & Chilton, N. B. (2001). Co-evolutionary relationships between the nematode subfamily Cloacininae and its macropodid marsupial hosts. *International Journal for Parasitology*, **21**, 976–996.

Bgytova, C. I. (1963). Data on the ecology of fleas. IV. Copulation and egg maturation in *Ctenophthalmus dolichus* under different environmental conditions. In *Proceedings of the Scientific Conference of the Natural Focality and Prophylaxis of Plague*, ed. M. A. Aikimbaev. Alma-Ata, USSR: The Middle Asian Scientific Anti-Plague Institute, pp. 15–16 (in Russian).

Bibikov, D. I. & Bibikova, V. A. (1955). Studying of the Isabelline wheatear and its ectoparasites. *Zoologicheskyi Zhurnal*, **34**, 399–407 (in Russian).

Bibikova, V. A. (1956). On the biology of fleas parasitic on marmots. *Proceedings of the Middle Asian Scientific Anti-Plague Institute*, **2**, 49–51 (in Russian).

Bibikova, V. A. (1963). Use of microscopy and spectral analysis when studying the hosts' blood in the fleas' stomachs. In *Proceedings of the 4th Scientific Conference on Natural Focality and Prophylaxis of Plague*, ed. Anonymous. Alma-Ata, USSR: Kainar, pp. 21–22 (in Russian).

Bibikova, V. A. (1965). Duration of metamorphosis in *Ceratophyllus trispinus balkhaschensis* Mik. 1958 under experimental conditions. In *Proceedings of the 4th Scientific Conference of the Anti-Plague Establishments of the Middle Asia and Kazakhstan*, ed. M. A. Aikimbaev. Alma-Ata, USSR: The Middle Asian Scientific Anti-Plague Institute and Kainar, pp. 31–32 (in Russian).

Bibikova, V. A. (1977). Contemporary views on the interrelationships between fleas and the pathogens of human and animal diseases. *Annual Review of Entomology*, **22**, 23–32.

Bibikova, V. A. & Gerasimova, N. G. (1967). On the biology of *Xenopsylla skrjabini* Ioff, 1928. II. Flea feeding under experimental conditions. *Zoologicheskyi Zhurnal*, **46**, 730–736 (in Russian).

Bibikova, V. A. & Gerasimova, N. G. (1973). Feeding of fleas *Xenopsylla nuttalli* Ioff, 1930 in the experiments. *Problems of Particularly Dangerous Diseases*, **29**, 122–125 (in Russian).

Bibikova, V. A. & Klassovsky, L. N. (1974). *Transmission of Plague by Fleas*. Moscow, USSR: Meditsina (in Russian).

Bibikova, V. A. & Zhovty, I. F. (1980). Review of certain studies of fleas in the USSR, 1967–1976. In *Fleas: Proceedings of the International Conference on Fleas*, Ashton Wold, Peterborough, UK, 21–25 June 1977, ed. R. Traub & H. Starcke. Rotterdam, the Netherlands: A. A. Balkema, pp. 257–272.

Bibikova, V. A., Ilyinskaya, V. L., Kaluzhenova, Z. P., Morozova, I. V. & Shmuter, M. F. (1963). On the biology of fleas of the genus *Xenopsylla* in the Sary-Ishik-Otrau Desert. *Zoologicheskyi Zhurnal*, **42**, 1045–1050 (in Russian).

Bibikova, V. A., Zolotova, S. I., Murzakhmetova, K. & Leonova, T. N. (1971). Fecundity and its dynamics in two fleas parasitic on the great gerbil under experimental conditions. In *Proceedings of the 7th Scientific Conference of the Anti-Plague Establishments of the Middle Asia and Kazakhstan*, ed. M. A. Aikimbaev. Alma-Ata, USSR: The Middle Asian Scientific Anti-Plague Institute, pp. 367–369 (in Russian).

Bidashko, F. G., Grazhdanov, A. K., Medzykhovsky, G. A., et al. (2001). Results of the studies of flea abundance in human houses in the plague focus of the West-Kazakhstan Region. *Quarantinable and Zoonotic Infections in Kazakhstan*, **4**, 89–93 (in Russian).

Bidashko, F. G., Grazhdanov, A. K., Tanitovsky, B. A., et al. (2004). Attack rate of fleas *Pulex irritans* on humans in various biotopes of human houses. *Quarantinable and Zoonotic Infections in Kazakhstan*, **9**, 58–61 (in Russian).

Biliński, S. M., Büning, J. & Simiczyjew, B. (1998). The ovaries of Mecoptera: basic similarities. *Folia Histochemica et Cytobiologica*, **36**, 189–196.

Bilyalov, Z. A., Ershova, L. S., Sviridov, G. G., Sokolov, P. N. & Arakelyants, V. S. (1989). A case of experimental transmission of the causative agent of plague from *Ornithodoros tartakovskyi* ticks to fleas of the great gerbil. *Parazitologiya*, **23**, 362–364 (in Russian).

Birtles, R. J., Harrison, T. G. & Molyneux, D. H. (1994). *Grahamella* in small woodland mammals in the UK: isolation, prevalence and host-specificity. *Annals of Tropical Medicine and Parasitology*, **88**, 317–327.

Bitam, I., Parola, P., Dittmar de la Cruz, K., et al. (2006). First molecular detection of *Rickettsia felis* in fleas from Algeria. *American Journal of Tropical Medicine and Hygiene*, **74**, 532–535.

Black, C. C. & Krishtalka, L. (1986). Rodents, bats, and insectivores from the Plio-Pleistocene sediments to the east of Lake Turkana, Kenya. *Contributions in Science, Natural History Museum of Los Angeles County*, **372**, 1–15.

Blackburn, T. M. & Gaston, K. J. (2001). Linking patterns in macroecology. *Journal of Animal Ecology*, **70**, 338–352.

Blanc, G. & Baltazard, M. (1941). Transmission du bacille de Withmore par la puce du rat *Xenopsylla cheopis*. *Comptes Rendus de l'Académie des Sciences*, **213**, 541–543.

Blanc, G. & Baltazard, M. (1942). Sur le mécanisme de la transmission de la peste par *Xenopsylla cheopis*. *Comptes Rendus de l'Académie des Sciences*, **136**, 645–647.

Blanc, G. & Baltazard, M. (1944a). Contribution à l'étude du comportement des microbes pathogènes chez les insectes hématophages. Premier mémoire. *Archives de l'Institut Pasteur du Maroc*, **3**, 21–49.

Blanc, G. & Balatazard, M. (1944b). Revue chronologique des recherches expérimentales sur la transmission et la conservation naturelle des typhus. *Archives de l'Institut Pasteur du Maroc*, **2**, 535–715.

Blank, S. M., Kutzscher, C., Masello, J. F., Pilgrim, R. L. C. & Quillfeldt, P. (2007). Stick-tight fleas in the nostrils and below the tongue: evolution of an extraordinary infestation site in *Hectopsylla* (Siphonaptera: Pulicidae). *Zoological Journal of the Linnean Society*, **149**, 117–137.

Blaski, M. (1989). Fleas occurring on rodents living in the area of coal-mine Boleslaw Smialy in Laziska (Poland). *Prace Naukowe Uniwersytetu Slaskiego w Katowicach*, **1035**, 92–96 (in Polish).

Blaski, M. (1991). Fleas (Siphonaptera) collected on the rodents from two industrial plants in Silesia. *Prace Naukowe Uniwersytetu Slaskiego w Katowicach*, **1184**, 87–88 (in Polish).

Blaski, M. (2004). Seasonal dynamics of *Ceratophyllus hirundinis* (Curtis, 1826) (Siphonaptera, Insecta) in the nests of *Delichon urbica* (L.). *Wiadomości Parazytologiczne*, **50**, 31–34 (in Polish).

Blount, J. D., Houston, D. C., Møller, A. P. & Wright, J. (2003). Do individual branches of immune defence correlate? A comparative case study of scavenging and non-scavenging birds. *Oikos*, **102**, 340–350.

Bock, C. E., Cruz, A., Grant, M. C., Aid, C. S. & Strong, T. R. (1992). Field experimental evidence for diffuse competition among southwestern riparian birds. *American Naturalist*, **140**, 515–528.

Bodrova, T. V. & Zhovty, I. F. (1961). Fleas of the Daurian ground squirrel in the area of the Zun-Torei Lake (S.-E. Trans-Baikalia). *Transactions of the Irkutsk State Scientific Anti-Plague Institute of Siberia and Far East*, **1**, 82–85 (in Russian).

Boeken, B. & Shachak, M. (1998). The dynamics of abundance and incidence of annual plant species richness during colonization in a desert. *Ecography*, **21**, 63–73.

Bogdanov, I. I., Malkova, M. G., Yakimenko, V. V. & Tantsev, A. K. (2001). Parasite–host associations between fleas and small mammals in the Omsk Region. *Parazitologiya*, **35**, 184–191 (in Russian).

Bohn, T. & Amundsen, P. A. (2001). The competitive edge of an invading specialist. *Ecology*, **82**, 2150–2163.

Bolger, D. T., Alberts, A. C. & Soulé, M. E. (1991). Occurrence patterns of bird species in habitat fragments: sampling, extinction, and nested species subsets. *American Naturalist*, **137**, 155–166.

Bossard, R. L. (2002). Speed and Reynolds number of jumping cat fleas (Siphonaptera: Pulicidae). *Journal of the Kansas Entomological Society*, **75**, 52–54.

Bossard, R. L. (2006). Mammal and flea relationships in the Great Basin Desert: from H. J. Egoscue's collections. *Journal of Parasitology*, **92**, 260–266.

Bossard, R. L., Broce, A. B. & Dryden, M. W. (2000). Effects of circadian rhythms and other bioassay factors on cat flea (Pulicidae: Siphonaptera) susceptibility to insecticides. *Journal of the Kansas Entomological Society*, **73**, 21–29.

Botelho, J. R. & Linardi, P. M. (1996). Interrelações entre ectoparasitos e roedores em ambientes silvestre e urbano de Belo Horizonte, Minas Gerais, Brasil. *Revista Brasileira de Entomologia*, **40**, 425–430.

Bouchet, F., Guidon, N., Dittmar, K., *et al.* (2003). Parasite remains in archeological sites. *Memórias do Instituto Oswaldo Cruz*, **98**, 39–47.

Boudreaux, H. B. (1979). *Arthropod Phylogeny with Special Reference to Insects*. New York: John Wiley.

Boughton, R. K., Atwell, J. W. & Schoech, S. J. (2006). An introduced generalist parasite, the sticktight flea (*Echidnophaga gallinacea*), and its pathology in the threatened Florida scrub-jay (*Aphelocoma coerulescens*). *Journal of Parasitology*, **92**, 941–948.

Boulouis, H.-J., Chang, C.-C., Henn, J. B., Kasten, R. W. & Chomel, B. B. (2005). Factors associated with the rapid emergence of zoonotic *Bartonella* infections. *Veterinary Research*, **36**, 383–410.

Bouslama, Z., Chabi, Y. & Lambrechts, M. M. (2001). Chicks resist high parasite intensities in an Algerian population of blue tits. *Ecoscience*, **8**, 320–324.

Bouslama, Z., Lambrechts, M. M., Ziane, N., Djenidi, R. D. & Chabi, Y. (2002). The effect of nest ectoparasites on parental provisioning in a north-African population of the blue tit *Parus caeruleus*. *Ibis*, **144**, E73–E78.

Boycott, A. E. (1913). The reaction of flea bites. *Journal of Pathology and Bacteriology*, **17**, 110.

Bozeman, F. M., Sonenshine, D. E., Williams, M. S., *et al.* (1981). Experimental infection of ectoparasitic arthropods with *Rickettsia prowazekii* (GvF-16 strain) and transmission to flying squirrels. *American Journal of Tropical Medicine and Hygiene*, **30**, 253–263.

Bradley, T. J. & Hetz, S. (2001). Specific dynamic action in the insect *Rhodnius prolixus*. *American Zoologist*, **41**, 1397.

Braks, M. A. H., Honorio, N. A., Lounibos, L. P., de Oliveira, R. L. & Juliano, S. A. (2004). Interspecific competition between two invasive species of container mosquitoes, *Aedes aegypti* and *Aedes albopictus* (Diptera: Culicidae), in Brazil. *Annals of the Entomological Society of America*, **97**, 130–139.

Brändle, M., Amarell, U., Auge, H., Klotz, S. & Brandl, R. (2001). Plant and insect diversity along a pollution gradient: understanding species richness across trophic levels. *Biodiversity and Conservation*, **10**, 1497–1511.

Brandt, C. A. (1992). Social factors in immigration and emigration. In *Animal Dispersal: Small Mammals as a Model*, ed. N. C. Stenseth & W. Z. Lidicker. London: Chapman & Hall, pp. 96–141.

Bray, J. R. & Curtis, J. T. (1957). An ordination of the upland forest communities of Southern Wisconsin. *Ecological Monographs*, **27**, 325–349.

Breitschwerdt, E. B. & Kordick D. L. (2000). *Bartonella* infection in animals: carriership, reservoir potential, pathogenicity, and zoonotic potential for human infection. *Clinical Microbiology Reviews*, **13**, 428–438.

Brinck, G. (1966). Siphonaptera from small mammals in natural foci of tick-borne encephalitis virus in Sweden. *Opuscula Entomologica*, **31**, 156–170.

Brinck, G. & Löfqvist, J. (1973). The hedgehog *Erinaceus europeus* and its flea *Archaeopsylla erinacei*. *Zoon (Supplement)*, **1**, 97–103.

Brinck-Lindroth, G. (1968). Host spectra and distribution of fleas of small mammals in Swedish Lapland. *Opuscula Entomologica*, **33**, 327–358.

Brinkerhoff, R. J., Markeson, A. B., Knouft, J. A., Gage, K. L. & Montenieri, J. A. (2006). Abundance patterns of two *Oropsylla* (Ceratophyllidae: Siphonaptera) species on black-tailed prairie dog (*Cynomys ludovicianus*) hosts. *Journal of Vector Ecology*, **31**, 355–363.

Brinkhof, M. W. G., Heeb, P., Kölliker, M. & Richner, H. (1999). Immunocompetence of nestling great tits in relation to rearing environment and parentage. *Proceedings of the Royal Society of London B*, **266**, 2315–2322.

Brooks, D. R. (1988). Macroevolutionary comparisons of host and parasite phylogenies. *Annual Review of Ecology and Systematics*, **19**, 235–259.

Brooks, D. R. & McLennan, D. A. (1991). *Phylogeny, Ecology, and Behaviour: A Research Program in Comparative Biology*. Chicago, IL: University of Chicago Press.

Brouwer, L. & Komdeur, J. (2004). Green nesting material has a function in mate attraction in the European starling. *Animal Behaviour*, **67**, 539–548.

Brown, C. R. & Brown, M. B. (1986). Ectoparasitism as a cost of coloniality in cliff swallows (*Hirundo pyrrhonota*). *Ecology*, **67**, 1206–1218.

Brown, C. R. & Brown, M. B. (1992). Ectoparasitism as a cause of natal dispersal in cliff swallows. *Ecology*, **73**, 1718–1723.

Brown, C. R., Brown, M. B. & Rannala, B. (1995). Ectoparasites reduce long-term survival of their avian host. *Proceedings of the Royal Society of London B*, **262**, 313–319.

Brown, J. H. (1975) Geographical ecology of desert rodents. In *Ecology and Evolution of Communities*, ed. M. L. Cody & J. M. Diamond. Cambridge, MA: Harvard University Press, pp. 315–341.

Brown, J. H. (1984). On the relationship between abundance and distribution of species. *American Naturalist*, **124**, 255–279.

Brown, J. H. (1995). *Macroecology*. Chicago, IL: University of Chicago Press.

Brown, J. H. & Lomolino, M. V. (1998). *Biogeography*. Sunderland, MA: Sinauer Associates.

Brown, J. H., Stevens, G. C. & Kaufman, D. M. (1996). The geographic range: size, shape, boundaries, and internal structure. *Annual Review of Ecology and Systematics*, **27**, 597–623.

Brown, J. S., Kotler, B. P. & Mitchell, W. M. (1994). Foraging theory, patch use, and the structure of a Negev Desert granivore community. *Ecology*, **75**, 2286–2300.

Brown, S. J. (1989). Pathological consequences of feeding by hematophagous arthropods: comparison of feeding strategies. In *Proceedings of a Symposium 'Physiological Interactions between Hematophagous Arthropods and their Vertebrate Hosts'*, ed. C. J. Johnston & R. E. Williams. Lanham, MD: Entomological Society of America, pp. 4–14.

Bruce, W. N. (1948). Studies on the biological requirements of the cat flea. *Annals of the Entomological Society of America*, **41**, 346–352.

Bryukhanova, L. V. (1966). Reproduction and feeding of fleas parasitic on ground squirrels. In *Particularly Dangerous Diseases in Caucasus*, ed. V. G. Pilipenko. Stavropol, USSR: Scientific Anti-Plague Institute of Caucasus and Trans-Caucasus, pp. 37–40 (in Russian).

Bryukhanova, L. V. & Myalkovskaya, C. A. (1974). On the duration of metamorphosis of fleas in the nests of the pygmy ground squirrel. In *Particularly Dangerous Diseases in Caucasus: Proceedings of the 3rd Scientific–Practical Conference of the Anti-Plague Establishments of Caucasus on Natural Focality, Epidemiology and Prophylaxis of Particularly Dangerous Diseases, 14–16 May 1974*, ed. V. G. Pilipenko. Stavropol, USSR: Scientific Anti-Plague Institute of Caucasus and Trans-Caucasus, pp. 121–124 (in Russian).

Bryukhanova, L. V. & Surkova, L. A. (1970). On the biology of *Frontopsylla semura* Wagn. et Ioff, 1926. In *Vectors of Particularly Dangerous Diseases and their Control*, ed. V. E. Tiflov. Stavropol, USSR: Scientific Anti-Plague Institute of Caucasus and Trans-Caucasus, pp. 247–252 (in Russian).

Bryukhanova, L. V. & Surkova, L. A. (1983). Digestion of the blood of ground squirrels by fleas. In *Prophylaxis of Diseases in the Natural Foci*, ed. I. F. Taran. Stavropol, USSR: Scientific Anti-Plague Institute of Caucasus and Trans-Caucasus, pp. 220–222 (in Russian).

Bryukhanova, L. V., Sardar, E. A. & Levi, M. I. (1961). On the method of measurement the amount of blood taken by fleas. *Proceedings of Scientific Anti-Plague Institute of Caucasus and Trans-Caucasus*, **5**, 28–32 (in Russian).

Bryukhanova, L. V., Darskaya, N. F. & Surkova, L. A. (1978). Blood digestion by fleas *Leptopsylla segnis* Schöncher. *Parazitologiya*, **12**, 383–386 (in Russian).

Bryukhanova, L. V., Darskaya, N. F., Surkova, L. A. & Karandina, R. S. (1983). Digestion of the blood of gerbils by fleas. In *Prophylaxis of Diseases in the Natural Foci*, ed. I. F. Taran. Stavropol, USSR: Scientific Anti-Plague Institute of Caucasus and Trans-Caucasus, pp. 219–220 (in Russian).

Buckland, P. C. & Sadler, J. P. (1989). A biogeography of the human flea, *Pulex irritans* L. (Siphonaptera, Pulicidae). *Journal of Biogeography*, **16**, 115–120.

Buckle, A. P. (1978). The mark, release and recapture of fleas in a wild population of woodmice, *Apodemus sylvaticus*. *Journal of Zoology*, **186**, 563–567.

Buechler, K., Fitze, P. S., Gottstein, B., Jacot, A. & Richner, H. (2002). Parasite-induced maternal response in a natural bird population. *Journal of Animal Ecology*, **71**, 247–252.

Buchman, K. (1991). Relationship between host size of *Anguilla anguilla* and the infection level of the monogeneans *Pseudodactylogyrus* spp. *Journal of Fish Biology*, **35**, 599–601.

Burdelov, A. S., Sabilayev, A. S., Musrepov, T., *et al.* (1999). Spatial distribution of fleas *Xenopsylla skrjabini* and *X. gerbilli minax* in colonies of the great gerbil in the north-western Pri-Balkhash area. *Parazitologiya*, **33**, 493–496 (in Russian).

Burdelov, L. A., Zhubanazarov, I. Z. & Rudenchik, N. F. (1989). The size of small mammals and the number of fleas parasitizing them. *Medical Parasitology and Parasitic Diseases [Meditsinskaya Parazitologiya i Parazitarnye Bolezni]*, **58**, 42–45 (in Russian).

Burdelov, S. A., Leiderman, M., Krasnov, B. R., Khokhlova, I. S. & Degen, A. A. (2007). Locomotor response to light and surface angle in three species of desert fleas. *Parasitology Research*, **100**, 973–982.

Burdelova, N. V. (1996). Flea fauna of some small mammals in the Dzhungarskyi Ala-Tau Ridge. In *Proceedings of the Conference 'Ecological Aspects of Epidemiology and Epizootology of Plague and Other Dangerous Diseases'*, ed. L. A. Burdelov. Almaty, Kazakhstan: Kazakh Scientific Center for Quarantine and Zoonotic Diseases, pp. 119–120 (in Russian).

Burdelova, N. V. & Burdelov, V. A. (1983). Dynamics of flea species composition at high and low density of the great gerbil in the Akdala Valley. In *Prophylaxis of Diseases in the Natural Foci*, ed. I. F. Taran. Stavropol, USSR: Scientific Anti-Plague Institute of Caucasus and Trans-Caucasus, pp. 222–223 (in Russian).

Burroughs, A. L. (1947). Sylvatic plague studies: the vector efficiency of nine species of fleas compared with *Xenopsylla cheopis*. *Journal of Hygiene*, **43**, 371–396.

Burrows, M. & Wolf, H. (2002). Jumping and kicking in the false stick insect *Prosarthria teretrirostris*: kinematics and motor control. *Journal of Experimental Biology*, **205**, 1519–1530.

Bursell, E. (1974). Environmental aspects: humidity. In *The Physiology of Insects*, vol. 2, ed. M. Rockstein. New York: Academic Press, pp. 44–84.

Bursten, S. N., Kimsey, R. B. & Owings, D. H. (1997). Ranging of male *Oropsylla montana* fleas via male California ground squirrel (*Spermophilus beecheyi*) juveniles. *Journal of Parasitology*, **83**, 804–809.

Busalaeva, N. N. & Fedosenko, A. K. (1964). Fleas parasitic on small mammals in the high mountain areas of the Trans-Ili Ala-Tau Ridge. *Proceedings of the Institute of Zoology of Academy of Sciences of the Kazakh SSR*, **22**, 177–183 (in Russian).

Bush, A. O. & Holmes, J. C. (1986). Intestinal helminths of lesser scaup ducks: patterns of association. *Canadian Journal of Zoology*, **64**, 132–141.

Bush, A. O., Lafferty, K. D., Lotz, J. M. & Shostak, A. W. (1997). Parasitology meets ecology on its own terms: Margolis *et al.* revisited. *Journal of Parasitology*, **83**, 575–583.

Butler, J. M. & Roper, T. J. (1996). Ectoparasites and sett use in European badgers. *Animal Behaviour*, **52**, 621–629.

Buxton, P. A. (1948). Experiments with mice and fleas. I. The baby mouse. *Parasitology*, **39**, 119–124.

Byers, G. W. (1996). More on the origin of Siphonaptera. *Journal of the Kansas Entomological Society*, **69**, 274–277.

Cabrero-Sanudo, F. J. & Lobo, J. M. (2003). Estimating the number of species not yet described and their characteristics: the case of the Western Palaearctic dung beetle species (Coleoptera, Scarabaeoidea). *Biodiversity and Conservation*, **12**, 147–166.

Cadiergues, M. C., Joubert, C. & Franc, M. (2000). A comparison of jump performances of the dog flea, *Ctenocephalides canis* (Curtis, 1826) and the cat flea, *Ctenocephalides felis felis* (Bouché, 1835). *Veterinary Parasitology*, **92**, 239–241.

Cadiergues, M. C., Santamarta, D., Mallet, X. & Franc, M. (2001). First blood meal of *Ctenocephalides canis* (Siphonaptera: Pulicidae) on dogs: time to initiation of feeding and duration. *Journal of Parasitology*, **87**, 214–215.

Cai, W.-F., Fang, Z. & Yang, G.-R. (2000). Effects of temperature, humidity on the population emergence rate of *Xenopsylla cheopis* in laboratory. *Chinese Journal of Vector Biology and Control*, **11**, 358–361 (in Chinese).

Caley, M. J. & Schluter, D. (1997). The relationship between local and regional diversity. *Ecology*, **78**, 70–80.

Callaway, R. M. & King, L. (1996). Temperature-driven variation in substrate oxygenation and the balance of competition and facilitation. *Ecology*, **77**, 1189–1195.

Callaway, R. M. & Walker, L. R. (1997). Competition and facilitation: a synthetic approach to interactions in plant communities. *Ecology*, **78**, 1958–1965.

Calvete, C., Blanco-Aguiar, J. A., Virgós, E., Cabezas-Díaz, S. & Villafuerte, R. (2004). Spatial variation in helminth community structure in the red-legged partridge (*Alectoris rufa* L.): effects of definitive host density. *Parasitology*, **129**, 101–113.

Campos, E. G., Maupin, G. O., Barnes, A. M. & Eads, R. B. (1985). Seasonal occurrence of fleas (Siphonaptera) on rodents in a foothills habitat in Larimer County, Colorado, USA. *Journal of Medical Entomology*, **22**, 266–270.

Carlier, Y. & Truyens, C. (1995). Influence of maternal infection on offspring resistance towards parasites. *Parasitology Today*, **11**, 94–99.

Carney, J. P. & Dick, T. A. (2000). Helminth communities of yellow perch (*Perca flavescens* (Mitchill)): determinants of pattern. *Canadian Journal of Zoology*, **78**, 538–555.

Caro, A., Combes, C. & Euzet, L. (1997). What makes a fish a suitable host for Monogenea in the Mediterranean? *Journal of Helminthology*, **71**, 203–210.

Carson, H. L. (1959). Genetic conditions that promote or retard the formation of species. *Cold Spring Harbor Symposia on Quantitative Biology*, **24**, 87–103.

Case, T. J. & Taper, M. L. (2000). Interspecific competition, environmental gradients, gene flow, and the coevolution of species' borders. *American Naturalist*, **155**, 583–605.

Case, T. J., Holt, R. D., McPeek, M. A. & Keitt, T. H. (2005). The community context of species' borders: ecological and evolutionary perspectives. *Oikos*, **108**, 28–46.

Castleberry, S. B., Castleberry, N. L., Wood, P. B., Ford, W. M. & Mengak, M. T. (2003). Fleas (Siphonaptera) of the Allegheny woodrat (*Neotoma magister*) in West Virginia with comments on host specificity. *American Midland Naturalist*, **149**, 233–236.

Castro, J. M., Nolan, V. & Ketterson, E. D. (2001). Steroid hormones and immune function: experimental studies in wild and captive dark-eyed juncos (*Junco hyemalis*). *American Naturalist*, **157**, 408–420.

Cavitt, J. F., Pearse, A. T. & Miller, T. A. (1999). Brown thrasher nest reuse: a time saving resource, protection from search-strategy predators, or cues for nest-site selection? *Condor*, **101**, 859–862.

Chadee, D. D. (1994). Distribution patterns of *Tunga penetrans* within a community in Trinidad, West Indies. *Journal of Tropical Medicine and Hygiene*, **97**, 167–170.

Chandy, L. & Prasad, R. (1987). Behavioral resistance of hosts of flea infestation. *Proceedings of the Indian National Science Academy B*, **53**, 27–30.

Charleston, M. A. (1998). Jungles: a new solution to the host/parasite phylogeny reconciliation problem. *Mathematical Biosciences*, **149**, 191–223.

Chase, J. M. (1996). Differential competitive interactions and the included niche: an experimental analysis with grasshoppers. *Oikos*, **76**, 103–112.

Chastel, O. & Beaucournu, J. C. (1992). Specificity and ecoethology of bird fleas in Kerguelen Islands. *Annales de Parasitologie Humaine et Comparée*, **67**, 213–220.

Chekchak, T., Chapuis, J. L., Pisanu, B. & Bousses, P. (2000). Introduction of the rabbit flea, *Spilopsyllus cuniculi* (Dale), to a subantarctic island (Kerguelen Archipelago) and its assessment as a vector of myxomatosis. *Wildlife Research*, **27**, 91–101.

Chesson, P. & Huntly, N. (1997). The roles of harsh and fluctuating conditions in the dynamics of ecological communities. *American Naturalist*, **150**, 519–553.

Christe, P., Oppliger, A. & Richner, H. (1994). Ectoparasite affects choice and use of roost sites in the great tit (*Parus major*). *Animal Behaviour*, **47**, 895–898.

Christe, P., Richner, H. & Oppliger, A. (1996a). Begging, food provisioning, and nestling competition in great tit broods infested with ectoparasites. *Behavioral Ecology*, **7**, 127–131.

Christe, P., Richner, H. & Oppliger, A. (1996b). Of great tits and fleas: sleep baby sleep . . .. *Animal Behaviour*, **52**, 1087–1092.

Christe, P., Møller, A. P. & de Lope, F. (1998). Immunocompetence and nestling survival in the house martin: the tasty chick hypothesis. *Oikos*, **83**, 175–179.

Christe, P., Arlettaz, R. & Vogel, P. (2000). Variation in intensity of a parasitic mite (*Spinturnix myoti*) in relation to the reproductive cycle and immunocompetence of its bat host (*Myotis myotis*). *Ecology Letters*, **3**, 207–212.

Christe, P., Giorgi, M. S., Vogel, P. & Arlettaz, R. (2003). Differential species-specific ectoparasitic mite intensities in two intimately coexisting sibling bat species: resource-mediated host attractiveness or parasite specialization? *Journal of Animal Ecology*, **72**, 866–872.

Christe, P., Michaux, J. & Morand, S. (2006). Biological conservation and parasitism. In *Micromammals and Macroparasites: From Evolutionary Ecology to Management*, ed. S. Morand, B. R. Krasnov & R. Poulin. New York: Springer-Verlag, pp. 593–613.

Christian, K. A. & Bedford, G. S. (1995). Physiological consequences of filarial parasites in the frillneck lizard, *Chlamydosaurus kingii*, in Northern Australia. *Canadian Journal of Zoology*, **73**, 2302–2306.

Christodoulopoulos, G. & Theodoropoulos, G. (2003). Infestation of dairy goats with the human flea, *Pulex irritans*, in central Greece. *Veterinary Record*, **152**, 371–372.

Christodoulopoulos, G., Theodoropoulos, G., Kominakis, A. & Theis, J. H. (2006). Biological, seasonal and environmental factors associated with *Pulex irritans* infestation of dairy goats in Greece. *Veterinary Parasitology*, **137**, 137–143.

Clark, F. & McNeil, D. A. C. (1981). The variation in population densities of fleas in house martin nests in Leicestershire. *Ecological Entomology*, **6**, 379–386.

Clark, F. & McNeil, D. A. C. (1991). Temporal variation in the population densities of fleas in house martin nests (*Delichon u. urbica* (Linnaeus)) in Leicestershire, U.K. *Entomologist's Gazette*, **42**, 281–288.

Clark, F. & McNeil, D. A. C. (1993). A study of some factors affecting mortality overwinter in three congeneric species of bird fleas. *Entomologist*, **112**, 55–66.

Clark, F., Greenwood, M. T. & Smith, J. S. (1993a). The use of an insect activity monitor in behavioral studies of the flea, *Xenopsylla cheopis* (Rothschild). *Bulletin of the Society of Vector Ecology*, **18**, 26–32.

Clark, F., McNeil, D. A. C. & Hill, L. A. (1993b). Studies on the dispersal of three congeneric species of flea monoxenous to the house martin (*Delichon urbica* (L.)). *Entomologist*, **112**, 85–94.

Clark, F., Deadman, D., Greenwood, M. T. & Larsen, K. S. (1997). A circadian rhythm of locomotor activity in newly emerged *Ceratophyllus sciurorum*. *Medical and Veterinary Entomology*, **11**, 213–216.

Clark, F., Deadman, D., Greenwood, M. T., Larsen, K. S. & Pudney, S. (1999). Effects of feeding on the circadian rhythm of *Ceratophyllus s. sciurorum* (Siphonaptera: Ceratophyllidae). *Journal of Vector Ecology*, **24**, 78–82.

Clark, L. & Mason, J. R. (1988). Effect of biologically active plants used as nest material and the derived benefit to starling nestlings. *Oecologia*, **77**, 174–180.

Clarke, K. R. & Warwick, R. M. (1998). A taxonomic distinctness index and its statistical properties. *Journal of Applied Ecology*, **35**, 523–531.

Clarke, K. R. & Warwick, R. M. (1999). The taxonomic distinctness measure of biodiversity: weighting of step lengths between hierarchical levels. *Marine Ecology Progress Series*, **184**, 21–29.

Clarke, K. R. & Warwick, R. M. (2001). A further biodiversity index applicable to species lists: variation in taxonomic distinctness. *Marine Ecology Progress Series*, **216**, 265–278.

Clayton, D. H. & Cotgreave, P. (1994). Relationship of bill morphology to grooming behaviour in birds. *Animal Behaviour*, **47**, 195–201.

Clayton, D. H. & Moore, J. (1997). Introduction. In *Host–Parasite Evolution: General Principles and Avian Models*, ed. D. H. Clayton & J. Moore. Oxford, UK: Oxford University Press, pp. 1–6.

Clayton, D. H. & Tompkins, D. M. (1994). Ectoparasite virulence is linked to mode of transmission. *Proceedings of the Royal Society of London B*, **256**, 211–217.

Clayton, D. H. & Tompkins, D. M. (1995). Comparative effects of mites and lice on the reproductive success of rock doves (*Columba livia*). *Parasitology*, **110**, 195–206.

Clayton, D. H., Al-Tamimi, S. & Johnston, K. P. (2003). The ecological basis of coevolutionary history. In *Tangled Trees: Phylogeny, Cospeciation and Coevolution*, ed. R. D. M. Page. Chicago, IL: University of Chicago Press, pp. 310–342.

Clements, A. N. (1992). *The Biology of Mosquitoes: Development, Nutrition and Reproduction*, vol. 1. Wallingford, UK: CAB International.

Clements, J. F. (2007). *The Clements Checklist of Birds of the World*, 6th edn. London: Christopher Helm.

Cole, L. C. & Koepke, J. A. (1946). A study of rodent ectoparasites in Mobile, Alabama. *Public Health Reports*, **61**, 1469–1487.

Cole, L. C. & Koepke, J. A. (1947a). Problems of interpretation of the data of rodent–ectoparasite surveys. *Public Health Reports (Supplement)*, **202**, 1–24.

Cole, L. C. & Koepke, J. A. (1947b). A study of rodent ectoparasites in Honolulu. *Public Health Reports (Supplement)*, **202**, 25–41.

Cole, L. C. & Koepke, J. A. (1947c). A study of rodent ectoparasites in Savannah, Georgia. *Public Health Reports (Supplement)*, **202**, 42–59.

Cole, L. C. & Koepke, J. A. (1947d). A study of rodent ectoparasites in Dothan, Alabama. *Public Health Reports (Supplement)*, **202**, 61–71.

Collen, B., Purvis, A. & Gittleman, J. L. (2004). Biological correlates of description date in carnivores and primates. *Global Ecology and Biogeography*, **13**, 459–467.

Collinge, S. K., Johnson, W. C., Ray, C., et al. (2005). Landscape structure and plague occurrence in black-tailed prairie dogs on grasslands of the western USA. *Landscape Ecology*, **20**, 941–955.

Colwell, R. K. (2000). Rensch's rule crosses the line: convergent allometry of sexual size dimorphism in hummingbirds and flower mites. *American Naturalist*, **156**, 495–510.

Combes, C. (1997). Fitness of parasites: pathology and selection. *International Journal for Parasitology*, **27**, 1–10.

Combes, C. (2001). *Parasitism: The Ecology and Evolution of Intimate Interactions*. Chicago, IL: University of Chicago Press.

Combes, C. (2005). *The Art of Being a Parasite*. Chicago, IL: University of Chicago Press.

Cooke, B. D. (1990a). Rabbit burrows as environments for the European rabbit fleas, *Spilopsyllus cuniculi* (Dale), in arid South Australia. *Australian Journal of Zoology*, **38**, 317–325.

Cooke, B. D. (1990b). Notes on the comparative reproductive biology and the laboratory breeding of the rabbit flea *Xenopsylla cunicularis* Smit (Siphonaptera, Pulicidae). *Australian Journal of Zoology*, **38**, 527–534.

Cooke, B. D. (1999). Notes on the life history of the rabbit flea *Caenopsylla laptevi ibera* Beaucornu & Marquez, 1987 (Siphonaptera: Ceratophyllidae) in eastern Spain. *Parasite*, **6**, 347–354.

Cooke, B. D. & Skewes, M. A. (1988). The effect of temperature and humidity on the survival and development of the European rabbit flea *Spilopsyllus cuniculi* (Dale). *Australian Journal of Zoology*, **36**, 449–459.

Cooper, J. E. (1976). *Tunga penetrans* infestation in pigs. *Veterinary Record*, **98**, 472.

Cornell, H. V. (1993). Unsaturated patterns in species assemblages: the role of regional processes in setting local species richness. In *Species Diversity in Ecological Communities: Historical and Geographical Perspectives*, ed. R. F. Ricklefs & D. Schluter. Chicago, IL: University of Chicago Press, pp. 243–252.

Cornell, H. V. & Lawton, J. H. (1992). Species interactions, local and regional processes, and limits to richness of ecological communities: a theoretical perspective. *Journal of Animal Ecology*, **61**, 1–12.

Correia, T. R., de Souza, C. P., Fernandes, J. I., *et al.* (2003). Life cycle of *Ctenocephalides felis felis* (Bouché, 1835) (Siphonaptera, Pulicidae) from different artificial diets. *Revista Brasiliera de Zoociências, Juiz de Fora*, **5**, 153–160.

Côté, I. M. & Poulin, R. (1995). Parasitism and group size in social animals: a meta-analysis. *Behavioral Ecology*, **6**, 159–165.

Cotgreave, P. & Clayton, D. H. (1994). Comparative analysis of time spent grooming by birds in relation to parasite load. *Behaviour*, **131**, 171–187.

Cotton, M. J. (1970a). The reproductive biology of *Ctenophthalmus nobilis* (Rothschild) (Siphonaptera). *Proceedings of the Royal Entomological Society of London A*, **45**, 141–148.

Cotton, M. J. (1970b). The life history of the hen flea, *Ceratophyllus gallinae* (Schrank) (Siphonaptera, Ceratophyllidae). *Entomologist*, **103**, 45–48.

Cox, F. E. G. (2001). Concomitant infections, parasites and immune responses. *Parasitology*, **122**, S23–S38.

Cox, R., Stewart, P. D. & Macdonald, D. W. (1999). The ectoparasites of the European badger, *Meles meles*, and the behavior of the host-specific flea, *Paraceras melis*. *Journal of Insect Behavior*, **12**, 245–265.

Craw, R. C., Grehan, J. R. & Heads, M. J. (1999). *Panbiogeography: Tracking the History of Life*. New York: Oxford University Press.

Cresswell, J. E., Vidal-Martinez, V. M. & Crichton, N. J. (1995). The investigation of saturation in the species richness of communities: some comments on methodology. *Oikos*, **72**, 301–304.

Croll, N. A., Anderson, R. M., Gyerkos, T. W. & Ghardian, E. (1982). The population biology and control of *Ascaris lumbricoides* in a rural community in Iran. *Transactions of the Royal Society of Tropical Medicine and Hygiene*, **76**, 187–197.

Crooks, K. R., Scott, C. A., Angeloni, L., *et al.* (2001). Ectoparasites of the island fox on Santa Cruz Island. *Journal of Wildlife Diseases*, **37**, 189–193.

Crooks, K. R., Garcelon, D. K., Scott, C. A., *et al.* (2004). Ectoparasites of a threatened insular endemic mammalian carnivore: the island spotted skunk. *American Midland Naturalist*, **151**, 35–41.

Crowell, K. L. & Pimm, S. L. (1976). Competition and niche shifts introduced onto small islands. *Oikos*, **27**, 251–258.

Crum, G. E., Knapp, F. W. & White, G. M. (1974). Response of the cat flea, *Ctenocephalides felis* (Bouche), and the Oriental rat flea, *Xenopsylla cheopis* (Rothschild), to electromagnetic radiation in the 300–700 nanometer range. *Journal of Medical Entomology*, **11**, 88–94.

Cumming, G. S. & Bernard, R. T. (1997). Rainfall, food abundance and timing of parturition in African bats. *Oecologia*, **111**, 309–317.

Cyprich, D., Krumpál, M. & Rolníková, T. (1999). Occurrence and distribution of *Ceratophyllus gallinae* (Schrank, 1803) (Siphonaptera) in Slovakia. *Folia Faunistica Slovaca*, **4**, 79–88 (in Slovak).

Cyprich, D., Pinowski, J. & Krumpál, M. (2002). Seasonal changes in numbers of fleas (Siphonaptera) in nests of the house sparrow (*Passer domesticus*) and tree sparrow (*P. montanus*) in Warsaw surroundings (Poland). *Acta Parasitologica*, **47**, 58–65.

Cyprich, D., Štiavnická, L. & Kiefer, M. S. (2003). Information about occurrence and distribution specific flea species (Siphonaptera) of the ground squirrel (*Spermophilus citellus* L., 1758) in Slovakia: the genus *Citellophilus* and *Neopsylla*. *Folia Faunistica Slovaca*, **8**, 47–51 (in Slovak).

Cyprich, D., Krumpál, M. & Duda, M. (2006). Characteristic of flea fauna (Siphonaptera) of ecological group of birds (Aves) nesting on the ground. *Entomofauna Carpathica*, **18**, 1–4 (in Slovak).

Dallai, R., Lupetti, P., Afzelius, B. A. & Frati, F. (2003). Sperm structure of Mecoptera and Siphonaptera (Insecta) and the phylogenetic position of *Boreus hyemalis*. *Zoomorphology*, **122**, 211–220.

Dampf, A. (1911). *Palaeopsylla klebsiana* n. sp., eine fossiler Floh aus dem baltischen Bernstein. *Schriften der Physikalisch–Ökonomischen Gesellschaft zu Königsberg*, **51**, 248–259.

Darby, C., Ananth, S. L., Tan, L. & Hinnebusch, B. J. (2005). Identification of gmhA, a *Yersinia pestis* gene required for flea blockage, by using a *Caenorhabditis elegans* biofilm system. *Infection and Immunity*, **73**, 7236–7242.

Darskaya, N. F. (1954). Fleas of the Daurian ground squirrel. *Proceedings of the Irkutsk State Scientific Anti-Plague Institute of Siberia and Far East*, **12**, 245–257 (in Russian).

Darskaya, N. F. (1955). Ecological characteristics of *Xenopsylla gerbilli caspica*. I. Fleas parasitic on the great gerbil, associated with the ecological features of their host. In *Natural Focality of Human Diseases and Regional Epidemiology*, ed. Anonymous. Moscow, USSR: Medgiz, pp. 400–407 (in Russian).

Darskaya, N. F. (1964). On the comparative ecology of fleas belonging to the genus *Ceratophyllus*. *Ectoparasites*, **4**, 31–180 (in Russian).

Darskaya, N. F. (1970). Ecological comparisons of some fleas of the USSR fauna. *Zoologicheskyi Zhurnal*, **49**, 729–745 (in Russian).

Darskaya, N. F. & Besedina, K. P. (1961). On the possibility of fleas feeding on reptiles. *Proceeding of Scientific Anti-Plague Institute of Caucasus and Trans-Caucasus*, **5**, 33–39 (in Russian).

Darskaya, N. F. & Karandina, P. C. (1974). Observations on pre-imaginal development of *Neopsylla setosa* Wagn., 1898: fleas parasitic on ground squirrels. In *Particularly Dangerous Diseases in Caucasus: Proceedings of the 3rd Scientific–Practical Conference of the Anti-Plague Establishments of Caucasus on Natural Focality, Epidemiology and Prophylaxis of Particularly Dangerous Disieases*, 14–16 May 1974, ed. V. G. Pilipenko.

Stavropol, USSR: Scientific Anti-Plague Institute of Caucasus and Trans-Caucasus, pp. 134–137 (in Russian).

Darskaya, N. F., Brukhanova, L. V. & Kunitskaya, N. T. (1965). On the method of investigation of flea oviposition. In *Problems of General Zoology and Medical Entomology*, ed. Anonymous. Moscow, USSR: Moscow University Press, pp. 6–9.

Darskaya, N. F., Bragina, Z. S. & Petrov, V. G. (1970). On fleas of the common vole and shrews in dependence of sharp density fluctuations of these mammals. In *Vectors of Particularly Dangerous Diseases and their Control*, ed. V. E. Tiflov. Stavropol, USSR: Scientific Anti-Plague Institute of Caucasus and Trans-Caucasus, pp. 132–152 (in Russian).

Davidson, D. W. (1980). Some consequences of diffuse competition in a desert ant community. *American Naturalist*, **116**, 92–105.

Davidson, D. W. (1985). An experimental study of diffuse competition in harvester ants. *American Naturalist*, **125**, 500–506.

Davis, D. E. (1951). Observations on rat ectoparasites and typhus fever in San Antonio, Texas. *Public Health Reports*, **66**, 1717–1726.

Davis, R. M., Smith, R. T., Madon, M. B. & Sitko-Cleugh, E. (2002). Flea, rodent and plague ecology at Chichupate Campground, Ventura County, California. *Journal of Vector Ecology*, **27**, 107–127.

Dawson, L. H. J., Renaud, F., Guégan, J. F. & de Meeûs, T. (2000). Experimental evidence of asymmetrical competition between two species of parasitic copepods. *Proceedings of the Royal Society of London B*, **267**, 1973–1978.

Dawson, R. D. (2004). Does fresh vegetation protect avian nests from ectoparasites? An experiment with tree swallows. *Canadian Journal of Zoology*, **82**, 1005–1010.

Dawson, R. D. & Bortolotti, G. R. (1997). Ecology of parasitism of nestling American kestrels by *Carnus hemapterus* (Diptera, Carnidae). *Canadian Journal of Zoology*, **75**, 2021–2026.

Day, J. F. & Benton, A. H. (1980). Population dynamics and coevolution of adult siphonapteran parasites of the southern flying squirrel (*Glaucomys volans volans*). *American Midland Naturalist*, **103**, 333–338.

de Albuquerque Cardoso, V. & Linardi, P. M. (2006). Scanning electron microscopy studies of sensilla and other structures of the head of *Polygenis* (*Polygenis*) *tripus* (Siphonapera: Rhopalopsyllidae). *Micron*, **37**, 557–565.

Dean, S. R. & Meola, R. W. (1997). Effect of juvenile hormone and juvenile hormone mimics on sperm transfer from the testes of the male cat flea (Siphonaptera: Pulicidae). *Journal of Medical Entomology*, **34**, 485–488.

Dean, S. R. & Meola, R. W. (2002a). Factors influencing sperm transfer and insemination in cat fleas (Siphonaptera: Pulicidae) fed on an artificial membrane system. *Journal of Medical Entomology*, **39**, 475–479.

Dean, S. R. & Meola, R. W. (2002b). Effect of diet composition on weight gain, sperm transfer, and insemination in the cat flea (Siphonaptera: Pulicidae). *Journal of Medical Entomology*, **39**, 370–375.

de Carvalho, R. W., de Almeida, A. B., Barbosa-Silva, S. C., *et al.* (2003). The patterns of tungiasis in Araruama township, state of Rio de Janeiro, Brazil. *Memórias do Instituto Oswaldo Cruz*, **98**, 31–36.

Degen, A. A. (1997). *Ecophysiology of Small Desert Mammals*. New York: Springer-Verlag.

Degen, A. A. (2006). Effect of macroparasites on the energy budget of small mammals. In *Micromammals and Macroparasites: From Evolutionary Ecology to Management*, ed. S. Morand, B. R. Krasnov & R. Poulin. New York: Springer-Verlag, pp. 371–399.

Degen, A. A., Kam, M., Khokhlova, I. S., Krasnov, B. R. & Barraclough, T. (1998). Average daily metabolic rate of rodents: habitat and dietary comparisons. *Functional Ecology*, **12**, 63–73.

Delahay, R. J., Speakman, J. R. & Moss, R. (1995). The energetic consequences of parasitism: effects of a developing infection of *Trichostongylus tenius* (Nematoda) of red grouse (*Lagopus lagopus scoticus*) energy balance, body weight and condition. *Parasitology*, **110**, 473–482.

de Lope, F., Møller, A. P. & de la Cruz, C. (1998). Parasitism, immune response and reproductive success in the house martin *Delichon urbica*. *Oecologia*, **114**, 188–193.

Demas, G. E. & Nelson, R. J. (1998). Photoperiod, ambient temperature, and food availability interact to affect reproductive and immune function in adult male deer mice (*Peromyscus maniculatus*). *Journal of Biological Rhythms*, **13**, 253–262.

Demin, E. P., Zagniborodova, E. N., Sageev, M. T., *et al.* (1970). Some features of ecology of fleas parasitic on *Rhombomys opimus* in western Turkmenistan in relation to their importance in the epizootology of plague. *Problems of Particularly Dangerous Diseases*, **11**, 49–56 (in Russian).

de Moraes, L. B., Bossi, D. E. P. & Linhares, A. X. (2003). Siphonaptera parasites of wild rodents and marsupials trapped in three mountain ranges of the Atlantic Forest in Southeastern Brazil. *Memórias do Instituto Oswaldo Cruz*, **98**, 1071–1076.

den Hollander, N. & Allen, J. R. (1986). Cross-reactive antigens between a tick *Dermacentor variabilis* (Acari: Ixodidae) and a mite *Prosoptes cuniculi* (Acari: Psoroptidae). *Journal of Medical Entomology*, **23**, 44–50.

Deoras, P. J. & Prasad, R. S. (1967). Feeding mechanisms of Indian fleas *X. cheopis* (Roths.) and *X. astia* (Roths.). *Indian Journal of Medical Research*, **55**, 1041–1050.

de Pedro, N., Delgado, M. J., Gancedo, B. & Alonso-Bedate, M. (2003). Changes in glucose, glycogen, thyroid activity and hypothalamic catecholamines in tench by starvation and refeeding. *Journal of Comparative Physiology B*, **173**, 475–481.

Desdevises, Y., Jovelin, R., Jousson, O. & Morand, S. (2000). Comparison of ribosomal DNA sequences of *Lamellodiscus* spp. (Monogenea, Diplectanidae) parasitising *Pagellus* (Sparidae, Teleostei) in the north Mediterranean Sea: species divergence and coevolutionary interactions. *International Journal for Parasitology*, **30**, 741–746.

Desdevises, Y., Morand, S., Jousson, O. & Legendre, P. (2002a). Coevolution between *Lamellodiscus* (Monogenea: Diplectanidae) and Sparidae (Teleostei): the study of a complex host–parasite system. *Evolution*, **56**, 2459–2471.

Desdevises, I., Morand, S. & Legendre, P. (2002b). Evolution and determinants of host specificity in the genus *Lamellodiscus* (Monogenea). *Biological Journal of the Linnean Society*, **77**, 431–443.

Devi, S. G. & Prasad, R. S. (1985). Phagostimulants in artificial feeding systems of rat fleas *Xenopsylla cheopis* and *Xenopsylla astia*. *Proceedings of the Indian National Science Academy B*, **51**, 566–573.

Dezfuli, B. S., Giari, L., De Biaggi, S. & Poulin, R. (2001). Associations and interactions among intestinal helminths of the brown trout, *Salmo trutta*, in northern Italy. *Journal of Helminthology*, **75**, 331–336.

Dick, C. W. & Patterson, B. D. (2006). Bat flies: obligate parasites of bats. In *Micromammals and Macroparasites: From Evolutionary Ecology to Management*, ed. S. Morand, B. R. Krasnov & R. Poulin. New York: Springer-Verlag, pp. 179–194.

Dick, C. W., Gannon, M. R., Little, W. E. & Patrick, M. J. (2003). Ectoparasite associations of bats from central Pennsylvania. *Journal of Medical Entomology*, **40**, 813–819.

Diamond, J. M. (1975). Assembly of species communities: chance or competition? In *Ecology and Evolution of Communities*, ed. M. L. Cody & J. M. Diamond. Cambridge, MA: Harvard University Press, pp. 342–444.

Dittmar de la Cruz, K. & Whiting, M. (2003). Genetic and phylogeographic structure of populations of *Pulex simulans* (Siphonaptera) in Peru inferred from two genes (CytB and CoII). *Parasitology Research*, **91**, 55–59.

Dittmar de la Cruz, K., Mamat, U., Whiting, M., *et al.* (2003). Techniques of DNA-studies on prehispanic ectoparasites (*Pulex* sp., Pulicidae, Siphonaptera) from animal mummies of the Chiribaya Culture, Southern Peru. *Memórias do Instituto Oswaldo Cruz*, **98**, 53–58.

Dobson, A. P. (1985). The population dynamics of competition between parasites. *Parasitology*, **91**, 317–347.

Dobson, A. P. (1990). Models for multi-species parasite–host communities. In *Parasite Communities: Patterns and Processes*, ed. G. Esch, A. O. Bush & J. M. Aho. London: Chapman & Hall, pp. 261–288.

Dobson, A. P. & Roberts, M. (1994). The population dynamics of parasitic helminth communities. *Parasitology*, **109**, S97–S108.

Dogiel, V. A., Petrushevski, G. K. & Polyanski, Y. I. (1961). *Parasitology of Fishes*. Edinburgh, UK: Oliver & Boyd.

Dong, B. (1991). Experimental studies on the transmission of hemorrhagic fever with renal syndrome virus by gamasid mites and fleas. *Chinese Medical Journal*, **71**, 502–504 (in Chinese).

Dowling, A. P. G. (2006). Mesostigmatid mites as parasites of small mammals: systematics, ecology, and the evolution of parasitic associations. In *Micromammals and Macroparasites: From Evolutionary Ecology to Management*, ed. S. Morand, B. R. Krasnov & R. Poulin. New York: Springer-Verlag, pp. 103–117.

Downing, J. A. (1986). Spatial heterogeneity: evolved behaviour or mathematical artefact. *Nature*, **323**, 255–257.

Drobney, R. D., Train, C. T. & Gredrickson, L. H. (1983). Dynamics of the platyhelminth fauna of wood ducks in relation to food habits and reproductive state. *Journal of Parasitology*, **69**, 375–380.

Dryden, M. W. (1989). Host association, on-host longevity and egg production of *Ctenocephalides felis felis*. *Veterinary Parasitology*, **34**, 117–122.

Dryden, M. W. & Blakemore, J. (1989). A review of flea allergy dermatitis in the dog and cat. *Companion Animal Practice*, **19**, 10–16.

Dryden, M. W. & Broce, A. B. (1993). Development of a trap for collecting newly emerged *Ctenocephalides felis* (Siphonaptera: Pulicidae) in homes. *Journal of Medical Entomology*, **30**, 901–906.

Dubinina, V. B. & Dubinin, M. N. (1951). Parasite fauna of mammals of the Dauric Steppe. *Fauna and Ecology of Rodents*, **4**, 98–156 (in Russian).

Dubyansky, M. A., Dubyanskaya. L. D., Zhubanazarov, I. Z. & Filipchenko, V. E. (1975). Alternative method for prognosis of abundance of *Xenopsylla* fleas. In *Proceedings of the 10th Scientific Conference of the Anti-Plague Establishments of the Middle Asia and Kazakhstan*, vol. 2, ed. M. A. Aikimbaev. Alma-Ata, USSR: The Middle Asian Scientific Anti-Plague Institute, pp. 93–96 (in Russian).

Duffy, J. E., Richardson, J. P. & France, K. E. (2005). Ecosystem consequences of diversity depend on food chain length in estuarine vegetation. *Ecology Letters*, **8**, 301–309.

Dufva, R, & Allander, K. (1996). Variable effects of the hen flea *Ceratophyllus gallinae* on the breeding success of the great tit *Parus major* in relation to weather conditions. *Ibis*, **138**, 772–777.

Dunnet, G. M. & Mardon, D. K. (1974). A monograph of Australian fleas (Siphonaptera). *Australian Journal of Zoology (Supplemental Series)*, **30**, 1–273.

Durden, L. A. (1995). Fleas (Siphonaptera) of cotton mice on a Georgia Barrier Island: a depauperate fauna. *Journal of Parasitology*, **81**, 526–529.

Durden, L. A. & Beaucournu, J. C. (2002). *Gymnomeropsylla* n. gen. (Siphonaptera: Pygiopsyllidae) from Sulawesi, Indonesia, with the description of two new species. *Parasite*, **9**, 225–232.

Durden, L. A. & Kollars, T. M. (1997). The fleas (Siphonaptera) of Tennessee. *Journal of Vector Ecology*, **22**, 13–22.

Durden, L. A., Ellis, B. A., Banks, C. W., Crowe, J. D. & Oliver, J. H. (2004). Ectoparasites of gray squirrels in two different habitats and screening of selected ectoparasites for bartonellae. *Journal of Parasitology*, **90**, 485–489.

Durden, L. A., Judy, T. N., Martin, J. E. & Spedding, L. S. (2005). Fleas parasitizing domestic dogs in Georgia, USA: species composition and seasonal abundance. *Veterinary Parasitology*, **130**, 157–162.

Durden, L. A., Cunningham, M. W., McBride, R. & Ferree, B. (2006). Ectoparasites of free-ranging pumas and jaguars in the Paraguayan Chaco. *Veterinary Parasitology*, **137**, 189–193.

Dynesius, M. & Jansson, R. (2000). Evolutionary consequences of changes in species' geographical distributions driven by Milankovitch climate oscillations. *Proceedings of the National Academy of Sciences of the USA*, **97**, 9115–9120.

Ebert, D. (1994). Virulence and local adaptation of a horizontally transmitted parasite. *Science*, **265**, 1084–1086.

Eckstein, R. A. & Hart, B. L. (2000a). The organization and control of grooming in cats. *Applied Animal Behaviour Science*, **68**, 131–140.

Eckstein, R. A. & Hart, B. L. (2000b). Grooming and control of fleas in cats. *Applied Animal Behaviour Science*, **68**, 141–150.

Edney, E. B. (1945). Laboratory studies on the bionomics of the rat fleas, *Xenopsylla brasiliensis* Baker and *X. cheopis* Roths. I. Certain effects of light, temperature and humidity on the rate of development and on adult longevity. *Bulletin of Entomological Research*, **35**, 399–416.

Edney, E. B. (1947a). Laboratory studies on the bionomics of the rat fleas, *Xenopsylla brasiliensis* Baker and *X. cheopis* Rothsch. II. Water relations during the cocoon period. *Bulletin of Entomological Research*, **38**, 263–280.

Edney, E. B. (1947b). Laboratory studies on the bionomics of the rat fleas, *Xenopsylla brasiliensis* Baker and *X. cheopis* Roths. III. Further factors affecting adult longevity. *Bulletin of Entomological Research*, **38**, 389–404.

Eeley, H. A. C. & Foley, R. A. (1999). Species richness, species range size and ecological specialization among African primates: geographical patterns and conservation implications. *Biodiversity and Conservation*, **8**, 1033–1056.

Eeva, T., Lehikoinen, E. & Nurmi, J. (1994). Effects of ectoparasites on breeding success of great tits (*Parus major*) and pied flycatchers (*Ficedula hypoleuca*) in an air-pollution gradient. *Canadian Journal of Zoology*, **72**, 624–635.

Egoscue, H. J. (1976). Flea exchange between deer mice and some associated small mammals in western Utah. *Great Basin Naturalist*, **36**, 475–480.

Eisele, M., Heukelbach, J., van Marck, E., *et al.* (2003). Investigations on the biology, epidemiology, pathology and control of *Tunga penetrans* in Brazil. I. Natural history of tungiasis in man. *Parasitology Research*, **90**, 87–99.

Eiseman, C. H. & Binnington, K. C. (1994). The peritrophic membrane: its formation, structure, chemical composition and permeability in relation to vaccination against ectoparasitic arthropods. *International Journal for Parasitology*, **24**, 15–26.

Elliot, J. M. (1977). *Some Methods for Statistical Analysis of Samples of Benthic Invertebrates*, 2nd edn, Freshwater Biological Association Sceintific Publications No. 25. Ambleside, UK: Titus Wilson & Son.

Ellis, B. A., Regnery, R. L., Beati, L., *et al.* (1999). Rats of the genus *Rattus* are reservoir hosts for pathogenic *Bartonella* species: an Old World origin for a New World disease. *Journal of Infectious Diseases*, **180**, 220–224.

Elmes, G. W., Barr, B., Thomas, J. A. & Clarke, R. T. (1999). Extreme host specificity by *Microdon mutabilis* (Diptera: Syrphidae), a social parasite of ants. *Proceedings of the Royal Society of London B*, **266**, 447–453.

Elshanskaya, N. I. & Popov, M. N. (1972). Zoologico-parasitological characteristics of the River Kenkeme valley (Central Yakutia). In *Theriology*, vol. 1, ed. L. D. Kolosova & I. V. Lukyanova. Novosibirsk, USSR: Nauka, Siberian Branch, pp. 368–372 (in Russian).

Emelianova, N. D. & Shtilmark, F. R. (1967). Fleas of insectivores, rodents and lagomorphs of the central part of Western Sayan. *Proceedings of the Irkutsk State Scientific Anti-Plague Institute of Siberia and Far East*, **27**, 241–253 (in Russian).

Engelthaler, D. M., Hinnebusch, B. J., Rittner, C. M. & Gage, K. L. (2000). Quantitative competitive PCR as a technique for exploring flea–*Yersina pestis* dynamics. *American Journal of Tropical Medicine and Hygiene*, **62**, 552–560.

Enquist, B. J., Haskell, J. P. & Tiffney, B. H. (2002). General patterns of taxonomic and biomass partitioning in extant and fossil plant communities. *Nature*, **419**, 610–613.

Esbérard, C. (2001). Infestation of *Rhynchopsyllus pulex* (Siphonaptera: Tungidae) on *Molossus molossus* (Chiroptera) in Southeastern Brazil. *Memórias do Instituto Oswaldo Cruz*, **96**, 1169–1170.

Euzet, L. & Combes, C. (1980). Les problèmes de l'espèce chez les animaux parasites. *Bulletin de la Société Zoologique de France*, **40**, 239–285.

Evans, F. G. & Freeman, R. B. (1950). On the relationship of some mammal fleas to their hosts. *Annals of the Entomological Society of America*, **43**, 320–333.

Ezenwa, V. O. (2004). Host social behavior and parasitic infection: a multifactorial approach. *Behavioral Ecology*, **15**, 446–454.

Fain, A. & Hyland, K. W. (1985). Evolution of astigmatid mites on mammals. In *Coevolution of Parasitic Arthropods and Mammals*, ed. K. C. Kim. New York: John Wiley, pp. 641–658.

Fairbairn, D. J. (1997). Allometry for sexual size dimorphism: pattern and process in the coevolution of body size in males and females. *Annual Review of Ecology and Systematics*, **28**, 659–687.

Fairbairn, D. J. (2005). Allometry for sexual size dimorphism: testing two hypotheses for Rensch's rule in the water strider *Aquarius remigis*. *American Naturalist*, **166**, S69–S84.

Farhang-Azad, A., Traub, R. & Wisseman, C. L. (1983). *Rickettsia mooseri* infection in the fleas *Leptopsylla segnis* and *Xenopsylla cheopis*. *American Journal of Tropical Medicine and Hygiene*, **32**, 1392–1400.

Farhang-Azad, A., Traub, R., Sofi, M. & Wisseman, C. L. (1984). Experimental murine typhus infection in the cat fleas, *Ctenocephalides felis* (Siphonaptera: Pulicidae). *Journal of Medical Entomology*, **21**, 675–680.

Farhang-Azad, A., Traub, R. & Boqar, S. (1985). Transovarial transmission of murine typhus rickettsia in *Xenopsylla cheopis* fleas. *Science*, **227**, 543–545.

Farhang-Azad, A., Radulovic, S., Higgins, J. A., Noden, B. H. & Troyer, J. M. (1997). Flea-borne rickettsioses: ecologic considerations. *Emerging Infectious Diseases*, **3**, 319–327.

Faulkenberry, G. D. & Robbins, R. G. (1980). Statistical measures of interspecific association between the fleas of the gray-tailed vole, *Microtus canicaudus* Miller. *Entomological News*, **91**, 93–101.

Faust, E. C. & Maxwell, T. A. (1930). The finding of the larva of the chigo, *Tunga penetrans*, in scrapings from human skin. *Archives of Dermatology and Syphilology*, **22**, 94–97.

Fauth, P. T., Krementz, D. G. & Hines, J. E. (1991). Ectoparasitism and the role of green nesting material in the European starling. *Oecologia*, **88**, 22–29.

Fedorov, Y. V., Igolkin, N. I. & Tyushnikova, M. K. (1959). Some data on virus-carrying fleas in areas of tick-borne encephalitis and lymphocytic choriomenengitis. *Medical Parasitology and Parasitic Diseases* [Meditsinskaya Parazitologiya i Parazitarnye Bolezni], **28**, 149–152 (in Russian).

Feingold, B. F., Benjamini, E. & Michaeli, D. (1968). The allergic responses to insect bites. *Annual Review of Entomology*, **13**, 138–158.

Feliu, C., Renaud, F., Catzeflis, F., *et al.* (1997). Comparative analysis of parasite species richness of Iberian rodents. *Parasitology*, **115**, 453–466.

Fellis, K. J., Negovetich, N. J., Esch, G. W., Horak, I. G. & Boomker, J. (2003). Patterns of association, nestedness, and species co-occurrence of helminth parasites in the greater kudu, *Tragelaphus strepsiceros*, in the Kruger National Park, South Africa, and the Etosha National Park, Namibia. *Journal of Parasitology*, **89**, 899–907.

Felsenstein, J. (1985). Phylogenies and the comparative method. *American Naturalist*, **125**, 1–15.

Feng, X.-Y., Ni, E.-J., Cao, P.-G., Zheng, C.-J. & Yang, G.-H. (2004). Investigation of the distributive features of flea population during epidemic outbreak of plague in Longlin County, Guangxi. *Acta Parasitologica et Medica Entomologica Sinica*, **11**, 235–237 (in Chinese).

Feng, Y.-M., Li, W. & Wang, Z.-Y. (2003). Observation on the life cycle of *Ischnopsyllus octactenus* at laboratory. *Chinese Journal of Vector Biology and Control*, **14**, 202–203 (in Chinese).

Fenner, F. & Ratcliff, F. N. (1965). *Myxomatosis*. Cambridge, UK: Cambridge University Press.

Feoktistov, A. Z., Vasiliev, G. I. & Kraminsky, N. N. (1968). Passage of the virus of tick-borne encephalitis via fleas *Xenopsylla cheopis* Roths. In *Problems of Epidemiology and Epizootology of Paricularly Dangerous Diseases*, ed. Anonymous. Kyzyl, USSR: The Tuva Anti-Plague Station, pp. 317–320.

Ferkin, M. H., Sorokin, E. S. & Johnston, R. E. (1996). Self-grooming as a sexually dimorphic communicative behaviour in meadow voles, *Microtus pennsylvanicus*. *Animal Behaviour*, **51**, 801–810.

Ferrari, N., Cattadori, I. M., Nespereira, J., Rizzoli, A. & Hudson, P. J. (2004). The role of host sex in parasite dynamics: field experiments on the yellow-necked mouse *Apodemus flavicollis*. *Ecology Letters*, **7**, 88–94.

Ferrari, N., Rosá, R., Pugliese, A. & Hudson, P. J. (2007). The role of sex in parasite dynamics: model simulations on transmission of *Heligmosomoides polygyrus* in populations of yellow-necked mice, *Apodemus flavicollis*. *International Journal for Parasitology*, **37**, 341–349.

Fichet-Calvet, E., Jomaa, I., Ben Ismail, R. & Ashford, R. W. (2000). Pattern of infection of haemoparasites in the fat sand rat, *Psammomys obesus*, in Tunisia and effect on the host. *Annals of Tropical Medicine and Parasitology*, **94**, 55–68.

Fieberg, J. & Ellner, S. P. (2000). When is it meaningful to estimate an extinction probability? *Ecology*, **81**, 2040–2047.

Fielden, L. J., Rechav, Y. & Bryson, N. R. (1992). Acquired immunity to larvae of *Amblyomma marmoreum* and *A. hebraeum* by tortoises, guinea-pigs and guinea-fowl. *Medical and Veterinary Entomology*, **6**, 251–254.

Fielden, L. J., Jones, R. M., Goldberg, M. & Rechav, Y. (1999). Feeding and respiratory gas exchange in the American dog tick, *Dermacentor variabilis*. *Journal of Insect Physiology*, **45**, 297–304.

Fielden, L. J., Krasnov, B. R. & Khokhlova, I. S. (2001). Respiratory gas exchange in the flea *Xenopsylla conformis* (Siphonaptera: Pulicidae). *Journal of Medical Entomology*, **38**, 735–739.

Fielden, L. J., Krasnov, B. R., Still, K. & Khokhlova, I. S. (2002). Water balance in two species of desert fleas, *Xenopsylla ramesis* and *X. conformis* (Siphonaptera: Pulicidae). *Journal of Medical Entomology*, **39**, 875–881.

Fielden, L. J., Krasnov, B. R., Khokhlova, I. S. & Arakelyan, M. S. (2004). Respiratory gas exchange in the desert flea *Xenopsylla ramesis* (Siphonaptera: Pulicidae): response to temperature and blood-feeding. *Comparative Biochemistry and Physiology A*, **137**, 557–565.

Filimonova, S. A. (1986). Changes in the ultra-structure of the intestinal epithelium of *Xenopsylla cheopis* (Siphonaptera) after emerging from cocoons and beginning of feeding. *Parazitologiya*, **20**, 99–105 (in Russian).

Filimonova, S. A. (1989). A morphologic analysis of digestion in *Leptopsylla segnis* (Siphonaptera: Leptopsyllidae) fleas. *Parazitologiya*, **23**, 480–488 (in Russian).

Fiorello, C. V., Robbins, R. G., Maffei, L. & Wade, S. E. (2006). Parasites of free-ranging small canids and felids in the Bolivian Chaco. *Journal of Zoo and Wildlife Medicine*, **37**, 130–134.

Fisher, R. A. (1930). *The Genetical Theory of Natural Selection*. Oxford, UK: Oxford University Press.

Fitze, P. S. & Richner, H. (2002). Differential effects of a parasite on ornamental structures based on melanins and carotenoids. *Behavioral Ecology*, **13**, 401–407.

Fitze, P. S., Clobert, J. & Richner, H. (2004a). Long-term life-history consequences of ectoparasite-modulated growth and development. *Ecology*, **85**, 2018–2026.

Fitze, P. S., Tschirren, B. & Richner, H. (2004b). Life history and fitness consequences of ectoparasites. *Journal of Animal Ecology*, **73**, 216–226.

Fleming, T. H., Breitwisch, R. L. & Whitesides, G. W. (1987). Patterns of tropical vertebrate frugivore diversity. *Annual Review of Ecology and Systematics*, **18**, 91–109.

Flux, J. E. C. (1972). Seasonal and regional abundance of fleas on hares in Kenya. *Journal of East African Natural History Society*, **29**, 1–8.

Folstad, I. & Karter, A. J. (1992). Parasites, bright males, and the immunocompetence handicap. *American Naturalist*, **139**, 603–622.

Forbes, M. R., Alisauskas, R. T., McLaughlin, J. D. & Cuddington, K. M. (1999). Explaining co-occurrence among helminth species of lesser snow geese (*Chen caerulescens*) during their winter and spring migration. *Oecologia*, **120**, 613–620.

Foster, W. A. & Olkowski, W. (1968). Natural invasion of artificial cliff swallow nests by *Oeciacus vicarius* (Hemiptera: Cimicidae) and *Ceratophyllus petrochelidoni* (Siphonaptera: Ceratophyllidae). *Journal of Medical Entomology*, **5**, 488–491.

Fowler, J. A., Cohen, S. & Greenwood, M. T. (1983). Seasonal variation in the infestation of blackbirds by fleas. *Bird Study*, **30**, 240–242.

Fox, B. J. & Brown, J. H. (1993). Assembly rules for the functional groups in North American desert rodent communities. *Oikos*, **67**, 358–370.

Fox, B. J. & Luo, J. (1996). Estimating competition coefficients from census data: a re-examination of the regression technique. *Oikos*, **77**, 291–300.

Fox, I., Fox, R. I. & Bayona, I. G. (1966). Fleas feed on the lizards in the laboratory in Puerto Rico. *Journal of Medical Entomology*, **2**, 395–396.

Fox, J. W., McGrady-Steed, J. & Petchey, O. L. (2000). Testing for local species saturation with nonindependent regional species pools. *Ecology Letters*, **3**, 198–206.

Fox, L. R. (1975). Cannibalism in natural populations. *Annual Review of Ecology and Systematics*, **6**, 87–106.

Fox, L. R. & Morrow, P. A. (1981). Specialization: species property or local phenomenon? *Science*, **211**, 887–893.

Franc, M., Choquart, P. & Cadiergues, M. C. (1998). Species of fleas found on dogs in France. *Revue de Médecine Vétérinaire*, **149**, 135–140.

Freeman, R. B. & Madsen, H. (1949). A parasitic flea larva. *Nature*, **164**, 187–188.

Fretwell, S. D. & Lucas, H. L. (1970). On territorial behavior and other factors influencing habitat distribution in birds. I. Theoretical development. *Acta Biotheoretica*, **19**, 16–36.

Frigessi, A., Holden, M., Marshall, C., *et al.* (2005). A Bayesian model for the population dynamics of two interacting species, with application to great gerbils and fleas in south-eastern Kazakhstan. *Biometrics*, **61**, 231–239.

Fry, J. D. (1996). The evolution of host specialization: are trade-offs overrated? *American Naturalist*, **148**, S84–S107.

Fuller, G. K. (1974). Observations on flea attachment at low hair densities on man. *Journal of Natural History*, **8**, 207–213.

Futuyma, D. J. & Moreno, G. (1988). The evolution of ecological specialization. *Annual Review of Ecology and Systematics*, **19**, 207–233.

Futuyma, D. J. & Slatkin, M. (1983). *Coevolution*. Sunderland, MA: Sinauer Associates.

Gabbutt, P. D. (1961). The distribution of some small mammals and their associated fleas from central Labrador. *Ecology*, **42**, 518–525.

Gäde, G. (2002). Sexual dimorphism in the pyrgomorphid grasshopper *Phymateus morbillosus*: from wing morphometry and flight behaviour to flight physiology and endocrinology. *Physiological Entomology*, **27**, 51–57.

Gage, K. L. & Kosoy, M. Y. (2005). Natural history of plague: perspectives from more than a century of research. *Annual Review of Entomology*, **50**, 505–528.

Gage, K. L., Ostfeld, R. S. & Olson, J. G. (1995). Nonviral vector-borne zoonoses associated with mammals in the United States. *Journal of Mammalogy*, **76**, 695–715.

Galaktionov, K. V. (1996). Life cycles and distribution of seabird helminths in Arctic and subArctic regions. *Bulletin of the Scandinavian Society for Parasitology*, **6**, 31–49.

Galbe, J. & Oliver, J. H. (1992). Immune response of lizards and rodents to larval *Ixodes scapularis* (Acari, Ixodidae). *Journal of Medical Entomology*, **29**, 774–783.

Gallivan, G. J. & Horak, I. G. (1997). Body size and habitat as determinants of tick infestations of wild ungulates in South Africa. *South African Journal of Wildlife Research*, **27**, 63–70.

Galun, R. (1975). Research into alternative arthropod control measures against livestock pests (part 1). In *Workshop on the Ecology and Control of External Parasites of Economic Importance on Bovines in Latin America*, ed. K. C. Thompson. Cali, Colombia: Centro Internacional de Agricultura Tropical (CIAT), pp. 155–161.

Garland, T., Harvey, P. H. & Ives, A. R. (1992). Procedures for the analysis of comparative data using phylogenetically independent contrasts. *American Naturalist*, **41**, 18–32.

Garland, T., Dickerman, A. W. C., Janis, M. & Jones, J. A. (1993). Phylogenetic analysis of covariance by computer simulation. *Systematic Biology*, **42**, 265–292.

Gaston, K. J. (1996). Spatial covariance in the species richness of higher taxa. In *Aspects of the Genesis and Maintenance of Biological Diversity*, ed. M. E. Hochberg, J. Clobert & R. Barbault. Oxford, UK: Oxford University Press, pp. 221–242.

Gaston, K. J. (2003). *The Structure and Dynamics of Geographic Ranges*. Oxford, UK: Oxford University Press.

Gaston, K. J. & Blackburn, T. M. (2000). *Pattern and Process in Macroecology*. Oxford, UK: Blackwell Science.

Gaston, K. J. & Blackburn, T. M. (2003). Dispersal and the interspecific abundance–occupancy relationship in British birds. *Global Ecology and Biogeography*, **12**, 373–379.

Gaston, K. J., Blackburn, T. M. & Loder, N. (1995). Which species are described first? The case of North American butterflies. *Biodiversity and Conservation*, **4**, 119–127.

Gaston, K. J., Blackburn, T. M. & Lawton, J. H. (1997). Interspecific abundance–range size relationships: an appraisal of mechanisms. *Journal of Animal Ecology*, **66**, 579–601.

Gauzshtein, D. M., Kunitsky, V. N., Kunitskaya, N. T. & Filimonov, V. I. (1965). On the time spent on the host body in fleas parasitic on the great gerbil. In *Proceedings of the 4th Scientific Conference of the Anti-Plague Establishments of the Middle Asia and Kazakhstan*, ed. M. A. Aikimbaev. Alma-Ata, USSR: The Middle Asian Scientific Anti-Plague Institute and Kainar, pp. 66–68 (in Russian).

Gauzshtein, D. M., Kunitsky, V. N., Gubaidullina, V. S., *et al.* (1967). On the phenology of reproduction in some fleas parasitic on the great gerbil in the southern Balkhash Region. In *Proceedings of the 5th Scientific Conference of the Anti-Plague Establishments of the Middle Asia and Kazakhstan*, ed. M. A. Aikimbaev. Alma-Ata, USSR: The Middle Asian Scientific Anti-Plague Institute, pp. 160–163 (in Russian).

Geigy, R. & Herbig, A. (1949). Die Hypertrophie der Organe beim Weibchen von *Tunga penetrans*. *Acta Tropica*, **6**, 246–262.

Gerasimova, N. G. (1970). Metamorphosis of fleas *Xenopsylla nuttalli* Ioff, 1930. In *Vectors of Dangerous Diseases and Their Control*, ed. V. E. Tiflov. Stavropol, USSR: Scientific Anti-Plague Institute of Caucasus and Trans-Caucasus, pp. 316–322 (in Russian).

Gerasimova, N. G. (1973). Some reproductive parameters in *Xenopsylla skrjabini* and *X. nuttalli*. *Problems of Particularly Dangerous Diseases*, **29**, 117–121 (in Russian).

Gerasimova, N. G., Denisova, N. G., Denisov, P. S., Knyazeva, T. V. & Lavrovsky, A. A. (1977). Species composition and population dynamics of fleas on the pygmy ground squirrel in the stable foci of plague in the Ergeni Upland. *Parazitologiya*, **11**, 446–452 (in Russian).

Gillespie, R. D., Mbow, M. L. & Titus, R. G. (2000). The immunomodulatory factors of bloodfeeding arthropod saliva. *Parasite Immunology*, **22**, 319–331.

Gillespie, S. H., Smith, G. L. & Osbourn, A. (2004). *Microbe–Vector Interactions in Vector-Borne Diseases*. Cambridge, UK: Cambridge University Press.

Giorgi, M. S., Arlettaz, R., Christe, P. & Vogel, P. (2001). The energetic grooming costs imposed by a parasitic mite (*Spinturnix myoti*) upon its bat host (*Myotis myotis*). *Proceedings of the Royal Society of London B*, **268**, 2071–2075.

Gliwicz, J. (1992). Patterns of dispersal in non-cyclic populations of small rodents. In *Animal Dispersal: Small Mammals as a Model*, ed. N. C. Stenseth & W. Z. Lidicker. London: Chapman & Hall, pp. 147–159.

Goater, C. P. & Ward, P. I. (1992). Negative effects of *Rhabdias bufonis* (Nematoda) on the growth and survival of toads (*Bufo bufo*). *Oecologia*, **89**, 161–165.

Gobel, E. & Krampitz, H. E. (1982). Histologische Untersuchungen zur Gamogonie und Sporogonie von *Hepatozoon erhardovae* in experimentell infizierten Rattenflöhen (*Xenopsylla cheopis*). *Zeitschrift für Parasitenkunde*, **67**, 261–271.

Gong, Y.-L., Li, Z.-L. & Ma, L.-M. (2004). Further research of the bloodsucking activities of the flea *Citellophilus tesquorum sungaris*. *Acta Parasitologica et Medica Entomologica Sinica*, **11**, 47–49 (in Chinese).

Gong, Z.-D., Xie, B.-Q. & Ling, J.-B. (1996). Ecology and fauna of fleas on Mt. Gaoligong of Yunnan. *Zoological Research*, **17**, 59–67 (in Chinese).

Gong, Z.-D., Duan, X.-D., Feng, X.-G., Wu, X.-Y. & Liu, Q. (1999). The fauna and ecology of fleas in Cangshan Mountain and Erhai Lake Nature Reserve, Dali. *Zoological Research*, **20**, 451–456 (in Chinese).

Gong, Z.-D., Wu, H.-Y., Duan, X.-D., Feng, X.-G. & Yang, G.-R. (2000). Fauna and community ecology of fleas in Lincang region, Yunnan province. *Acta Parasitologica et Medica Entomologica Sinica*, **7**, 160–169 (in Chinese).

Gong, Z.-D., Zheng, D., Wu, H.-Y, et al. (2001). The relationship between the geographical distribution trends of flea species diversity and the important environmental factor in the Hengduan Mountains, Yunnan. *Biodiversity Science*, **9**, 319–328 (in Chinese).

Gong, Z.-D., Wu, H.-Y., Duan, X.-D., et al. (2004). Vertical distribution pattern and fauna characteristics of flea communities in the Mt. Wuliang Nature Reserve, Jingdong, Yunnan. *Chinese Journal of Vector Biology and Control*, **15**, 344–348 (in Chinese).

Gong, Z.-D., Wu, H.-Y., Duan, X.-D., et al. (2005). Species richness and vertical distribution pattern of flea fauna in Hengduan Mountains of western Yunnan, China. *Biodiversity Science*, **13**, 279–289 (in Chinese).

Gong, Z.-D., Zhang, L.-Y., Duan, X.-D., et al. (2007). Species richness and fauna of fleas along a latitudinal gradient in the Three Parallel Rivers landscape, China. *Biodiversity Science*, **15**, 61–69 (in Chinese).

González, M. T. & Poulin, R. (2005). Spatial and temporal predictability of the parasite community structure of a benthic marine fish along its distributional range. *International Journal for Parasitology*, **35**, 1369–1377.

Gotelli, N. J. (2000). Null model analysis of species co-occurrence patterns. *Ecology*, **81**, 2606–2621.

Gotelli, N. J. & Arnett, A. E. (2000). Biogeographic effects of red fire ant invasion. *Ecology Letters*, **3**, 257–261.

Gotelli, N. J. & Entsminger, G. L. (2001). Swap and fill algorithms in null model analysis: rethinking the Knight's Tour. *Oecologia*, **129**, 281–291.

Gotelli, N. J. & Entsminger, G. L. (2006). *EcoSim: Null Models Software for Ecology. Version 7*. Jericho, VT: Acquired Intelligence Inc. & Kesey-Bear. Available online at http://garyentsminger.com/ecosim.htm.

Gotelli, N. J. & Graves, G. R. (1996). *Null Models in Ecology*. Washington, DC: Smithsonian Institution Press.

Gotelli, N. J. & McCabe, D. J. (2002). Species co-occurrence: a meta-analysis of J. M. Diamond's assembly rules model. *Ecology*, **83**, 2091–2096.

Gotelli, N. J. & Rohde, K. (2002). Co-occurrence of ectoparasites of marine fishes: a null model analysis. *Ecology Letters*, **5**, 86–94.

Goüy de Bellocq, J., Sarà, M., Casanova, J. C., Feliu, C. & Morand, S. (2003). A comparison of the structure of helminth communities in the woodmouse, *Apodemus sylvaticus*, on islands of the western Mediterranean and continental Europe. *Parasitology Research*, **90**, 64–70.

Goüy de Bellocq, J., Krasnov, B. R., Khokhlova, I. S., Ghazaryan, L. & Pinshow, B. (2006a). Immunocompetence and flea parasitism in a desert rodent. *Functional Ecology*, **20**, 637–646.

Goüy de Bellocq, J., Krasnov, B. R., Khokhlova, I. S. & Pinshow, B. (2006b). Temporal dynamics of a T-cell mediated immune response in desert rodents. *Comparative Biochemistry and Physiology A*, **145**, 554–559.

Gracia, M. J., Lucientes, J., Castillo, J. A., et al. (2000). *Pulex irritans* infestation in dogs. *Veterinary Record*, **147**, 748–749.

Grafen, A. (1989). The phylogenetic regression. *Philosophical Transactions of the Royal Society of London B*, **326**, 119–157.

Gray, C. A., Gray, P. N. & Pence, D. B. (1989). Influence of social status on the helminth community of late-winter mallards. *Canadian Journal of Zoology*, **67**, 1937–1944.

Grazhdanov, A. K., Bidashko, F. G., Tanitovky, V. A., et al. (2002). Comparative fecundity of fleas parasitic on *Meriones* gerbils in the laboratory. *Quarantinable and Zoonotic Infections in Kazakhstan*, **6**, 39–43 (in Russian).

Grebenyuk, R. V. (1951). *Sheep Vermipsylleses and their Control*. Frunze, USSR: Ylym (in Russian).

Greene, W. K., Carnegie, R. L., Shaw, S. E., Thompson, R. C. A. & Penhale, W. J. (1993). Characterization of allergens of the cat flea, *Ctenocephalides felis*: detection and frequency of IgE antibodies in canine sera. *Parasite Immunology*, **15**, 69–74.

Greenwood, M. T., Clark, F. & Smith, J. S. (1991). Automatic recording of flea activity. *Medical and Veterinary Entomology*, **5**, 93–100.

Gregory, R. D., Montgomery, S. S. J. & Montgomery, W. I. (1992). Population biology of *Heligmosomoides polygyrus* (Nematoda) in the wood mouse. *Journal of Animal Ecology*, **61**, 749–757.

Gregory, R. D., Keymer, A. E. & Harvey, P. H. (1996). Helminth parasite richness among vertebrates. *Biodiversity and Conservation*, **5**, 985–997.

Greives, T. J., McGlothlin, J. W., Jawor, J. M., Demas, G. E. & Ketterson, E. D. (2006). Testosterone and innate immune function inversely covary in a wild population of breeding dark-eyed juncos (*Junco hyemalis*). *Functional Ecology*, **20**, 812–818.

Grenfell, B. T. (1992). Parasitism and the dynamics of ungulate grazing systems. *American Naturalist*, **139**, 907–929.

Grenfell, B. T. & Dobson, A. P. (eds.) (1995). *Ecology of Infectious Diseases in Natural Populations*. Cambridge, UK: Cambridge University Press.

Grenfell, B. T., Dietz, K. & Roberts, M. G. (1995). Modelling the immuno-epidemiology of macroparasites in naturally-fluctuated host populations. In *Ecology of Infectious Diseases in Natural Populations*, ed. B. T. Grenfell & A. P. Dobson. Cambridge, UK: Cambridge University Press, pp. 362–383.

Griffiths, D. (1999). On investigation local–regional species richness relationships. *Journal of Animal Ecology*, **68**, 1051–1055.

Grodzinski, W. & Wunder, B. A. (1975). Ecological energetics of small mammals. In *Small Mammals: Their Productivity and Population Dynamics*, ed. F. B. Golley, K. Petrusewitz & L. Ryszkowski. Cambridge, UK: Cambridge University Press, pp. 173–204.

Gromov, V. S., Krasnov, B. R. & Shenbrot, G. I. (2000). Space use in Wagner's gerbil *Gerbillus dasyurus* in the Negev Highlands, Israel. *Acta Theriologica*, **45**, 175–182.

Gruen, J. R. & Weissman, S. M. (1997). Evolving views of the major histocompatibility complex. *Blood*, **90**, 4252–4265.

Gubareva, N. P., Akiev, A. K., Zemelman, B. M. & Abdurakhmanov, G. A. (1976). Effect of some factors on formation of the plague blockage in *Ceratophyllus tesquorum* and *Neopsylla setosa setosa*. *Parazitologiya*, **10**, 315–319 (in Russian).

Guégan, J.-F. & Hugueny, B. A. (1994). A nested parasite species subset pattern in tropical fish host as major determinant of parasite infracommunity structure. *Oecologia*, **100**, 184–189.

Guégan, J.-F. & Kennedy, C. R. (1996). Parasite richness/sampling effort/host range: the fancy three-piece jigsaw puzzle. *Parasitology Today*, **12**, 367–369.

Guégan, J.-F. & Morand, S. (1996). Polyploid hosts: strange attractors for parasites! *Oikos*, **7**, 366–370.

Guégan, J.-F., Lambert, A., Leveque, C. & Euzet, L. (1992). Can host body size explain the parasite species richness in tropical freshwater fishes? *Oecologia*, **90**, 197–204.

Guégan, J.-F., Morand, S. & Poulin, R. (2005). Are there general laws in parasite community ecology? The emergence of spatial parasitology and epidemiology. In *Parasitism and Ecosystems*, ed. F. Thomas, J.-F. Guégan & F. Renaud. Oxford, UK: Oxford University Press, pp. 22–42.

Guerrero, O. M., Chinchilla, M. & Abrahams, E. (1997). Increasing of *Toxoplasma gondii* (Coccidia: Sarcocystidae) infections by *Trypanosoma lewisi* (Kinetoplastida, Trypanosomatidae) in white rats. *Revista de Biologia Tropical*, **45**, 877–882.

Gulland, F. M. D. (1995). The impact of infectious diseases on wild animal populations: a review. In *Ecology of Infectious Diseases in Natural Populations*, ed. B. T. Grenfell & A. P. Dobson. Cambridge, UK: Cambridge University Press, pp. 20–51.

Guo, T. & Xu, R. (1999). Trophic niche of flea in the southern slope of the Himalaya Mountains. *Chinese Journal of Applied Ecology*, **10**, 67–70 (in Chinese).

Guo, X.-G., Gong, Z.-D., Qian, T.-J., et al. (2000). Flea fauna investigation in some foci of human plague in Yunnan, China. *Acta Zootaxonomica Sinica*, **25**, 291–297 (in Chinese).

Gurevitch, J., Morrow, L. L., Wallace, A. & Walsh, J. S. (1992). A meta-analysis of competition in field experiments. *American Naturalist*, **140**, 539–572.

Gurtler, R. E., Cohen, J. E., Cecere, M. C. & Chuit, R. (1997). Shifting host choices of the vector of Chagas disease, *Triatoma infestans*, in relation to the availability of hosts in houses in north-west Argentina. *Journal of Applied Ecology*, **34**, 699–715.

Gusev, V. M., Petrosyan, E. A., Guseva, A. A., Eigelis, Y. K. & Tchernyavsky, A. M. (1962). Wild birds: carriers of ectoparasites in the Trans-Caucasus. *Proceedings of the Azerbaijanian Anti-Plague Station*, **3**, 177–184 (in Russian).

Guseva, A. A. & Kosminsky, R. B. (1974). Feeding and reproduction of *Frontopsylla elata caspica* Ioff et Arg., 1934 (Ceratophyllidae, Siphonaptera) in experiments. In *Particularly Dangerous Diseases in Caucasus: Proceedings of the 3rd Scientific–Practical Conference of the Anti-Plague Establishments of Caucasus on Natural Focality, Epidemiology and Prophylaxis of Particularly Dangerous Diseases*, 14–16 May 1974, ed. V. G. Pilipenko. Stavropol, USSR: Scientific Anti-Plague Institute of Caucasus and Trans-Caucasus, pp. 132–134 (in Russian).

Gustafson, C. R., Bickford, A. A., Cooper, G. L. & Charlton, B. R. (1997). Sticktight fleas associated with fowl pox in a backyard chicken flock in California. *Avian Diseases*, **41**, 1006–1009.

Gwinner, H. & Berger, S. (2005). European starlings: nestling condition, parasites and green nest material during the breeding season. *Journal of Ornithology*, **146**, 365–371.

Haag-Wackernagel, D. & Spiewak, R. (2004). Human infestation by pigeon fleas (*Ceratophyllus columbae*) from feral pigeons. *Annals of Agricultural and Environmental Medicine*, **11**, 343–346.

Haas, G. E. (1965). Comparative suitability of the four murine rodents of Hawaii as hosts for *Xenopsylla vexabilis* and *X. cheopis* (Siphonaptera). *Journal of Medical Entomology*, **2**, 75–83.

Haas, G. E. (1966). A technique for estimating the total number of rodent fleas in cane fields in Hawaii. *Journal of Medical Entomology*, **2**, 392–394.

Haas, G. E. (1969). Quantitative relationships between fleas and rodents in a Hawaiian cane field. *Pacific Science*, **23**, 70–82.

Haas, G. E. & Kucera, J. R. (2004). Fleas (Siphonaptera) in nests of voles (*Microtus* spp.) in montane habitats of three regions of Utah. *Western North American Naturalist*, **64**, 346–352.

Haas, G. E. & Wilson, N. (1985). Rodent fleas (Siphonaptera) in tree cavities of woodpeckers in Alaska, USA. *Canadian Field Naturalist*, **100**, 554–556.

Haas, G. E. & Wilson, N. (1998). *Polygenis martinezbaezi* (Siphonaptera: Rhopalopsyllidae) reared from a rodent nest found in the Peloncillo Mountains of southwestern New Mexico. *Journal of Medical Entomology*, **35**, 431–432.

Haas, G. E., Johnson, L. & Wilson, N. (1980). Siphonaptera from mammals in Alaska, USA. Supplement 2. Southeastern Alaska. *Journal of the Entomological Society of British Columbia*, **77**, 43–46.

Haas, G. E., Rumfelt, T. & Wilson, N. (1981). Fleas (Siphonaptera) from nests of the tree swallow *Iridoprocne bicolor* and the violet-green swallow *Tachycineta thalassina* in Alaska, USA. *Wasmann Journal of Biology*, **39**, 37–41.

Haas, G. E., Kucera, J. R., Runck, A. M., Macdonald, S. O. & Cook, J. A. (2005). Mammal fleas (Siphonaptera: Ceratophyllidae) new for Alaska and the southeastern mainland collected during seven years of a field survey of small mammals. *Journal of the Entomological Society of British Columbia*, **102**, 65–75.

Haeselbarth, E., Segerman, J. & Zumpt, F. (1966). The arthropod parasites of vertebrates in Africa south of the Sahara (Ethiopian region). III. Insecta excl. Phthiraptera. *Publications of the South African Institute for Medical Research*, **13**, 1–283.

Hafner, M. S. & Nadler, S. A. (1988). Phylogenetic trees support the coevolution of parasites and their hosts. *Nature*, **332**, 258–259.

Hafner, M. S. & Nadler, S. A. (1990). Cospeciation in host–parasite assemblages: comparative analysis of rates of evolution and timing of cospeciation events. *Systematic Zoology*, **39**, 192–204.

Hafner, M. S. & Page, R. D. M. (1995). Molecular phylogenies and host–parasite cospeciation: gophers and lice as a model system. *Philosophical Transactions of the Royal Society of London B*, **349**, 77–83.

Haftorn, S. (1994). The act of tremble-thrusting in tit nests, performance and possible function. *Fauna Norvegica C*, **17**, 55–74.

Haitlinger, R. (1970). Die Flöhe (Siphonaptera) der Kleinsäuger aus den West- und Mittelsudeten. *Polskie Pismo Entomologiczne*, **40**, 749–762.

Haitlinger, R. (1971). Aphanipterofauna drobnych gryzoni i owadożernych Wrocławia. *Zootechnika*, **30**, 9–22.

Haitlinger, R. (1973). The parasitological investigation of small mammals of the Góry Sowie (Middle Sudetes). I. Siphonaptera (Insecta). *Polskie Pismo Entomologiczne*, **43**, 499–519.

Haitlinger, R. (1974). Fleas (Siphonaptera) of small mammals of the Pieniny, Poland. *Polskie Pismo Entomologiczne*, **44**, 765–788.

Haitlinger, R. (1975). The parasitological investigation of small mammals of the Góry Sowie (Middle Sudetes). II. Siphonaptera (Insecta). *Polskie Pismo Entomologiczne*, **45**, 373–396.

Haitlinger, R. (1977). The parasitological investigation of small mammals of the Góry Sowie (Middle Sudetes). VI. Siphonaptera, Anoplura, Acarina. *Polskie Pismo Entomologiczne*, **47**, 429–492.

Haitlinger, R. (1981). Structure of Arthropod communities occurring on *Microtus arvalis* (Pall.) in various habitats. I. Faunistic differentiation, dominance structure, arthropod infestation intensiveness in relation to habitats and host population dynamics. *Polish Ecological Studies*, **7**, 271–292.

Haitlinger, R. (1989). Arthropods (Acari, Anoplura, Siphonaptera, Coleoptera) of small mammals of the Babia Góra Mts. *Acta Zoologica Cracoviensia*, **32**, 15–56.

Hallas, T. & Bang, P. (1976). Fleas caught on small mammals at seven locations in eastern Denmark. *Flora og Fauna*, **82**, 11–18.

Halliwell, R. E. W. & Longino, S. J. (1985). IgE and IgG antibodies to flea antigen in differing dog populations. *Veterinary Immunology and Immunopathology*, **8**, 215–223.

Halliwell, R. E. W. & Schemmer, K. R. (1987). The role of basophils in the immunopathogenesis of hypersensitivity to fleas (*Ctenocephalides felis*) in dogs. *Veterinary Immunology and Immunopathology*, **15**, 203–213.

Halvorsen, O. (1985). On the relationship between social status of a host and risk of parasitic infection. *Oikos*, **47**, 71–74.

Hamilton, W. D. & Zuk, M. (1982). Heritable true fitness and bright birds: a role for parasites? *Science*, **218**, 384–387.

Hamilton, P. B., Stevens, J. R., Holz, P., *et al.* (2005). The inadvertent introduction into Australia of *Trypanosoma nabiasi*, the trypanosome of the European rabbit (*Oryctolagus cuniculus*), and its potential for biocontrol. *Molecular Ecology*, **14**, 3167–3175.

Hanski, I. (1982). Communities of bumblebees: testing the core–satellite hypothesis. *Annales Zoologici Fennici*, **19**, 65–73.

Hanski, I. (1998). Metapopulation dynamics. *Nature*, **396**, 41–49.

Hanski, I., Kouki, J. & Halkka, A. (1993). Three explanations of the positive relationship between distribution and abundance of species. In *Species Diversity in Ecological Communities: Historical and Geographical Perspectives*, ed. R. E. Ricklefs & D. Schluter. Chicago, IL: University of Chicago Press, pp. 108–116.

Hanski, I., Moilanen, A. & Gyllenberg, M. (1996). Minimum viable metapopulation size. *American Naturalist*, **147**, 527–541.

Harper, G. H., Marchant, A. & Boddington, D. G. (1992). The ecology of the hen flea *Ceratophyllus gallinae* and the moorhen flea *Dasypsyllus gallinulae* in nestboxes. *Journal of Animal Ecology*, **61**, 317–327.

Harrison, J. F. & Roberts, S. P. (2000). Flight respiration and energetics. *Annual Review of Physiology*, **62**, 179–205.

Hart, B. L. (1990). Behavioral adaptations to pathogens and parasites: five strategies. *Neuroscience and Biobehavioural Reviews*, **14**, 273–294.

Hart, B. L. & Pryor, P. A. (2004). Developmental and hair-coat determinants of grooming behaviour in goats and sheep. *Animal Behaviour*, **67**, 11–19.

Hart, B. L., Hart, L. A., Mooring, M. S. & Olubayo, R. (1992). Biological basis of grooming behaviour in antelope: the body size, vigilance and habitat principles. *Animal Behaviour*, **44**, 615–631.

Hartley, S. & Shorrocks, B. (2002). A general framework for the aggregation model of coexistence. *Journal of Animal Ecology*, **71**, 651–662.

Hartwell, W. V., Quan, S. F., Scott, K. G. & Kartman, L. (1958). Observations on flea transfer between hosts: a mechanism in the spread of bubonic plague. *Science*, **127**, 814.

Harvey, P. H. & Pagel, M. D. (1991). *The Comparative Method in Evolutionary Biology*. Oxford, UK: Oxford University Press.

Hasibender, G. & Dye, C. (1988). Population dynamics of mosquito-borne disease: persistence in a completely heterogenous environment. *Theoretical Population Biology*, **33**, 31–53.

Hastriter, M. W. (1997). Establishment of the tungid flea, *Tunga monositus* (Siphonaptera: Pulicidae), in the United States. *Great Basin Naturalist*, **57**, 281–282.

Hastriter, M. W. (2000a). *Jordanopsylla becki* (Siphonaptera: Ctenophthalmidae), a new species of flea from the Nevada Test Site. *Proceedings of the Entomological Society of Washington*, **102**, 135–141.

Hastriter, M. W. (2000b). *Echidnophaga suricatta* (Siphonaptera: Pulicidae), a new species of flea from the Northern Cape Province, South Africa. *African Zoology*, **35**, 77–83.

Hastriter, M. W. (2001a). Five new species and new subgenus of fleas (Siphonaptera: Chimaeropsyllidae, Ctenophthalmidae) from Africa. *Proceedings of the Entomological Society of Washington*, **103**, 832–848.

Hastriter, M. W. (2001b). Fleas (Siphonaptera: Ctenophthalmidae and Rhopalopsyllidae) from Argentina and Chile with two new species from the rock rat, *Aconaemys fuscus*, in Chile. *Annals of Carnegie Museum*, **70**, 169–178.

Hastriter, M. W. (2004). Revision of the flea genus *Jellisonia* Traub, 1944 (Siphonaptera: Ceratophyllidae). *Annals of Carnegie Museum*, **73**, 213–238.

Hastriter, M. W. & Eckerlin, R. P. (2003). *Jellisonia painteri* (Siphonaptera: Ceratophyllidae), a new species of flea from Guatemala. *Annals of Carnegie Museum*, **72**, 215–221.

Hastriter, M. W. & Haas, G. E. (2005). Bionomics and distribution of species of *Hystrichopsylla* in Arizona and New Mexico, with a description of *Hystrichopsylla dippiei oblique*, n. ssp. (Siphonaptera: Hystrichopsyllidae). *Journal of Vector Ecology*, 30, 251–262.

Hastriter, M. W. & Tipton, V. J. (1975). Fleas (Siphonaptera) associated with small mammals of Morocco. *Journal of the Egyptian Public Health Association*, **50**, 79–169.

Hastriter, M. W. & Whiting, M. E. (2002). *Macropsylla novaehollandiae* (Siphonaptera: Hystrichopsyllidae), a new species of flea from Tasmania. *Proceedings of the Entomological Society of Washington*, **104**, 663–671.

Hastriter, M. W. & Whiting, M. F. (2003). Siphonaptera (fleas). In *Encyclopedia of Insects*, ed. V. H. Resh & R. Carde. Orlando, FL: Elsevier Science, pp. 1039–1045.

Hastriter, M. W., Egoscue, H. J. & Traub, R. (1998). A description of the male of *Jordanopsylla allredi* Traub and Tipton, 1951, and characterization of the tribes within Anomiopsyllinae (Siphonaptera: Ctenophthalmidae). *Proceedings of the Entomological Society of Washington*, **100**, 141–146.

Hastriter, M. W., Zyzak, M. D., Soto, R., *et al.* (2002). Fleas (Siphonaptera) from Ancash Department, Peru with the description of a new species, *Ectinorus alejoi* (Rhopalopsyllidae), and the description of the male of *Plocopsylla pallas* (Rothschild, 1914) (Sephanocircidae). *Annals of Carnegie Museum*, **71**, 87–106.

Hastriter, M. W., Haas, G. E. & Wilson, N. (2006). New distribution records for *Stenoponia americana* (Baker) and *Stenoponia ponera* Traub and Johnson (Siphonaptera: Ctenophthalmidae) with a review of records from the southwestern United States. *Zootaxa*, **1253**, 51–59.

Haukisalmi, V. & Henttonen, H. (1990). The impact of climatic factors and host density on the long-term population dynamics of vole helminths. *Oecologia*, **83**, 309–315.

Haukisalmi, V. & Henttonen, H. (1993). Coexistence in helminths of the bank vole *Clethrionomys glareolus*. I. Patterns of co-occurrence. *Journal of Animal Ecology*, **62**, 221–229.

Hawlena, H., Abramsky, Z. & Krasnov, B. R. (2005). Age-biased parasitism and density-dependent distribution of fleas (Siphonaptera) on a desert rodent. *Oecologia*, **146**, 200–208.

Hawlena, H., Krasnov, B. R., Abramsky, Z., *et al.* (2006a). Flea infestation and energy requirements of rodent hosts: are there general rules? *Functional Ecology*, **20**, 1028–1036.

Hawlena, H., Khokhlova, I. S., Abramsky, Z. & Krasnov, B. R. (2006b). Age, intensity of infestation by flea parasites and body mass loss in a rodent host. *Parasitology*, **133**, 187–193.

Hawlena, H., Abramsky, Z. & Krasnov, B. R. (2006c). Ectoparasites and age-dependent survival in a desert rodent. *Oecologia*, **148**, 30–39.

Hawlena, H., Abramsky, Z., Krasnov, B. R. & Saltz, D. (2007a). Host defence versus intraspecific competition in the regulation of infrapopulations of the flea *Xenopsylla conformis* on its rodent host *Meriones crassus*. *International Journal for Parasitology*, **37**, 919–925.

Hawlena, H., Bashary, D., Abramsky, Z. & Krasnov, B. R. (2007b). Benefits, costs and constraints of anti-parasitic grooming in adult and juvenile rodents. *Ethology*, **113**, 394–402.

He, J.-H., Liang, Y. & Zhang, H.-Y. (1997). A study on the transmission of plague though seven kinds of fleas in rat type and wild rodent type plague foci in Yunnan. *Chinese Journal of Epidemiology*, **18**, 236–240 (in Chinese).

Heath, A. W., Arfsten, A., Yamanaka, M., Dryden, M. W. & Dale, B. (1994). Vaccination against the cat flea *Ctenocephalides felis felis*. *Parasite Immunology*, **16**, 187–191.

Hecht, O. (1943). La reacciones da la piel contra las picaduras de insectos como fenómenas alergicos. *Revista de Sanidad y Asistencia Social*, **8**, 945–959.

Heckmann, R., Gansen, B. & Hom, M. (1967). Maternal transfer of immunity to rat coccidiosis: *Eimeria nieschulzi* (Dieben – 1924). *Journal of Protozoology*, **S14**, 35.

Hecnar, S. J. & M'Closkey, R. T. (1997). Patterns of nestedness and species association in a pond-dwelling amphibian fauna. *Oikos*, **80**, 371–381.

Heeb, P., Werner, I., Richner, H. & Kölliker, M. (1996). Horizontal transmission and reproductive rates of hen fleas in great tit nests. *Journal of Animal Ecology*, **65**, 474–484.

Heeb, P., Werner, I., Kölliker, M. & Richner, H. (1998). Benefits of induced host responses against an ectoparasite. *Proceedings of the Royal Society of London B*, **265**, 51–56.

Heeb, P., Werner, I., Mateman, A. C., et al. (1999). Ectoparasite infestation and sex-biased local recruitment of hosts. *Nature*, **400**, 63–65.

Heeb, P., Kölliker, M. & Richner, H. (2000). Bird–ectoparasite interactions, nest humidity, and ectoparasite community structure. *Ecology*, **81**, 958–968.

Heino, J., Muotka, T. & Paavola, R. (2003). Determinants of macroinvertebrate diversity in headwater streams: regional and local influences. *Journal of Animal Ecology*, **72**, 425–434.

Heitman, T. L., Koski, K. G. & Scott, M. E. (2003). Energy deficiency alters behaviours involved in transmission of *Heligmosomoides polygyrus* (Nematoda) in mice. *Canadian Journal of Zoology*, **81**, 1767–1773.

Heller-Haupt, A., Kagaruki, L. K. & Varma, M. G. R. (1996). Resistance and cross-resistance in rabbits to adults of three species of African ticks (Acari: Ixodidae). *Experimental and Applied Acarology*, **20**, 155–165.

Hemmes, R. B., Alvarado, A. & Hart, B. L. (2002). Use of California bay foliage by wood rats for possible fumigation of nest-borne ectoparasites. *Behavioral Ecology*, **13**, 381–385.

Henry, P. Y., Poulin, B., Rousset, F., Renaud, F. & Thomas, F. (2004). Infestation by the mite *Harpirhynchus nidulans* in the bearded tit *Panurus biarmicus*. *Bird Study*, **51**, 34–40.

Heukelbach, J., de Oliveira, F. A. S., Hesse, G. & Feldmeier, H. (2001). Tungiasis: a neglected health problem of poor communities. *Tropical Medicine and International Health*, **6**, 267–272.

Heukelbach, J., Wilcke, T. & Feldmeier, H. (2004). Cutaneous larva migrans (creeping eruption) in an urban slum in Brazil. *International Journal of Dermatology*, **43**, 511–515.

Hillebrand, H. (2005). Regressions of local on regional diversity do not reflect the importance of local interactions or saturation of local diversity. *Oikos*, **110**, 195–198.

Hillebrand, H. & Blenckner, T. (2002). Regional and local impact on species diversity: from pattern to processes. *Oecologia*, **132**, 479–491.

Hinaidy, H. K. (1991). The biology of *Dipylidium caninum*. *Zentralblatt für Veterinärmedizin B*, **38**, 329–336.

Hinkle, N. C., Koehler, P. G. & Kern, W. H. (1991). Hematophagous strategies of the cat flea (Siphonaptera: Pulicidae). *Florida Entomologist*, **74**, 377–385.

Hinkle, N. C., Koehler, P. G. & Patterson, R. S. (1998). Host grooming efficiency for regulation of cat flea (Siphonaptera: Pulicidae) populations. *Journal of Medical Entomology*, **35**, 266–269.

Hinnebusch, B. J., Gage, K. L. & Schwan, T. G. (1998). Estimation of vector infectivity rates for plague by means of a standard curve-based competitive polymerase chain reaction method to quantify *Yersinia pestis* in fleas. *American Journal of Tropical Medicine and Hygiene*, **58**, 562–569.

Hinton, H. E. (1958). The phylogeny of the Panorpoid orders. *Annual Review of Entomology*, **3**, 181–206.

Hoberg, E. P., Brooks, D. R. & Siegel-Causey, D. (1997). Host–parasite co-speciation: history, principles, and prospects. In *Host–Parasite Evolution: General Principles and Avian Models*, ed. D. H. Clayton & J. Moore. Oxford, UK: Oxford University Press, pp. 212–235.

Holland, C. (1984). Interactions between *Moniliformis* (Acanthocephala) and *Nippostrongylus* (Nematoda) in the small intestine of laboratory rats. *Parasitology*, **88**, 303–315.

Holland, G. P. (1955). Primary and secondary sexual characteristics of some Ceratophyllinae, with notes on the mechanism of copulation (Siphonaptera). *Transactions of the Royal Entomological Society of London*, **107**, 233–248.

Holland, G. P. (1964). Evolution, classification and host relationships of Siphonaptera. *Annual Review of Entomology*, **9**, 123–146.

Holland, G. P. (1985). The fleas of Canada, Alaska and Greenland (Siphonaptera). *Memoirs of the Entomological Society of Canada*, **130**, 1–631.

Holmes, J. C. & Price P. W. (1986). Communities of parasites. In *Community Ecology: Patterns and Processes*, ed. J. Kikkawa & D. J. Anderson. Oxford, UK: Blackwell Science, pp. 187–213.

Holmstad, P. R., Hudson, P. J. & Skorping, A. (2005). The influence of a parasite community on the dynamics of a host population: a longitudinal study on willow ptarmigan and their parasites. *Oikos*, **111**, 377–391.

Holt, R. D., Dobson, A. P., Begon, M., Bowers, R. G. & Schauber, E. M. (2003). Parasite establishment in host communities. *Ecology Letters*, **6**, 837–842.

Honnay, O., Hermy, M. & Coppin, P. (1999). Nested plant communities in deciduous forest fragments: species relaxation or nested habitats? *Oikos*, **84**, 119–129.

Hoogland, J. L. & Sherman, P. W. (1976). Advantages and disadvantages of bank swallow (*Riparia riparia*) coloniality. *Ecological Monographs*, **46**, 33–58.

Hopkins, G. H. E. (1957). Host associations of Siphonaptera. In *1st Symposium on Host Specificity amongst Parasites of Vertebrates*, ed. J. G. Baer. Neuchâtel, Switzerland: Institut de Zoologie, Université de Neuchâtel, pp. 64–87.

Hopkins, G. H. E. & Rothschild, M. (1953). *An Illustrated Catalogue of the Rothschild Collection of Fleas (Siphonaptera) in the British Museum (Natural History)*, vol. 1, *Tungidae and Pulicidae*. London: Trustees of the British Museum.

Hopkins, G. H. E. & Rothschild, M. (1956). *An Illustrated Catalogue of the Rothschild Collection of Fleas (Siphonaptera) in the British Museum (Natural History)*, vol. 2, *Coptopsyllidae, Vermipsyllidae, Stephanocircidae, Ischnopsyllidae, Hypsophthalmidae and Xiphiopsyllidae*. London: Trustees of the British Museum.

Hopkins, G. H. E. & Rothschild, M. (1962). *An Illustrated Catalogue of the Rothschild Collection of Fleas (Siphonaptera) in the British Museum (Natural History)*, vol. 3, *Hystrichopsyllidae*. London: Trustees of the British Museum.

Hopkins, G. H. E. & Rothschild, M. (1966). *An Illustrated Catalogue of the Rothschild Collection of Fleas (Siphonaptera) in the British Museum (Natural History)*, vol. 4, *Hystrichopsyllidae*. London: Trustees of the British Museum.

Hopkins, G. H. E. & Rothschild, M. (1971). *An Illustrated Catalogue of the Rothschild Collection of Fleas (Siphonaptera) in the British Museum (Natural History)*, vol. 5, *Leptopsyllidae and Ancistropsyllidae*. London: Trustees of the British Museum.

Hopla, C. E. (1980). Fleas as vectors of tularemia in Alaska. In *Fleas: Proceedings of the International Conference on Fleas*, Ashton Wold, Peterborough, UK, 21–25 June 1977, ed. R. Traub & H. Starcke. Rotterdam, the Netherlands: A. A. Balkema, pp. 287–300.

Hõrak, P., Ots, I., Vellau, H., Spottiswoode, C. & Møller, A. P. (2001). Carotenoid-based plumage coloration reflects hemoparasite infection and local survival in breeding great tits. *Oecologia*, **126**, 166–173.

Hsu, M.-H. & Wu, W.-J. (2000). Effects of multiple mating on female reproductive output in the cat flea (Siphonaptera: Pulicidae). *Journal of Medical Entomology*, **37**, 828–834.

Hsu, M.-H. & Wu, W.-J. (2001). Off-host observations of mating and postmating behaviors in the cat flea (Siphonaptera: Pulicidae). *Journal of Medical Entomology*, **38**, 352–360.

Hsu, M.-H., Hsu, T.-C. & Wu, W.-J. (2002). Distribution of cat fleas (Siphonaptera: Pulicidae) on the cat. *Journal of Medical Entomology*, **39**, 685–688.

Hu, X.-L., He, J.-H., Zhang, H.-Y., Zhao W.-H. & Liang, Y. (1996). Evaluation on qualities of house rat fleas from laboratory breeding. *Endemic Diseases Bulletin*, **11**, 14–17 (in Chinese).

Hu, X.-L., He, J.-H., Zhang, H.-Y., *et al.* (1998). Experimental breeding and life history of the flea *Ctenophthalmus quadratus*. *Endemic Diseases Bulletin*, **13**, 26–28 (in Chinese).

Hu, X.-L., He, J.-H. & Zhang, H.-Y. (2001). Body weight and quantity of blood-sucking of six species of fleas in Yunnan province, China. *Chinese Journal of Pest Control*, **17**, 393–395 (in Chinese).

Hudson, B. W. & Prince, F. M. (1958a). A method for large-scale rearing of the cat flea, *Ctenocephalides felis felis* (Bouché). *Bulletin of World Health Organization*, **19**, 1126–1129.

Hudson, B. W. & Prince, F. M. (1958b). Culture methods for fleas *Pulex irritans* (L.) and *Pulex simulans* Baker. *Bulletin of World Health Organization*, **19**, 1129–1133.

Hudson, B. W., Feingold, B. F. & Kartman, L. (1960a). Allergy to flea bites. I. Experimental induction of flea-bite sensitivity in guinea pigs. *Experimental Parasitology*, **9**, 18–24.

Hudson, B. W., Feingold, B. F. & Kartman, L. (1960b). Allergy to flea bites. II. Investigations of flea bite sensitivity in humans. *Experimental Parasitology*, **9**, 264–270.

Hudson, P. J. & Dobson, A. P. (1995). Macroparasites: observed patterns. In *Ecology of Infectious Diseases in Natural Populations*, ed. B. T. Grenfell & A. P. Dobson. Cambridge, UK: Cambridge University Press, pp. 144–176.

Hudson, P. J. & Dobson, A. P. (1997). Host–parasite processes and demographic consequences. In *Host–Parasite Evolution: General Principles and Avian Models*, ed. D. H. Clayton & J. M. Moore. Oxford, UK: Oxford University Press, pp. 128–154.

Hudson, P. J. & Greenman, J. (1998). Competition mediated by parasites: biological and theoretical progress. *Trends in Ecology and Evolution*, **13**, 387–390.

Hughes, J. B. (2000). The scale of resource specialization and the distribution and abundance of lycaenid butterflies. *Oecologia*, **123**, 375–383.

Hughes, T. P., Baird, A. H., Dinsdale, E. A., *et al.* (2000). Supply-side ecology works both ways: the link between benthic adults, fecundity, and larval recruits. *Ecology*, **81**, 2241–2249.

Hughes, V. L. & Randolph, S. E. (2001). Testosterone depresses innate and acquired resistance to ticks in natural rodent hosts: a force for aggregated distributions of parasites. *Journal of Parasitology*, **87**, 49–54.

Hugueny, B. A. & Guégan, J.-F. (1997). Community nestedness and the proper way to assess statistical significance by Monte-Carlo tests: some comments on Worthen and Rohde's (1996) paper. *Oikos*, **80**, 572–574.

Humphreys, N. E. & Grencis, R. K. (2002). Effects of ageing on the immunoregulation of parasitic infection. *Infection and Immunity*, **70**, 5148–5157.

Humphries, D. A. (1966). The function of combs in fleas. *Entomological Monthly Magazine*, **102**, 232–236.

Humphries, D. A. (1967a). The mating behaviour of the hen flea *Ceratophyllus gallinae* (Schrank) (Siphonaptera: Insecta). *Animal Behaviour*, **15**, 82–90.

Humphries, D. A. (1967b). The action of the male genitalia during the copulation of the hen flea, *Ceratophyllus galinnae* (Schrank). *Proceedings of the Royal Entomological Society of London A*, **42**, 101–106.

Humphries, D. A. (1967c). Function of combs in ectoparasites. *Nature*, **215**, 319.

Humphries, D. A. (1968). The host-finding behavior of the hen flea, *Ceratophyllus gallinae* (Schrank) (Siphonaptera). *Parasitology*, **59**, 403–414.

Humphries, D. A. (1969). Behavioral aspects of the ecology of the sand-martin flea *Ceratophyllus styx jordani* Smit (Siphonaptera). *Parasitology*, **59**, 311–334.

Hunter, M. D. & Price, P. W. (1992). Playing chutes and ladders: heterogeneity and the relative roles of bottom–up and top–down forces in natural communities. *Ecology*, **73**, 724–732.

Hůrka, K. (1963a). Bat fleas (Aphaniptera, Ischnopsyllidae) of Czechoslovakia: contribution to the distribution, morphology, bionomy, ecology and systematics. I. Subgenus *Ischnopsyllus* Westw. *Acta Faunistica Entomologica Musei Nationalis Pragae*, **9**, 57–120.

Hůrka, K. (1963b). Bat fleas (Aphaniptera, Ischnopsyllidae) of Czechoslovakia: contribution to the distribution, morphology, bionomy, ecology and systematics. II. Subgenus *Hexactenopsylla* Oud., genus *Rhinolophopsylla* Oud., subgenus *Nycteridopsylla* Oud., subgenus *Dinycteropsylla* Ioff. *Acta Universitatis Carolinae, Biologica*, **3**, 1–73.

Ioff, I. G. (1941). *Ecology of Fleas in Relevance to their Medical Importance*. Pyatygorsk, USSR: Pyatygorsk Publishers (in Russian).

Ioff, I. G. (1949). Aphaniptera of Kyrgyzstan. *Ectoparasites*, **1**, 5–212 (in Russian).

Ioff, I. G. (1950). The alakurt. *Materials to Knowledge of Fauna and Flora of the USSR* [*Materialy k Poznaniju Fauny i Flory SSSR*], **2**, 4–29 (in Russian).

Ioff, I. G. & Tiflov, V. E. (1954). *The Key to Identification of Aphaniptera of the South-East of the USSR*. Stavropol, USSR: Stavropol Publishers (in Russian).

Ioff, I. G., Tiflov, V. E., Argyropulo, A. I., et al. (1946). News species of fleas (Aphaniptera). *Medical Parasitology* [*Meditsinskaya Parazitologiya*], **15**, 85–94 (in Russian).

Ioff, I. G., Mikulin, M. A. & Scalon, O. N. (1965). *The Key to Identification of Fleas of the Middle Asia and Kazakhstan*. Moscow, USSR: Meditsina (in Russian).

Iqbal, Q. L. (1973). On the presence of mating pheromone in the rat flea *Nosopsyllus fasciatus* (Bosc.). *Pakistan Journal of Zoology*, **5**, 123–125.

Iqbal, Q. L. (1974). Host-finding behaviour of the rat flea *Nosopsyllus fasciatus* (Bosc.). *Biologia, Lahore*, **20**, 147–150.

Iqbal, Q. J. & Humphries, D. A. (1970). Temperature as a critical factor in the mating behavior of the rat flea, *Nosopsyllus fasciatus* (Bosc.). *Parasitology*, **61**, 375–380.

Iqbal, Q. J. & Humphries, D. A. (1974). The mating behavior of the rat flea *Nosopsyllus fasciatus* Bosc. *Pakistan Journal of Zoology*, **6**, 163–174.

Iqbal, Q. J. & Humphries, D. A. (1976). Remating in the rat flea *Nosopsyllus fasciatus* (Bosc.). *Pakistan Journal of Zoology*, **8**, 39–41.

Iqbal, Q. J. & Humphries, D. A. (1983). Feeding behavior of the rat flea *Nosopsyllus fasciatus*. *Pakistan Journal of Zoology*, **14**, 71–74.

Ives, A. R. (1988a). Aggregation and the coexistence of competitors. *Annales Zoologici Fennici*, **25**, 75–88.

Ives, A. R. (1988b). Covariance, coexistence and the population dynamics of two competitors using a patchy resource. *Journal of Theoretical Biology*, **133**, 345–361.

Ives, A. R. (1991). Aggregation and coexistence in a carrion fly community. *Ecological Monographs*, **61**, 75–94.

Iwao, K. & Ohsaki, N. (1996). Inter- and intraspecific interactions among larvae of specialist and generalist parasitoids. *Research on Population Ecology*, **38**, 265–273.

Izsák, J. & Papp, L. (1995). Application of the quadratic entropy index for diversity studies on drosophilid species assemblages. *Environmental and Ecological Statistics*, **2**, 213–224.

Jackson, T. P. (2000). Adaptation to living in an open arid environment: lessons from the burrow structure of two South African whistling rats, *Parotomys brantsii* and *P. littledalei*. *Journal of Arid Environments*, **46**, 345–355.

Jaenike, J. (1990). Host specialization in phytophagous insects. *Annual Review of Ecology and Systematics*, **21**, 243–273.

Jaenike, J. & James, A. C. (1991). Aggregation and the coexistence of mycophagous *Drosophila*. *Journal of Animal Ecology*, **60**, 913–928.

Jameson, E. W. (1985). Pleioxenous host-restriction in fleas. *Journal of Natural History*, **19**, 861–876.

Jameson, E. W. (1999). Host–ectoparasite relationships among North American chipmunks. *Acta Theriologica*, **44**, 225–231.

Jamieson, B. G. M. (1987). *The Ultrastructure and Phylogeny of Insect Spermatozoa*. Cambridge, UK: Cambridge University Press.

Janeway, C., Travers, P., Walport, M. & Capra, J. (1999). *Immunobiology: The Immune System in Health and Disease*. New York: Garland.

Janion, S. M. (1962). Flea infestation of three rodent species: *Apodemus agrarius*, *Apodemus flavicollis* and *Clethrionomys glareolus* at the period of *Apodemus agrarius* mass occurrence. *Bulletin of the Polish Academy of Sciences, Series of Biological Sciences*, **10**, 361–366.

Janion, S. M. (1968). Certain host–parasite relationships between rodents (Muridae) and fleas (Aphaniptera). *Ekologia Polska*, **16**, 561–606.

Jarrett, W. F. (1975). Cat leukemia and its viruses. *Advances in Veterinary Science and Comparative Medicine*, **19**, 165–193.

Jell, P. A. & Duncan, P. M. (1986). Invertebrates, mainly insects, from the freshwater, Lower Cretaceous, Koonwarra Fossil Bed (Korumburra Group), South Gippsland, Victoria. *Memoir of the Association of Australasian Palaeontologists*, **3**, 111–205.

Jellison, W. L. (1959). Fleas and disease. *Annual Review of Entomology*, **4**, 389–414.

Johnsen, T. S. & Zuk, M. (1999). Parasites and tradeoffs in the immune response of female red jungle fowl. *Oikos*, **86**, 487–492.

Johnson, K. P., Adams, R. J. & Clayton, D. H. (2002). The phylogeny of the louse genus *Brueelia* does not reflect host phylogeny. *Biological Journal of the Linnean Society*, **77**, 233–247.

Johnson, K. P., Bush, S. E. & Clayton, D. H. (2005). Correlated evolution of host and parasite body size: tests of Harrison's rule using birds and lice. *Evolution*, **59**, 1744–1753.

Johnston, C. M. & Brown, S. J. (1985). *Xenopsylla cheopis*: cellular expression of hypersensitivity to guinea pigs. *Experimental Parasitology*, **59**, 81–89.

Johnston, M., Johnston, D. & Richardson, A. (2005). Digestive capabilities reflect the major food sources in three species of talitrid amphipods. *Comparative Biochemistry and Physiology B*, **140**, 251–257.

Jokela, J., Schmid-Hempel, P. & Rigby, M. C. (2000). Dr Pangloss restrained by the Red Queen: steps towards a unified defence theory. *Oikos*, **89**, 267–274.

Jones, C. J. (1996). Immune responses to fleas, bugs and sucking lice. In *The Immunology of Host–Ectoparasitic Arthropod Relationships*, ed. S. K. Wikel. Wallingford, UK: CAB International, pp. 150–174.

Jordan, K. (1945). On the deciduous frontal tubercle of some genera of Siphonaptera. *Proceedings of the Royal Entomological Society of London B*, **14**, 113–116.

Jordan, K. (1962). Notes on the *Tunga caecigena* (Siphonaptera: Tungidae). *Bulletin of the British Museum of Natural History, Entomology*, **12**, 353–364.

Jordan, K. & Rothschild, N. C. (1915a). On some Siphonaptera collected by W. Ruckbeil in East Turkestan. *Ectoparasites*, **1**, 1–24.

Jordan, K. & Rothschild, N. C. (1915b). Contribution to our knowledge of American Siphonaptera. *Ectoparasites*, **1**, 45–60.

Joseph, S. A. (1974). Incidence of the flea *Ancistropsylla nepalensis* Lewis, 1968 on the barking deer (*Muntiacus muntjak aureus* Smith, 1827) in India. *Indian Veterinary Journal*, **51**, 356–358.

Joseph, S. A. & Mani, K. R. (1980). *Cervus unicolor niger*: the Indian sambar, a new host for *Ancistropsylla nepalensis* Lewis, 1968. *Cheiron*, **9**, 200–202.

Joy, J. E. & Briscoe, N. J. (1994). Parasitic arthropods of white-footed mice at McClintock Wildlife Station, West Virginia. *Journal of the American Mosquito Control Association*, **10**, 108–111.

Juricova, Z., Halouzka, J. & Hubalek, Z. (2002). Serologic survey for antibodies to *Borrelia burgdorferi* in rodents and detection of spirochaetes in ticks and fleas in South Moravia (Czech Republic). *Biologia, Bratislava*, **57**, 383–387.

Jurík, M. (1974). Bionomics of fleas in birds' nests in the territory of Czechoslovakia. *Acta Scientiarum Naturalium Brno*, **8**, 1–54.

Jurík, M. (1983a). To the knowledge of ecological conditions affecting the occurrence of specific and non-specific flea species on their hosts (*Talpa europaea* – Siphonaptera). *Biologia, Bratislava*, **38**, 949–957.

Jurík, M. (1983b). *Ceratophyllus vagabundus insularis* Rothschild, 1906 and *Ceratophyllus rossittensis* Dampf, 1913 in Czechoslovakia (Siphonaptera). *Folia Parasitologica*, **30**, 169–173.

Kaal, J. F., Baker, K. & Torgerson, P. R. (2006). Epidemiology of flea infestation of ruminants in Libya. *Veterinary Parasitology*, **141**, 313–318.

Kadatskaya, K. P. (1983). Facultative imaginal diapause in fleas *Xenopsylla conformis* (Siphonaptera). *Parazitologiya*, **17**, 370–374 (in Russian).

Kadatskaya, K. P. & Kadatsky, N. G. (1983). Comparative data on abundance of fleas parasitic on *Meriones erythrourus* in the western and eastern parts of the Apsheron Peninsula in relation to the plague epizootic in 1976–1978. In *Prophylaxis of Diseases in the Natural Foci*, ed. I. F. Taran. Stavropol, USSR: Scientific Anti-Plague Institute of Caucasus and Trans-Caucasus, pp. 237–238 (in Russian).

Kadatskaya, K. P. & Shirova, L. F. (1983). Seasonal changes of reproduction in fleas *Xenopsylla conformis* in Azerbaijan. In *Prophylaxis of Diseases in the Natural Foci*, ed. I. F. Taran. Stavropol, USSR: Scientific Anti-Plague Institute of Caucasus and Trans-Caucasus, pp. 238–240 (in Russian).

Kam, M. & Degen, A. A. (1993). Energetics of lactation and growth in the fat sand rat, *Psammomys obesus*: new perspectives of resource partitioning and the effect of litter size. *Journal of Theoretical Biology*, **162**, 353–369.

Kam, M. & Degen, A. A. (1997). Energy requirements and the efficiency of utilization of metabolizable energy in free-living animals: evaluation of existing theories and generation of a new model. *Journal of Theoretical Biology*, **184**, 101–104.

Kam, M., Khokhlova, I. S. & Degen, A. A. (1997). Granivory and plant selection by desert gerbils of different body size. *Ecology*, **78**, 2218–2229.

Kamala Bai, M. & Prasad, R. S. (1979). Influence of nutrition on maturation of male rat fleas, *Xenopsylla cheopis* and *X. astia*. *Journal of Medical Entomology*, **16**, 164–165.

Kaňuch, P., Krištín, A. & Krištofik, J. (2005). Phenology, diet, and ectoparasites of Leisler's bat (*Nyctalus leisleri*) in the western Carpathians (Slovakia). *Acta Chiropterologica*, **7**, 249–257.

Karandina, P. C. & Darskaya, N. F (1974). Observations on the pre-imaginal development in fleas parasitic on ground squirrels: *Ceratophyllus (Citellophilus) tesquorum* Wagn., 1898. In *Particularly Dangerous Diseases in Caucasus: Proceedings of the 3rd Scientific–Practical Conference of the Anti-Plague Establishments of Caucasus on Natural Focality, Epidemiology and Prophylaxis of Particularly Dangerous Diseases*, 14–16 May 1974, ed. V. G. Pilipenko. Stavropol, USSR: Scientific Anti-Plague Institute of Caucasus and Trans-Caucasus, pp. 143–144 (in Russian).

Kareiva, P. & Wennergren, U. (1995). Connecting landscape patterns to ecosystem and population processes. *Nature*, **373**, 299–302.

Karlson, R. H., Cornell, H. V. & Hughes, T. P. (2004). Coral communities are regionally enriched along an oceanic biodiversity gradient. *Nature*, **429**, 867–870.

Kartman, L., Prince, F. M., Quan, S. F. & Stark, H. E. (1958). New knowledge of the ecology of sylvatic plague. *Annals of the New York Academy of Sciences*, **70**, 668–711.

Kavaliers, M. & Colwell, D. (1994). Parasite infection attenuates nonopioid mediated predator-induced analgesia in mice. *Physiology and Behavior*, **55**, 505–510.

Kavaliers, M. & Colwell, D. D. (1995). Reduced spatial learning in mice infected with the nematode *Heligmosomoides polygyrus*. *Parasitology*, **110**, 591–597.

Kavaliers, M., Colwell, D. D. & Choleris, E. (1998). Parasitized female mice display reduced aversive responses to the odours of infected males. *Proceedings of the Royal Society of London B*, **265**, 1111–1118.

Kędra, A. H., Kruszewicz, A. G., Mazgajski, T. D. & Modlińska, E. (1996). The effects of the presence of fleas in nestboxes on fledglings of pied flycatchers and great tits. *Acta Parasitologica*, **41**, 211–213.

Keeling, M. G. & Gilligan, C. A. (2000a). Metapopulation dynamics of bubonic plague. *Nature*, **407**, 903–906.

Keeling, M. G. & Gilligan, C. A. (2000b). Bubonic plague: a metapopulation model of a zoonosis. *Proceedings of the Royal Society of London B*, **267**, 2219–2230.

Kehr, J. D., Heukelbach, J., Mehlhorn, H. & Feldmeier, H. (2007). Morbidity assessment in sand flea disease (tungiasis). *Parasitology Research*, **100**, 413–421.

Kelly, D. W. & Thompson, C. E. (2000). Epidemiology and optimal foraging: modeling the ideal free distribution of insect vectors. *Parasitology*, **120**, 319–327.

Kelly, D. W., Mustafa, Z. & Dye, C. (1996). Density-dependent feeding success in a field population of the sandfly *Lutzomyia longipalpis*. *Journal of Animal Ecology*, **65**, 517–527.

Kennedy, C. R. & Bush, A. O. (1994). The relationship between pattern and scale in parasite communities: a stranger in a strange land. *Parasitology*, **109**, 187–196.

Kern, W. H. (1993). The autecology of the cat flea (*Ctenocephalides felis felis* Bouché) and the synecology of the cat flea and its domestic hosts (*Felis catus*). Unpublished Ph.D. thesis, University of Florida, Gainesville, FL.

Kern, W. H., Koehler, P. G. & Patterson, R. S. (1992). Diel patterns of cat flea (Siphonaptera, Pulicidae) egg and fecal deposition. *Journal of Medical Entomology*, **29**, 203–206.

Kern, W. H., Richman, D. L., Koehler, P. G. & Brenner, R. J. (1999). Outdoor survival and development of immature cat fleas (Siphonaptera: Pulicidae) in Florida. *Journal of Medical Entomology*, **36**, 207–211.

Kettle, D. S. (1995). *Medical and Veterinary Entomology*. Wallingford, UK: CAB International.

Key, B. H. & Kemp, D. H. (1994). Vaccines against arthropods. *American Journal of Tropical Medicine and Hygiene*, **50**, 87–96.

Khalid, M. L., Morsy, T. A., El Shennawy, S. F., et al. (1992). Studies on flea fauna in El Fayoum Governorate, Egypt. *Journal of the Egyptian Society of Parasitology*, **22**, 783–799.

Kharlamov, V. P. (1965). Changes in feeding activity and mobility of the flea *Xenopsylla cheopis* marked with radioactive ^{32}P. *Zoologicheskyi Zhurnal*, **44**, 547–550 (in Russian).

Khokhlova, I. S. & Knyazeva, T. V. (1983). The effect of spatial and social structure of the house mouse populations on flea assemblages. In *Prophylaxis of Diseases in the Natural Foci*, ed. I. F. Taran. Stavropol, USSR: Scientific Anti-Plague Institute of Caucasus and Trans-Caucasus, pp. 165–167 (in Russian).

Khokhlova, I. S., Krasnov, B. R., Shenbrot, G. I. & Degen, A. A. (1994). Seasonal body mass changes and habitat distribution in several rodent species from the Ramon erosion cirque, Negev Highlands, Israel. *Zoologicheskyi Zhurnal*, **73**, 115–121 (in Russian).

Khokhlova, I. S., Krasnov, B. R., Shenbrot, G. I. & Degen, A. A. (2001). Body mass and environment: a study in Negev rodents. *Israel Journal of Zoology*, **47**, 1–14.

Khokhlova, I. S., Krasnov, B. R., Kam, M., Burdelova, N. V. & Degen, A. A. (2002). Energy cost of ectoparasitism: the flea *Xenopsylla ramesis* on the desert gerbil *Gerbillus dasyurus*. *Journal of Zoology*, **258**, 349–354.

Khokhlova, I. S., Spinu, M., Krasnov, B. R. & Degen, A. A. (2004a). Immune response to fleas in a wild desert rodent: effect of parasite species, parasite burden, sex of host and host parasitological experience. *Journal of Experimental Biology*, **207**, 2725–2733.

Khokhlova, I. S., Spinu, M., Krasnov, B. R. & Degen, A. A. (2004b). Immune responses to fleas in two rodent species differing in natural prevalence of infestation and diversity of flea assemblages. *Parasitology Research*, **94**, 304–311.

Khokhlova, I. S., Hovhanyan, A., Krasnov, B. R. & Degen, A. A. (2007). Reproductive success in two species of desert fleas: density-dependence and host effect. *Journal of Experimental Biology*, **210**, 2121–2127.

Khrustselevsky, V. P., Sokolova, A. A. & Balabas, N. G. (1971). Materials on the reproduction of *Xenopsylla gerbilli* in the Moyynkum Desert. In *Proceedings of the 7th Scientific Conference of the Anti-Plague Establishments of the Middle Asia and Kazakhstan*, ed. M. A. Aikimbaev. Alma-Ata, USSR: The Middle Asian Scientific Anti-Plague Institute, pp. 436–439 (in Russian).

Khudyakov, I. S. (1965). Fleas (Aphaniptera) of the coastal zone of southern Primorie Region. *Entomological Review*, **44**, 117–122 (in Russian).

Kiefer, M., Klimaszewski, S. M. & Krumpál, M. (1982). Zoogeographical regionalization of Mongolia on the basis of flea fauna (Siphonaptera). *Polskie Pismo Entomologiczne*, **52**, 13–29.

Kiefer, M., Krumpál, M., Cendsuren, N., Lobachev, V. S. & Chotolchu, N. (1984). Checklist, distribution and bibliography of Mongolian Siphonaptera. In *Erforschung biologischer Ressourcen der mongolischen Volksrepublik*, vol. 4, ed. M. Stubbe, W. Hilbig & N. Dawaa. Halle, Germany: Wissenschaftliche Beiträge Universität Halle-Wittenberg, pp. 91–123.

Kilpatrick, A. M. & Ives, A. R. (2003). Species interactions can explain Taylor's power law for ecological time series. *Nature*, **422**, 65–68.

Kim, K. C. (1985a). Evolution and host associations of Anoplura. In *Coevolution of Parasitic Arthropods and Mammals*, ed. K. C. Kim. New York: John Wiley, pp. 197–232.

Kim, K. C. (1985b). Evolutionary relationships of parasitic arthropods and mammals. In *Coevolution of Parasitic Arthropods and Mammals*, ed. K. C. Kim. New York: John Wiley, pp. 3–82.

King, C. M. (1976). The fleas of a population of weasels in Wytham Wood, Oxford. *Journal of Zoology*, **180**, 525–535.

King, C. M. & Moody, J. E. (1982). The biology of the stoat (*Mustela erminea*) in the National Parks of New Zealand. VII. Fleas. *New Zealand Journal of Zoology*, **9**, 141–144.

Kings, R. C. & Teasly, M. (1980). Insect oogenesis: some generalities and their bearing on the ovarian development of fleas. In *Fleas: Proceedings of the International Conference on Fleas*, Ashton Wold, Peterborough, UK, 21–25 June 1977, ed. R. Traub & H. Starcke. Rotterdam, the Netherlands: A. A. Balkema, pp. 337–340.

Kingsolver, J. G. (1987). Mosquito host choice and the epidemiology of malaria. *American Naturalist*, **130**, 811–827.

Kiriakova, A. N., Koptzev, L. A. & Koptzeva, Z. G. (1970). Annual number of generations of *Xenopsylla* fleas in the northern Kyzylkum Desert. *Parazitologiya*, **6**, 528–536 (in Russian).

Kirillova, N. Y., Kirillov, A. A. & Ivashkina, V. A. (2006). Ectoparasites of the edible dormouse *Glis glis* L. of the Samarskaya Luka Peninsula (Russia). *Polish Journal of Ecology*, **54**, 387–390.

Kirk, W. D. J. (1991). The size relationship between insects and their hosts. *Ecological Entomology*, **16**, 351–359.

Kisielewska, K. (1970). Ecological organization of intestinal helminth groupings in *Clethrionomys glareolus* (Schreb.) (Rodentia). III. Structure of helminth groupings in *C. glareolus* populations of various forest biocoenoses in Poland. *Acta Parasitologica*, **18**, 163–176.

Klasing, K. C. (1998). Nutritional modulation of resistance to infectious diseases. *Poultry Science*, **77**, 1119–1125.

Klassen, G. J. (1992). Coevolution: a history of the macroevolutionary approach to studying host–parasite associations. *Journal of Parasitology*, **78**, 573–587.

Kleiber, M. (1961). *The Fire of Life: An Introduction to Animal Energetics*. New York: John Wiley.

Klein, J. (1990). *Immunology*. Oxford, UK: Blackwell Science.

Klein, J. M. (1966). Données écologiques et biologiques sur *Synopsyllus fonquerniei* Wagner et Roubaud, 1932 (Siphonaptera) puce du rat péridomestique, dans la région de Tananarive. *Cahiers ORSTOM, Série Entomologie Médicale et Parasitologie*, **4**, 3–29.

Klein, J. M., Simonkovich, E., Alonso, J. M. & Baranton, G. (1975). Observations écologiques dans une zone epizootique de peste en Mauritanie. II. Les puces de rongeurs (Insecta, Siphonaptera). *Cahiers ORSTOM, Série Entomologie Médicale et Parasitologie*, **13**, 29–39.

Klein, S. L. & Nelson, R. J. (1998a). Sex and species differences in cell-mediated immune responses in voles. *Canadian Journal of Zoology*, **76**, 1394–1398.

Klein, S. L. & Nelson, R. J. (1998b). Adaptive immune responses are linked to the mating system of arvicoline rodents. *American Naturalist*, **151**, 59–67.

Klompen, J. S. H., Black, W. C., Keirans, J. E. & Oliver, J. H. (1996). Evolution of ticks. *Annual Review of Entomology*, **41**, 141–161.

Knopf, P. M. & Coghlan, R. L. (1989). Maternal transfer of resistance to *Schistosoma mansoni*. *Journal of Parasitology*, **75**, 398–404.

Knülle, W. (1967). Physiological properties and biological implications of the water vapour sorption mechanism in larvae of the oriental rat flea, *Xenopsylla cheopis* (Roths). *Journal of Insect Physiology*, **13**, 333–357.

Koehler, P. G., Leppla, N. C. & Patterson, S. (1990). Circadian rhythm in the cat flea *Ctenocephalides felis* (Siphonaptera: Pulicidae). In *Chronobiology: Its Role in Clinical Medicine, General Biology, and Agriculture*, part B, ed. D. K. Hayes, J. E. Pauly & R. J. Reiter. New York: Wiley-Liss, pp. 661–665.

Kolpakova, S. A. (1950). Migration of fleas from the burrows of the great gerbil. *Ectoparasites*, **2**, 115–128 (in Russian).

Kondrashkina, K. I. & Dudnikova, A. F. (1962). Oxygen consumption in fleas parasitic on the ground squirrels as a physiological test of their viability. In *Particularly Dangerous and Natural Diseases*, ed. Anonymous. Moscow, USSR: Medgiz, pp. 63–69 (in Russian).

Kondrashkina, K. I. & Dudnikova, A. F. (1968). Oxygen consumption in fleas parasitic on the Norway rats. In *Rodents and their Ectoparasites*, ed. B. K. Fenyuk. Saratov, USSR: Saratov University Press, pp. 87–91 (in Russian).

Kondrashkina, K. I. & Gerasimova, N. G. (1971). Dependence of the metabolic rate of two common fleas parasitic on the great gerbil *Rhombomys opimus* on their physiological conditions. *Problems of Particularly Dangerous Diseases*, **17**, 32–37 (in Russian).

Konkova, K. V. & Timofeeva, A. A. (1970). Studies of fleas (Siphonaptera) of the Sakhalin and Kuril Islands. In *Vectors of Dangerous Diseases and their Control*, ed. V. E. Tiflov. Stavropol, USSR: Scientific Anti-Plague Institute of Caucasus and Trans-Caucasus, pp. 371–390 (in Russian).

Konnov, N. P., Demchenko, T. A., Anisimov, P. I., Kondrashkina, K. I. & Lukyanova, A. D. (1986). Pathogenic effect of the plague microbe on the flea *Xenopsylla cheopis* and ultrastructure of the agent at various periods of its stay in the vector. *Parazitologiya*, **20**, 19–22 (in Russian).

Korallo, N. P., Vinarski, M. V., Krasnov, B. R., et al. (2007). Are there general rules governing parasite diversity? Small mammalian hosts and gamasid mite assemblages. *Diversity and Distributions*, **13**, 353–360.

Korneeva, L. A. & Sadovenko, E. V. (1990). Patterns of development of the fleas of the great gerbils in the laboratory. In *Advantages of Medical Entomology and Acarology in the USSR*, ed. G. S. Medvedev. Leningrad, USSR: All-Union Entomological Society and Zoological Institute, Academy of Sciences of the USSR, pp. 12–14 (in Russian).

Korovin, E. P. (1961). *Vegetation of Middle Asia and Southern Kazakhstan*. Tashkent, USSR: Academy of Sciences of the Uzbek SSR Press (in Russian).

Korpimaki, E., Tolonen, P. & Bennett, G. F. (1995). Blood parasites, sexual selection and reproductive success of European kestrels. *Ecoscience*, **2**, 335–343.

Koshkin, S. M. (1966). Materials on flea fauna in Sovetskaya Gavan. *Proceedings of the Irkutsk State Scientific Anti-Plague Institute of Siberia and Far East*, **26**, 242–248 (in Russian).

Kosminsky, R. B. (1960). Reproduction of fleas parasitic on mice in the field and experimental conditions. In *Parasitological Problems: Proceedings of the 3rd Parasitological Conference of the Ukrainian SSR*, ed. Anonymous. Kiev, USSR: Insitute of Zoology of the Academy of Sciences of the Ukrainian SSR, pp. 326–328 (in Russian).

Kosminsky, R. B. (1961). Study of bionomics of the fleas parasitic on the house mice. Unpublished Ph.D. thesis, Scientific Anti-Plague Institute of Caucasus and Trans-Caucasus, Stavropol, USSR (in Russian).

Kosminsky, R. B. (1962). Characteristics of the life cycles of fleas parasitic on the house mice in the buildings of a steppe settlement in the Stavropol Region. In

Problems of Ecology, vol. 4, ed. Anonymous. Kiev, USSR: Kiev University Press, pp. 65–66 (in Russian).

Kosminsky, R. B. (1965). Feeding and reproduction in fleas parasitic on the house mice in the field and under experiment. *Zoologicheskyi Zhurnal*, **44**, 1372–1375 (in Russian).

Kosminsky, R. B. & Guseva, A. A. (1974a). On gonotrophic activity of some fleas belonging to the genus *Ctenophthalmus* Kolenati (Siphonaptera). In *Particularly Dangerous Diseases in Caucasus: Proceedings of the 3rd Scientific–Practical Conference of the Anti-Plague Establishments of Caucasus on Natural Focality, Epidemiology and Prophylaxis of Particularly Dangerous Diseases, 14–16 May 1974*, ed. V. G. Pilipenko. Stavropol, USSR: Scientific Anti-Plague Institute of Caucasus and Trans-Caucasus, pp. 154–156 (in Russian).

Kosminsky, R. B. & Guseva, A. A. (1974b). On biology of *Amphipsylla rossica* Wagn., 1912 (Ceratophyllidae, Siphonaptera): a flea with all-year-round activity. In *Particularly Dangerous Diseases in Caucasus: Proceedings of the 3rd Scientific–Practical Conference of the Anti-Plague Establishments of Caucasus on Natural Focality, Epidemiology and Prophylaxis of Particularly Dangerous Diseases, 14–16 May 1974*, ed. V. G. Pilipenko. Stavropol, USSR: Scientific Anti-Plague Institute of Caucasus and Trans-Caucasus, pp. 156–158 (in Russian).

Kosminsky, R. B. & Guseva, A. A. (1975a). Age-related changes in the imago of *Ctenophthalmus wagneri* Tifl, 1928 (Ctenophthalmidae, Siphonaptera). *Medical Parasitology and Parasitic Diseases* [*Meditsinskaya Parazitologiya i Parazitarnye Bolezni*], **44**, 96–100 (in Russian).

Kosminsky, R. B. & Guseva, A. A. (1975b). Feeding behavior and reproduction of *Ctenophthalmus wagneri* Tifl., 1928 (Ctenophthalmidae, Siphonaptera) under experimental conditions. *Parazitologiya*, **9**, 265–270 (in Russian).

Kosminsky, R. B. & Udovitskaya, E. Y. (1975). Upper temperature boundary of life of imago of some species of fleas (Siphonaptera). *Proceedings of the Academy of Sciences of the USSR* [*Doklady Akademii Nauk SSSR*], **222**, 500–503 (in Russian).

Kosminsky, R. B., Avetisyan, G. A. & Talybov, A. N. (1970). Annual cycle of *Ctenophthalmus wladimiri* Isajeva-Gurvich, 1948: common flea species of the common vole in the south-east of Trans-Caucasian Upland. In *Vectors of Dangerous Diseases and their Control*, ed. V. E. Tiflov. Stavropol, USSR: Scientific Anti-Plague Institute of Caucasus and Trans-Caucasus, pp. 79–107 (in Russian).

Kosminsky, R. B., Bryukhanova, L. V., Darskaya, N. F., *et al.* (1974). Annual cycle of *Ceratophyllus* (*Nosopsyllus*) *consimilis* Wagn., 1898 (Ceratophyllidae, Siphonaptera) in the Stavropol Upland. In *Particularly Dangerous Diseases in Caucasus: Proceedings of the 3rd Scientific–Practical Conference of the Anti-Plague Establishments of Caucasus on Natural Focality, Epidemiology and Prophylaxis of Particularly Dangerous Diseases, 14–16 May 1974*, ed. V. G. Pilipenko. Stavropol, USSR: Scientific Anti-Plague Institute of Caucasus and Trans-Caucasus, pp. 152–154 (in Russian).

Kosoy, M. Y., Regnery, R., Tzianabos, T., *et al.* (1997). Distribution, diversity, and host specificity of *Bartonella* in rodents from the southern United States. *American Journal of Tropical Medicine and Hygiene*, **57**, 578–588.

Kosoy, M. Y., Mandel, E., Green, D., Marston, E. L. & Childs, J. E. (2004). Prospective studies of *Bartonella* of rodents. I. Demographic and temporal patterns in population dynamics. *Vector-Borne and Zoonotic Diseases*, **4**, 285–95.

Kotler, B. P., Brown, J. S. & Subach, A. (1993). Mechanisms of species coexistence of optimal foragers: temporal partitioning by two species of sand dune gerbils. *Oikos*, **67**, 548–556.

Kotti, B. K. & Kovalevsky, I. (1995). Fleas parasitic on small mammals in the area between the Amur and Bureya Rivers. *Zoologicheskyi Zhurnal*, **74**, 70–76 (in Russian).

Kouki, J. & Hanski, I. (1995). Population aggregation facilitates coexistence of many competing carrion fly species. *Oikos*, **72**, 223–227.

Kozlovskaya, O. L. (1958). Flea (Aphaniptera) fauna of rodents of the valley of the Ussury River in the Khabarovsk Region. *Proceedings of the Irkutsk State Scientific Anti-Plague Institute of Siberia and Far East*, **17**, 109–116 (in Russian).

Kozlovskaya, O. L. & Demidova, A. A. (1958). Data on the ecology of fleas parasitic on the field mouse in the Khabarovsk Region. *Proceedings of the Irkutsk State Scientific Anti-Plague Institute of Siberia and Far East*, **17**, 59–64 (in Russian).

Krämer, F. & Mencke, N. (2001). *Flea Biology and Control: The Biology of the Cat Flea – Control and Prevention with Imidacloprid in Small Animals*. New York: Springer-Verlag.

Krampitz, H. E. (1964). Über Vorkomen und Verhalten von Haemococcidien der Gattung *Hepatozoon* Miller, 1908 (Protozoa, Adeloidea) in mittel- und südeuropäischen Säugern. *Acta Tropica*, **21**, 114–154.

Krampitz, H. E. (1981). Development of *Hepatozoon erhardovae* Krampitz, 1964, (Protozoa: Haemogregarinidae) in experimental mammalian and arthropod hosts. II. Sexual development in fleas and sporozoite indices in xenodiagnosis. *Transactions of the Royal Society of Tropical Medicine and Hygiene*, **75**, 155–157.

Krasnov, B. R. & Khokhlova, I. S. (2001). The effect of behavioural interactions on the exchange of flea (Siphonaptera) between two rodent species. *Journal of Vector Ecology*, **26**, 181–190.

Krasnov, B. R. & Knyazeva, T. V. (1983). Ectoparasite exchange between the midday jird, *Meriones meridianus*, and the house mouse, *Mus musculus*, under experiment. In *Prophylaxis of Diseases in the Natural Foci*, ed. I. F. Taran. Stavropol, USSR: Scientific Anti-Plague Institute of Caucasus and Trans-Caucasus, pp. 243–244 (in Russian).

Krasnov, B. R. & Shenbrot, G. I. (2002). Coevolutionary events in history of association of jerboas (Rodentia: Dipodidae) and their flea parasites. *Israel Journal of Zoology*, **48**, 331–350.

Krasnov, B. R., Shenbrot, G. I., Khokhlova, I. S., Degen, A. A. & Rogovin, K. V. (1996a). On the biology of Sundevall's jird (*Meriones crassus* Sundevall) in Negev Highlands, Israel. *Mammalia*, **60**, 375–391.

Krasnov, B. R., Shenbrot, G. I., Khokhlova, I. S. & Ivanitskaya, E. Y. (1996b). Spatial structure of rodent community in the Ramon erosion cirque, Negev Highlands (Israel). *Journal of Arid Environments*, **32**, 319–327.

Krasnov, B. R., Shenbrot, G. I., Medvedev, S. G., Vatschenok, V. S. & Khokhlova, I. S. (1997). Host–habitat relations as an important determinant of spatial distribution of flea assemblages (Siphonaptera) on rodents in the Negev Desert. *Parasitology*, **114**, 159–173.

Krasnov, B. R., Shenbrot, G. I., Medvedev, S. G., Khokhlova, I. S. & Vatschenok, V. S. (1998). Habitat-dependence of a parasite–host relationship: flea assemblages in two gerbil species of the Negev Desert. *Journal of Medical Entomology*, **35**, 303–313.

Krasnov, B. R., Hastriter, M., Medvedev, S. G., *et al.* (1999). Additional records of fleas (Siphonaptera) on wild rodents in the southern part of Israel. *Israel Journal of Zoology*, **45**, 333–340.

Krasnov, B. R., Khokhlova, I. S., Fielden, L. J. & Burdelova, N. V. (2001a). The effect of temperature and humidity on the survival of pre-imaginal stages of two flea species (Siphonaptera: Pulicidae). *Journal of Medical Entomology*, **38**, 629–637.

Krasnov, B. R., Khokhlova, I. S., Fielden, L. J. & Burdelova, N. V. (2001b). Development rates of two *Xenopsylla* flea species in relation to air temperature and humidity. *Medical and Veterinary Entomology*, **15**, 249–258.

Krasnov, B. R., Khokhlova, I. S., Oguzoglu, I. & Burdelova, N. V. (2002a). Host discrimination by two desert fleas using an odour cue. *Animal Behaviour*, **64**, 33–40.

Krasnov, B. R., Khokhlova, I. S., Fielden, L. J. & Burdelova, N. V. (2002b). The effect of substrate on survival and development of two species of desert fleas (Siphonaptera: Pulicidae). *Parasite*, **9**, 135–142.

Krasnov, B. R., Burdelova, N. V., Shenbrot, G. I. & Khokhlova, I. S. (2002c). Annual cycles of four flea species (Siphonaptera) in the central Negev Desert. *Medical and Veterinary Entomology*, **16**, 266–276.

Krasnov, B. R., Khokhlova, I. S., Fielden, L. J. & Burdelova, N. V. (2002d). Time to survival under starvation in two flea species (Siphonaptera: Pulicidae) at different air temperatures and relative humidities. *Journal of Vector Ecology*, **27**, 70–81.

Krasnov, B. R., Khokhlova, I. S. & Shenbrot, G. I. (2002e). The effect of host density on ectoparasite distribution: an example with a desert rodent parasitized by fleas. *Ecology*, **83**, 164–175.

Krasnov, B. R., Burdelov, S. A., Khokhlova, I. S. & Burdelova, N. V. (2003a). Sexual size dimorphism, morphological traits and jump performance in seven species of desert fleas (Siphonaptera). *Journal of Zoology*, **261**, 181–189.

Krasnov, B. R., Sarfati, M., Arakelyan, M. S., *et al.* (2003b). Host-specificity and foraging efficiency in blood-sucking parasite: feeding patterns of a flea *Parapulex chephrenis* on two species of desert rodents. *Parasitology Research*, **90**, 393–399.

Krasnov, B. R., Khokhlova, I. S. & Shenbrot, G. I. (2003c). Density-dependent host selection in ectoparasites: an application of isodar theory to fleas parasitizing rodents. *Oecologia*, **134**, 365–373.

Krasnov, B. R., Khokhlova, I. S., Burdelova, N. V., Mirzoyan, N. S. & Degen, A. A. (2004a). Fitness consequences of density-dependent host selection in ectoparasites: testing reproductive patterns predicted by isodar theory in fleas parasitizing rodents. *Journal of Animal Ecology*, **73**, 815–820.

Krasnov, B. R., Khokhlova, I. S., Burdelov, S. A. & Fielden, L. J. (2004b). Metabolic rate and jumping performance in seven species of desert fleas. *Journal of Insect Physiology*, **50**, 149–156.

Krasnov, B. R., Shenbrot, G. I., Khokhlova, I. S. & Poulin, R. (2004c). Relationships between parasite abundance and the taxonomic distance among a parasite's host species: an example with fleas parasitic on small mammals. *International Journal for Parasitology*, **34**, 1289–1297.

Krasnov, B. R., Shenbrot, G. I. & Khokhlova, I. S. (2004d). Sampling fleas: the reliability of host infestation data. *Medical and Veterinary Entomology*, **18**, 232–240.

Krasnov, B. R., Mouillot, D., Shenbrot, G. I., Khokhlova, I. S. & Poulin, R. (2004e). Geographical variation in host specificity of fleas (Siphonaptera) parasitic on small mammals: the influence of phylogeny and local environmental conditions. *Ecography*, **27**, 787–797.

Krasnov, B. R., Poulin, R., Shenbrot, G. I., Mouillot, D. & Khokhlova, I. S. (2004f). Ectoparasitic 'jacks-of-all-trades': relationship between abundance and host specificity in fleas (Siphonaptera) parasitic on small mammals. *American Naturalist*, **164**, 506–515.

Krasnov, B. R., Shenbrot, G. I., Khokhlova, I. S. & Degen, A. A. (2004g). Flea species richness and parameters of host body, host geography and host 'milieu'. *Journal of Animal Ecology*, **73**, 1121–1128.

Krasnov, B. R., Shenbrot, G. I., Khokhlova, I. S. & Degen, A. A. (2004h). Relationship between host diversity and parasite diversity: flea assemblages on small mammals. *Journal of Biogeography*, **31**, 1857–1866.

Krasnov, B. R., Poulin, R., Shenbrot, G. I., Mouillot, D. & Khokhlova, I. S. (2005a). Host specificity and geographic range in haematophagous ectoparasites. *Oikos*, **108**, 449–456.

Krasnov, B. R., Shenbrot, G. I., Khokhlova, I. S. & Poulin, R. (2005b). Diversification of ectoparasite assemblages and climate: an example with fleas parasitic on small mammals. *Global Ecology and Biogeography*, **14**, 167–175.

Krasnov, B. R., Morand, S., Hawlena, H., Khokhlova, I. S. & Shenbrot, G. I. (2005c). Sex-biased parasitism, seasonality and sexual size dimorphism in desert rodents. *Oecologia*, **146**, 209–217.

Krasnov, B. R., Khokhlova, I. S., Arakelyan, M. S. & Degen, A. A. (2005d). Is a starving host tastier? Reproduction in fleas parasitizing food limited rodents. *Functional Ecology*, **19**, 625–631.

Krasnov, B. R., Burdelova, N. V., Khokhlova, I. S., Shenbrot, G. I. & Degen, A. A. (2005e). Pre-imaginal interspecific competition in two flea species parasitic on the same rodent host. *Ecological Entomology*, **30**, 146–155.

Krasnov, B. R., Shenbrot, G. I., Mouillot, D., Khokhlova, I. S. & Poulin, R. (2005f). What are the factors determining the probability of discovering a flea species (Siphonaptera)? *Parasitology Research*, **97**, 228–237.

Krasnov, B. R., Mouillot, D., Shenbrot, G. I., Khokhlova, I. S. & Poulin, R. (2005g). Abundance patterns and coexistence processes in communities of fleas parasitic on small mammals. *Ecography*, **28**, 453–464.

Krasnov, B. R., Morand, S., Khokhlova, I. S., Shenbrot, G. I. & Hawlena, H. (2005h). Abundance and distribution of fleas on desert rodents: linking Taylor's power law to ecological specialization and epidemiology. *Parasitology*, **131**, 825–837.

Krasnov, B. R., Stanko, M., Miklisova, D. & Morand, S. (2005i). Distribution of fleas (Siphonaptera) among small mammals: mean abundance predicts prevalence via simple epidemiological model. *International Journal for Parasitology*, **35**, 1097–1101.

Krasnov, B. R., Shenbrot, G. I., Khokhlova, I. S. & Poulin, R. (2005j). Nested pattern in flea assemblages across the host's geographic range. *Ecography*, **28**, 475–484.

Krasnov, B. R., Shenbrot, G. I., Mouillot, D., Khokhlova, I. S. & Poulin, R. (2005k). Spatial variation in species diversity and composition of flea assemblages in small mammalian hosts: geographic distance or faunal similarity? *Journal of Biogeography*, **32**, 633–644.

Krasnov, B. R., Mouillot, D., Khokhlova, I. S., Shenbrot, G. I. & Poulin, R. (2005l). Covariance in species diversity and facilitation among non-interactive parasite taxa: all against the host. *Parasitology*, **131**, 557–568.

Krasnov, B. R., Stanko, M. & Morand, S. (2006a). Age-dependent flea (Siphonaptera) parasitism in rodents: a host's life history matters. *Journal of Parasitology*, **92**, 242–248.

Krasnov, B. R., Morand, S., Mouillot, D., et al. (2006b). Resource predictability and host specificity in fleas: the effect of host body mass. *Parasitology*, **133**, 81–88.

Krasnov, B. R., Shenbrot, G. I., Mouillot, D., Khokhlova, I. S. & Poulin, R. (2006c). Ecological characteristics of flea species relate to their suitability as plague vectors. *Oecologia*, **149**, 474–481.

Krasnov, B. R., Shenbrot, G. I., Khokhlova, I. S. & Poulin, R. (2006d). Is abundance a species attribute? An example with haematophagous ectoparasites. *Oecologia*, **150**, 132–140.

Krasnov, B. R., Stanko, M., Miklisova, D. & Morand, S. (2006e). Host specificity, parasite community size and the relation between abundance and its variance. *Evolutionary Ecology*, **20**, 75–91.

Krasnov, B. R., Stanko, M., Khokhlova, I. S., et al. (2006f). Aggregation and species coexistence in fleas parasitic on small mammals. *Ecography*, **29**, 159–168.

Krasnov, B. R., Shenbrot, G. I., Khokhlova, I. S., Hawlena, H. & Degen, A. A. (2006g). Temporal variation in parasite infestation of a host individual: does a parasite-free host remain uninfested permanently? *Parasitology Research*, **99**, 541–545.

Krasnov, B. R., Stanko, M. & Morand, S. (2006h). Are ectoparasite communities structured? Species co-occurrence, temporal variation and null models. *Journal of Animal Ecology*, **75**, 1330–1339.

Krasnov, B. R., Stanko, M., Khokhlova, I. S., et al. (2006i). Relationships between local and regional species richness in flea communities of small mammalian hosts: saturation and spatial scale. *Parasitology Research*, **98**, 403–413.

Krasnov, B. R., Morand, S. & Poulin, R. (2006j). Patterns of macroparasite diversity in small mammals. In *Micromammals and Macroparasites: From Evolutionary Ecology to Management*, ed. S. Morand, B. R. Krasnov & R. Poulin. New York: Springer-Verlag, pp. 197–232.

Krasnov, B. R., Stanko, M., Miklisova, D. & Morand, S. (2006k). Habitat variation in species composition of flea assemblages on small mammals in central Europe. *Ecological Research*, **21**, 460–469.

Krasnov, B. R., Korine, C., Burdelova, N. V., Khokhlova, I. S. & Pinshow, B. (2007a). Between-host phylogenetic distance and feeding efficiency in haematophagous ectoparasites: rodent fleas and a bat host. *Parasitology Research*, **101**, 365–371.

Krasnov, B. R., Hovhanyan, A., Khokhlova, I. S. & Degen, A. A. (2007b). Density-dependence and feeding success in haematophagous ectoparasites. *Parasitology*, **134**, 1379–1386.

Krasnov, B. R., Shenbrot, G. I., Khokhlova, I. S. & Poulin, R. (2007c). Geographic variation in the 'bottom-up' control of diversity: fleas and their small mammalian hosts. *Global Ecology and Biogeography*, **16**, 179–186.

Krebs, C. J. (1994). *Ecology: The Experimental Analysis of Distribution and Abundance*, 4th edn. New York: HarperCollins.

Krebs, C. J. (1996). Population cycles revisited. *Journal of Mammalogy*, **77**, 8–24.

Kristan, D. M. (2002). Maternal and direct effects of the intestinal nematode *Heligmosomoides polygyrus* on offspring growth and susceptibility to infection. *Journal of Experimental Biology*, **205**, 3967–3977.

Kristensen, N. P. (1975). The phylogeny of hexapod 'orders': a critical review of recent accounts. *Zeitschrift für zoologische Systematik und Evolutionsforschung*, **13**, 1–44.

Kristensen, N. P. (1981). Phylogeny of insect orders. *Annual Review of Entomology*, **26**, 135–157.

Krivokhatsky, V. A. (1984). Seasonal changes in the distribution of fleas in burrow passages as indicator of their migration activity. *Parazitologiya*, **18**, 150–153 (in Russian).

Krylov, D. G. (1986). On fauna and ecology of fleas parasitic on small mammals from the Moscow Region, Russian Federation, USSR. *Parazitologiya*, **20**, 356–363 (in Russian).

Kucheruk, V. V. (1983). Mammal burrows: their structure, topology and use. *Fauna and Ecology of Rodents*, **15**, 5–54 (in Russian).

Kucheruk, V. V., Kulik, I. L. & Dubrovsky, Y. A. (1972). The great gerbil as a desert life form. *Fauna and Ecology of Rodents*, **11**, 5–70 (in Russian).

Kulakova, Z. G. (1962). On the role of fleas in transmission of the tick-borne encephalitis (experimental results). *Bulletin of the Moscow Naturalist Society, Series Biology*, **67**, 144–145 (in Russian).

Kulakova, Z. G. (1964). Feeding of *Xenopsylla gerbilli caspica* and some other flea species. *Ectoparasites*, **4**, 205–219 (in Russian).

Kumar, R. & Kumar, R. (1996). Cross-resistance to *Hyalomma anatolicum anatolicum* ticks in rabbits immunized with midgut antigens of *Hyalomma dromedarii*. *Indian Journal of Animal Sciences*, **66**, 657–661.

Kunitskaya, N. T. (1960). Study of the reproductive organs of female fleas and identification of their physiological age. *Medical Parasitology and Parasitic Diseases* [*Meditsinskaya Parazitologiya i Parazitarnye Bolezni*], **29**, 688–701 (in Russian).

Kunitskaya, N. T. (1970). Structure of the ovaries in fleas. *Parazitologiya*, **4**, 444–450 (in Russian).

Kunitskaya, N. T., Gauzshtein, D. M., Kunitsky, V. N., Rodionov, I. A. & Filimonov, V. I. (1965a). Feeding activity of fleas parasitic on the great gerbil in experiments. In *Proceedings of the 4th Scientific Conference on Natural Focality and Prophylaxis of Plague*, ed. Anonymous. Alma-Ata, USSR: Kainar, pp. 135–137 (in Russian).

Kunitskaya, N. T., Kunitsky, V. N. & Gauzshtein, D. M. (1965b). On the reproduction of fleas parasitic on the great gerbil. In *Proceedings of the 4th Scientific Conference on Natural Focality and Prophylaxis of Plague*, ed. Anonymous. Alma-Ata, USSR: Kainar, pp. 137–138 (in Russian).

Kunitskaya, N. T., Kunitsky, V. N. & Gauzshtein, D. M. (1969). Reproduction and age structure of populations of fleas from genera *Coptopsylla* and *Paradoxopsyllus* in the southern Balkhash region. In *Proceedings of the 6th Scientific Conference of the Anti-Plague Establishments of the Middle Asia and Kazakhstan*, vol. 2, ed. M. A. Aikimbaev. Alma-Ata, USSR: The Middle Asian Scientific Anti-Plague Institute, pp. 74–76 (in Russian).

Kunitskaya, N. T., Kunitsky, V. N., Gauzshtein, D. M., Morozova, I. V. & Savelova, N. M. (1971). Age composition of imago in populations of *Xenopsylla gerbilli* and *Xenopsylla hirtipes* in the southern Pri-Balkhashie. In *Proceedings of the 7th Scientific Conference of the Anti-Plague Establishments of the Middle Asia and Kazakhstan*, ed. M. A. Aikimbaev. Alma-Ata, USSR: The Middle Asian Scientific Anti-Plague Institute, pp. 387–389 (in Russian).

Kunitskaya, N. T., Kunitsky, V. N. & Gauzshtein, D. M. (1974). On fecundity of fleas *Xenopsylla gerbilli* in experiments. In *Proceedings of the 8th Scientific Conference of the Anti-Plague Establishments of the Middle Asia and Kazakhstan*, ed. M. A. Aikimbaev. Alma-Ata, USSR: The Middle Asian Scientific Anti-Plague Institute, pp. 323–325 (in Russian).

Kunitskaya, N. T., Kunitsky, V. N. & Gauzshtein, D. M. (1977). Phenological age of fleas and an attempt to analyze age composition of natural populations of *Xenopsylla gerbilli* Wagn. *Parazitologiya*, **11**, 202–210 (in Russian).

Kunitskaya, N. T., Kunitsky, V. N., Gauzshtein, D. M. & Savelova, N. M. (1979). Spatial distribution of the larvae of fleas parasitic on the great gerbil in the host's burrow. *Proceedings of the 10th Scientific Conference of the Anti-Plague Establishments of the Middle Asia and Kazakhstan*, vol. 2, ed. M. A. Aikimbaev. Alma-Ata, USSR: The Middle Asian Scientific Anti-Plague Institute, pp. 107–110 (in Russian).

Kunitsky, V. N. (1961). On the environmental conditions of fleas parasitic on gerbils in the southeastern Azerbaijan SSR. *Zoologicheskyi Zhurnal*, **40**, 848–858 (in Russian).

Kunitsky, V. N. (1970). Essay on the comparative ecology of fleas parasitic on gerbils in southwestern Azerbaijan. In *Vectors of Dangerous Diseases and their Control*, ed. V. E. Tiflov. Stavropol, USSR: Scientific Anti-Plague Institute of Caucasus and Trans-Caucasus, pp. 153–227 (in Russian).

Kunitsky, V. N. & Kunitskaya, N. T. (1962). Fleas of southwestern Azerbaijan. *Proceedings of the Azerbaijanian Anti-Plague Station*, 3, 156–169 (in Russian).

Kunitsky, V. N., Kunitskaya, N. T., Gauzshtein, D. M. & Podkovyrova, T. S. (1963). Reproduction and development of *Ceratophyllus laeviceps* Wagn. in natural and experimental conditions. In *Proceedings of Scientific Conference of the Anti-Plague Establishments of the Middle Asia and Kazakhstan*, ed. Anonymous. Alma-Ata, USSR: The Middle Asian Scientific Anti-Plague Institute, pp. 126–128 (in Russian).

Kunitsky, V. N., Gauzshtein, D. M. & Kunitskaya, N. T. (1971a). On the effect of the soil moisture of the longevity of fleas under low ambient temperatures. In *Proceedings of the 7th Scientific Conference of the Anti-Plague Establishments of the Middle Asia and Kazakhstan*, ed. M. A. Aikimbaev. Alma-Ata, USSR: The Middle Asian Scientific Anti-Plague Institute, pp. 395–397 (in Russian).

Kunitsky, V. N., Volkov, V. M., Lelikova, Z. F., *et al.* (1971b). Changes in the numbers of fleas in the artificially decreased populations of the great gerbil in the northeastern part of the Caspian Lowland. In *Proceedings of the 7th Scientific Conference of the Anti-Plague Establishments of the Middle Asia and Kazakhstan*, ed. M. A. Aikimbaev. Alma-Ata, USSR: The Middle Asian Scientific Anti-Plague Institute, pp. 392–395 (in Russian).

Kunitsky, V. N., Volkov, V. M., Lelikova, Z. F. & Agunkova, O. S. (1974). On annual number of generations of *Xenopsylla skrjabini* in the Caspian Lowlands. In *Proceedings of the 8th Scientific Conference of the Anti-Plague Establishments of the Middle Asia and Kazakhstan*, ed. M. A. Aikimbaev. Alma-Ata, USSR: The Middle Asian Scientific Anti-Plague Institute, pp. 328–3330 (in Russian).

Kuris, A. M., Blaustein, A. R. & Aho, J. J. (1980). Hosts as islands. *American Naturalist*, 116, 570–586.

Kusiluka, L. J. M., Kambarage, D. M., Matthewman, R. W., Daborn, C. J. & Harrison, L. J. S. (1995). Prevalence of ectoparasites of goats in Tanzania. *Journal of Applied Animal Research*, 7, 69–74.

Kuznetsov, A. A. & Matrosov, A. N. (2003). The use of individual marking of fleas (Siphonaptera) for studies of their dispersal by hosts. *Zoologicheskyi Zhurnal*, 82, 964–972 (in Russian).

Kuznetsov, A. A., Matrosov, A. N., Nikitin, P. N. & Eigelis, S. Y. (1993). Method of individual marking of fleas and the results of the application of this method for the study of the dispersal of ectoparasites of gerbils in the Volga-Ural Sands. *Problems of Particularly Dangerous Diseases*, 73, 58–64 (in Russian).

Kuznetsov, A. A., Matrosov, A. N., Chyong, L. T. V. & Dat, D. T. (1999). Movements of the synantropous rats and their fleas in the settlements of the southern Vietnam. *Problems of Particularly Dangerous Diseases*, 79, 59–65 (in Russian).

Kyriazakis, I., Tolkamp, B. J. & Hutchings, M. R. (1998). Towards a functional explanation for the occurrence of anorexia during parasitic infections. *Animal Behaviour*, **56**, 265–274.

Labandeira, C. C. (1997). Insect mouthparts: ascertaining the paleobiology of insect feeding strategies. *Annual Review of Ecology and Systematics*, **28**, 153–193.

Labunets, N. F. (1967). Zoogeographic characteristics of the western Khangay. *Proceedings of the Irkutsk State Scientific Anti-Plague Institute of Siberia and Far East*, **27**, 231–240 (in Russian).

Lafferty, K. D. & Kuris, A. M. (2000). Parasite–host modelling meets reality: adaptive peaks and their ecological attributes. In *Evolutionary Biology of Host–Parasite Relationships: Theory Meets Reality*, ed. R. Poulin, S. Morand & A. Skorping. Amsterdam, the Netherlands: Elsevier Science, pp. 9–26.

Lafferty, K. D. & Kuris, A. M. (2002). Trophic strategies, animal diversity and body size. *Trends in Ecology and Evolution*, **17**, 507–513.

Lambrechts, M. M. & Dos Santos, A. (2000). Aromatic herbs in Corsican blue tit nests: the 'Potpourri' hypothesis. *Acta Oecologica*, **21**, 175–178.

Lang, J. D. (1996). Factors affecting the seasonal abundance of ground squirrel and wood rat fleas (Siphonaptera) in San Diego County, California. *Journal of Medical Entomology*, **33**, 790–804.

Lapchin, L. & Guillemaud, T. (2005). Asymmetry in host and parasitoid diffuse coevolution: when the Red Queen has to keep a finger in more than one pie. *Frontiers in Zoology*, **2**, 4. doi:10.1186/1742-9994-2-4.

Lareschi, M. (2006). The relationship of sex and ectoparasite infestation in the water rat *Scapteromys aquaticus* (Rodentia: Cricetidae) in La Plata, Argentina. *Revista de Biologia Tropical*, **54**, 673–679.

Larrivee, D. H., Benjamini, E., Feingold, B. F. & Shimizu, M. (1964). Histologic studies of guinea pig skin: different stages of allergic reactivity to flea bites. *Experimental Parasitology*, **15**, 491–502.

Larsen, K. S. (1995). Laboratory rearing of the squirrel flea *Ceratophyllus sciurorum sciurorum* with notes on its biology. *Entomologia Experimentalis et Applicata*, **76**, 241–245.

Larson, O. R. (1973). North Dakota fleas. IV. Cold tolerance in the bird flea *Ceratophillus idius* (Jordan and Rothschild). *Proceedings of the North Dakota Academy of Sciences*, **26**, 51–55.

Lauer, D. M. & Sonenshine, D. E. (1978). Bionomics of the squirrel flea, *Orchopeas howardi* (Siphonaptera: Ceratophyllidae), in laboratory and field colonies of the southern flying squirrel, *Glaucomys volans*, using radiolabelling technique. *Journal of Medical Entomology*, **15**, 1–10.

Launay, H. (1989). Facteurs écologiques influençant la répartition et la dynamique des populations de *Xenopsylla cunicularis* Smit, 1957 (Insecta: Siphonaptera) puce inféodée au lapin de garenne, *Oryctolagus cuniculus* (L.). *Vie et Milieu*, **39**, 111–120.

Lavoipierre, M. M. J., Radovsky, F. J. & Budwiser, P. D. (1979). The feeding process of a tungid flea, *Tunga monositus* (Siphonaptera: Tungidae), and its relationship to the host inflammatory and repair response. *Journal of Medical Entomology*, **15**, 187–217.

Lawrence, W. & Foil, L. D. (2002). The effect of diet upon pupal development and cocoon formation by the cat flea (Siphonaptera: Pulicidae). *Journal of Vector Ecology*, **27**, 39–43.

Lawton, J. H. & Hassell, M. P. (1981). Asymmetrical competition in insects. *Nature*, **289**, 793–795.

Layene, J. N. (1954). The biology of the red squirrel, *Tamiasciurus hudsonicus loquax* (Bangs), in central New York. *Ecological Monographs*, **24**, 227–267.

Layene, J. N. (1963). A study of the parasites of the Florida mouse, *Peromyscus floridanus*, in relation to host and environmental factors. *Tulane Studies in Zoology and Botany*, **11**, 1–27.

Lee, C. Y., Alexander, P. S., Yang, V. V. C. & Yu, J. Y. L. (2001). Seasonal reproductive activity of male formosan wood mice (*Apodemus semotus*): relationships to androgen levels. *Journal of Mammalogy*, **82**, 700–708.

Lee, P. L. M. & Clayton, D. H. (1995). Population biology of swift (*Apus apus*) ectoparasites in relation to host reproductive success. *Ecological Entomology*, **20**, 43–50.

Lee, S. E., Johnstone, I. P., Lee, R. P. & Opdebeeck, J. P. (1999). Putative salivary allergens of the cat flea, *Ctenocephalides felis*. *Veterinary Immunology and Immunopathology*, **69**, 229–237.

Lee, W. B. & Houston, D. C. (1993). The effect of diet quality on gut anatomy in British voles (Microtinae). *Journal of Comparative Physiology B*, **163**, 337–339.

Leeson, H. S. (1936). Further experiments upon the longevity of *Xenopsylla cheopis* Roths. (Siphonaptera). *Parasitology*, **28**, 403–409.

Legendre, P., Galzin, R. & Harmelin-Vivien, M. L. (1997). Relating behavior to habitat: solutions to the fourth-corner problem. *Ecology*, **78**, 547–562.

Legendre, P., Desdevises, Y. & Bazin, E. (2002). A statistical test for host–parasite coevolution. *Systematic Biology*, **51**, 217–234.

Lehane, M. (2005). *The Biology of Blood-Sucking in Insects*, 2nd edn. Cambridge, UK: Cambridge University Press.

Lehmann, T. (1992). Reproductive activity of *Synosternus cleopatrae* (Siphonaptera: Pulicidae) in relation to host factors. *Journal of Medical Entomology*, **29**, 946–952.

Lehmann, T. (1993). Ectoparasites: direct impact on host fitness. *Parasitology Today*, **9**, 8–13.

Lehmann, T. (1994). Reinfestation analysis to estimate ectoparasite population size, emergence and mortality. *Journal of Medical Entomology*, **31**, 257–264.

Leonov, Y. A. (1958). Fleas parasitic on rodents in the southern part of Primorie (Far East). *Proceedings of the Irkutsk State Scientific Anti-Plague Institute of Siberia and Far East*, **17**, 147–152 (in Russian).

Lepage, D. (2006). *Avibase: The World Bird Database*. Available online at www.bsc-eoc.org/avibase/avibase.jsp?pg=home&lang=EN

Letov, G. S., Emelianova, N. D., Letova, G. I. & Sulimov, A. D. (1966). Rodents and their ectoparasites in the settlements of Tuva. *Proceedings of the Irkutsk State Scientific Anti-Plague Institute of Siberia and Far East*, **26**, 270–276 (in Russian).

Letova, G. I., Letov, G. S. & Mamontova, E. V. (1969). Parasitological description of the Mungun-Taigin part of the Altai plague focus. *Problems of Particularly Dangerous Infections*, **10**, 55–60 (in Russian).

Levenbook, L. (1985). Storage proteins. In *Comprehensive Insect Biochemistry, Physiology and Pharmacology*, ed. L. I. Gilbert & G. Kerkut. Oxford, UK: Pergamon Press, pp. 307–346.

Levine, J. M. (1999). Indirect facilitation: evidence and predictions from a riparian community. *Ecology*, **80**, 1762–1769.

Levine, J. M. & Rees, M. (2002). Coexistence and relative abundance in annual plant assemblages: the roles of competition and colonization. *American Naturalist*, **160**, 452–467.

Levins, R. (1968). *Evolution in Changing Environments*. Princeton, NJ: Princeton University Press.

Lewis, R. E. (1998). Résumé of the Siphonaptera (Insecta) of the world. *Journal of Medical Entomology*, **35**, 377–389.

Lewis, R. E. & Eckerlin, R. P. (2004). A new species of *Hystrichopsylla* Taschenberg, 1880 (Siphonaptera: Hystrichopsyllidae) from Guatemala. *Proceedings of the Entomological Society of Washington*, **106**, 757–760.

Lewis, R. E. & Grimaldi, D. (1997). A pulicid flea in Miocene amber from the Dominican Republic (Insecta: Siphonaptera: Pulicidae). *American Museum Novitates*, **3205**, 1–9.

Lewis, R. E. & Haas, G. E. (2001). A review of the North American *Catallagia* Rothschild, 1915, with the description of a new species (Siphonaptera: Ctenophthalmidae: Neopsyllinae: Phalacropsyllini). *Journal of Vector Ecology*, **26**, 51–69.

Lewis, R. E. & Lewis, J. H. (1990). An annotated checklist of the fleas (Siphonaptera) of the Middle East. *Fauna of Saudi Arabia*, **11**, 251–277.

Lewis, R. E. & Stone, E. (2001). *Psittopsylla mexicana*, a new genus and species of bird flea from Chihuahua, Mexico (Siphonaptera: Ceratophyllidae: Ceratophyllinae). *Journal of the New York Entomological Society*, **109**, 360–366.

Lewis, R. E., Lewis, J. H. & Maser, C. (1988). *The Fleas of the Pacific Northwest*. Corvallis, OR: Oregon State University Press.

Li, B.-G., Zhang, P., Watanabe, K. & Tan, C.-L. (2002). Does allogrooming serve a hygienic function in the Sichuan snub-nosed monkey (*Rhinopithecus roxellana*)? *Acta Zoologica Sinica*, **48**, 705–715 (in Chinese).

Li, W., Wang, Z.-Y., Wang, C., Zhang, Z.-J. & Ye, R.-Y. (2004). Observations on the life cycle of *Neopsylla teratura* Rothschild, 1913 in the laboratory. *Endemic Diseases Bulletin*, **19**, 8–9 (in Chinese).

Li, Z.-L. & Ma, L.-M. (1999). Aggregation degree of distribution of two fleas, *Citellophilus tesquorum sungaris* (Jordan) and *Neopsylla bidentatiformis* Wagner, on host body. *Entomological Knowledge*, **36**, 89–91 (in Chinese).

Li, Z.-L. & Zhang, Y.-X. (1997). Analysis on the yearly dynamics relation between body flea index and population of *Citellus dauricus*. *Acta Entomologica Sinica*, **40**, 166–170 (in Chinese).

Li, Z.-L. & Zhang, Y.-X. (1998). The yearly dynamics relationship between burrow nest flea index and population of *Citellus dauricus*. *Acta Entomologica Sinica*, **41**, 77–81 (in Chinese).

Li, Z.-L., Zhang, W.-R. & Ma, L.-M. (1995). Analysis of the relations among flea index, populations of *Meriones unguiculatus* and meteorological factors. *Acta Entomologica Sinica*, **38**, 442–447 (in Chinese).

Li, Z.-L., Zhang, W.-R. & Yan, W.-L. (2000). Studies on dynamics of body and burrow nest fleas of *Meriones unguiculatus*. *Acta Entomologica Sinica*, **43**, 58–63 (in Chinese).

Li, Z.-L., Yang, Y. & Chen, S.-G. (2001a). An analysis on the population dynamics of fleas in the man-made plague focuses of Harbin suburbs. *Acta Entomologica Sinica*, **44**, 507–511 (in Chinese).

Li, Z.-L., Liu, T.-C. & Niu, Y. (2001b). Studies on dynamics of body and burrow fleas of *Microtus brandti* and succession of their community. *Acta Entomologica Sinica*, **44**, 327–331 (in Chinese).

Li, Z.-L., Yu, G.-J. & Chen, D. (2002). Statespace model between *Nosopsyllus laeviceps kuzenkovi* and population of *Meriones unguiculatus*. *Acta Entomologica Sinica*, **45**, 132–133 (in Chinese).

Liao, H.-R. & Lin, D.-H. (1993). Laboratorial observation on some biological characters of two rat fleas in south China. *Endemic Diseases Bulletin*, **8**, 61–64 (in Chinese).

Lighton, J. R. B. (1991). Measurements in insects. In *Concise Encyclopedia on Biological and Biomedical Measurement Systems*, ed. C. A. Payne. Oxford, UK: Pergamon Press, pp. 201–208.

Lighton, J. R. B., Fielden, L. J. & Rechav, Y. (1993). Discontinuous ventilation in a non-insect, the tick *Amblyomma marmoreum* (Acari, Ixodidae): characterization and metabolic modulation. *Journal of Experimental Biology*, **180**, 229–245.

Liker, A., Markus, M., Vazar, A., Zemankovics, E. & Rózsa, L. (2001). Distribution of *Carnus hemapterus* in a starling colony. *Canadian Journal of Zoology*, **79**, 574–580.

Linardi, P. M. & Guimarães, L. R. (2000). *Sifonapteros do Brasil*. São Paulo, Brazil: Museu de Zoologia da Universidade de São Paulo.

Linardi, P. M., Botelho, J. R. & Cunha, H. C. (1985). Ectoparasites of rodents of the urban region of Belo Horizonte, MG. II. Variations of the infestation indices in *Rattus norvegicus norvegicus*. *Memórias do Instituto Oswaldo Cruz*, **80**, 227–232.

Linardi, P. M., Gomes, A. F., Botelho, J. R. & Lopes, C. M. L. (1994). Some ectoparasites of commensal rodents from Huambo, Angola. *Journal of Medical Parasitology*, **31**, 754–756.

Linardi, P. M., DeMaria, M. & Botelho, J. R. (1997). Effects of larval nutrition on the postembryonic development of *Ctenocephalides felis felis* (Siphonaptera: Pulicidae). *Journal of Medical Entomology*, **34**, 494–497.

Lindén, M. (1991). Divorce in great tits: chance or choice? An experimental approach. *American Naturalist*, **138**, 1039–1048.

Lindsay, L. R. & Galloway, T. D. (1997). Seasonal activity and temporal separation of four species of fleas (Insecta: Siphonaptera) infesting Richardson's ground squirrels, *Spermophilus richardsoni* (Rodentia: Sciuridae), in Manitoba, Canada. *Canadian Journal of Zoology*, **75**, 1310–1322.

Lindsay, L. R. & Galloway, T. D. (1998). Reproductive status of four species of fleas (Insecta: Siphonaptera) on Richardson's ground squirrel (Rodentia: Sciuridae) in Manitoba, Canada. *Journal of Medical Entomology*, **35**, 423–430.

Linley, J. R., Benton, A. H. & Day, J. F. (1994). Ultrastructure of the eggs of seven flea species (Siphonaptera). *Journal of Medical Entomology*, **31**, 813–827.

Linsdale, J. M. (1947). *The California Ground Squirrel*. Berkeley, CA: University of California Press.

Linsdale, J. M. & Davis, B. S. (1956). Taxonomic appraisal and occurrence of fleas at the Hastings Reservation in Central California. *University of California Publications in Zoology*, **54**, 293–370.

Linsdale, J. M. & Tevis, L. P. (1951). *The Dusky-Footed Wood Rat*. Berkeley, CA: University of California Press.

Litvinova, E. A. (2004). On biology and ecology of *Ctenophthalmus congeneroides* Wagner, 1929 (Siphonaptera: Hystrichopsyllidae): one of the most abundant fleas of rodents in the Primorie Region. *A. I. Kurentsov's Annual Memorial Meetings*, **15**, 104–107 (in Russian).

Liu, J., Li, S.-J., Amin, O. M. & Zhang, Y.-M. (1993). Blood-feeding of the gerbil flea *Nosopsyllus laeviceps kuzenkovi* (Yagubyants), vector of plague in Inner Mongolia, China. *Medical and Veterinary Entomology*, **7**, 54–58.

Liu, Z.-Y., Wu, H.-Y., Li, G.-Z., et al. (1986). *Fauna Sinica. Insecta. Siphonaptera*. Beijing: Science Press (in Chinese).

Lively, C. M. (1989). Adaptation by a parasitic nematode to local populations of its snail host. *Evolution*, **50**, 1663–1671.

Lloyd, M. (1967). Mean crowding. *Journal of Animal Ecology*, **36**, 1–30.

Lo, C. M., Morand, S. & Galzon, R. (1998). Parasite diversity/host age and size relationship in three coral-reef fish from French Polynesia. *International Journal for Parasitology*, **28**, 1695–1708.

Lochmiller, R. L. & Dabbert, C. B. (1993). Immunocompetence, environmental stress, and the regulation of animal populations. *Trends in Comparative Biochemistry and Physiology*, **1**, 823–855.

Lochmiller, R. L. & Deerenberg, C. (2000). Trade-offs in evolutionary immunology: just what is the cost of immunity? *Oikos*, **88**, 87–98.

Lochmiller, R. L., Vestey, M. R. & Boren, J. C. (1993). Relationship between protein nutritional status and immunocompetence in northern bobwhite chicks. *Auk*, **110**, 503–510.

Lombardero, M. J., Ayres, M. P., Hofstetter, R. W., Moser, J. C. & Lepzig, K. D. (2003). Strong indirect interactions of *Tarsonemus* mites (Acarina: Tarsonemidae) and *Dendroctonous frontalis*. *Oikos*, **102**, 243–252.

Lomnicki, A. (1988). *Population Ecology of Individuals*. Princeton, NJ: Princeton University Press.

López-Sepulcre, A. & Kokko, H. (2005). Territorial defense, territory size, and population regulation. *American Naturalist*, **166**, 317–329.

Lorange, E. A., Race, B. L., Sebbane, F. & Hinnebusch, B. J. (2005). Poor vector competence of fleas and the evolution of hypervirulence in *Yersinia pestis*. *Journal of Infectious Diseases*, **191**, 1907–1912.

Loreau, M. (2000). Are communities saturated? On the relationship of S, J and T diversity. *Ecology Letters*, **3**, 73–76.

Losos, J. B., Leal, M., Glor, R. E., et al. (2003). Niche lability in the evolution of a Caribbean lizard community. *Nature*, **424**, 542–545.

Lott, D. F. (1991). *Intraspecific Variation in the Social Systems of Wild Vertebrates*. Cambridge, UK: Cambridge University Press.

Louw, J. P., Horak, I. G. & Braack, L. E. O. (1993). Fleas and lice on scrub hares (*Lepus saxatilis*). *Onderstepoort Journal of Veterinary Research*, **60**, 95–101.

Louw, J. P., Horak, I. G., Horak, M. L. & Braack, L. E. O. (1995). Fleas, lice and mites on scrub hares (*Lepus saxatilis*) in Northern and Eastern Transvaal and in KwaZulu-Natal, South Africa. *Onderstepoort Journal of Veterinary Research*, **62**, 133–137.

Lu, L. & Wu, H. (2001). The molecular phylogeny of some species of the *bidentatiformis* group of the genus *Neopsylla* based on 16s rRNA gene. *Acta Entomologica Sinica*, **44**, 548–554 (in Chinese).

Lu, L. & Wu, H. (2002). The variation of rDNA ITS2 sequences in nine species of *Neopsylla* (Siphonaptera: Ctenophthalmidae). *Acta Parasitologica et Medica Entomologica Sinica*, **9**, 106–113 (in Chinese).

Lu, L. & Wu, H. (2005). Morphological phylogeny of *Geusibia* Jordan, 1932 (Siphonaptera: Leptopsyllidae) and the host–parasite relationships with pikas. *Systematic Parasitology*, **61**, 65–78.

Luchetti, A., Mantovani, B., Pampiglione, S. & Trentini, M. (2005). Molecular characterization of *Tunga trimamillata* and *T. penetrans* (Insecta, Siphonaptera, Tungidae): taxonomy and genetic variability. *Parasite*, **12**, 123–129.

Ludwig, D. (1999). Is it meaningful to estimate a probability of extinction? *Ecology*, **80**, 298–310.

Lukashevich, E. D. & Mostovsky, M. B. (2003). Haematophagous insects in the fossil record. *Paleontologicheskyi Zhurnal*, **37**, 48–56 (in Russian).

Lundqvist, L. (1988). Reproductive strategies of ectoparasites on small mammals. *Canadian Journal of Zoology*, **66**, 774–781.

Lundqvist, L. & Brinck-Lindroth, G. (1990). Patterns of coexistence: ectoparasites on small mammals in northern Fennoscandia. *Holarctic Ecology*, **13**, 39–49.

Ma, L.-M. (1983). Distribution of fleas in the hair coat of the host. *Acta Entomologica Sinica*, **26**, 409–412 (in Chinese).

Ma, L.-M. (1988). Abundance of fleas in relation to population fluctuations of their hosts. *Acta Entomologica Sinica*, **31**, 50–54 (in Chinese).

Ma, L.-M. (1989). The distribution of fleas on the host body in relation to temperature and the number of fleas. *Acta Entomologica Sinica*, **32**, 68–73 (in Chinese).

Ma, L.-M. (1990). Observations on longevity of adult *Neopsylla bidentatiformis* Wagner and *Citellophilus tesquorum sungaris* (Jordan) under different conditions. *Entomological Knowledge*, **27**, 358–359 (in Chinese).

Ma, L.-M. (1993a). Resistance of fleas *Neopsylla bidentatiformis* and *Citellophilus* (*Citellophilus*) *tesquorum sungaris* to low temperature. *Endemic Diseases Bulletin*, **8**, 71–73 (in Chinese).

Ma, L.-M. (1993b). The sex ratios of some fleas in north China. *Acta Entomologica Sinica*, **36**, 63–66 (in Chinese).

Ma, L.-M. (1994a). Observation on survival of fleas and their hosts under high temperature in field of northern China. *Acta Zoologica Sinica*, **40**, 100–104 (in Chinese).

Ma, L.-M. (1994b). Laboratory studies on fleas *Neopsylla bidentatiformis* and *Citellophilus tesquorum sungaris* attacking and leaving from hosts. *Acta Entomologica Sinica*, **36**, 63–66 (in Chinese).

Ma, L.-M. (1995). Activity of *Neopsylla bidentatiformis* Wagner and *Citellophilus tesquorum sungaris* Jordan. *Entomological Knowledge*, **32**, 225–227 (in Chinese).

Ma, L.-M. (1997). The monstrosities of fleas reared in laboratory. *Entomological Journal of East China*, **6**, 107–109 (in Chinese).

Ma, L.-M. (2000). Body length of fleas in relation to some factors, and influence of host nutrition on fleas. *Acta Parasitologica et Medica Entomologica Sinica*, **7**, 235–240 (in Chinese).

Ma, L.-M. (2002). Relationship of hunger tolerance to some environmental and physiologic factors in fleas *Neopsylla bidentatiformis* and *Citellophilus tesquorum sungaris*. *Acta Parasitologica et Medica Entomologica Sinica*, **9**, 246–248 (in Chinese).

MacArthur, R. H. (1955). Fluctuations in animal populations, and a measure of community stability. *Ecology*, **36**, 533–536.

MacArthur, R. H. (1972). *Geographical Ecology*. New York: Harper & Row.

MacArthur, R. H. & Levins, R. (1964). Competition, habitat selection and character displacement in a patchy environment. *Proceedings of the National Academy of Sciences of the USA*, **51**, 1207–1210.

MacArthur, R. H. & Levins, R. (1967). The limiting similarity of convergence and divergence of coexisting species. *American Naturalist*, **101**, 377–385.

MacArthur, R. H. & Wilson, E. O. (1967). *The Theory of Island Biogeography*. Princeton, NJ: Princeton University Press.

MacInnis, A. J. (1976). How parasites find hosts: some thoughts on the inception of host–parasite integration. In *Ecological Aspects of Parasitology*, ed. C. R. Kennedy. Amsterdam, the Netherlands: North Holland, pp. 3–20.

Madhavi, R. & Anderson, R. M. (1985). Variability in the susceptibility of the fish host, *Poecilia reticulata*, to infection with *Gyrodactylus bullatarudis* (Monogenea). *Parasitology*, **91**, 531–544.

Main, A. J. (1983). Fleas (Siphonaptera) on small mammals in Connecticut, USA. *Journal of Medical Entomology*, **20**, 33–39.

Maizels, R. M., Balic, A., Gomez-Escobar, N., *et al.* (2004). Helminth parasites: masters of regulation. *Immunological Reviews*, **201**, 89–116.

Makundi, R. H. & Kilonzo, B. S. (1994). Seasonal dynamics of rodent fleas and its implication on control strategies in Lushoto district, north-eastern Tanzania. *Journal of Applied Entomology*, **118**, 165–171.

Manhert, V. (1972). Zum Auftreten von Kleinsäuger-Flöhen auf ihren Wirten in Abhängigkeit von Jahreszeit und Höhenstufen. *Oecologia*, **8**, 400–418.

Mans, B. J., Louw, A. I. & Neitz, A. W. H. (2002). Evolution of hematophagy in ticks: common origins for blood coagulation and platelet aggregation inhibitors from soft ticks of the genus *Ornithodoros*. *Molecular Biology and Evolution*, **19**, 1695–1705.

Mappes, T., Mappes, J. & Kotiaho, J. (1994). Ectoparasites, nest-site choice and breeding success in the pied flycatcher. *Oecologia*, **98**, 147–149.

Margalit, Y. & Shulov, A. S. (1972). Effect of temperature on development of prepupa and pupa of the rat flea, *Xenopsylla cheopis* Rothschild. *Journal of Medical Entomology*, **9**, 117–125.

Margolis, L., Esch, G. W., Holmes, J. C., Kuris, A. M. & Schad, G. A. (1982). The use of ecological terms in parasitology (report of an *ad hoc* committee of the American Society of Parasitologists). *Journal of Parasitology*, **68**, 131–133.

Marie, J.-L., Fournier, P.-E., Rolain, J.-M., et al. (2006). Molecular detection of *Bartonella quintana, B. elizabethae, B. koehlerae, B. doshiae, B. taylorii,* and *Rickettsia felis* in rodent fleas collected in Kabul, Afghanistan. *American Journal of Tropical Medicine and Hygiene*, **74**, 436–439.

Marshall, A. G. (1981a). *The Ecology of Ectoparasitic Insects*. London: Academic Press.

Marshall, A. G. (1981b). Sex ratio in ectoparasitic insects. *Ecological Entomology*, **6**, 155–174.

Martin, T. E., Møller, A. P., Merino, S. & Clobert, J. (2001). Does clutch size evolve in response to parasites and immunocompetence? *Proceedings of the National Academy of Sciences of the USA*, **98**, 2071–2076.

Mashek, H., Licznerski, B. & Pincus, S. (1997). Tungiasis in New York. *International Journal of Dermatology*, **36**, 276–278.

Mashtakov, V. I. (1969). Density dynamics of gerbils and their fleas in different landscape-ecological regions. *Problems of Particularly Dangerous Infections*, **8**, 100–105 (in Russian).

Maslennikova, Z. P., Bibikova, V. A. & Morozova, I. V. (1967). Abundance of *Xenopsylla* fleas in winter micropopulations in the association with the abundance of the great gerbil. In *Proceedings of the 5th Scientific Conference of the Anti-Plague Establishments of the Middle Asia and Kazakhstan*, ed. M. A. Aikimbaev. Alma-Ata, USSR: The Middle Asian Scientific Anti-Plague Institute, pp. 176–178 (in Russian).

Matejusová, I., Morand, S. & Gelnar, M. (2000). Nestedness in assemblages of gyrodactylids (Monogenea: Gyrodactylidea) parasitising two species of cyprinid: with reference to generalists and specialists. *International Journal for Parasitology*, **30**, 1153–1158.

Matthews, J. W. (2004). Effects of site and species characteristics on nested patterns of species composition in sedge meadows. *Plant Ecology*, **174**, 271–278.

May, R. M. & Anderson, R. M. (1978). Regulation and stability of host–parasite population interactions. II. Destabilizing processes. *Journal of Animal Ecology*, **47**, 455–461.

May, R. M. & Anderson, R. M. (1979). Population biology of infectious diseases. II. *Nature*, **280**, 455–461.

Mayevsky, M. P., Bazanova L. P. & Popkov, A. F. (1999). Winter survival of the causative agent of plaque in the long-tailed ground squirrel in the Tuva natural focus. *Medical Parasitology and Parasitic Diseases* [*Meditsinskaya Parazitologiya i Parazitarnye Bolezni*], **68**, 55–58 (in Russian).

Mazgajski, T. D., Kedra, A. H., Modlińska, E. & Samborski, J. (1997). Siphonaptera influence the condition of starling *Sturnus vulgaris* nestlings. *Acta Ornithologica*, **32**, 185–190.

McCallum, H. & Dobson, A. P. (1995). Detecting disease and parasite threats to endangered species and ecosystems. *Trends in Ecology and Evolution*, **10**, 190–194.

McCoy, G. W. & Mitzmain, M. B. (1909). The regional distribution of fleas on rodents. *Parasitology*, **2**, 297–304.

McDermott, M. J., Weber, E., Hunter, S., et al. (2000). Identification, cloning, and characterization of a major cat flea salivary allergen (Cte f 1). *Molecular Immunology*, **37**, 361–375.

McKenzie, A. A. (1990). The ruminant dental grooming apparatus. *Zoological Journal of the Linnean Society*, **99**, 117–128.

McKinney, P. & McDonald, L. C. (2001). Tungiasis. *Medical Journal*, **2**, 3–10.

McTier, T. L., George, J. E. & Bennet, S. N. (1981). Resistance and cross-resistance of guinea pigs to *Dermacentor andersoni* Stiles, *D. variabilis* (Say), *Amblyomma americanum* (Linnaeus) and *Ixodes scapularis* Say. *Journal of Parasitogy*, **67**, 813–822.

Mead-Briggs, A. R. (1964). The reproductive biology of the rabbit flea *Spilopsyllus cuniculi* (Dale) and the dependence of this species upon the breeding of its host. *Journal of Experimental Biology*, **41**, 371–402.

Mead-Briggs, A. R. & Rudge, A. J. B. (1960). Breeding of the rabbit flea, *Spilopsyllus cuniculi* (Dale): requirement of a 'factor' from a pregnant rabbit for ovarian maturation. *Nature*, **187**, 1136–1137.

Mead-Briggs, A. R. & Vaughan, J. A. (1969). Some requirements for mating in the rabbit flea, *Spilopsyllus cuniculi* (Dale). *Journal of Experimental Biology*, **51**, 495–511.

Mead-Briggs, A. R., Vaughan, J. A. & Rennison, B. D. (1975). Seasonal variation in numbers of the rabbit flea on the wild rabbit. *Parasitology*, **70**, 103–118.

Mears, S., Clark, F., Greenwood, M. & Larsen, K. S. (2002). Host location, survival and fecundity of the Oriental rat flea *Xenopsylla cheopis* (Siphonaptera: Pulicidae) in relation to black rat *Rattus rattus* (Rodentia: Muridae) host sex and age. *Bulletin of Entomological Research*, **92**, 375–384.

Medvedev, S. G. (1989a). Structure of the head capsule in fleas (Siphonaptera). I. *Entomological Review*, **68**, 1–18.

Medvedev, S. G. (1989b). Ecological characteristics and distribution of fleas of the family Ischnopsyllidae (Siphonaptera). *Parasitological Collection*, **36**, 21–43 (in Russian).

Medvedev, S. G. (1990). Evolution of fleas parasitic on Chiroptera. *Parazitologiya*, 24, 457–465 (in Russian).

Medvedev, S. G. (1994). Morphological basis of classification of the order Siphonaptera. *Entomological Review*, 73, 22–43 (in Russian).

Medvedev, S. G. (1996). Geographical distribution of families of fleas (Siphonaptera). *Entomological Review*, 76, 978–992.

Medvedev, S. G. (1997a). Host–parasite relations in fleas (Siphonaptera). I. *Entomological Review*, 77, 318–337.

Medvedev, S. G. (1997b). Host–parasite relations in fleas (Siphonaptera). II. *Entomological Review*, 77, 511–521.

Medvedev, S. G. (1998). Fauna and host–parasite relations of fleas (Siphonaptera) in the Palaearctic. *Entomological Review*, 78, 292–308.

Medvedev, S. G. (2000a). Fauna and host–parasite associations of fleas (Siphonaptera) in different zoogeographical regions of the world. I. *Entomological Review*, 80, 409–435.

Medvedev, S. G. (2000b). Fauna and host–parasite associations of fleas (Siphonaptera) in different zoogeographical regions of the world. II. *Entomological Review*, 80, 640–655.

Medvedev, S. G. (2001). Morphological structure of thoracic and abdominal ctenidia of fleas (Siphonaptera). *Parazitologiya*, 35, 291–306 (in Russian).

Medvedev, S. G. (2002). Specific features of the distribution and host associations of fleas (Siphonaptera). *Entomological Review*, 82, 1165–1177.

Medvedev, S. G. (2003a). Morphological adaptations of fleas (Siphonaptera) to parasitism. I. *Entomological Review*, 82, 40–62 (in Russian).

Medvedev, S. G. (2003b). Morphological adaptations of fleas (Siphonaptera) to parasitism. II. *Entomological Review*, 82, 820–835 (in Russian).

Medvedev, S. G. (2004). Morphological adaptations of fleas (Siphonaptera) to parasitism. III. *Entomological Review*, 83, 313–333 (in Russian).

Medvedev, S. G. (2005). *An Attempted System Analysis of the Evolution of the Order of Fleas (Siphonaptera)*, Lectures in Memoriam N. A. Kholodkovsky, No. 57. St Petersburg, Russia: Russian Entomological Society and Zoological Institute of Russian Academy of Sciences (in Russian).

Medvedev, S. G. & Lobanov, A. L. (1999). Information-analytical system of the World fauna of fleas (Siphonaptera): results and prospects. *Entomological Review*, 79, 654–665.

Medvedev, S. G. & Krasnov, B. R. (2006). Fleas: permanent satellites of small mammals. In *Micromammals and Macroparasites: From Evolutionary Ecology to Management*, ed. S. Morand, B. R. Krasnov & R. Poulin. New York: Springer-Verlag, pp. 161–177.

Medvedev, S. G., Khabilov, T. K. & Rybin, S. N. (1984). On the biology of the bat fleas (Ischnopsyllidae: Siphonaptera) in the Middle Asia and Kazakhstan. *Parazitologiya*, 18, 140–149 (in Russian).

Medvedev, S. G., Lobanov, A. L. & Lyangouzov, I. A. (2005). World Database of Fleas (Nov 2004 Version). In *Species 2000 and ITIS Catalogue of Life: 2005 Annual Checklist*, ed. F. A. Bisby, M. A. Ruggiero, K. L. Wilson, *et al.* Reading, MA: Species 2000 (CD-ROM).

Medzykhovsky, G. A. (1971). On the method of measurement of the duration of uninterrupted stay of fleas on a host. In *Proceedings of the 7th Scientific Conference of the Anti-Plague Establishments of the Middle Asia and Kazakhstan*, ed. M. A. Aikimbaev. Alma-Ata, USSR: The Middle Asian Scientific Anti-Plague Institute, pp. 404–406 (in Russian).

Mellanby, K. (1933). The influence of the temperature and humidity on the pupation of *Xenopsylla cheopis*. *Bulletin of Entomological Research*, **24**, 197–202.

Méndez, E. (1977). Mammalian–siphonapteran associations, the environment, and biogeography of mammals of southwestern Colombia. *Quaestiones Entomologicae*, **13**, 91–182.

Meng, F.-X., Feng, Y.-L., Chen, J.-Q., Song, X.-P. & Liu, Q.-Y. (2006). The sex ratio and adult eclosion of *Xenopsylla cheopis* in laboratory. *Chinese Journal of Vector Biology and Control*, **17**, 15–16 (in Chinese).

Menier, K. (2003). Infestation of dairy goats with *Pulex irritans*. *Veterinary Record*, **153**, 128–128.

Merila, J. & Allander, K. (1995). Do great tits (*Parus major*) prefer ectoparasite-free roost sites? – An experiment. *Ethology*, **99**, 53–60.

Merino, S. & Potti, J. (1996). Weather dependent effects of nest ectoparasites on their bird hosts. *Ecography*, **19**, 107–113.

Merino, S., Minguez, E. & Belliure, B. (1999). Ectoparasite effects on nestling European storm-petrels. *Waterbirds*, **22**, 297–301.

Metzger, M. E. & Rust, M. K. (1992). Egg production and emergence of adult cat fleas (Siphonaptera: Pulicidae) exposed to different photoperiods. *Journal of Medical Entomology*, **33**, 651–655.

Metzger, M. E. & Rust, M. K. (1997). Effect of temperature on cat flea (Siphonaptera: Pulicidae) development and overwintering. *Journal of Medical Entomology*, **34**, 173–178.

Meyer, K. F. (1947). The prevention of plague in the light of newer knowledge. *Annals of the New York Academy of Sciences*, **48**, 429–467.

Michelsen, V. (1997). A revised interpretation of the mouthparts in adult fleas (Insecta, Siphonaptera). *Zoologischer Anzeiger*, **235**, 217–223.

Miklisova, D. & Stanko, M. (1992). Negative binomial distribution as a model for fleas on small rodents. *Biologia, Bratislava*, **52**, 647–652.

Mikulin, M. A. (1956). Data on fleas of the Middle Asia. II. Fauna and some characteristic of geographic distribution of fleas parasitic on the great gerbil in deserts of the southern Trans-Balkhash Desert. *Proceedings of the Middle Asian Scientific Anti-Plague Institute*, **2**, 95–107 (in Russian).

Mikulin, M. A. (1958). Data on fleas of the Middle Asia and Kazakhstan. V. Fleas of the Tarbagatai Mountains. *Proceedings of the Middle Asian Scientific Anti-Plague Institute*, **4**, 227–240 (in Russian).

Mikulin, M. A. (1959a). Data on fleas of the Middle Asia and Kazakhstan. VIII. Fleas of the Akmolinsk Region. *Proceedings of the Middle Asian Scientific Anti-Plague Institute*, **5**, 237–245 (in Russian).

Mikulin, M. A. (1959b). Data on fleas of the Middle Asia and Kazakhstan. X. Fleas of the eastern Balkhash Desert, Trans-Alakul Desert and Sungorian Gates. *Proceedings of the Middle Asian Scientific Anti-Plague Institute*, **6**, 205–220 (in Russian).

Milazzo, C., Goüy de Bellocq, J., Cagnin, M., et al. (2003). Helminths and ectoparasites of *Rattus rattus* and *Mus musculus* from Sicily, Italy. *Comparative Parasitology*, **70**, 199–204.

Milinski, M. (1990). Parasites and host decision-making. In *Parasitism and Host Behaviour*, ed. C. J. Barnard & J. M. Behnke. London: Taylor & Francis, pp. 95–116.

Milinski, M. & Bakker, T. C. M. (1990). Female sticklebacks use male coloration in mate choice and hence avoid parasitized males. *Nature*, **344**, 330–333.

Miller, D. H. & Benton, A. H. (1970). Cold tolerance of some adult fleas (Ceratophyllidae: Siphonaptera). *Canadian Field Naturalist*, **84**, 396–397.

Miller, R. A. (1996). The aging immune system: primer and prospectus. *Science*, **273**, 70–74.

Mineur, Y. S., Prasol, D. J., Belzung, C. & Crusio, W. E. (2003). Agonistic behavior and unpredictable chronic mild stress in mice. *Behavior Genetics*, **33**, 513–519.

Mironov, A. N. & Pasyukov, V. V. (1987). Observations on the construction of cocoons by fleas *Nosopsyllus fasciatus*. *Parazitologiya*, **21**, 10–15 (in Russian).

Mitchell-Jones, A. J., Amori, G., Boganowicz, W., et al. (1999). *The Atlas of European Mammals*. London: T. & A. D. Poyser.

Miyamoto, K. & Hashimoto, Y. (2000). Outbreak of the sparrow flea bite cases in Hokkaido, Japan. *Medical Entomology and Zoology*, **51**, 111 (in Japanese).

Moeller, D. A. (2004). Facilitative interactions among plants via shared pollinators. *Ecology*, **85**, 3289–3301.

Mohr, C. O. (1958). Relation to mean number of fleas to prevalence of infestation on rats. *American Journal of Tropical Medicine and Hygiene*, **7**, 519–522.

Moll, A. A. & O'Leary, S. B. (1945). *Plague in the Americas: Historical and Quasi-Epidemiological Survey*. Washington, DC: The Pan American Sanitary Bureau.

Møller, A. P. (1989). Parasites, predators and nest boxes: facts and artefacts in the nest boxes studies of birds? *Oikos*, **56**, 421–423.

Møller, A. P. (1991). Ectoparasite loads affect optimal clutch size in swallows. *Functional Ecology*, **5**, 351–359.

Møller, A. P. (1993). Parasites differentially increase the degree of fluctuating asymmetry in secondary sexual characters. *Journal of Evolutionary Biology*, **5**, 691–699.

Møller, A. P. (1997). Parasitism and the evolution of host life history. In *Host–Parasite Evolution: General Principles and Avian Models*, ed. D. H. Clayton & J. Moore. Oxford, UK: Oxford University Press, pp. 105–127.

Møller, A. P. & de Lope, F. (1999). Senescence in a short-lived migratory bird: age-dependent morphology, migration, reproduction and parasitism. *Journal of Animal Ecology*, **68**, 163–171.

Møller, A. P., Christe, P. & Garamszegi, L. Z. (2005). Coevolutionary arms races: increased host immune defense promotes specialization by avian fleas. *Journal of Evolutionary Biology*, **18**, 46–59.

Moore, J. (2002). *Parasites and the Behavior of Animals*. New York: Oxford University Press.

Moore, S. L. & Wilson, K. (2002). Parasites as a viability cost of sexual selection in natural populations of mammals. *Science*, **297**, 2015–2018.

Mooring, M. S. (1995). The effect of tick challenge on grooming rate by impala. *Animal Behaviour*, **50**, 377–392.

Mooring, M. S. & Hart, B. L. (1995). Costs of allogrooming in impala: distraction from vigilance. *Animal Behaviour*, **49**, 1414–1416.

Mooring, M. S. & Hart, B. L. (1997). Self grooming in impala mothers and lambs: testing the body size and tick challenge principles. *Animal Behaviour*, **53**, 925–934.

Mooring, M. S. & Samuel, W. M. (1999). Premature loss of winter hair in free-ranging moose (*Alces alces*) infested with winter ticks (*Dermacentor albipictus*) is correlated with grooming rate. *Canadian Journal of Zoology*, **77**, 148–156.

Mooring, M. S., Benjamin, J. E., Harte, C. R. & Herzog, N. B. (2000). Testing the interspecific body size principle in ungulates: the smaller they come, the harder they groom. *Animal Behaviour*, **60**, 35–45.

Mooring, M. S., Reisig, D. D., Niemeyer, J. M. & Osborne, E. R. (2002). Sexually and developmentally dimorphic grooming: a comparative survey of the Ungulata. *Ethology*, **108**, 911–934.

Mooring, M. S., Blumstein, D. T. & Stoner, C. J. (2004). The evolution of parasite-defence grooming in ungulates. *Biological Journal of the Linnean Society*, **81**, 17–37.

Morand, S. (1996). Life-history traits in parasitic nematodes: a comparative approach for the search of invariants. *Functional Ecology*, **10**, 210–218.

Morand, S. (2000). Wormy world: comparative tests of theoretical hypotheses on parasite species richness. In *Evolutionary Biology of Host–Parasite Relationships: Theory Meets Reality*, ed. R. Poulin, S. Morand & A. Skorping. Amsterdam, the Netherlands: Elsevier Science, pp. 63–79.

Morand, S. & Guégan, J.-F. (2000). Distribution and abundance of parasite nematodes: ecological specialization, phylogenetic constraints or simply epidemiology? *Oikos*, **88**, 563–573.

Morand, S. & Harvey, P. H. (2000). Mammalian metabolism, longevity and parasite species richness. *Proceedings of the Royal Society of London B*, **267**, 1999–2003.

Morand, S. & Poulin, R. (1998). Density, body mass and parasite species richness of terrestrial mammals. *Evolutionary Ecology*, **12**, 717–727.

Morand, S. & Poulin, R. (2000). Nematode parasite species richness and the evolution of spleen size in birds. *Canadian Journal of Zoology*, **78**, 1356–1360.

Morand, S. & Poulin, R. (2003). Phylogenies, the comparative method and parasite evolutionary ecology. *Advances in Parasitology*, **54**, 281–302.

Morand, S., Pointier, J.-P., Borel, G. & Theron, A. (1993). Pairing probability of schistosomes related to their distribution among the host population. *Ecology*, **74**, 2444–2449.

Morand, S., Poulin, R., Rohde, K. & Hayward, C. (1999). Aggregation and species coexistence of ectoparasites of marine fishes. *International Journal for Parasitology*, **29**, 663–672.

Morand, S., Hafner, M. S., Page, R. D. M. & Reed, D. L. (2000). Comparative body size relationships in pocket gophers and their chewing lice. *Biological Journal of the Linnean Society*, **70**, 239–249.

Morand, S., Goüy de Bellocq, J., Stanko, M. & Miklisova, D. (2004). Is sex-biased ectoparasitism related to sexual size dimorphism in small mammals of central Europe? *Parasitology*, **129**, 505–510.

Moret, Y. & Schmid-Hempel, P. (2000). Survival for immunity: the price of immune system activation for bumblebee workers. *Science*, **290**, 1166–1168.

Morlan, H. B. (1955). Mammal fleas of Santa Fe County, New Mexico. *Texas Reports on Biology and Medicine*, **13**, 93–125.

Morozkina, E. A., Lysenko, L. S. & Kafarskaya, D. G. (1970). Materials on the ecology of fleas of the red marmot in the eastern Pamir Mountains. In *Vectors of Particularly Dangerous Diseases and their Control*, ed. V. E. Tiflov. Stavropol, USSR: Scientific Anti-Plague Institute of Caucasus and Trans-Caucasus, pp. 337–341 (in Russian).

Morozkina, E. A., Lysenko, L. S. & Kafarskaya, D. G. (1971). Fleas of the red marmot (*Marmota caudata*) and other animals inhabiting the Gissar Ridge. *Problems of Particularly Dangerous Infections*, **17**, 38–44 (in Russian).

Morozov, Y. A. (1974). The level of flea infestation in the great gerbils of different age. In *Proceedings of the 8th Scientific Conference of the Anti-Plague Establishments of the Middle Asia and Kazakhstan*, ed. M. A. Aikimbaev. Alma-Ata, USSR: The Middle Asian Scientific Anti-Plague Institute, pp. 337–338 (in Russian).

Morozov, Y. A., Rapoport, L. P. & Kovtun, I. P. (1972). Parasitological contacts between the great gerbils (*Rhombomys opimus* Licht.) and the yellow ground squirrels (*Citellus fulvus* Licht.) in the Tchu-Moyynkum Desert. *Parazitologiya*, **6**, 334–337 (in Russian).

Morris, D. W. (1987a). Ecological scale and habitat use. *Ecology*, **68**, 362–369.

Morris, D. W. (1987b). Spatial scale and the cost of density-dependent habitat selection. *Evolutionary Ecology*, **1**, 379–388.

Morris, D. W. (1988). Habitat-dependent population regulation and community structure. *Evolutionary Ecology*, **2**, 253–269.

Morris, D. W. (1990). Temporal variation, habitat selection and community structure. *Oikos*, **59**, 303–312.

Morris, D. W. (2003). Toward an ecological synthesis: a case for habitat selection. *Oecologia*, **136**, 1–13.

Morrison, M. L., Marcot, B. G. & Mannan, R. W. (1992). *Wildlife–Habitat Relationships: Concepts and Applications*. Madison, WI: University of Wisconsin Press.

Morrone, J. J. & Acosta, R. (2006). A synopsis of the fleas (Insecta: Siphonaptera) parasitizing New World species of Soricidae (Mammalia: Insectivora). *Zootaxa*, **1354**, 1–30.

Morrone, J. J. & Gutiérrez, A. (2004). Do fleas (Insecta: Siphonaptera) parallel their mammal host diversification in the Mexican transition zone? *Journal of Biogeography*, **32**, 1315–1325.

Morrone, J. J., Acosta, R. & Gutiérrez, A. (2000). Cladistics, biogeography, and host relationships of the flea subgenus *Ctenophthalmus* (*Alloctenus*), with the description of a new Mexican species (Siphonaptera: Ctenophthalmidae). *Journal of the New York Entomological Society*, **108**, 1–12.

Morsy, T. A., El Kady, G. A., Salama, M. M. & Sabry, A. H. (1993). The seasonal abundance of *Gerbillus pyramidum* and their flea ectoparasites in Al Arish, North Sinai Governorate, Egypt. *Journal of the Egyptian Society of Parasitology*, **23**, 269–276.

Moser, B. A., Koehler, P. G. & Patterson, R. S. (1991). Effect of larval diet on cat flea (Siphonaptera: Pulicidae) developmental times and adult emergence. *Journal of Economic Entomology*, **84**, 1257–1261.

Moskalenko, V. V. (1958). On the effect of temperature on the behaviour of fleas after deaths of their host. *Proceedings of the Irkutsk State Scientific Anti-Plague Institute of Siberia and Far East*, **17**, 181–184 (in Russian).

Moskalenko, V. V. (1963a). On the longevity of several flea species parasitic on rodents in the Primorie Region. *Transactions of the Irkutsk State Scientific Anti-Plague Institute of Siberia and Far East*, **6**, 166–169 (in Russian).

Moskalenko, V. V. (1963b). On the effect of temperature on reproduction of some flea species from the Primorie Region under laboratory conditions. *Transactions of the Irkutsk State Scientific Anti-Plague Institute of Siberia and Far East*, **6**, 162–165 (in Russian).

Moskalenko, V. V. (1966). On the frequency of feeding in fleas from the Primorie Region. *Proceedings of the Irkutsk State Scientific Anti-Plague Institute of Siberia and Far East*, **26**, 349–354 (in Russian).

Mouillot, D. & Poulin, R. (2004). Taxonomic partitioning shedding light on the diversification of parasite communities. *Oikos*, **104**, 205–207.

Mouillot, D., Krasnov, B. R., Gaston, K., Shenbrot, G. I. & Poulin, R. (2006). Conservatism of host specificity in parasites. *Ecography*, **29**, 596–602.

Moura, M. O., Bordignon, M. & Graciolli, G. (2003). Host characteristics do not affect community structure of ectoparasites on the fishing bat *Noctilio leporinus* (L., 1758) (Mammalia: Chiroptera). *Memórias do Instituto Oswaldo Cruz*, **98**, 811–815.

Moynahan, E. J. (1987). Kawasaki disease: a novel feline virus transmitted by fleas. *Lancet*, **i** (8526), 195.

Muehlen, M., Heukelbach, J., Wilcke, T., et al. (2003). Investigations on the biology, epidemiology, pathology and control of *Tunga penetrans* in Brazil. II. Prevalence, parasite load and topographic distribution of lesions in the population of a traditional fishing village. *Parasitology Research*, **90**, 449–455.

Muirhead-Thomson, R. C. (1968). *Ecology of Insect Vector Populations*. New York: Academic Press.

Mulenga, A., Sugimoto, C. & Onuma, M. (2000). Issues in tick vaccine development: identification and characterization of potential candidate vaccine antigens. *Microbes and Infection*, **2**, 1353–1361.

Mules, M. W. (1940). Notes on the life history and artificial breeding of the Australian 'stickfast' flea *Echidnophaga mirmecobii* Rothschild. *Australian Journal of Experimental Biological and Medical Science*, **18**, 385–390.

Muller, G. H., Kirk, R. W., Scott, D. W., *et al.* (2001). *Muller and Kirk's Small Animal Dermatology*, 6th edn. Philadelphia, PA: Saunders College Publishing.

Muñoz, L., Milenko, A. & Casanueva, M. E. (2003). Prevalencia e intensidad de ectoparasitos asociados a *Tadarida brasiliensis* (Geoffroy Saint-Hilaire, 1824) (Chiroptera: Molossidae) en Concepción. *Gayana*, **67**, 1–8.

Murzakhmetova, K. (1958). On studying physiological activity of fleas parasitic on gerbils. *Proceedings of the Middle Asian Scientific Anti-Plague Institute*, **4**, 223–226 (in Russian).

Myalkovskaya, S. A. (1983). Some ecological characteristics of fleas parasitic on the pygmy ground squirrels in Dagestan. In *Prophylaxis of Diseases in the Natural Foci*, ed. I. F. Taran. Stavropol, USSR: Scientific Anti-Plague Institute of Caucasus and Trans-Caucasus, pp. 254–255 (in Russian).

Nakazawa, K., Oisi, I. & Kumi, S. (1957). Seasonal prevalence of rodent fleas. *Medical Entomology and Zoology*, **8**, 11–13 (in Japanese).

Naumov, R. L. & Gutova, V. P. (1984). Experimental study of the participation of gamasid mites and fleas in circulation of the tick-borne encephalitis virus (a review). *Parazitologiya*, **18**, 106–115 (in Russian).

Navarro, C., Marzal, A., de Lope, F. & Møller, A. P. (2003). Dynamics of an immune response in house sparrows *Passer domesticus* in relation to time of day, body condition and blood parasite infection. *Oikos*, **101**, 291–298.

Nayak, N. C. & Bhowmik, M. K. (1990). Goat flea (order Siphonaptera) as a possible vector for the transmission of caprine mycoplasmal polyarthritis with septicaemia. *Preventive Veterinary Medicine*, **9**, 259–266.

Nazarova, I. V. (1981). *Fleas of the Volga–Kama Region*. Moscow, USSR: Nauka (in Russian).

Neal, E. G. & Roper, T. J. (1991). The environmental impact of badgers (*Meles meles*) and their setts. *Symposia of the Zoological Society of London*, **63**, 89–106.

Nee, S. (1994). How populations persist. *Nature*, **367**, 123–124.

Nee, S., Hassell, M. P. & May, R. M. (1997). Two-species metapopulation models. In *Metapopulation Biology: Ecology, Genetics, and Evolution*, ed. I. Hanski & M. E. Gilpin. San Diego, CA: Academic Press, pp. 123–147.

Nekola, J. C. & White, P. S. (1999). The distance decay of similarity in biogeography and ecology. *Journal of Biogeography*, **26**, 867–878.

Nelson, R. J. & Demas, G. E. (1996). Seasonal changes in immune function. *Quarterly Review of Biology*, **71**, 511–548.

Nelzina, E. N., Danilova, G. M. & Tchernova, G. I. (1963). Abundance and spatial distribution of micropopulations of hematophagous arthropods in microhabitats of *Citellus pygmaeus*. *Medical Parasitology and Parasitic Diseases* [*Meditsinskaya Parazitologiya i Parazitarnye Bolezni*], **32**, 45–54 (in Russian).

Němec, F. (1993). Flea community inhabiting nests of the common vole (*Microtus arvalis* Pallas, 1779) in west Bohemian farmland. *Folia Musei Rerum Naturalium Bohemiae Occidentalis Zoologica*, **38**, 1–37.

Neter, J., Wasserman, W. & Kutner, M. H. (1990). *Applied Linear Statistical Models: Regression, Analysis of Variance, and Experimental Design*. Homewood, IL: Richard D. Irwin.

Neuhaus, P. (2003). Parasite removal and its impact on litter size and body condition in Columbian ground squirrels (*Spermophilus columbianus*). *Proceedings of the Royal Society of London B*, **270**, S213–S215.

Newton, I. (1998). *Population Limitation in Birds*. London: Academic Press.

Nijssen, A. (1985). Body temperature regulation by care of the fur in rats *Rattus norvegicus albinos*. *Netherlands Journal of Zoology*, **35**, 423–437.

Nikitina, N. A. (1961). Results of marking and recapturing of small mammals in the Komi Soviet Republic. *Bulletin of the Moscow Naturalist Society, Series Biology*, **66**, 15–25 (in Russian).

Nikitina, N. A. & Nikolaeva, G. (1979). Study of the ability of some rodents to get rid of fleas. *Zoologicheskyi Zhurnal*, **58**, 931–933 (in Russian).

Nikitina, N. A. & Nikolaeva, G. (1981). Ability of rodents to clean themselves of specific and non-specific fleas. *Zoologicheskyi Zhurnal*, **60**, 165–167 (in Russian).

Nikulshin, S. V. (1980). Descriptions of the annual cycles of fleas (Aphaniptera) parasitic on *Citellus musicus* from the Baksan Valley. *Parazitologiya*, **14**, 134–141 (in Russian).

Nikulshin, S. V. & Shinkareva, V. N. (1983). Seasonal patterns of ecology of fleas *Citellophilus tesquorum* in the Enikol Tract. In *Prophylaxis of Diseases in the Natural Foci*, ed. I. F. Taran. Stavropol, USSR: Scientific Anti-Plague Institute of Caucasus and Trans-Caucasus, pp. 260–261 (in Russian).

Nilsson, J. A. (2003). Ectoparasitism in marsh tits: costs and functional explanations. *Behavioral Ecology*, **14**, 175–181.

Njau, B. C. & Nyindo, M. (1987). Detection of immune response in rabbits infested with *Rhipicephalus appendiculatus* and *Rhipicephalus evertsi evertsi*. *Research in Veterinary Science*, **43**, 217–221.

Njunwa, K. J., Mwaiko, G. L., Kilonzo, B. S. & Mhina, J. I. (1989). Seasonal patterns of rodents, fleas and plague status in the Western Usambara Mountains, Tanzania. *Medical and Veterinary Entomology*, **3**, 17–22.

Nordling, D., Andersson, M., Zohari, S. & Gustafsson, L. (1998). Reproductive effort reduces specific immune response and parasite resistance. *Proceedings of the Royal Society of London B*, **265**, 1291–1298.

Norman, R., Bowers, R. G., Begon, M. & Hudson, P. J. (1999). Persistence of tick-borne virus in the presence of multiple host species: tick reservoirs and parasite mediated competition. *Journal of Theoretical Biology*, **200**, 111–118.

Nosil, P. (2002). Transition rates between specialization and generalization in phytophagous insects. *Evolution*, **56**, 1701–1706.

Novokreshchenova, N. S. (1960). Materials on ecology of fleas of *Citellus pygmaeus* in relation to their epizootological importance. *Proceedings of the Anti-Plague 'Microb' Institute*, **4**, 444–456 (in Russian).

Novokreshchenova, N. S. (1962). Materials on comparative ecology of three species of fleas parasitic on the great gerbil. In *Particularly Dangerous and Natural Diseases*, ed. Anonymous. Moscow, USSR: Medgiz, pp. 53–63 (in Russian).

Novokreshchenova, N. S. & Kuznetsova, G. S. (1964). Ecological characteristic of fleas of the great gerbil in regions with permanent plague epizootics. *Zoologicheskyi Zhurnal*, **43**, 1638–1647 (in Russian).

Novokreshchenova, N. S., Soldatkin, I. S. & Levoshina, A. I. (1968). Comparative frequency of blood meals in various flea species, measured in the laboratory conditions using radioactive indicators. In *Rodents and their Ectoparasites*, ed. B. K. Fenyuk. Saratov, USSR: Saratov University Press, pp. 49–54 (in Russian).

Novokreshchenova, N. S., Zagniborodova, E. N., Zabegalova, M. N., et al. (1975). Multiannual dynamics of density in fleas parasitic on the great gerbils in Turkmenistan. *Problems of Particularly Dangerous Diseases*, **43/44**, 84–90 (in Russian).

Novozhilova, E. N. (1977). Ectoparasites of small mammals and inhabitants of their burrows in the Pre-Polar Ural. *Proceedings of the Komi Branch of the Academy of Sciences of the USSR*, **34**, 125–139 (in Russian).

Nuriev, K. K., Rapoport, L. P., Shokputov, T. M., Orlova, L. M. & Duisenbiev, D. M. (2004). Materials on rodent fleas from mountain regions of southern Kazakhstan. *Quarantinable and Zoonotic Infections in Kazakhstan*, **9**, 66–70 (in Russian).

Oberdorff, T., Hugueny, B., Compin, A. & Belkessam, D. (1998). Non-interactive fish communities in the coastal streams of north-western France. *Journal of Animal Ecology*, **67**, 472–484.

O'Brien, E. L. & Dawson, R. D. (2005). Perceived risk of ectoparasitism reduces primary reproductive investment in tree swallows *Tachycineta bicolor*. *Journal of Avian Biology*, **36**, 269–275.

O'Donnell, M. J. & Machin, J. (1988). Water vapor absorption by terrestrial organisms. *Advances in Comparative and Environmental Physiology*, **2**, 47–90.

Oguge, N., Rarieya, M. & Ondiaka, P. (1997). A preliminary survey of macroparasite community of rodents of Kahawa, central Kenya. *Belgian Journal of Zoology*, **127**, S113–S118.

O'Hare, N. (2005). Asking the right questions. *New Zealand Listener*, **197** (Jan. 29–Feb. 4), No. 3377.

Olsen, N. J. & Kovacs, W. J. (1996). Gonadal steroids and immunity. *Endocrine Review*, **17**, 369–384.

Olsson, K. & Allander, K. (1995). Do fleas, and/or old nest material, influence nest-site preference in hole-nesting passerines? *Ethology*, **101**, 160–170.

Olsufiev, N. G. (1975). *Taxonomy, Microbiology and Laboratory Diagnostics of the Tularaemia Pathogen*. Moscow, USSR: Meditsina (in Russian).

Olsufiev, N. G. & Dunaeva, T. N. (1960). Epizootology (natural focality) of tularaemia. In *Tularemia*, ed. N. G. Olsufiev & G. P. Rudnev. Moscow, USSR: Meditsina, pp. 136–206 (in Russian).

Oparina, O., Starozhitskaya, G. S. & Samurov, M. A. (1989). Human blood-sucking ability of fleas parasitic on rodents and small gerbils. *Medical Parasitology and Parasitic Diseases* [Meditsinskaya Parazitologiya i Parazitarnye Bolezni], **58**, 23–25 (in Russian).

Opdebeeck, J. P. & Slacek, B. (1993). An attempt to protect cats against infestation with *Ctenocephalides felis felis* using gut membrane antigens as a vaccine. *International Journal for Parasitology*, **23**, 1063–1067.

Oppliger, A., Richner, H. & Christe, P. (1994). Effect of an ectoparasite on lay date, nest-site choice, desertion, and hatching success in the great tit (*Parus major*). *Behavioral Ecology*, **5**, 130–134.

Orell, M., Rytkönen, S. & Ilomäki, K. (1993). Do pied flycatchers prefer nest boxes with old nest material? *Annales Zoologici Fennici*, **30**, 313–316.

Osacar-Jimenez, J. J., Lucientes-Curdi, J. & Calvete-Margolles, C. (2001). Abiotic factors influencing the ecology of wild rabbit fleas in north-eastern Spain. *Medical and Veterinary Entomology*, **15**, 157–166.

Osbrink, W. L. & Rust, M. (1984). Fecundity and longevity of the adult cat flea *Ctenocephalides felis felis* (Siphonaptera: Pulicidae). *Journal of Medical Entomology*, **21**, 727–731.

Osbrink, W. L. & Rust, M. (1985). Cat flea (Siphonaptera: Pulicidae): factors influencing host finding behaviour in the laboratory. *Annals of the Entomological Society of America*, **78**, 29–34.

Ostfeld, R. & Keesing, F. (2000). The function of biodiversity in the ecology of vector-borne zoonotic diseases. *Canadian Journal of Zoology*, **78**, 2061–2078.

Overal, W. L. (1980). Host-relations of the batfly *Megistopoda aranea* (Diptera: Streblidae) in Panama. *University of Kansas Scientific Bulletin*, **52**, 1–20.

Öztürk, L., Pelin, Z., Karadeniz, D., *et al.* (1999). Effects of 48 hours' sleep deprivation on human immune profile. *Sleep Research Online*, **2**, 107–111.

Pacala, S. W. & Dobson, A. P. (1988). The relation between the number of parasites/host and host age: population dynamic causes and maximum likelihood estimation. *Parasitology*, **96**, 197–210.

Pacejka, A. J., Gratton, C. M. & Thompson, C. F. (1998). Do potentially virulent mites affect house wren (*Troglodytes aedon*) reproductive success? *Ecology*, **79**, 1797–1806.

Page, R. M. D. (1990). Component analysis: a valiant failure? *Cladistics*, **6**, 119–136.

Page, R. M. D. (1993a). Parasites, phylogeny and cospeciation. *International Journal for Parasitology*, **23**, 499–506.

Page, R. M. D. (1993b). Genes, organisms, and areas: the problem of multiple lineages. *Systematic Biology*, **42**, 77–84.

Page, R. M. D. (1994a). Parallel phylogenies: reconstructing the history of host–parasite assemblages. *Cladistics*, **10**, 155–173.

Page, R. M. D. (1994b). Maps between trees and cladistic analysis of historical associations among genes, organisms, and areas. *Systematic Biology*, **43**, 58–77.

Page, R. D. M. (ed.) (2003). *Tangled Trees: Phylogeny, Cospeciation, and Coevolution*. Chicago, IL: University of Chicago Press.

Page, R. D. M. & Charleston, M. D. (1998). Trees within trees: phylogeny and historical associations. *Trends in Ecology and Evolution*, **13**, 356–359.

Palombit, R. A., Cheney, D. L. & Seyfarth, R. M. (2001). Female–female competition for male 'friends' in wild chacma baboons, *Papio cynocephalus ursinus*. *Animal Behaviour*, **61**, 1159–1171.

Pampiglione, S., Trentini, M., Fioravanti, M. L., Onore, G. & Rivasi, F. (2003). Additional description of a new species of *Tunga* (Siphonaptera) from Ecuador. *Parasite*, **10**, 9–15.

Panchenko, G. M. (1971). Ecological study of larvae of the rat fleas of Siberia and Far East. *Transactions of the Irkutsk State Scientific Anti-Plague Institute of Siberia and Far East*, **9**, 229–231 (in Russian).

Paramonov, B. B., Emelianova, N. D., Zarubina, V. N. & Kontrimavitchus, V. L. (1966). Materials for the study of ectoparasites of rodents and shrews of the Kamchatka Peninsula. *Proceedings of the Irkutsk State Scientific Anti-Plague Institute of Siberia and Far East*, **26**, 333–341 (in Russian).

Park, T. (1948). Experimental studies of interspecific competition. I. Competition between populations of the flour beetles, *Tribolium confusum* Duval and *Tribolium castaneum* Herbst. *Ecological Monographs*, **18**, 265–308.

Parman, D. C. (1923). Biological notes on the hen flea, *Echidnophaga gallinacea*. *Journal of Agricultural Research*, **23**, 1007–1009.

Parola, R. & Raoult, D. (2006). Tropical rickettsioses. *Clinics in Dermatology*, **24**, 191–200.

Parola, P., Davoust, B. & Raoult, D. (2005). Tick- and flea-borne rickettsial emerging zoonoses. *Veterinary Research*, **36**, 469–492.

Parsons, P. A. (1990). Fluctuating asymmetry: an epigenetic measure of stress. *Biological Reviews*, **17**, 391–421.

Pascal, M., Beaucournu, J. C. & Lorvelec, O. (2004). An enigma: the lack of Siphonaptera on wild rats and mice on densely populated tropical islands. *Acta Parasitologica*, **49**, 168–172.

Paterson, A. M. & Banks, J. (2001). Analytical approaches to measuring cospeciation of host and parasites: through a glass, darkly. *International Journal for Parasitology*, **31**, 1012–1022.

Paterson, A. M. & Gray, R. D. (1997). Host–parasite co-speciation, host switching, and missing the boat. In *Host–Parasite Evolution: General Principles and Avian Models*, ed. D. H. Clayton & J. Moore. Oxford, UK: Oxford University Press, pp. 236–250.

Paterson, A. M., Gray, R. D. & Wallis, G. P. (1993). Parasites, petrels and penguins: does louse presence reflect seabird phylogeny? *International Journal for Parasitology*, **23**, 515–526.

Paterson, A. M., Wallis, G. P., Wallis, L. J. & Gray, R. D. (2000). Seabird and louse coevolution: Complex histories revealed by 12S rRNA sequences and reconciliation analyses. *Systematic Biology*, **38**, 144–153.

Patrick, M. J. (1991). Distribution of enteric helminthes in *Glaucomys volans* L. (Sciuridae): a test for competition. *Ecology*, **72**, 755–758.

Patterson, B. D. (1990). On the temporal development of nested subset patterns of species composition. *Oikos*, **59**, 330–342.

Patterson, B. D. & Atmar, W. (1986). Nested subsets and the structure of insular mammalian faunas and archipelagos. *Biological Journal of the Linnean Society*, **28**, 65–82.

Patterson, B. D. & Brown, J. H. (1991). Regionally nested patterns of species composition in granivorous rodent assemblages. *Journal of Biogeography*, **18**, 395–402.

Pauller, O. F. & Tchipizubova, P. A. (1958). Ecology of fleas of the Daurian ground squirrel in the Trans-Baikalia. *Proceedings of the Irkutsk State Scientific Anti-Plague Institute of Siberia and Far East*, **17**, 161–179 (in Russian).

Pauller, O. F., Elshanskaya, N. I. & Shvetsova, I. V. (1966). Ecological and faunistical review of mammalian and bird ectoparasites in the tularemia focus of the Selenga River delta. *Proceedings of the Irkutsk State Scientific Anti-Plague Institute of Siberia and Far East*, **26**, 322–332 (in Russian).

Peach, W. J., Fowler, J. A. & Greenwood, M. T. (1987). Seasonal variation in the infestation of starlings *Sturnus vulgaris* by fleas (Siphonaptera). *Bird Study*, **34**, 251–252.

Pennings, S. C. & Silliman, B. R. (2005). Linking biogeography and community ecology: latitudinal variation in plant–herbivore interaction strength. *Ecology*, **86**, 2310–2319.

Perlman, S. J. & Jaenike, J. (2001). Competitive interactions and persistence of two nematode species that parasitize *Drosophila recens*. *Ecology Letters*, **4**, 577–584.

Perry, J. N. (1988). Some models for spatial variability of animal species. *Oikos*, **51**, 124–130.

Perry, J. N. & Taylor, L. R. (1986). Stability of real interacting populations in space and time: implications, alternatives and negative binomial k_c. *Journal of Animal Ecology*, **55**, 1053–1068.

Peters, R. H. (1983). *The Ecological Implications of Body Size*. Cambridge, UK: Cambridge University Press.

Peterson, A. T., Soberon, J. & Sanchez-Cordero, V. (1999). Conservatism of ecological niches in evolutionary time. *Science*, **285**, 1265–1267.

Petit, C., Hossaert-McKey, M., Perret, P., Blondel, J. & Lambrechts, M. M. (2002). Blue tits use selected plants and olfaction to maintain an aromatic environment for nestlings. *Ecology Letters*, **5**, 585–589.

Peus, F. (1968). Über die beiden Bernstein-Flöhe (Insecta, Siphonaptera). *Paläontologische Zeitschrift*, **42**, 62–72.

Peus, F. (1970). Zur Kenntnis der Flöhe Deutschlands (Insecta, Siphonaptera). IV. Faunistik und Ökologie der Säugeteierflöhe. *Zoologische Jahrbücher, Abteilung für Systematic Ökologie und Geographie der Tiere*, **99**, 400–418.

Piersma, T. & Drent, J. (2003). Phenotypic flexibility and the evolution of organismal design. *Trends in Ecology and Evolution*, **18**, 228–233.

Pigage, H. K. & Larson, O. R. (1983). The detection of glycerol in overwintering purple martin fleas, *Ceratophyllus idius*. *Comparative Biochemistry and Physiology A*, **75**, 593–595.

Pigage, H. K., Pigage, J. C. & Tillman, J. F. (2005). Fleas associated with the northern pocket gopher (*Thomomys talpoides*) in Elbert County, Colorado. *Western North American Naturalist*, **65**, 210–214.

Pilgrim, R. L. C. & Galloway, T. D. (2004). Descriptions of flea larvae (Siphonaptera: Ceratophyllidae, Leptopsyllidae) found in nests of the house martin, *Delichon urbica* (Aves: Hirundinidae), in Great Britain. *Journal of Natural History*, **38**, 473–502.

Pimm, S. L. (1979). The structure of food webs. *Theoretical Population Biology*, **16**, 144–158.

Pimm, S. L. & Rosenzweig, M. L. (1981). Competitors and habitat use. *Oikos*, **37**, 1–6.

Pollitzer, R. (1960). A review of recent literature on plague. *Bulletin of the World Health Organization*, **23**, 313–400.

Pollitzer, R. & Meyer, K. F. (1961). The ecology of plague. In *Studies of Disease Ecology*, ed. J. F. May. New York: Hafner, pp. 433–590.

Ponomarenko, A. G. (1976). The new insect from the Cretaceous of the Trans-Baikalia was a probable parasite of pterosaurs. *Paleontologicheskyi Zhurnal*, **3**, 102–106 (in Russian).

Ponomarenko, A. G. (1986). Scarabaeiformis *incertae sedis*. *Transactions of the Joint Soviet–Mongolian Paleontological Expedition*, **28**, 110–112 (in Russian).

Popov, V. N. & Verzhutsky, D. B. (1988). Characteristics of intrapopulation aggregations of the long-tailed ground squirrel (*Citellus undulatus* Pall.) under density decline. *Bulletin of the Moscow Naturalist Society, Series Biology*, **93**, 47–50 (in Russian).

Popova, A. S. (1968). Flea fauna of the Moyynkum Desert. In *Rodents and their Ectoparasites*, ed. B. K. Fenyuk. Saratov, USSR: Saratov University Press, pp. 402–406 (in Russian).

Poulin, R. (1993). The disparity between observed and uniform distibutions: a new look at parasite aggregation. *International Journal for Parasitology*, **23**, 937–944.

Poulin, R. (1995a). Clutch size and egg size in free-living and parasitic copepods: a comparative analysis. *Evolution*, **49**, 325–336.

Poulin, R. (1995b). Phylogeny, ecology, and the richness of parasite communities in vertebrates. *Ecological Monographs*, **65**, 283–302.

Poulin, R. (1996a) Sexual inequalities in helminth infections: a cost of being male? *American Naturalist*, **147**, 289–295.

Poulin, R. (1996b). Richness, nestedness and randomness in parasite infracommunity structure. *Oecologia*, **105**, 545–551.

Poulin, R. (1997). Parasite faunas of freshwater fish: the relationship between richness and the specificity of parasites. *International Journal for Parasitology*, **27**, 1091–1098.

Poulin, R. (1998). Large-scale patterns of host use by parasites of freshwater fishes. *Ecology Letters*, **1**, 118–128.

Poulin, R. (1999). Body size vs. abundance among parasite species: positive relationships? *Ecography*, **22**, 246–250.

Poulin, R. (2003). The decay of similarity with geographical distance in parasite communities of vertebrate hosts. *Journal of Biogeography*, **30**, 1609–1615.

Poulin, R. (2005). Relative infection levels and taxonomic distances among the host species used by a parasite: insights into parasite specialization. *Parasitology*, **130**, 109–115.

Poulin, R. (2006). Variation in infection parameters among populations within parasite species: intrinsic properties versus local factors. *International Journal for Parasitology*, **36**, 877–885.

Poulin, R. (2007a). *Evolutionary Ecology of Parasites: From Individuals to Communities*, 2nd edn. Princeton, NJ: Princeton University Press.

Poulin, R. (2007b). Are there general laws in parasite ecology? *Parasitology*, **134**, 763–776.

Poulin, R. & Guégan, J.-F. (2000). Nestedness, antinestedness, and relationship between prevalence and intensity in ectoparasite assemblages of marine fish: a spatial model of species co-existence. *International Journal for Parasitology*, **30**, 1147–1152.

Poulin, R. & Hamilton, W. J. (1997). Ecological correlates of body size and egg size in parasitic Ascothoracida and Rhizocephala (Crustacea). *Acta Oecologica*, **18**, 621–635.

Poulin, R. & Morand, S. (2004). *Parasite Biodiversity*. Washington, DC: Smithsonian Institution Press.

Poulin, R. & Mouillot, D. (2003). Parasite specialization from a phylogenetic perspective: a new index of host specificity. *Parasitology*, **126**, 473–480.

Poulin, R. & Mouillot, D. (2004a). The relationship between specialization and local abundance: the case of helminth parasites of birds. *Oecologia*, **140**, 372–378.

Poulin, R. & Mouillot, D. (2004b). The evolution of taxonomic diversity in helminth assemblages of mammalian hosts. *Evolutionary Ecology*, **18**, 231–247.

Poulin, R. & Mouillot, D. (2005a). Combining phylogenetic and ecological information into a new index of host specificity. *Journal of Parasitology*, **91**, 511–514.

Poulin, R. & Mouillot, D. (2005b). Host specificity and the probability of discovering species of helminth parasites. *Parasitology*, **130**, 709–715.

Poulin, R. & Valtonen, E. T. (2001). Nested assemblages resulting from host-size variation: the case of endoparasite communities in fish hosts. *International Journal for Parasitology*, **31**, 194–1204.

Poulin, R. & Valtonen, E. T. (2002). The predictability of helminth community structure in space: a comparison of fish populations from adjacent lakes. *International Journal for Parasitology*, **30**, 1235–1243.

Poulin, P., Brodeur, J. & Moore, J. (1994). Parasite manipulation of host behaviour: should hosts always lose? *Oikos*, **70**, 479–484.

Poulin, R., Mouillot, D. & George-Nascimento, M. (2003). The relationship between species richness and productivity in metazoan parasite communities. *Oecologia*, **137**, 277–285.

Poulin, R., Krasnov, B. R. & Morand, S. (2006a). Patterns of host specificity in parasites exploiting small mammals. In *Micromammals and Macroparasites: From Evolutionary Ecology to Management*, ed. S. Morand, B. R. Krasnov & R. Poulin. New York: Springer-Verlag, pp. 233–256.

Poulin, R., Krasnov, B. R., Shenbrot, G. I., Mouillot, D. & Khokhlova, I. S. (2006b). Evolution of host specificity in fleas: is it directional and irreversible? *International Journal for Parasitology*, **36**, 185–191.

Prasad, R. S. (1969). Influence of host on fecundity of the Indian rat flea, *Xenopsylla cheopis* (Roths.). *Journal of Medical Entomology*, **6**, 443–447.

Prasad, R. S. (1972). Different site selections by the rat fleas *Xenopsyllsa cheopis* and *Xenopsyllsa astia* (Siphonaptera, Pulicidae). *Entomologist's Gazette*, **108**, 63–64.

Prasad, R. S. (1973). Studies on host–flea relationships. II. Sex hormones of the host and fecundity of rat fleas *Xenopsylla astia* (Rothschild) and *Xenopsylla cheopis* (Rothschild) (Siphonaptera). *Indian Journal of Medical Research*, **61**, 38–44.

Prasad, R. S. (1976). Studies on host–flea relationships. IV. Progesterone and cortisone do not influence reproductive potentials of rat fleas *Xenopsylla astia* (Rothschild) and *Xenopsylla cheopis* (Rothschild) (Siphonaptera). *Parasitology Research*, **50**, 81–46.

Prasad, R. S. (1987). Host dependency among haematophagous insects: a case study on flea–host association. *Proceedings of the Indian Academy of Sciences*, **96**, 349–360.

Prasad, R. S. & Kamala Bai, M. (1976). Studies on host specificity in fleas. In *Insect and Host Specificity: Proceedings of the Symposium on Problems of Host Specificity*, ed. T. N. Ananthakrishnan. Madras, India: Loyola College, pp. 111–115.

Price, T., Lovette, I. J., Bermingham, E., Gibbs, H. L. & Richman, A. D. (2000). The imprint of history on communities of North American and Asian warblers. *American Naturalist*, **156**, 354–367.

Prokopiev, V. N. (1969). Morphological types of cocoons and mechanism of emergence of imago fleas. In *Proceedings of the 4th Scientific Conference of the Anti-Plague Establishments of the Middle Asia and Kazakhstan*, ed. M. A. Aikimbaev. Alma-Ata, USSR: The Middle Asian Scientific Anti-Plague Institute and Kainar, pp. 182–184 (in Russian).

Promislow, D. E. L. (1992). Costs of sexual selection in natural populations of mammals. *Proceedings of the Royal Society of London B*, **247**, 203–210.

Puchala, P. (2004). Detrimental effects of larval blow files (*Protocalliphora azurea*) on nestlings and breeding success of tree sparrows (*Passer montanus*). *Canadian Journal of Zoology*, **82**, 1285–1290.

Pullen, S. R. & Meola, R. W. (1995). Survival and reproduction of the cat flea (Siphonaptera: Pulicidae) fed human blood on an artificial membrane system. *Journal of Medical Entomology*, **32**, 467–4670.

Pulliam, H. R. (2000). On the relationship between niche and distribution. *Ecology Letters*, **3**, 349–361.

Punsky, E. E. & Zagniborodova, E. N. (1964). Role of fleas *Xenopsylla gerbilli gerbilli* in maintenance and transmission of the pathogen of erysepeloid. *Proceedings of the Turkmenian Scientific Institute of Epidemiology and Hygiene*, **6**, 345–347 (in Russian).

Pushnitsa, F. A., Shevchenko, S. F., Mironov, A. N., et al. (1978). Present status of abundance and habitat distribution of the ground squirrels and their fleas in the eastern parts of the Rostov Region. *Problems of Particularly Dangerous Diseases*, **60**, 48–52 (in Russian).

Qi, Y.-M. (1990a). Fine structures of the reproductive system of three flea species: development of female genitalia. *Acta Entomologica Sinica*, **33**, 182–188 (in Chinese).

Qi, Y.-M. (1990b). Fine structures of the reproductive system of three flea species: development of male genitalia. *Acta Entomologica Sinica*, **33**, 403–411 (in Chinese).

Qian, T.-J., Gong, Z.-D. & Guo, X.-G. (2000). Sex ratio analysis on dominant flea species of the flea community in the foci of human plague in Yunnan. *Medical Journal of Dali College*, **9**, 1–3 (in Chinese).

Radovsky, F. J. (1985). Evolution of mammalian mesostigmatid mites. In *Coevolution of Parasitic Arthropods and Mammals*, ed. K. C. Kim. New York: John Wiley, pp. 441–504.

Randolph, S. E. (1977). Changing spatial relationships in a population of *Apodemus sylvaticus* with the onset of breeding. *Journal of Animal Ecology*, **46**, 653–676.

Randolph, S. E. (1994). Density-dependent acquired resistance to ticks in natural hosts, independent of concurrent infection with *Babesia microti*. *Parasitology*, **108**, 413–419.

Rapoport, E. H. (1982). *Areography: Geographic Strategies of Species*. London: Pergamon Press.

Rapoport, L. P., Morozov, Y. A. & Korneyev, G. A. (1976). Intensity of parasitological contacts in the colonies of the great gerbils under different population densities of animals and their fleas. *Parazitologiya*, **10**, 392–396 (in Russian).

Rapoport, L. P., Kondratenko, L. P., Orlova, L. M., et al. (2007). Fleas in the Moyynkum Desert and their epozootological role. *Zoologicheskyi Zhurnal*, **86**, 44–51 (in Russian).

Rasnitsyn, A. P. (1992). *Strashila incredibilis*, a new enigmatic mecopteroid insect with possible siphonapteran affinities from upper Jurassic of Siberia. *Psyche*, **99**, 323–333.

Rasnitsyn, A. P. (2002a). Infraclass Gryllones Laicharting, 1781: the grylloneans. In *History of Insects*, ed. A. P. Rasnitsyn & D. L. J. Quicke. Dordrecht, the Netherlands: Kluwer, pp. 254–324.

Rasnitsyn, A. P. (2002b). Order Pulicida Billbergh, 1820: the fleas. In *History of Insects*, ed. A. P. Rasnitsyn & D. L. J. Quicke. Dordrecht, the Netherlands: Kluwer, pp. 239–242.

Rassokhina, O. S., Starozhitskaya, G. S. & Knyazeva, T. V. (1985). Fecundity of four flea species in the laboratory colonies. *Parazitologiya*, **19**, 488–490 (in Russian).

Raszl, S. M., Cabral, D. D. & Linardi, P. M. (1998). *Xenopsylla cheopis* on dogs from Brazil: first report. *Arquivo Brasileiro de Medicina Veterinaria e Zootecnia*, **50**, 211–212 (in Portugese).

Raszl, S. M., Cabral, D. D. & Linardi, P. M. (1999). Notas sobre Sifonápteros (Pulicidae, Tungidae e Rhopalopsyllidae) de carnívoros domésticos brasileiros. *Revista Brasiliera de Entomologia*, **43**, 95–97.

Ratovonjato, J., Duchemin, J. B. & Chanteau, S. (2000). Optimized method for rearing fleas (*Xenopsylla cheopis* and *Synopsyllus fonquerniei*). *Archives de l'Institut Pasteur de Madagascar*, **66**, 75–77.

Rausher, M. D. (1993). The evolution of habitat preference: avoidance and adaptation. In *Evolution of Insect Pests: Patterns of Variation*, ed. K. C. Kim & B. A. McPherson. New York: John Wiley, pp. 259–283.

Ravkin, Y. S. & Sapegina, V. F. (1990). Fleas of rodents of the southern taiga of the Pri-Angarie. *Bulletin of the Siberian Branch of the Academy of Sciences of the USSR, Series Biological Sciences*, **3**, 63–68 (in Russian).

Rea, J. G. & Irwin, S. W. B. (1994). The ecology of host-finding behaviour and parasite transmission: past and future perspectives. *Parasitology*, **109**, S31–S39.

Rechav, Y. (1992). Naturally acquired resistance to ticks: a global view. *Insect Science and its Application*, **13**, 495–504.

Rechav, Y. & Dauth, J. (1987). Development of resistance in rabbits to immature stages of the ixodid tick *Rhipicephalus appendiculatus*. *Medical and Veterinary Entomology*, **1**, 177–183.

Rechav, Y. & Fielden, L. J. (1995). The effect of host resistance on the metabolic rate of engorged females of *Rhipicephalus evertsi evertsi*. *Medical and Veterinary Entomology*, **9**, 289–292.

Rechav, Y & Fielden, L. J. (1997). The effect of various host species on the feeding performance of immature stages of the tick *Hyalomma truncatum* (Acari: Ixodidae). *Experimental and Applied Acarology*, **21**, 551–559.

Rechav, Y., Heller-Haupt, A. & Varma, M. G. R. (1989). Resistance and cross-resistance in guinea-pigs and rabbits to immature stages of ixodid ticks. *Medical and Veterinary Entomology*, **3**, 333–336.

Redford, K. H. & Eisenberg, J. F. (1992). *Mammals of the Neotropics: The Southern Cone*, vol. 2, *Chile, Argentine, Uruguay, Paraguay*. Chicago, IL: University of Chicago Press.

Reichardt, T. R. & Galloway, T. D. (1994). Seasonal occurrence and reproductive status of *Opisocrostis bruneri* (Siphonaptera, Ceratophyllidae), a flea on Franklin ground-squirrel, *Spermophilus franklinii* (Rodentia, Sciuridae) near Birds Hill Park, Manitoba. *Journal of Medical Entomology*, **31**, 105–113.

Reichman, O. J. & Smith, S. C. (1990). Burrows and burrowing behavior by mammals. In *Current Mammology*, vol. 2, ed. H. H. Genoways. New York: Plenum Press, pp. 197–244.

Reiczigel, J. & Rózsa, L. (1998). Host-mediated site-segregation of ectoparasites: an individual-based simulation study. *Journal of Parasitology*, **84**, 491–498.

Reinhold, K. (1999). Energetically costly behaviour and the evolution of resting metabolic rate in insects. *Functional Ecology*, **13**, 217–224.

Reiss, M. J. (1986). Sexual dimorphism in body size: are larger species more dimorphic? *Journal of Theoretical Biology*, **121**, 163–172.

Reiss, M. J. (1989). *The Allometry of Growth and Reproduction*. Cambridge, UK: Cambridge University Press.

Reitblat, A. G. & Belokopytova, A. M. (1974). On cannibalism and predatory habits of flea larvae. *Zoologicheskyi Zhurnal*, **53**, 135–137 (in Russian).

Řeháček, J. (1961). Transmission of tick-borne encephalitis virus by fleas. *Journal of Hygiene, Epidemiology and Immunology (Prague)*, **5**, 282–285 (in Russian).

Rementsova, M. M. (1962). *Brucellosis in Wildlife*. Alma-Ata, USSR: Kainar (in Russian).

Rendell, W. B. & Verbeek, N. A. M. (1996). Old nest material in nest boxes of tree swallows: effects on nest-site choice and nest building. *Auk*, **113**, 319–328.

Rensch, B. (1960). *Evolution above the Species Level*. New York: Columbia University Press.

Reshetnikova, P. I. (1959). Flea fauna of the Kustanai Region. *Proceedings of the Middle Asian Scientific Anti-Plague Institute*, **6**, 261–265 (in Russian).

Ribeiro, J. M. C. (1987). Role of saliva in blood feeding in arthropods. *Annual Review of Entomology*, **32**, 463–478.

Ribeiro, J. M. C. (1995). Blood-feeding arthropods: live syringes or invertebrate pharmacologists? *Infectious Agents and Disease – Reviews Issues and Commentary*, **4**, 143–152.

Ribeiro, J. M. C. (1996). Common problems of arthropod vectors of disease. In *The Biology of Disease Vectors*, ed. B. J. Beaty & W. C. Marquardt. Niwot, CO: University of Colorado Press, pp. 25–33.

Ribeiro, J. M. C., Vaughan, J. A. & Farhang-Azad, A. (1990). Characterization of the salivary apyrase activity of three rodent flea species. *Comparative Biochemistry and Physiology B*, **95**, 215–218.

Richards, P. A. & Richards, A. G. (1969). Acanthae: a new type of cuticular process in the proventriculus of Mecoptera and Siphonaptera. *Zoologische Jahrbücher, Abteilung für Anatomie und Ontogenie der Tiere*, **86**, 158–176.

Richner, H. (1996). Flohzirkus im Vogelnest: Wirt–Parasiten-Interaktionen in der Brutzeit. *Ornithologische Beobachter*, **93**, 103–110.

Richner, H. (1998). Host–parasite interactions and life-history evolution. *Zoology*, **101**, 333–344.

Richner, H. & Heeb, P. (1995). Are clutch and brood size patterns in birds shaped by ectoparasites? *Oikos*, **73**, 435–441.

Richner, H., Oppliger, A. & Christe, P. (1993). Effect of an ectoparasite on reproduction in great tits. *Journal of Animal Ecology*, **62**, 703–710.

Ricotta, C. (2004). A parametric diversity measure combining the relative abundances and taxonomic distinctiveness of species. *Diversity and Distributions*, **10**, 143–146.

Riddoch, B. J., Greenwood, M. T. & Ward, R. D. (1984). Aspects of the population structure of the sand martin flea, *Ceratophyllus styx*, in Britain. *Journal of Natural History*, **18**, 475–484.

Riek, E. F. (1970). Lower Cretaceous fleas. *Nature*, **227**, 746–747.

Ritter, R. C. & Epstein, A. N. (1974). Saliva lost by grooming: major item in rats' water economy. *Behavioral Biology*, **11**, 581–585.

Robbins, R. G. & Faulkenberry, G. D. (1982). A population model for fleas of the gray-tailed vole, *Microtus canicaudus* Miller. *Entomological News*, **93**, 70–74.

Roberts, M. G., Smith, G. & Grenfell, B. T. (1995). Matematical models for macroparasites of wildlife. In *Ecology of Infectious Diseases in Natural Populations*, ed. B. T. Grenfell & A. P. Dobson. Cambridge, UK: Cambridge University Press, pp. 177–208.

Robson, D. S. (1972). Appendix: Statistical tests of significance. *Journal of Theoretical Biology*, **34**, 350–352.

Rödl, P. (1979). Investigation of the transfer of fleas among small mammals using radioactive phosphorus. *Folia Parasitologica*, **26**, 265–274.

Rodrigues, A. F. S. F. & Daemon, E. (2004). Ixodideos e sifonapteros em *Cerdocyon thous* L. (Carnivora, Canidae) procedentes da zona da Mata Mineira, Brasil. *Arquivos do Instituto Biologico São Paulo*, **71**, 371–372.

Rodríguez, Z., Moreira, E. C., Linardi, P. M. & Santos, H. A. (1999). Notes on the bat flea *Hormopsylla fosteri* (Siphonaptera: Ischnopsyllidae) infesting *Molossops abrasus* (Chiroptera). *Memórias do Instituto Oswaldo Cruz*, **94**, 727–728.

Rodríguez-Gironés, M. A. & Santamaría, L. (2006). A new algorithm to calculate the nestedness temperature of presence–absence matrices. *Journal of Biogeography*, **33**, 924–935.

Roehrig, J. T., Piesman, J., Hunt, A. R., et al. (1992). The hamster immune-response to tick-transmitted *Borrelia burgdorferi* differs from the response to needle-inoculated, cultured organisms. *Journal of Immunology*, **149**, 3648–3653.

Rogovin, K. A., Randall, J., Kolosova, I. & Moshkin, M. (2003). Social correlates of stress in adult males of the great gerbil, *Rhombomys opimus*, in years of high and low population densities. *Hormones and Behavior*, **43**, 132–139.

Rogowitz, G. L. & Chappell, M. A. (2000). Energy metabolism of eucalyptus-boring beetles at rest and during locomotion: gender makes a difference. *Journal of Experimental Biology*, **203**, 1131–1139.

Rohde, K. (1979). A critical evaluation of intrinsic and extrinsic factors responsible for niche restriction in parasites. *American Naturalist*, **114**, 648–671.

Rohde, K. (1985). Increased viviparity of marine parasites at high latitudes. *Hydrobiologia*, **137**, 197–201.

Rohde, K. (1992). Latitudinal gradients in species diversity: the search for the primary cause. *Oikos*, **65**, 514–527.

Rohde, K. (1994). Niche restriction in parasites: proximate and ultimate causes. *Parasitology*, **109**, S69–S84.

Rohde, K. (1996). Rapoport's rule is a local phenomenon and cannot explain latitudinal gradients in species diversity. *Biodiversity Letters*, **3**, 10–13.

Rohde, K. (1998). Is there a fixed number of niches for endoparasites of fish? *International Journal for Parasitology*, **28**, 1861–1865.

Rohde, K. (1999). Latitudinal gradients in species diversity and Rapoport's rule re-visited: a review of recent work and what can parasites teach us about the causes of the gradients? *Ecography*, **22**, 593–613.

Rohde, K., Hayward, C. & Heap, M. (1995). Aspects of the ecology of metazoan ectoparasites of marine fishes. *International Journal for Parasitology*, **25**, 945–970.

Rohde, K., Worthen, W., Heap, M., Hugueny, B. A. & Guégan, J.-F. (1998). Nestedness in assemblages of metazoan ecto- and endoparasites of marine fish. *International Journal for Parasitology*, **28**, 543–549.

Rolff, J. (2002). Bateman's principle and immunity. *Proceedings of the Royal Society of London B*, **269**, 867–872.

Roman, E. & Pichot, J. (1975). Fleas of mammals in bird nests during winter. *Bulletin Mensuel de la Société Linnéenne de Lyon*, **44**, 53–57.

Ronquist, F. (1995). Reconstructing the history of host–parasite associations using generalized parsimony. *Cladistics*, **11**, 73–89.

Ronquist, F. (2001). *TreeFitter, Version 1.0.* Available online at www.ebc.uu.se/systzoo/research/treefitter/treefitter.html.

Ronquist, F. & Liljeblad, J. (2001). Evolution of the gall wasp–host plant association. *Evolution*, **55**, 2503–2522.

Roper, T. J., Ostler, J. R., Schmid, T. K. & Christian, S. F. (2001). Sett use in European badgers *Meles meles*. *Behaviour*, **138**, 173–187.

Roper, T. J., Jackson, T. P., Conradt, L. & Bennett, N. C. (2002). Burrow use and the influence of ectoparasites in Brants' whistling rat *Parotomys brantsii*. *Ethology*, **108**, 557–564.

Rosenfeld, J. S. (2002). Functional redundancy in ecology and conservation. *Oikos*, **98**, 156–162.

Rosenzweig, M. L. (1981). A theory of habitat selection. *Ecology*, **62**, 327–335.

Rosenzweig, M. L. (1989). Habitat selection as a source of biological diversity. *Evolutionary Ecology*, **1**, 315–330.

Rosenzweig, M. L. (1991). Habitat selection and population interactions: the search for mechanism. *American Naturalist*, **137**, 5–28.

Rosenzweig, M. L. (1992). Species diversity gradients: we know more and less than we thought. *Journal of Mammalogy*, **73**, 715–730.

Rosenzweig, M. L. (1995). *Species Diversity in Space and Time*. Cambridge, UK: Cambridge University Press.

Rosenzweig, M. L. & Abramsky, Z. (1985). Detecting density-dependent habitat selection. *American Naturalist*, **126**, 405–417.

Rosenzweig, M. L. & Ziv, Y. (1999). The echo pattern of species diversity: pattern and processes. *Ecography*, **22**, 614–628.

Rosenzweig, M. L., Abramsky, Z., Kotler, B. P. & Mitchell, W. A. (1985). Can interaction coefficients be determined from census data? *Oecologia*, **66**, 194–198.

Rossin, A. & Malizia, A. I. (2002). Relationship between helminth parasites and demographic attributes of a population of the subterranean rodent *Ctenomys talarum* (Rodentia: Octodontidae). *Journal of Parasitology*, **88**, 1268–1270.

Rothschild, N. C. (1915a). A synopsis of the British Siphonaptera. *Entomological Monthly Magazine*, **51**, 49–112.

Rothschild, N. C. (1915b). On *Neopsylla* and some allied genera of Siphonaptera. *Ectoparasites*, **1**, 30–44.

Rothschild, M. (1965a). Fleas. *Scientific American*, **213**, 44–53.

Rothschild, M. (1965b). The rabbit flea and hormones. *Endeavour*, **24**, 162–168.

Rothschild, M. (1969). Notes of fleas: with the first record of the mermithid nematode from the order. *Proceedings of the British Entomological and Natural History Society*, **1**, 1–8.

Rothschild, M. (1973). Note given at meeting on *Pulex irritans* found in Viking pit excavations. *Proceedings of the Royal Entomological Society of London A*, **38**, 29.

Rothschild, M. (1975). Recent advances in our knowledge of the order Siphonaptera. *Annual Review of Entomology*, **20**, 241–259.

Rothschild, M. (1992). Neosomy in fleas, and the sessile life-style. *Journal of Zoology*, **226**, 613–629.

Rothschild, M. & Clay, T. (1952). *Fleas, Flukes and Cuckoos: A Study of Bird Parasites*, 3rd edn. London: Collins.

Rothschild, M. & Ford, R. (1966). Hormones of the vertebrate host controlling ovarian regression and copulation of the rabbit flea. *Nature*, **211**, 261–266.

Rothschild, M. & Ford, R. (1969). Does a pheromone-like factor from the nestling rabbit stimulate impregnation and maturation in the rabbit flea? *Nature*, **221**, 1169–1170.

Rothschild, M. & Ford, R. (1972). Breeding cycle of the flea *Cediopsylla simplex* is controlled by breeding cycle of host. *Science*, **178**, 625–626.

Rothschild, M. & Ford, R. (1973). Factors influencing the breeding of the rabbit flea (*Spilopsyllus cuniculi*): a spring-time accelerator and a kairomone in nestling rabbit urine (with notes on *Cediopsylla simplex*, another 'hormone bound' species). *Journal of Zoology*, **170**, 87–137.

Rothschild, M. & Hinton, H. E. (1968). Holding organs on the antennae of male fleas. *Proceedings of the Royal Entomological Society of London A*, **43**, 105–107.

Rothschild, M. & Neville, C. (1967). Fleas: insects which fly with their legs. *Proceedings of the Royal Entomological Society of London C*, **32**, 9–10.

Rothschild, M. & Schlein, J. (1975). The jumping mechanism of *Xenospylla cheopis*. I. Exoskeletal structures and musculature. *Philosophical Transactions of the Royal Society of London B*, **271**, 457–489.

Rothschild, M., Ford, B. & Hughes, M. (1970). Maturation of the rabbit flea (*Spilopsyllus cuniculi*) and the oriental flea (*Xenopsylla cheopis*): some effects of mammalian hormones on development and impregnation. *Transactions of the Zoological Society of London*, **32**, 105–188.

Rothschild, M., Schlein, J., Parker, K. & Sternberg, S. (1972). Jump of the oriental rat flea *Xenopsylla cheopis* (Roths.). *Nature*, **239**, 45–48.

Rothschild, M., Schlein, J., Parker, K., Neville, C. & Sternberg, S. (1973). The flying leap of the flea. *Scientific American*, **229**, 92–100.

Rothschild, M., Schlein, J., Parker, K., Neville, C. & Sternberg, S. (1975). The jumping mechanism of *Xenospylla cheopis*. III. Execution of the jump and activity. *Philosophical Transactions of the Royal Society of London B*, **271**, 499–515.

Roulin, A., Jeanmonod, J. & Blanc, T. (1997). Green plant material on common buzzard's (*Buteo buteo*) nests during the rearing of chicks. *Alauda*, **65**, 251–257.

Roulin, A., Riols, C., Dijkstra, C. & Ducrest, A. (2001). Female plumage spottiness and parasite resistance in the barn owl (*Tyto alba*). *Behavioral Ecology*, **12**, 103–110.

Rousset, F., Thomas, F., de Meeûs, T. & Renaud, F. (1996). Inference of parasite-induced host mortality from distribution of parasite loads. *Ecology*, **77**, 2203–2211.

Roy, B. A. (2001). Patterns of association between crucifers and their flower-mimic pathogens: host jumps are more common than coevolution or cospeciation. *Evolution*, **55**, 41–53.

Rózsa, L., Rekasi, J. & Reiczigel, J. (1996). Relationship of host coloniality to the population ecology of avian lice (Insecta: Phthiraptera). *Journal of Animal Ecology*, **65**, 242–248.

Rudenchik, Y. V., Soldatkin, I. S., Klimova, Z. I. & Severova, Z. A. (1967). On the relationships between abundance of fleas and abundance of the great gerbils. In *Proceedings of the 5th Scientific Conference of the Anti-Plague Establishments of the Middle Asia and Kazakhstan*, ed. M. A. Aikimbaev. Alma-Ata, USSR: The Middle Asian Scientific Anti-Plague Institute, pp. 181–183 (in Russian).

Rudolph, D. & Knülle, W. (1982). Novel uptake systems for atmospheric water vapor among insects. *Journal of Experimental Zoology*, **222**, 321–333.

Ruffer, D. G. (1965). Burrows and burrowing behavior of *Onychomys leucogaster*. *Journal of Mammalogy*, **46**, 241–247.

Rust, M. K. (1992). Influence of photoperiod on egg production of cat fleas (Siphonaptera: Pulicidae) infesting cats. *Journal of Medical Entomology*, **29**, 242–245.

Rust, M. K. (1994). Interhost movement of adult cat fleas (Siphonaptera: Pulicidae). *Journal of Medical Entomology*, **31**, 486–489.

Ryba, J., Rodl, P., Bartos, L., Daniel, M. & Cerny, V. (1986). Some features of the ecology of fleas inhabiting the nests of the suslik (*Citellus citellus* (L.). I. Population dynamics, sex ratio, feeding, reproduction. *Folia Parasitologica*, **33**, 265–275.

Ryba, J., Rodl, P., Bartos, L., Daniel, M. & Cerny, V. (1987). Some features of the ecology of fleas inhabiting the nests of the suslik (*Citellus citellus* (L.). II. The influence of mesostigmatid mites on fleas. *Folia Parasitologica*, **34**, 61–68.

Ryckman, R. E. (1971). Plague vector studies. I. The rate of transfer of fleas among *Citellus*, *Rattus* and *Sylvilagus* under field conditions in southern California. *Journal of Medical Entomology*, **8**, 535–540.

Rytkönen, S., Lehtonen, R. & Orell, M. (1998). Breeding great tits *Parus major* avoid nestboxes infested with fleas. *Ibis*, **140**, 687–690.

Rzhevskaya, A. E., Rapoport, L. P., Orlova, L. M., Nuriev, K. K. & Suslova, L. P. (1991). Fleas in the eastern Kyzylkum Desert and their epizootic importance. *Parazitologiya*, **25**, 504–511 (in Russian).

Sabilaev, A. S., Davydova, V. N. & Pole, D. S. (2003). About rodents' flea fauna on the left bank of the Ili River. *Quarantinable and Zoonotic Infections in Kazakhstan*, **7**, 148–149 (in Russian).

Saino, N., Calza, S. & Møller, A. P. (1998). Effects of a dipteran ectoparasite on immune response and growth trade-offs in barn swallow, *Hirundo rustica*, nestlings. *Oikos*, **81**, 217–228.

Sakaguti, K. & Jameson, E. W. (1962). The Siphonaptera of Japan. *Pacific Insects Monographs*, **3**, 1–169.

Salas, V. & Herrera, E. A. (2004). Intestinal helminths of capybaras, *Hydrochoerus hydrochaeris*, from Venezuela. *Memórias do Instituto Oswaldo Cruz*, **99**, 563–566.

Salkeld, D. J. & Stapp, P. (2006). Seroprevalence rates and transmission of plague (*Yersinia pestis*) in mammalian carnivores. *Vector-Borne and Zoonotic Diseases*, **6**, 231–239.

Samarina, G. P., Alekseev, A. N. & Shiranovich, P. I. (1968). Study of fecundity of the rat fleas (*Xenopsylla cheopis* Rothschild and *Ceratophyllus fasciatus* Bosc.) when fed on different host species. *Zoologicheskyi Zhurnal*, **47**, 261–268 (in Russian).

Samurov, M. A. (1985). Life history and the prognosis of abundance of fleas parasitizing gerbils of the genus *Meriones* in the Volga–Ural Sands. Unpublished Ph.D. thesis, Institute of Zoology of Academy of Science of the Kazakh SSR, Alma-Ata, USSR (in Russian).

Samurov, M. A. & Ageyev, V. S. (1983). Annual number of generations of fleas *Xenopsylla conformis* Wagn. (Siphonaptera, Pulicidae) in the Volga–Ural Sands. *Entomological Review*, **62**, 226–269 (in Russian).

Samurov, M. A. & Yakunin, B. M. (1979). On the annual number of generations of fleas *Ceratophyllus laeviceps* in the Volga–Ural Sands. In *Proceedings of the 10th Scientific Conference of the Anti-Plague Establishments of the Middle Asia and Kazakhstan*, vol. 1, ed. M. A. Aikimbaev. Alma-Ata, USSR: The Middle Asian Scientific Anti-Plague Institute, pp. 128–130 (in Russian).

Samurov, M. A. & Yakunin, B. M. (1980). Age composition and cycle of reproductive activity in female fleas *Ceratophyllus laeviceps* Wagn. (Siphonaptera) in the Volga–Ural Sands. *Entomological Review*, **59**, 510–512 (in Russian).

Sapegina, V. F. (1976). Distribution of fleas parasitic on small mammals and birds in the southern taiga of the Pri–Irtyshie Region. *Parazitologiya*, **10**, 397–400 (in Russian).

Sapegina, V. F. (1988). Fleas of small mammals and birds in the forest-park area of the city of Novosibirsk. *Parazitologiya*, **22**, 132–136 (in Russian).

Sapegina, V. F. & Kharitonova, N. N. (1969). On the ability of bird fleas to transmit virus of the Omsk haemorrhagic fever in the experiment. In *Migrating Birds and their Role in the Circulation of Arboviruses*, ed. A. I. Tcherepanov. Novosibirsk, USSR: Nauka, Siberian Branch, pp. 263–267 (in Russian).

Sapegina, V. F., Yudin, B. S. & Dudareva, G. V. (1980a). Materials on the biology of fleas of the Taimyr and Gydanskyi Penunsulae. In *Parasitic Insects and Ticks of Siberia*, ed. M. S. Davydova. Novosibirsk, USSR: Nauka, Siberian Branch, pp. 225–231 (in Russian).

Sapegina, V. F., Ravkin, Y. S., Lukianova, I. V. & Sebeleva, G. G. (1980b). Fleas of the forest zone of western Siberia. In *Problems of Zoogeography and Faunal History*, ed. B. F. Belyshev & Y. S. Ravkin. Novosibirsk, USSR: Nauka, Siberian Branch, pp. 94–166 (in Russian).

Sapegina, V. F., Lukianova, I. V. & Fomin, B. N. (1981a). Fleas of small mammals in northern foothills of the Altai Mountains and Upper Ob River Region. In *Biological Problems of Natural Foci*, ed. A. A. Maximov. Novosibirsk, USSR: Nauka, Siberian Branch, pp. 167–176 (in Russian).

Sapegina, V. F., Yudin, B. S. & Yudina, S. A. (1981b). Fleas of small mammals in the northern taiga of the southern Taimyr Peninsula. *Bulletin of the Siberian Branch of the Academy of Sciences of the USSR, Series Biological Sciences*, **1**, 96–104 (in Russian).

Sapegina, V. F., Vartapetov, L. G. & Pokrovskaya, I. V. (1990). The fleas of small mammals in the northern taiga of western Siberia. *Parazitologiya*, **24**, 56–62 (in Russian).

Sarfati, M., Krasnov, B. R., Ghazaryan, L., et al. (2005). Energy costs of blood digestion in a host-specific haematophagous parasite. *Journal of Experimental Biology*, **208**, 2489–2496.

Sasal, P., Trouvé, S., Müller-Graf, C. & Morand, S. (1999). Specificity and host predictability: a comparative analysis among monogenean parasites of fish. *Journal of Animal Ecology*, **68**, 437–444.

Saxena, V. K. (1987). Rodent–ectoparasite association in selected biotopes of Mirzapur and Varanasi districts of Uttar Pradesh. *Journal of Communicable Diseases*, **19**, 310–316.

Saxena, V. K. (1999). Mesostigmatid mite infestations of rodents in diverse biotopes of central and southern India. *Journal of Parasitology*, **85**, 147–149.

Scalon, O. I. (1981). On fleas from eastern Mongolia with description of male and female of *Echidnophaga tiscadaea* Smit, 1967 (Siphonaptera). *Parazitologiya*, **15**, 280–287 (in Russian).

Schall, J. J. (1990). The ecology of lizard malaria. *Parasitology Today*, **6**, 264–269.

Scharf, W. C. (1991). Geographic distribution of Siphonaptera collected from small mammals on Lake Michigan islands. *Great Lakes Entomologist*, **24**, 39–43.

Scharf, W. C. (1998). Fleas (Siphonaptera) from migrating owls: passengers on the journey. *Michigan Birds and Natural History*, **5**, 167–171.

Scheidt, V. J. (1988). Flea allergy dermatitis. In *The Veterinary Clinics of North America: Small Animal Practice*, ed. S. D. White. Philadelphia, PA: Saunders College Publishing, pp. 1023–1042.

Schelhaas, D. P. & Larson, O. R. (1989). Cold hardiness and winter survival in the bird flea, *Ceratophyllus idius*. *Journal of Insect Physiology*, **35**, 149–153.

Schemmer, K. R. & Halliwell, R. E. (1987). Efficacy of alum-precipitated flea antigen for hyposensitization of flea-allergic dogs. *Seminars in Veterinary Medicine and Surgery (Small Animal)*, **2**, 195–198.

Schlein, Y. (1980). Morphological similarities between the skeletal structures of Siphonaptera and Mecoptera. In *Fleas: Proceedings of the International Conference on Fleas, Ashton Wold, Peterborough, UK, 21–25 June 1977*, ed. R. Traub & H. Starcke. Rotterdam, the Netherlands: A. A. Balkema, pp. 359–367.

Schluter, D. (1984). A variance test for detecting species associations, with some example applications. *Ecology*, **65**, 998–1005.

Schmid-Hempel, P. (2003). Variation in immune defence as a question of evolutionary ecology. *Proceeding of the Royal Society of London B*, **270**, 357–366.

Schmid-Hempel, P. & Ebert, D. (2003). On the evolutionary ecology of specific immune defence. *Trends in Ecology and Evolution*, **18**, 27–32.

Schmidt-Nielsen, K. (1990). *Animal Physiology: Adaptation and Environment*, 4th edn. Cambridge, UK: Cambridge University Press.

Schoener, T. W. (1974). Competition and the form of the habitat shift. *Theoretical Population Biology*, **6**, 265–307.

Schofield, S. & Torr, S. J. (2002). A comparison of feeding behaviour of tsetse and stable flies. *Medical and Veterinary Entomology*, **16**, 177–185.

Schönrogge, K., Gardner, M. G., Elmes, G. W., et al. (2006). Host propagation permits extreme local adaptation in a social parasite of ants. *Ecological Letters*, **9**, 1032–1040.

Schradin, C. & Pillay, N. (2005). Demography of the striped mouse (*Rhabdomys pumilio*) in the succulent karoo. *Mammalian Biology*, **70**, 84–92.

Schwan, T. G. (1975). Flea reinfestation on the California meadow vole (*Microtus californicus*). *Journal of Medical Entomology*, **21**, 760.

Schwan, T. G. (1986). Seasonal abundance of fleas (Siphonaptera) on grassland rodents in Lake Nakuru National Park, Kenya, and potential for plague transmission. *Bulletin of Entomological Research*, **76**, 633–648.

Schwan, T. G. (1993). Sex ratio and phoretic mites of fleas (Siphonaptera, Pulicidae and Hystrichopsyllidae) on the Nile grass rat (*Arvicanthis niloticus*) in Kenya. *Journal of Medical Entomology*, **30**, 122–135.

Schwan, T. G. & Schwan, V. R. (1980). Observations on the fleas *Xenopsylla debilis* and *Xenopsylla difficilis* (Siphonaptera) infesting the gerbil *Tatera nigricauda* in Southern Kenya. *African Journal of Ecology*, **18**, 267–272.

Scofield, A., Riera, M. D. F., Eliseri, C., Massard, C. L. & Linardi, P. M. (2005). Ocorrência de *Rhopalopsyllus lutzi lutzi* (Baker) (Siphonaptera, Rhopalopsyllidae) em *Canis familiaris* (Linnaeus) de zona rural do município de Piraí, Rio de Janeiro, Brasil. *Revista Brasileira de Entomologia*, **49**, 159–161.

Seal, S. C. & Bhattacharji, L. M. (1961). Epidemiological studies of plague in Calcutta. I. Bionomics of two species of rat fleas and distribution, densities and resistance of rodents in relation to the epidemiology of plague in Calcutta. *Indian Journal of Medical Research*, **49**, 974–1007.

Segerman, J. (1995). Siphonaptera of Southern Africa: handbook for the identification of fleas. *Publications of the South African Institute for Medical Research*, **57**, 1–264.

Serzhan, O. S. (2002). Pathways of evolution of the faunistic complexes of rodent fleas in Kazakhstan, Middle Asia and adjacent regions and their role in the endemism of plague. *Quarantinable and Zoonotic Diseases in Kazakhstan*, **6**, 83–90 (in Kazakh).

Serzhan, O. S. & Ageyev, V. S. (2000). Geographical distribution and host complexes of plague-infected fleas in relation to some problems of paleogenesis of plague enzootics. *Quarantinable and Zoonotic Diseases in Kazakhstan*, **2**, 183–192 (in Russian).

Sevenster, J. G. (1996). Aggregation and coexistence. I. Theory and analysis. *Journal of Animal Ecology*, **65**, 297–307.

Sgonina, K. (1935). Die Reizphysiologie des Igelflöhs (*Archaeopsylla erinacei* Bouché) und seiner Larve. *Zeitschrift für Parasitenkunde*, **7**, 539–571.

Shafi, M. M., Ali, R., Ghazi, R. R. & Noor, U. N. (1988). Flea index studies of synanthropic rats in Karachi, Pakistan. *Acta Parasitologica Polonica*, **33**, 185–194.

Shaftesbury, A. D. (1934). The Siphonaptera (fleas) of North Carolina, with special reference to sex ratios. *Journal of the Elisha Mitchell Scientific Society*, **49**, 247–263.

Shargal, E., Kronfeld-Schor, N. & Dayan, T. (2000). Population biology and spatial relationships of coexisting spiny mice (*Acomys*) in Israel. *Journal of Mammalogy*, **81**, 1046–1052.

Sharif, M. (1949). Effects of constant temperature and humidity on the development of the larvae and the pupae of the three Indian species of *Xenopsylla* (Insecta: Siphonaptera). *Philosophical Transactions of the Royal Society of London B*, **233**, 581–633.

Shaw, D. J. & Dobson, A. P. (1995). Patterns of macroparasite abundance and aggregation in wildlife populations: a quantitative review. *Parasitology*, **111**, S111–S127.

Shaw, D. J., Grenfell, B. T. & Dobson, A. P. (1998). Patterns of macroparasite aggregation in wildlife host populations. *Parasitology*, **117**, 597–610.

Shaw, J. L. & Moss, R. (1990). Effect of the caecal nematode *Trichostrongylus tenius* on egg-laying by captive red grouse. *Research in Veterinary Science*, **48**, 253–258.

Shaw, S. E., Kenny, M. J., Tasker, S. & Birtles, R. J. (2004). Pathogen carriage by the cat flea *Ctenocephalides felis* (Bouché) in the United Kingdom. *Veterinary Microbiology*, **102**, 183–188.

Shchedrin, V. I. (1974). Morphological and histochemical data on the blood digestion in some flea species: vectors of plague. Unpublished Ph.D. thesis, All-Union Scientific Anti-Plague Institute 'Microb', Saratov, USSR (in Russian).

Shchedrin, V. I., Loktev, N. A. & Lunina, E. A. (1974). Morphological and histochemical data on digestion in fleas *P. irritans*. In *Particularly Dangerous Diseases in Caucasus: Proceedings of the 3rd Scientific–Practical Conference of the Anti-Plague Establishments of Caucasus on Natural Focality, Epidemiology and Prophylaxis of Particularly Dangerous*

Diseases, 14–16 May 1974, ed. V. G. Pilipenko. Stavropol, USSR: Scientific Anti-Plague Institute of Caucasus and Trans-Caucasus, pp. 281–283 (in Russian).

Sheldon, B. C. & Verhulst, S. (1996). Ecological immunology: costly parasite defenses and trade-offs in evolutionary ecology. *Trends in Ecology and Evolution*, **11**, 317–321.

Shenbrot, G. I. & Krasnov, B. R. (2002). Can interaction coefficients be determined from census data? Testing two estimation methods with Negev Desert rodents. *Oikos*, **99**, 47–58.

Shenbrot, G. I., Krasnov, B. R. & Khokhlova, I. S. (1994). On the biology of *Gerbillus henleyi* (Rodentia: Gerbillidae) in the Negev Highlands, Israel. *Mammalia*, **58**, 581–589.

Shenbrot, G. I., Sokolov, V. E., Geptner, V. G. & Kovalskaya, Y. M. (1995). *The Mammals of Russia and Adjacent Regions: Dipodoidea*. Moscow, Russia: Nauka (in Russian).

Shenbrot, G. I., Krasnov, B. R. & Khokhlova, I. S. (1997). On the biology of Wagner's gerbil *Gerbillus dasyurus* (Wagner, 1842) (Rodentia: Gerbillidae) in the Negev Highlands, Israel. *Mammalia*, **61**, 467–486.

Shenbrot, G. I., Krasnov, B. R. & Rogovin, K. A. (1999a). *Spatial Ecology of Desert Rodent Communities*. New York: Springer-Verlag.

Shenbrot, G. I., Krasnov, B. R. & Khokhlova, I. S. (1999b). Notes on the biology of the bushy-tailed jird, *Sekeetamys calurus*, in the central Negev, Israel. *Mammalia*, **63**, 374–377.

Shenbrot, G. I., Krasnov, B. R., Khokhlova, I. S., Demidova, T. & Fielden, L. J. (2002). Habitat-dependent differences in architecture and microclimate of the Sundevall's jird (*Meriones crassus*) burrows in the Negev Desert, Israel. *Journal of Arid Environments*, **51**, 265–279.

Shenbrot, G. I., Krasnov, B. R. & Lu, L. (2007). Geographic range size and host specificity in ectoparasites: A case study with *Amphipsylla* fleas and rodent hosts. *Journal of Biogeography*, **34**, 1679–1690.

Shepherd, R. C. H. & Edmonds, J. W. (1979). The distribution of the stickfast fleas, *Echidnophaga myrmecobii* Rothschild and *E. perilis* Jordan, on the wild rabbit, *Oryctolagus cuniculus* (L.). *Australian Journal of Zoology*, **27**, 261–271.

Shevchenko, V. L., Samurov, M. A., Kaimashnikov, V. I. & Polyakov, V. K. (1971). Some patterns of variation in the abundance of the midday jirds and fleas *Xenopsylla conformis* in the Volga–Ural Sands. In *Proceedings of the 7th Scientific Conference of the Anti-Plague Establishments of the Middle Asia and Kazakhstan*, ed. M. A. Aikimbaev. Alma-Ata, USSR: The Middle Asian Scientific Anti-Plague Institute, pp. 449–450 (in Russian).

Shevchenko, V. L., Grazhdanov, A. K., Zharinova, L. K. & Andreeva, T. A. (1976). Abilities of avian fleas *Frontopsylla frontalis alatau* Fed., 1946 to infect rodents with plague. *Medical Parasitology and Parasitic Diseases* [*Meditsinskaya Parazitologiya i Parazitarnye Bolezni*], **45**, 49–52 (in Russian).

Shields, W. M. & Crook, J. R. (1987). Barn swallow coloniality: a net cost for group breeding in the Adirondacks. *Ecology*, **68**, 1373–1386.

Shinozaki, Y., Shiibashi, T., Yoshizawa, K., et al. (2004). Ectoparasites of the Pallas squirrel, *Callosciurus erythraeus*, introduced to Japan. *Medical and Veterinary Entomology*, **18**, 61–63.

Shorrocks, B. (1996). Local diversity: a problem with too many solutions. In *The Genesis and Maintenance of Biological Diversity*, ed. M. Hochberg, J. Clobert & R. Barbault. Oxford, UK: Oxford University Press, pp. 104–122.

Shorrocks, B. & Rosewell, J. (1986). Guild size in drosophilids: a simulation model. *Journal of Animal Ecology*, **55**, 527–541.

Shryock, J. A. & Houseman, R. M. (2005). A comparison of fecal protein content in male and female cat fleas, *Ctenocephalides felis* (Bouché) (Siphonaptera: Pulicidae). *Florida Entomologist*, **88**, 335–337.

Shubber, A. H., Lloyd, S. & Soulsby, E. J. L. (1981). Infection with gastrointestinal helminths: effect of lactation and maternal transfer of immunity. *Parasitology Research*, **65**, 181–189.

Shudo, E. & Iwasa, Y. (2001). Inducible defense against pathogens and parasites: optimal choice among multiple options. *Journal of Theoretical Biology*, **209**, 233–247.

Shulov, A. & Naor, D. (1964). Experiments on the olfactory responses and host specificity of the Oriental rat flea (*Xenopsylla cheopis*). *Parasitology*, **54**, 225–231.

Shurin, J. B. & Allen, E. G. (2001). Effects of competition, predation, and dispersal on species richness at local and regional scales. *American Naturalist*, **158**, 624–637.

Shurin, J. B., Havel, J. E., Leibold, M. A. & Pinel-Alloul, B. (2000). Local and regional zooplankton species richness: a scale-independent test for saturation. *Ecology*, **81**, 3062–3073.

Shutler, D. & Campbell, A. A. (2007). Experimental addition of greenery reduces flea loads in nests of a non-greenery using species, the tree swallow *Tachycineta bicolor*. *Journal of Avian Biology*, **38**, 7–12.

Shutler, D., Petersen, S. D., Dawson, R. D. & Campbell, A. (2003). Sex ratios of fleas (Siphonaptera: Ceratophyllidae) in nests of tree swallows (Passeriformes: Hirundinidae) exposed to different chemicals. *Environmental Entomology*, **32**, 1045–1048.

Shutler, D., Mullie, A. & Clark, R. G. (2004). Tree swallow reproductive investment, stress, and parasites. *Canadian Journal of Zoology*, **82**, 442–448.

Shwartz, E. A., Berendyaeva, E. L. & Grebenyuk, R. V. (1958). Fleas parasitic on rodents in the Frunze Region. *Proceedings of the Middle Asian Scientific Anti-Plague Institute*, **4**, 255–261 (in Russian).

Siemann, E. (1998). Experimental tests of effects of plant productivity and diversity on grassland arthropod diversity. *Ecology*, **79**, 2057–2070.

Silverman, J. & Appel, A. G. (1994). Adult cat flea (Siphonaptera, Pulicidae) excretion of host blood proteins in relation to larval nutrition. *Journal of Medical Entomology*, **31**, 265–271.

Silverman, J. & Rust, M. K. (1983). Some abiotic factors affecting the survival of the cat flea, *Ctenocephalides felis* (Siphonaptera: Pulicidae). *Environmental Entomology*, **12**, 490–495.

Silverman, J. & Rust, M. K. (1985). Extended longevity of the pre-emerged adult of the cat flea (Siphonaptera: Pulicidae) and factors stimulating emergence from the pupal cocoon. *Annals of the Entomological Society of America*, **78**, 763–768.

Silverman, J., Rust, M. K. & Reierson, D. A. (1981). Influence of temperature and humidity on survival and development of the cat flea, *Ctenocephalides felis* (Siphonaptera: Pulicidae). *Journal of Medical Entomology*, **18**, 78–83.

Silvertown, J., Franco, M. & Harper, J. L. (eds.) (1997). *Plant Life Histories: Ecology, Phylogeny and Evolution*. New York: Cambridge University Press.

Simiczyjew, B. & Margas, W. (2001). Ovary structure in the bat flea *Ischnopsyllus* spp. (Siphonaptera: Ischnopsyllidae): phylogenetic implications. *Zoologica Poloniae*, **46**, 5–14.

Šimková, A., Desdevises, Y., Gelnar, M. & Morand, S. (2000). Coexistence of nine gill ectoparasites (*Dactylogyrus*: Monogenea) parasitizing the roach (*Rutilus rutilus* L.): history and present ecology. *International Journal for Parasitology*, **30**, 1077–1088.

Šimková, A., Desdevises, Y., Gelnar, M. & Morand, S. (2001). Morphometric correlates of host specificity in *Dactylogyrus* species (Monogenea) parasites of European cyprinid fish. *Parasitology*, **123**, 169–177.

Šimková, A., Kadlec, D., Gelnar, M. & Morand, S. (2002). Abundance–prevalence relationship of gill congeneric ectoparasites: testing the core–satellite hypothesis and ecological specialization. *Parasitology Research*, **88**, 682–686.

Šimková, A., Sitko, J., Okulewicz, J. & Morand, S. (2003). Occurrence of intermediate hosts and structure of digenean communities of the black-headed gull, *Larus ridibundus* (L.). *Parasitology*, **126**, 69–78.

Šimková, A., Verneau, O., Gelnar, M. & Morand, S. (2006). Specificity and specialization of congeneric monogeneans parasitizing cyprinid fish. *Evolution*, **60**, 1023–1037.

Sinelshchikov, V. A. (1956). Study of flea fauna of the Pavlodar Region. *Proceedings of the Middle Asian Scientific Anti-Plague Institute*, **2**, 147–153 (in Russian).

Singh, S. K. & Girschick, H. J. (2003). Tick–host interactions and their immunological implications in tick-borne diseases. *Current Science*, **85**, 1284–1298.

Skinner, J. D. & Smithers, H. N. (1990). *The Mammals of the Southern African Subregion*, 2nd edn. Pretoria, South Africa: University of Pretoria Press.

Skorping, A., Read, A. F. & Keymer, A. E. (1991). Life history covariation in intestinal nematodes of mammals. *Oikos*, **60**, 365–372.

Skuratowicz, W. (1960). Pchly (Aphaniptera) ptaków i ssaków Bialowieskiego Parku Narodowego. *Annales Zoologici (Warszawa)*, **19**, 1–32.

Skuratowicz, W. (1967). *Pchly – Siphonaptera (Aphaniptera): Klucze do Oznacznia Owadow Polski 29*. Warsaw: Panstwowe Wydawnictwo Naukowe.

Slomczyński, R., Kaliński, A., Wawrzyńiak, J., *et al.* (2006). Effects of experimental reduction in nest micro-parasite and macro-parasite loads on nestling hemoglobin level in blue tits *Parus caeruleus*. *Acta Oecologica*, **30**, 223–227.

Slonov, M. N. (1965). On the biology of a flea *Ceratophyllus tamias* Wagn., 1927. *Medical Parasitology and Parasitic Diseases* [*Meditsinskaya Parazitologiya i Parazitarnye Bolezni*], **34**, 485–487 (in Russian).

Smetana, A. (1965). On the transmission of tick-borne encephalitis by fleas. *Acta Virologica*, **9**, 375–379.

Smit, F. G. A. M. (1954). *Lopper: Danmarks Fauna, 60*. Copenhagen, Denmark: G. E. C. Gads Forlag.

Smit, F. G. A. M. (1962a). *Neotunga euloidea* gen. n., sp. n. (Siphonaptera, Pulicidae). *Bulletin of the British Museum of Natural History, Entomology*, **12**, 365–378.

Smit, F. G. A. M. (1962b). Siphonaptera collected from moles and their nests at Wilp, Netherlands, by Jhr. W. C. Van Heurn. *Tijdschrift voor Entomologie*, **105**, 29–44.

Smit, F. G. A. M. (1972). On some adaptive structures in Siphonaptera. *Folia Parasitologica*, **19**, 5–17.

Smit, F. G. A. M. (1974). Siphonaptera collected by Dr J. Martens in Nepal. *Senckenbergiana Biologica*, **55**, 357–398.

Smit, F. G. A. M. (1977). An unusual form of the stick-tight flea *Echidnophaga gallinacea*. *Revue Zoologique Africaine*, **91**, 198–199.

Smit, F. G. A. M. (1978). Fossil 'fleas'. *Flea News*, **14**, 1–2.

Smit, F. G. A. M. (1982). Classification of the Siphonaptera. In *Synopsis and Classification of Living Organisms*, vol. 2, ed. S. Parker. New York: McGraw-Hill, pp. 557–563.

Smit, F. G. A. M. (1987). *An Illustrated Catalogue of the Rothschild Collection of Fleas (Siphonaptern) in the British Museum (Natural History)*, vol. 7, *Malacopsylloidea (Malacopsyllidae and Rhopalopsyllidae)*. London: Oxford University Press.

Smith, A. (1951). The effect of relative humidity on the activity of the tropical rat flea *Xenopsylla cheopis* (Roths.). *Bulletin of Entomological Research*, **42**, 585–600.

Smith, A. (1980). Lack of interspecific interactions of Everglades rodents on two spatial scales. *Acta Theriologica*, **25**, 61–70.

Smith, A., Telfer, S., Burthe, S., Bennett, M. & Begon, M. (2005). Trypanosomes, fleas and field voles: ecological dynamics of a host–vector–parasite interaction. *Parasitology*, **131**, 355–365.

Smith, F. A., Brown, J. H., Haskell, J. P., et al. (2004). Similarity of mammalian body size across the taxonomic hierarchy and across space and time. *American Naturalist*, **163**, 672–691.

Smith, S. A. & Clay, M. E. (1985). Morphology of the antennae of the bat flea *Myodopsylla insignis* (Siphonaptera: Ischnopsyllidae). *Journal of Medical Entomology*, **22**, 64–71.

Smits, J. E., Bortolotti, G. R. & Tella, J. L. (1999). Simplifying the phytohaemagglutinin skin-testing technique in studies of avian immunocompetence. *Functional Ecology*, **13**, 567–572.

Snodgrass, R. E. (1944). The feeding apparatus of biting and sucking insects affecting man and animals. *Smithsonian Miscellaneous Collections*, **104**, 1–107.

Sobey, W. R., Menzies, W. & Conolly, D. (1974). Myxomatosis: some observations on breeding the European rabbit flea *Spilopsyllus cuniculi* (Dale) in an animal house. *Journal of Hygiene*, **71**, 453–465.

Sokolova, A. A. & Popova, A. S. (1969). On the biology of fleas *Coptopsylla lamellifer*. In *Proceedings of the 6th Scientific Conference of the Anti-Plague Establishments of the*

Middle Asia and Kazakhstan, vol. 2, ed. M. A. Aikimbaev. Alma-Ata, USSR: The Middle Asian Scientific Anti-Plague Institute, pp. 90–92 (in Russian).

Sokolova, A. A., Balabas, N. G. & Trofimenko, I. P. (1971). Data on reproduction of *Coptopsylla lamellifer* in the Moyynkum Desert. In *Proceedings of the 7th Scientific Conference of the Anti-Plague Establishments of the Middle Asia and Kazakhstan*, ed. M. A. Aikimbaev. Alma-Ata, USSR: The Middle Asian Scientific Anti-Plague Institute, pp. 416–417 (in Russian).

Soliman, S., Marzouk, A. S., Main, A. J. & Montasser, A. A. (2001). Effect of sex, size, and age of commensal rat hosts on the infestation parameters of their ectoparasites in a rural area of Egypt. *Journal of Parasitology*, **87**, 1307–1316.

Soloshenko, I. Z. (1958). Role of haematophagous arthropods in the maintenance of the epizootics of leptospiroses in the foci of these diseases. In *Proceedings of the 10th Conference on Parasitological Problems and Diseases with Natural Focality*, ed. Anonymous. Moscow–Leningrad, USSR, pp. 139–140 (in Russian).

Soloshenko, I. Z. (1962). Role of haematophagous arthropods in transmission and maintanence of pathogenous leptospires. II. Relationships between haematophagous arthropods and leptospiroses. *Journal of Microbiology* [*Zhurnal Mikrobiologii*], **4**, 31–34 (in Russian).

Solovieva, A. V., Alania, I. I. & Kosminsky, R. B. (1976). On the ecology of fleas *Ctenophthalmus* (*Euctenophthalmus*) *strigosus* Rostigaev et Zolotova, 1964 (Ctenophthalmidae, Siphonaptera) in southern Trans-Caucasus. *Problems of Particularly Dangerous Diseases*, **51**, 46–49 (in Russian).

Sorci, G. (1996). Patterns of haemogregarine load, aggregation and prevalence as a function of host age in the lizard *Lacerta vivipara*. *Journal of Parasitology*, **82**, 676–678.

Sorci, G., Defraipont, M. & Clobert, J. (1997). Host density and ectoparasite avoidance in the common lizard (*Lacerta vivipara*). *Oecologia*, **11**, 183–188.

Soshina, Y. F. (1973). The rate of flea infestation in common myomorph rodents in the forest zone of the Krym Mountains. *Parazitologiya*, **7**, 31–35 (in Russian).

Sosnina, E. F. (1967a). The dependence of the infestation and species composition of rodent ectoparasites on the host's habitat (on the example of *Rattus turkestanicus*). *Wiadomosci Parazytologiczne*, **13**, 637–641.

Sosnina, E. F. (1967b). An attempt of biocoenotical analysis of the assemblages of arthropods collected from rodents. *Parasitological Collection*, **23**, 61–69 (in Russian).

Sotnikova, A. N. & Soldatov, G. M. (1969). Extraction of the virus of the tick-borne encephalitis from fleas *Ceratophyllus tamias* Wagn. *Medical Parasitology and Parasitic Diseases* [*Meditsinskaya Parazitologiya i Parazitarnye Bolezni*], **33**, 622–624 (in Russian).

Southwood, T. R. E. (1966). *Ecological Methods*. London: Chapman & Hall.

Souza, W. P. (1994). Patterns and processes in communities of helminth parasites. *Trends in Ecology and Evolution*, **9**, 52–57.

Sreter-Lancz, Z., Tornyai, K., Szell, Z., Sreter, T. & Marialigeti, K. (2006). Bartonella infections in fleas (Siphonaptera: Pulicidae) and lack of Bartonellae in ticks (Acari: Ixodidae) from Hungary. *Folia Parasitologica*, **53**, 313–316.

Srivastava, D. (1999). Using local–regional richness plots to test for species saturation: pitfalls and potentials. *Journal of Animal Ecology*, **68**, 1–16.

Stanko, M. (1987). Siphonaptera of small mammals in the northern part of the Krupina Plain. *Stredné Slovensko, Zbornk Stredoslovenského Muzea, Banska Bystrica*, **6**, 108–117 (in Slovak).

Stanko, M. (1988). Fleas (Siphonaptera) of small mammals in eastern part of the Volovské Vrchy Mountains. *Acta Rerum Naturalium Musei Nationalis Slovaci Bratislava*, **34**, 29–40 (in Slovak).

Stanko, M. (1994). Fleas synusy (Siphonaptera) of small mammals from the central part of the East-Slovakian lowlands. *Biologia, Bratislava*, **49**, 239–246.

Stanko, M., Miklisova, D., Goüy de Bellocq, J. & Morand, S. (2002). Mammal density and patterns of ectoparasite species richness and abundance. *Oecologia*, **131**, 289–295.

Stanko, M., Krasnov, B. R. & Morand, S. (2006). Relationship between host density and parasite distribution: inferring regulating mechanisms from census data. *Journal of Animal Ecology*, **75**, 575–583.

Stanko, M., Krasnov, B. R., Miklisova, D. & Morand, S. (2007). Simple epidemiological model predicts the relationships between prevalence and abundance in ixodid ticks. *Parasitology*, **134**, 59–68.

Starikov, V. P. & Sapegina, V. F. (1987). Ectoparasites of small mammals in the forest-steppe Trans-Ural Region. In *Ecology and Geography of Arthropods in Siberia*, ed. A. I. Tcherepanov. Novosibirsk, USSR: Nauka, Siberian Branch, pp. 76–83 (in Russian).

Stark, H. E. (2002). Population dynamics of adult fleas (Siphonaptera) on hosts and in nests of the California vole. *Journal of Medical Entomology*, **39**, 818–824.

Stark, H. E. & Miles, V. I. (1962). Ecological studies of wild rodent plague in the San Francisco Bay area of California. VI. The relative abundance of the certain flea species and their host relationships on coexisting wild and domestic rodents. *American Journal of Tropical Medicine and Hygiene*, **11**, 525–534.

Starozhitskaya, G. S. (1968). Effect of gonotrophic cycle of fleas on the duration of their uninterrupted stay on a host. In *Rodents and their Ectoparasites*, ed. B. K. Fenyuk. Saratov, USSR: Saratov University Press, pp. 59–64 (in Russian).

Starozhitskaya, G. S. (1970). Diapause in fleas of the genus *Xenopsylla* and its epizootological importance. *Problems of Particularly Dangerous Diseases*, **13**, 148–155 (in Russian).

Statzner, B., Dolédec, S. & Hugueny, B. (2004). Biological trait composition of European stream invertebrate communities: assessing the effects of various trait filter types. *Ecography*, **27**, 470–488.

Stearns, S. C. (1992). *The Evolution of Life Histories*. New York: Oxford University Press.

Steele, W. K., Pilgrim, R. L. C. & Palma, R. L. (1997). Occurrence of the flea *Glaciopsyllus antarcticus* and avian lice in central Dronning Maud Land. *Polar Biology*, **18**, 292–294.

Stenseth, N. C, Samia, N. I., Viljugrein, H., *et al.* (2006). Plague dynamics are driven by climate variation. *Proceedings of the National Academy of Sciences of the USA*, **103**, 13110–13115.

Stepanova, N. A. & Mitropolsky, O. V. (1971). Relationships between co-occurring fleas *Xenopsylla hirtipes* and *Xenopsylla gerbilli*. In *Proceedings of the 7th Scientific Conference of the Anti-Plague Establishments of the Middle Asia and Kazakhstan*, ed. M. A. Aikimbaev. Alma-Ata, USSR: The Middle Asian Scientific Anti-Plague Institute, pp. 418–420 (in Russian).

Stepanova, N. A. & Mitropolsky, O. V. (1977). Spatial distribution of two sympatric species of fleas parasitic on the great gerbil in the Kyzyl-Kum Desert. *Parazitologiya*, **11**, 147–152 (in Russian).

Stevens, G. C. (1989). The latitudinal gradient in geographical range: how so many species coexist in the tropics. *American Naturalist*, **133**, 240–256.

Stevenson, H. L., Bai, Y., Kosoy, M. Y., *et al.* (2003). Detection of novel *Bartonella* strains and *Yersinia pestis* in pairie dogs and their fleas (Siphonaptera: Ceratophyllidae and Pulicidae) using multiplex polymerase chain reaction. *Journal of Medical Entomology*, **40**, 329–337.

Stevenson, H. L., Labruna, M. B., Montenieri, J. A., *et al.* (2005). Detection of *Rickettsia felis* in a New World flea species, *Anomiopsyllus nudata* (Siphonaptera: Ctenophthalmidae). *Journal of Medical Entomology*, **42**, 163–167.

Stewart, M. A. & Evans, F. C. (1941). A comparative study of rodent and burrow flea populations. *Proceedings of the Society for Experimental Biology and Medicine*, **47**, 140–142.

Stewart, P. D. & MacDonald, D. W. (2003). Badgers and badger fleas: strategies and counter-strategies. *Ethology*, **109**, 751–764.

Stireman, J. O. (2005). The evolution of generalization? Parasitoid flies and the perils of inferring host range evolution from phylogenies. *Journal of Evolutionary Biology*, **18**, 325–336.

Stone, L. & Roberts, A. (1991). Conditions for a species to gain advantage from the presence of competitors. *Ecology*, **72**, 1964–1972.

Strahan, R. (ed.) (1983). *The Complete Book of Australian Mammals*. North Ryde, Australia: Collins, Angus & Robertson.

Streilein, J. W. (1990). Skin associated lymphoid tissues (SALT): the next generation. In *Skin Immune System (SIS)*, ed. J. D. Bos. Boca Raton, FL: CRC Press, pp. 25–48.

Studdert, V. P. & Arundel, J. H. (1988). Dermatitis of the pinnae of cats in Australia associated with the European rabbit flea (*Spilopsyllus cuniculi*). *Veterinary Record*, **123**, 624–625.

Štys, P. & Bilinski, S. M. (1990). Ovariole types and the phylogeny of hexapods. *Biology Review*, **65**, 401–429.

Sukhanova, V. I., Tchernikina, M. A., Sosnovtseva, V. P., *et al.* (1978). Multiannual dynamics of abundance in fleas parasitic on the great gerbil in northern Turkmenistan. *Problems of Particularly Dangerous Diseases*, **60**, 53–57 (in Russian).

Sukhdeo, M. V. K. (1997). Earth's third environment: the worm's eye view. *BioScience*, **47**, 141–149.

Sukhdeo, M. V. K. (2000). Inside the vertebrate host: ecological strategies by parasites living in the third environment. In *Evolutionary Biology of Host–Parasite Relationships: Theory Meets Reality*, ed. R. Poulin, S. Morand & A. Skorping. Amsterdam, the Netherlands: Elsevier Science, pp. 43–62.

Sukhdeo, M. V. K. & Bansemir, A. D. (1996). Critical resources that influence habitat selection decisions by gastrointestinal helminth parasites. *International Journal for Parasitology*, **26**, 483–498.

Sukhdeo, M. V. K., Sukhdeo, S. C. & Bansemir, A. D. (2002). Interactions between intestinal nematodes and vertebrate hosts. In *The Behavioural Ecology of Parasites*, ed. E. E. Lewis, J. F. Campbell & M. V. K. Sukhdeo. Wallingford, UK: CAB International, pp. 223–242.

Suleimenov, B. M. (2004). *Mechanism of Plague Enzootic*. Almaty, Kazakhstan: Almaty (in Russian).

Suntsov, V. V. & Suntsova, N. I. (2003). Origin and genesis of natural and anthropogenic plague foci: ecological, geographical and social aspects. In *Ecological and Epizootological Aspects of Plague in Vietnam*, ed. L. P. Korzun & V. V. Suntsov. Moscow (Russia), Ho Chi Minh, Buonmathuot (Vietnam): GEOS, pp. 109–149 (in Russian).

Suntsov, V. V. & Suntsova, N. I. (2006). *Plague: Origin and Evolution of Epizootic System (Ecological, Geographical and Social Aspects)*. Moscow, Russia: KMK Scientific Press (in Russian).

Suntsov, V. V., Li Thi Vi, K. & Suntsova, N. I. (1992a). Some features of the flea (Insecta, Siphonaptera) fauna of small mammals in Vietnam. *Zoologicheskyi Zhurnal*, **71**, 88–94 (in Russian).

Suntsov, V. V., Huong, L. T. & Suntsova, N. I. (1992b). Notes on fleas (Siphonaptera) in the plague foci on the Tay Nguyen Plateau (Vietnam). *Parazitologiya*, **26**, 516–520 (in Russian).

Suter, P. R. (1964). Biologie von *Echidnophaga gallinacea* (Westw.) und Vergleich mit andern Verhaltenstypen bei Flöhen. *Acta Tropica*, **21**, 193–238.

Sutherland, W. J. (1983). Aggregation and the 'ideal free' distribution. *Journal of Animal Ecology*, **52**, 821–828.

Sutherland, W. J. (1996). *From Individual Behaviour to Population Ecology*. Oxford, UK: Oxford University Press.

Sviridov, G. G. (1963). Application of radioactive isotopes for the study of some problems of flea ecology. II. The contact of animals and intensity of the exchange of ectoparasites in the population of *Rhombomis opimus*. *Zoologicheskyi Zhurnal*, **42**, 546–550 (in Russian).

Syrvatcheva, N. G. (1964). Data on flea fauna of the Kabardino-Balkarian ASSR. *Proceedings of the Armenian Anti-Plague Station*, **3**, 389–405 (in Russian).

Szidat, L. (1940). Beiträge zum Aubfau eines natürlichen Systems der Trematoden. I. Die Entwicklung von *Echinocercaria choanophila* U. Szidat zu *Cathaemasia hians* und die Ableitung der Fasciolidae von den Echinostomidae. *Zeitschrift für Parasitenkunde*, **11**, 239–283.

Tabor, S. P., Williams, D. F., Germano, D. J. & Thomas, R. E. (1993). Fleas (Siphonaptera) infesting giant kangaroo rats (*Dipodomys ingens*) on the Elkhorn and Carrizo plains, San Luis Obispo County, California. *Journal of Medical Entomology*, **30**, 291–294.

Takahashi, K., Tuno, N. & Kagaya, T. (2005). The relative importance of spatial aggregation and resource partitioning on the coexistence of mycophagous insects. *Oikos*, **109**, 125–134.

Talybov, A. N. (1974). Some data on the lifespan of fleas parasitic on the common vole in the Trans-Caucasus mountains. In *Particularly Dangerous Diseases in Caucasus: Proceedings of the 3rd Scientific–Practical Conference of the Anti-Plague Establishments of Caucasus on Natural Focality, Epidemiology and Prophylaxis of Particularly Dangerous Diseases, 14–16 May 1974*, ed. V. G. Pilipenko. Stavropol, USSR: Scientific Anti-Plague Institute of Caucasus and Trans-Caucasus, pp. 183–185 (in Russian).

Talybov, A. N. (1975). Life expectancy of *Ctenophthalmus wladimiri* Is.-Gurv., 1948 (Siphonaptera, Ctenophthalmidae) under laboratory conditions. *Parazitologiya*, **9**, 354–358 (in Russian).

Talybov, A. N. (1976). Development of the pre-imaginal phases of flea *Ctenophthalmus wladimiri* Is.-Gurv., 1948. *Parazitologiya*, **10**, 320–324 (in Russian).

Tanitovsky, V. A., Bidashko, F. G., Grazhdanov, A. K. & Dauletova, S. B. (2004). Species structure and number of fleas parasitizing small mammals in the middle part of the Ural River valley. *Quarantinable and Zoonotic Infections in Kazakhstan*, **9**, 76–80 (in Russian).

Tarasevich, L. N., Tagiltsev, A. A. & Malkov, G. B. (1969). Results of virological examination of ixodid ticks and fleas in the southern Omsk Region. *Medical Parasitology and Parasitic Diseases [Meditsinskaya Parazitologiya i Parazitarnye Bolezni]*, **38**, 705–707 (in Russian).

Tarshis, I. B. (1956). Feeding techniques for blood-sucking arthropods. *Proceedings of the 10th International Congress of Entomology*, **3**, 767–784.

Taylor, J. & Purvis, A. (2003). Have mammals and their chewing lice diversified in parallel? In *Tangled Trees: Phylogeny, Cospeciation, and Coevolution*, ed. R. D. M. Page. Chicago, IL: University of Chicago Press, pp. 240–261.

Taylor, L. H., Mackinnon, M. J. & Read, A. F. (1998). Virulence of mixed-clone and single-clone infections of the rodent malaria *Plasmodium chabaudi*. *Evolution*, **52**, 583–591.

Taylor, L. R. (1961). Aggregation, variance and the mean. *Nature*, **189**, 732–735.

Taylor, L. R. & Taylor, R. A. J. (1977). Aggregation, migration and population dynamics. *Nature*, **265**, 415–421.

Taylor, L. R. & Woiwod, I. P. (1980). Temporal stability as a density-dependent species characteristic. *Journal of Animal Ecology*, **49**, 209–224.

Taylor, L. R., Woiwod, I. P. & Perry, J. N. (1979). The negative binomial as a dynamic ecological model and density-dependence of k. *Journal of Animal Ecology*, **48**, 289–304.

Taylor, R. A. J., Lindquist, R. K. & Shipp, J. L. (1998). Variation and consistency in spatial distribution as measured by Taylor's power law. *Environmental Entomology*, **27**, 191–201.

Taylor, S. D., Dittmar de la Cruz, K., Porter, M. L. & Whiting, M. F. (2005). Characterization of the long-wavelength opsin from Mecoptera and Siphonaptera: does a flea see? *Molecular Biology and Evolution*, **22**, 1165–1174.

Tchernova, N. A. (1971). Reproduction of *Xenopsylla skrjabini* and their preferences for different elements of the great gerbil burrow in the Mangyshlak Peninsula. In *Proceedings of the 7th Scientific Conference of the Anti-Plague Establishments of the Middle Asia and Kazakhstan*, ed. M. A. Aikimbaev. Alma-Ata, USSR: The Middle Asian Scientific Anti-Plague Institute, pp. 443–444 (in Russian).

Tchimanina, B. M. & Kozlovskaya, O. L. (1971a). Experimental study of the role of fleas *Ctenophthalmus congeneroides* Wagn. and *Neopsylla bidentatiformis* Wagn. in the circulation of the tick-borne encephalitis virus. *Transactions of the Irkutsk State Scientific Anti-Plague Institute of Siberia and Far East*, **9**, 235–236 (in Russian).

Tchimanina, B. M. & Kozlovskaya, O. L. (1971b). Experimental study of the circulation of the tick-borne encephalitis virus in the nests of the forest voles via fleas *Frontopsylla elata botis*, Jord., 1929. *Transactions of the Irkutsk State Scientific Anti-Plague Institute of Siberia and Far East*, **9**, 237–238 (in Russian).

Tchumakova, I. V. & Kozlov, M. P. (1983). Quantitative parameters of mortality and survival in fleas *Nosopsyllus consimilis* at different stages of the metamorphosis. In *Prophylaxis of Diseases in the Natural Foci*, ed. I. F. Taran. Stavropol, USSR: Scientific Anti-Plague Institute of Caucasus and Trans-Caucasus, pp. 280–281 (in Russian).

Tchumakova, I. V., Tokanev, F. I. & Kozlov, M. P. (1978). Dependence of the reproduction capacity of fleas (Aphaniptera) on the recurrence of mating. *Parazitologiya*, **12**, 292–296 (in Russian).

Tchumakova, I. V., Kozlov, M. P. & Belokopytova, A. (1981). Estimation of the dependence of the reproduction of rodent fleas (Siphonaptera) on feeding by experimental breeding of fleas on different hosts. *Entomological Review*, **60**, 562–569 (in Russian).

Tchumakova, I. V., Ermolova, N. V. & Shaposhnikova, L. I. (2002). Principles for prediction of population densities of fleas parasitic on rodents. *Medical Parasitology and Parasitic Diseases [Meditsinskaya Parazitologiya i Parazitarnye Bolezni]*, **72**, 45–48 (in Russian).

Telfer, S., Bown, K. J., Sekules, R., et al. (2005). Disruption of a host–parasite system following the introduction of an exotic host species. *Parasitology*, **130**, 661–668.

Tella, J. L. (2002). The evolutionary transition to coloniality promotes higher blood parasitism in birds. *Journal of Evolutionary Biology*, **15**, 32–41.

Tella, J. L., Scheuerlein, A. & Ricklefs, R. E. (2002). Is cell-mediated immunity related to the evolution of life-history strategies in birds? *Proceedings of the Royal Society of London B*, **269**, 1059–1066.

Tellam, R. L., Smith, D., Kemp, D. H. & Willadsen, P. (1992). Vaccination against ticks. In *Animal Parasite Control Utilizing Biotechnology*, ed. W. K. Yong. Boca Raton, FL: CRC Press, pp. 303–331.

Tenquist, J. D. & Charleston, W. A. G. (2001). A revision of the annotated checklist of ectoparasites of terrestrial mammals in New Zealand. *Journal of the Royal Society of New Zealand*, **31**, 481–542.

Teplinskaya, T. A., Labunetz, N. F. & Kuliev, M. T. (1983). Seasonal dynamics of age structure and reproduction of *Xenopsylla conformis* in the Caspian natural plague focus. In *Prophylaxis of Diseases in the Natural Foci*, ed. I. F. Taran. Stavropol, USSR: Scientific Anti-Plague Institute of Caucasus and Trans-Caucasus, pp. 277–278 (in Russian).

Terborgh, J. W. & Faaborg, J. (1980). Saturation of bird communities in the West Indies. *American Naturalist*, **116**, 178–195.

ter Hofstede, H. M. & Fenton, M. B. (2005). Relationships between roost preferences, ectoparasite density and grooming behaviour of neotropical bats. *Journal of Zoology*, **266**, 333–340.

Theodor, O. & Costa, M. (1967). *A Survey of the Parasites of Wild Mammals and Birds in Israel*. vol. 1, *Ectoparasites*. Jerusalem: Israel Academy of Science and Humanities.

Théron, A. & Combes, C. (1995). Asynchrony of infection timing, habitat preference, and sympatric speciation of schistosome parasites. *Evolution*, **49**, 372–375.

Thomas, C. D. & Hanski, I. (1997). Butterfly metapopulations. In *Metapopulation Biology: Ecology, Genetics, and Evolution*, ed. I. Hanski & M. E. Gilpin. San Diego, CA: Academic Press, pp. 359–386.

Thomas, K. & Shutler, D. (2001). Ectoparasites, nestling growth, parental feeding rates, and begging intensity of tree swallows. *Canadian Journal of Zoology*, **79**, 346–353.

Thomas, R. (1988). A review of flea collection records from *Onychomys leucogaster* with observations on the role of grasshopper mice in the epizootiology of wild rodent plague. *Great Basin Naturalist*, **48**, 83–95.

Thomas, R. (1996). Fleas and the agents they transmit. In *The Biology of Disease Vectors*, ed. B. J. Beaty & W. C. Marquardt. Niwot, CO: University of Colorado Press, pp. 146–159.

Thompson, C. F. & Neill, A. J. (1991). House wrens do not prefer clean nestboxes. *Animal Behaviour*, **42**, 1022–1024.

Thompson, C. W., Hillgarth, N., Leu, M. & McClure, H. E. (1997). High parasite load in house finches (*Carpodacus mexicanus*) is correlated with reduced expression of a sexually selected trait. *American Naturalist*, **149**, 270–294.

Thompson, J. N. (1994). *The Coevolutionary Process*. Chicago, IL: University of Chicago Press.

Thompson, J. N. (2005). *The Geographic Mosaic of Coevolution*. Chicago, IL: University of Chicago Press.

Tian, J. (1995). Niches of 27 flea species in the natural focus of plague in Jianchuan, Yunnan Province. *Endemic Diseases Bulletin*, **10**, 27–32 (in Chinese).

Tiflov, V. E. (1959). Role of fleas in epizootology of tularemia. *Proceedings of the Scientific Anti-Plague Institute of Caucasus and Trans-Caucasus*, **2**, 363–392 (in Russian).

Tiflov, V. E. (1964). Destiny of the bacterial cultures in the organism of a flea. *Ectoparasites*, **4**, 181–198 (in Russian).

Tiflov, V. E. & Ioff, I. G. (1932). Observations on the biology of fleas. *Herald of Microbiology and Epidemiology [Vestnik Mikrobiologii i Epidemiologii]*, **11**, 95–117 (in Russian).

Tillyard, R. J. (1926). *The Insects of Australia and New Zealand*. Sydney Australia: Angus and Robertson.

Tillyard, R. J. (1935). The evolution of the scorpion-flies and their derivatives (order Mecoptera). *Annals of the Entomological Society of America*, **28**, 1–45.

Timi, J. T. & Poulin, R. (2003). Parasite community structure within and across host populations of a marine pelagic fish: how repeatable is it? *International Journal for Parasitology*, **33**, 1353–1362.

Tipton, V. J. & Machado-Allison, C. E. (1972). Fleas of Venezuela. *Brigham Young University Scientific Bulletin, Biological Series*, **17**, 1–115.

Tipton, V. J. & Méndez, E. (1966). The fleas (Siphonaptera) of Panama. In *Ectoparasites of Panama*, ed. R. L. Wenzel & V. J. Tipton. Chicago, IL: Field Museum of Natural History, pp. 289–385.

Tipton, V. J. & Méndez, E. (1968). New species of fleas (Siphonaptera) from Cerro Potosi, Mexico, with notes on ecology and host–parasite relationships. *Pacific Insects*, **10**, 177–214.

Toft, C. A. & Karter, A. J. (1990). Parasite–host coevolution. *Trends in Ecology and Evolution*, **5**, 326–329.

Tofts, R. & Silvertown, J. (2000). A phylogenetic approach to community assembly from a local species pool. *Proceedings of the Royal Society of London B*, **267**, 363–369.

Tokeshi, M. (1999). *Species Coexistence: Ecological and Evolutionary Perspectives*. Oxford, UK: Blackwell Science.

Tokmakova, E. G., Verzhutsky, D. B. & Bazanova, L. P. (2006). Formation of the proventriculus blockage, alimentary activity and mortality in fleas *Amphipsylla primaris primaris* infected with *Yersinia pestis*. *Parazitologiya*, **40**, 215–224 (in Russian).

Trager, W. (1939). Acquired immunity to ticks. *Journal of Parasitology*, **25**, 57–81.

Tränkle, S. B. (1989). Wirtspecifizität und Wanderaktivität des Katzenflohes *Ctenocephalides felis* (Bouché). Unpublished M.Sc. thesis, Albert Ludwigs Universität, Freiburg im Beisgau, Germany.

Traub, R. (1972a). The Gunong Benom Expedition 1967. XII. Notes on zoogeography, convergent evolution and taxonomy of fleas (Siphonaptera), based on collection from Gunong Benom and elsewhere in South-East Asia. 2. Convergent evolution. *Bulletin of the British Museum Natural History, Zoology*, **23**, 309–387.

Traub, R. (1972b). The relationship between the spines, combs and other skeletal features of fleas (Siphonaptera) and the vesture, affinities and habits of their hosts. *Journal of Medical Entomology*, **9**, 601.

Traub, R. (1980). The zoogeography and evolution of some fleas, lice and mammals. In *Fleas: Proceedings of the International Conference on Fleas,* Ashton Wold, Peterborough, UK, 21–25 June 1977, ed. R. Traub & H. Starcke. Rotterdam, the Netherlands: A. A. Balkema, pp. 93–172.

Traub, R. (1985). Coevolution of fleas and mammals. In *Coevolution of Parasitic Arthropods and Mammals,* ed. K. C. Kim. New York: John Wiley, pp. 295–437.

Traub, R., Wisseman, C. L. & Farhang-Azad, A. (1978). The ecology of murine typhus: a critical review. *Tropical Diseases Bulletin,* **75,** 237–317.

Traub, R., Rothschild, M. & Haddow, J. F. (1983). *The Ceratophyllidae: Key to the Genera and Host Relationships.* Cambridge, UK: Cambridge University Press.

Tripet, F. & Richner, H. (1997a). Host responses to ectoparasites: food compensation by parent blue tits. *Oikos,* **78,** 557–561.

Tripet, F. & Richner, H. (1997b). The coevolutionary potential of a 'generalist' parasite, the hen flea *Ceratophyllus gallinae. Parasitology,* **115,** 419–427.

Tripet, F. & Richner, H. (1999a). Density-dependent processes in the population dynamics of a bird ectoparasite *Ceratophyllus gallinae. Ecology,* **80,** 1267–1277.

Tripet, F. & Richner, H. (1999b). Demography of the hen flea *Ceratophyllus gallinae* in blue tit *Parus caeruleus* nests. *Journal of Insect Behavior,* **12,** 159–174.

Tripet, F., Christe, P. & Møller, A. P. (2002a). The importance of host spatial distribution for parasite specialization and speciaton: a comparative study of bird fleas (Siphonaptera: Ceratophyllidae). *Journal of Animal Ecology,* **71,** 735–748.

Tripet, F., Glaser, M. & Richner, H. (2002b). Behavioural responses to ectoparasites: time-budget adjustments and what matters to blue tits *Parus caeruleus* infested by fleas. *Ibis,* **144,** 461–469.

Tripet, F., Jacot, A. & Richner, H. (2002c). Larval competition affects the life histories and dispersal behaviour of an avian ectoparasite. *Ecology,* **83,** 935–945.

Trivers, R. L. & Willard, D. E. (1973). Natural selection of parental ability to vary sex ratio of offspring. *Science,* **179,** 90–92.

Trudeau, W. L., Fernandez-Caldas, E. & Fox, R. W. (1993). Allergenicity of the cat flea (*Ctenocephalides felis felis*). *Clinical and Experimental Allergy,* **23,** 377–383.

Trukhachev, N. N. (1971). Effect of host on the offspring of fleas *Xenopsylla cheopis*. In *Proceedings of the 7th Scientific Conference of the Anti-Plague Establishments of the Middle Asia and Kazakhstan,* ed. M. A. Aikimbaev. Alma-Ata, USSR: The Middle Asian Scientific Anti-Plague Institute, pp. 423–425 (in Russian).

Tschirren, B., Fitze, P. S. & Richner, H. (2003). Sexual dimorphism in susceptibility to parasites and cell-mediated immunity in great tit nestlings. *Journal of Animal Ecology,* **72,** 839–845.

Tschirren, B., Richner, H. & Schwabl, H. (2004). Ectoparasite-modulated deposition of maternal androgens in great tit eggs. *Proceedings of the Royal Society of London B,* **271,** 1371–1375.

Tschirren, B., Saladin, V., Fitze, P. S., Schwabl, H. & Richner, H. (2005). Maternal yolk testosterone does not modulate parasite susceptibility or immune function in great tit nestlings. *Journal of Animal Ecology,* **74,** 675–682.

Tschirren, B., Bischoff, L. L., Saladin, V. & Richner, H. (2007a). Host condition and host immunity affect parasite fitness in a bird–ectoparasite system. *Functional Ecology*, **21**, 372–378.

Tschirren, B., Fitze, P. S. & Richner, H. (2007b). Maternal modulation of natal dispersal in a passerine bird: an adaptive strategy to cope with parasitism? *American Naturalist*, **169**, 87–93.

Uchikawa, K., Sato, A. & Kugimoto, M. (1967). A report on the flea fauna on the Oki Islands. *Medical Entomology and Zoology*, **18**, 14–17 (in Japanese).

Ulmanen, I. & Myllymäki, A. (1971). Species composition and numbers of fleas (Siphonaptera) in a local population of the field vole, *Microtus agrestis* (L.). *Annales Zoologici Fennici*, **8**, 374–384.

Uvarov, B. P. (1931). Insects and climate. *Transactions of the Royal Entomological Society of London*, **79**, 1–247.

Vainikka, A., Jokinen, E. I., Kortet, R. & Taskinen, J. (2004). Gender- and season-dependent relationships between testosterone, oestradiol and immune functions in wild roach. *Journal of Fish Biology*, **64**, 227–240.

Valone, T. J. & Hoffman, C. D. (2002). Effects of regional pool size on local diversity in small-scale annual plant communities. *Ecology Letters*, **5**, 477–480.

Valtonen, E. T., Pulkkinen, K., Poulin, R. & Julkunen, M. (2001). The structure of parasite component communities in brackish water fishes of the northeastern Baltic Sea. *Parasitology*, **122**, 471–481.

Vandermeer, J. (1990). Indirect and diffuse interactions: complicated cycles in a population embedded in a large community. *Journal of Theoretical Biology*, **142**, 429–442.

Vansulin, S. A. (1961). Ecology of fleas parasitic on the great gerbils. In *Proceedings of the Interdisciplinary Conference Dedicated to the 40th Anniversary of the Kazakh Soviet Socialist Republic*, ed. Anonymous. Alma-Ata, USSR: Kainar, pp. 47–49 (in Russian).

Vansulin, S. A. (1965). On the ecology of fleas (Aphaniptera) of the great gerbil. *Entomological Review*, **44**, 307–314 (in Russian).

Vansulin, S. A. & Volkova, L. A. (1962). Fur structure in *Rhombomys opimus* Licht. and its effect on the abundance of fleas parasitizing these rodents in different seasons. *Zoologicheskyi Zhurnal*, **41**, 147–150 (in Russian).

Van Vuren, D. (1996). Ectoparasites, fitness, and social behaviour of yellow-bellied marmots. *Ethology*, **102**, 686–694.

Varma, M. G. R. & Page, R. J. C. (1966). The epidemiology of louping ill in Ayshire, Scotland: ectoparasites of small mammals. I. (Siphonaptera). *Journal of Medical Entomology*, **3**, 331–335.

Varma, M. G. R., Hellerhaupt, A., Trinder, P. K. E. & Langi, A. O. (1990). Immunization of guinea-pigs against *Rhipicephalus appendiculatus* adult ticks using homogenates from unfed immature ticks. *Immunology*, **71**, 133–138.

Vashchenok, V. S. (1966a). Histological description of the oogenesis in fleas *Echidnophaga oschanini* Wagn. (Pulicidae, Aphaniptera). *Zoologicheskyi Zhurnal*, **45**, 1821–1831 (in Russian).

Vashchenok, V. S. (1966b). Morphophysiological changes in the organism of fleas *Echidnophaga oschanini* Wagn. (Aphaniptera, Pulicidae) during feeding and reproduction. *Entomological Review*, **45**, 715–727 (in Russian).

Vashchenok, V. S. (1967a). Gonotrophic relationships in fleas *Ceratophyllus consimilis* Wagn. (Aphaniptera, Ceratophyllidae). *Parasitological Collection*, **13**, 222–235 (in Russian).

Vashchenok, V. S. (1967b). On the ecology of fleas *Echidnophaga oshanini* Wagn. (Pulicidae, Aphaniptera) in the Tuva ASSR. *Parazitologiya*, **1**, 27–35 (in Russian).

Vashchenok, V. S. (1974). Activity of blood digestion in fleas. In *Proceedings of the 7th Meeting of the All-Union Entomological Society*, ed. Anonymous. Leningrad, USSR: Zoological Institute of the Academy of Sciences of the USSR, p. 209 (in Russian).

Vashchenok, V. S. (1979). Maintanence of the causative agent of the yersiniosis in fleas *Xenopsylla cheopis*. *Parazitologiya*, **13**, 19–25 (in Russian).

Vashchenok, V. S. (1984). Fleas and agents of bacterial diseases of humans and animals. *Parasitological Collection*, **32**, 79–123 (in Russian).

Vashchenok, V. S. (1988). *Fleas: Vectors of Pathogens Causing Diseases in Humans and Animals*. Leningrad, USSR: Nauka (in Russian).

Vashchenok, V. S. (1993). Factors regulating egg production in fleas *Leptopsylla segnis* (Leptopsyllidae: Siphonaptera). *Parazitologiya*, **27**, 382–388 (in Russian).

Vashchenok, V. S. (1995). The dependence of the egg-laying activity in the fleas *Leptopsylla segnis* (Siphonaptera: Leptopsyllidae) on their abundance on a host. *Parazitologiya*, **29**, 267–271.

Vashchenok, V. S. (2001). Age changes of fecundity in fleas *Leptopsylla segnis* (Siphonaptera: Leptopsyllidae). *Parazitologiya*, **35**, 460–464 (in Russian).

Vashchenok, V. S. (2006). Species composition, host preferences and niche differentiation in fleas (Siphonaptera) parasitic on small mammals in the Ilmen–Volkhov Lowland. *Parazitologiya*, **40**, 425–437 (in Russian).

Vashchenok, V. S. & Solina, L. T. (1969). On blood digestion in fleas *Xenopsylla cheopis* Roths. (Aphaniptera, Pulicidae). *Parazitologiya*, **3**, 451–460 (in Russian).

Vashchenok, V. S. & Solina, L. T. (1972). Age-related changes in the fat tissue of female fleas *Xenopsylla cheopis*. *Zoologicheskyi Zhurnal*, **60**, 79–85 (in Russian).

Vashchenok, V. S. & Tchirov, P. A. (1976). Histological study of fleas *Ceratophyllus consimilis* Wagn. infected with the causative agent of listeriosis (*Listeria monocytogenes*). *Parazitologiya*, **10**, 61–66 (in Russian).

Vashchenok, V. S. & Tretiakov, K. A. (2003). Seasonal dynamic of flea (Siphonaptera) abundance on *Clethrionomys glareolus* in the northern part of the Novgorod Region. *Parazitologiya*, **37**, 177–189 (in Russian).

Vashchenok, V. S. & Tretiakov, K. A. (2004). Seasonal dynamic of flea (Siphonaptera) abundance on the common shrew *Sorex araneus* in the northern part of the Novgorod Region. *Parazitologiya*, **38**, 503–514 (in Russian).

Vashchenok, V. S. & Tretiakov, K. A. (2005). Seasonal dynamic of flea (Siphonaptera) abundance on the pygmy woodmouse *Apodemus uralensis* in the northern part of the Novgorod Region. *Parazitologiya*, **39**, 270–277 (in Russian).

Vashchenok, V. S., Solina L, T. & Zhirnov, A. E. (1976). Digestion of blood of different animals by fleas *Xenopsylla cheopis*. *Parazitologiya*, **10**, 544–549 (in Russian).

Vashchenok, V. S., Bryukhanova, L. V. & Shchedrin, V. I. (1985). Characteristics of feeding and digestion in fleas. *Parasitological Collection*, **33**, 134–148 (in Russian).

Vashchenok, V. S., Karandina, R. S. & Bryukhanova, L. V. (1988). Amount of blood consumed by different flea species in the experiments. *Parazitologiya*, **22**, 312–320 (in Russian).

Vashchenok, V. S., Sheikin, A. O. & Serzhanov, O. S. (1992). Morphophysiological characteristics of fleas *Xenopsylla gerbilli* during the autumn–winter diapause. *Parasitological Collection*, **37**, 5–15 (In Russian).

Vasiliev, G. I. (1961). Observations on flea breeding in the laboratory. *Transactions of the Irkutsk State Scientific Anti-Plague Institute of Siberia and Far East*, **2**, 97–99 (in Russian).

Vasiliev, G. I. (1966). On ectoparasites and their hosts in relation to the plague epizootic in the Bajan-Khongor Aimak (Mongolian People's Republic). *Proceedings of the Irkutsk State Scientific Anti-Plague Institute of Siberia and Far East*, **26**, 277–281 (in Russian).

Vasiliev, G. I. (1971). Fleas of the long-tailed ground squirrel (species composition, ecology, epizootological importance for plague). Unpublished Ph.D. thesis, Irkutsk State Scientific Anti-Plague Institute of Siberia and Far East, Irkutsk, USSR (in Russian).

Vasiliev, G. I. & Zhovty, I. F. (1961). An attempt to investigate the rules of the distribution of micropopulations of fleas in a microhabitat. *Transactions of the Irkutsk State Scientific Anti-Plague Institute of Siberia and Far East*, **1**, 88–91 (in Russian).

Vasiliev, G. I. & Zhovty, I. F. (1971). On the annual cycle of *Oropsylla asiatica* Wagn., 1929 (Siphonaptera) parasitic on *Spermophilus ungulatus* in the Siberian Cis-Baikalia. *Transactions of the Irkutsk State Scientific Anti-Plague Institute of Siberia and Far East*, **9**, 227–229 (in Russian).

Vaughan, J. A. & Coombs, M. E. (1979). Laboratory breeding of the European rabbit flea, *Spilopsyllus cuniculi* (Dale). *Journal of Hygiene*, **83**, 521–530.

Vaughan, J. A. & Mead-Briggs, A. R. (1970). Host-finding behaviour of the rabbit flea, *Spilopsyllus cuniculi* with special reference to the significance of urine as an attractant. *Parasitology*, **61**, 397–409.

Vaughan, J. A., Jerse, A. E. & Farhang-Azad, A. (1989). Rat leucocyte's response to the bites of rat fleas (Siphonaptera: Pulicidae). *Journal of Medical Entomology*, **26**, 449–453.

Vázquez, D. P. & Aizen, M. A. (2003). Null model analyses of specialization in plant–pollinator interactions. *Ecology*, **84**, 2493–2501.

Vázquez, D. P. & Stevens, R. D. (2004). The latitudinal gradient in niche breadth: concepts and evidence. *American Naturalist*, **164**, E1–E19.

Vázquez, D. P., Poulin, R., Krasnov, B. R. & Shenbrot, G. I. (2005). Species abundance patterns and the distribution of specialization in host–parasite interaction networks. *Journal of Animal Ecology*, **74**, 946–955.

Verts, B. J. (1961). Observations on the fleas (Siphonaptera) of some small mammals in northwestern Illinois. *American Midland Naturalist*, **66**, 471–476.

Verzhutsky, D. B., Zonov, G. B. & Popov, V. V. (1990). Epizootological importance of flea accumulation in the aggregations of female long-tailed ground squirrels in the Tuva plague focus. *Parazitologiya*, **24**, 186–92 (in Russian).

Via, S. (2001). Sympatric speciation in animals: the ugly duckling grows up. *Trends in Ecology and Evolution*, **16**, 381–390.

Vidal-Martinez, V. M. & Poulin, R. (2003). Spatial and temporal repeatability in parasite community structure of tropical fish hosts. *Parasitology*, **127**, 387–398.

Viitala, J., Hakkarainen, H. & Ylönen, H. (1994). Different dispersal in *Clethrionomys* and *Microtus*. *Annales Zoologici Fennici*, **31**, 411–415.

Violovich, N. A. (1969). Landscape and geographic distribution of fleas. In *Biological Regionalization of the Novosibirsk Region*, ed. A. A. Maximov. Novosibirsk, USSR: Nauka, Siberian Branch, pp. 211–221 (in Russian).

Visser, M., Rehbein, S. & Wiedemann, C. (2001). Species of flea (Siphonaptera) infesting pets and hedgehogs in Germany. *Journal of Veterinary Medicine B*, **48**, 197–202.

Vobis, M., D'Haese, J., Mehlhorn, H., Heukelbach, J., Mencke, N. & Feldmeier, H. (2005). Molecular biological investigations of Brazilian *Tunga* sp. isolates from man, dogs, cats, pigs and rats. *Parasitological Research*, **96**, 107–112.

Volfertz, A. A. & Kolpakova, S. A. (1946). On epizootology of tularemia: role of fleas *Ctenophthalmus orientalis* Wagn. in epizootology of tularemia. *Medical Parasitology and Parasitic Diseases* [*Meditsinskaya Parazitologiya i Parazitarnye Bolezni*], **16**, 83–84 (in Russian).

Volis, S., Mendlinger, S., Olswig-Whittaker, L., Safriel, U. N. & Orlovsky, N. (1998). Phenotypic variation and stress resistance in core and peripheral populations of *Hordeum spontaneum*. *Biodiversity and Conservation*, **7**, 799–813.

Vysotskaya, S. O. (1967). Biocoenotical relationships between ectoparasites of rodents and other inhabitants of their nests. *Parasitological Collection*, **23**, 19–60 (in Russian).

Waage, J. K. (1979). The evolution of insect/vertebrate associations. *Biological Journal of the Linnean Society*, **12**, 187–224.

Wade, S. E. & Georgi, J. R. (1988). Survival and reproduction of artificially fed cat fleas, *Ctenocephalides felis* Bouché (Siphonaptera: Pulicidae). *Journal of Medical Entomology*, **25**, 186–190.

Waeber, P. O. & Hemelrijk, C. K. (2003). Female dominance and social structure in Alaotran gentle lemurs. *Behaviour*, **140**, 1235–1246.

Wagner, J. (1929). About new species of Palaearctic fleas (Aphaniptera). II. *Annual Reports of the Zoological Museum of the Academy of Sciences of the USSR*, **30**, 531–547 (in Russian).

Wakelin, D. (1996). *Immunity to Parasites: How Parasitic Infections Are Controlled*, 2nd edn. Cambridge, UK: Cambridge University Press.

Walker, M., Steiner, S., Brinkhof, M. W. G. & Richner, H. (2003). Induced responses of nestling great tits reduce hen flea reproduction. *Oikos*, **102**, 67–74.

Wall, R. & Shearer, D. (2001). *Veterinary Ectoparasites: Biology, Pathology and Control*, 2nd edn. Oxford: Blackwell Science.

Walshe, B. M. (1948). The oxygen requirements and thermal resistance of chironomid larvae from flowing and from still waters. *Journal of Experimental Biology*, **25**, 35–44.

Walton, D. W. & Hong, H.-K. (1976). Fleas of small mammals from the endemic haemorrhagic fever zones of Kyonggi and Kangwon Provinces of the Repubic of Korea. *Korean Journal of Parasitology*, **14**, 17–24.

Walton, D. W. & Tun, U. M. (1978). Fleas of small mammals from Rangoon, Burma. *Southeast Asian Journal of Tropical Medicine and Public Health*, **9**, 369–377.

Wang, G.-L., Xi, N., Adily, S., et al. (2004a). Studies on some characteristics of bioecology and morphology of *Vermipsylla alakurt*. *Endemic Diseases Bulletin*, **19**, 25–27 (in Chinese).

Wang, G.-L., Xi, N., Dang, X.-S., et al. (2004b). Pathogen identification of vermipsyllosis of domestic animal in Bazhou, Xinjiang. *Chinese Journal of Veterinary Parasitology*, **12**, 2–3 (in Chinese).

Warburg, A., Saraiva, E., Lanzaro, G. C., Titus, R. G. & Neva, F. (1994). Saliva of *Lutzomyia longipalpis* sibling species differs in its composition and capacity to enhance leishmaniasis. *Philosophical Transactions of the Royal Society of London B*, **345**, 223–230.

Ward, S. A. (1992). Assessing functional explanations of host specificity. *American Naturalist*, **139**, 883–891.

Warwick, R. M. & Clarke, K. R. (2001). Practical measures of marine biodiversity based on relatedness of species. *Oceanography and Marine Biology*, **39**, 207–231.

Watkins, R. A., Moshier, S. E. & Pinter, A. J. (2006). The flea *Megabothris abantis*: an invertebrate host of *Hepatozoon* sp. and a likely definitive host in *Hepatozoon* infections of the montane vole, *Microtus montanus*. *Journal of Wildlife Diseases*, **42**, 386–390.

Watt, C., Dobson, A. P. & Grenfell, B. T. (1995). Glossary. In *Ecology of Infectious Diseases in Natural Populations*, ed. B. T. Grenfell & A. P. Dobson. Cambridge, UK: Cambridge University Press, pp. 510–521.

Watts, M. M., Pascoe, D. & Carroll, K. (2002). Population responses of the freshwater amphipod *Gammarus pulex* (L.) to an environmental estrogen, 17 alpha-ethinylestradiol. *Environmental Toxicology and Chemistry*, **21**, 445–450.

Webb, C. T., Brooks, C. P., Gage, K. L. & Antolin, M. F. (2006). Classic flea-borne transmission does not drive plague epizootics in prairie dogs. *Proceedings of the National Academy of Sciences of the USA*, **103**, 6236–6241.

Webb, D. R., Porter, W. P. & Mcclure, P. A. (1990). Development of insulation in juvenile rodents: functional compromise in insulation. *Functional Ecology*, **4**, 251–256.

Webber, L. A. & Edman, J. D. (1972). Anti-mosquito behaviour of ciconiiform birds. *Animal Behaviour*, **20**, 228–232.

Wedekind, C. (1992). Detailed information about parasites revealed by sexual ornamentation. *Proceedings of the Royal Society of London B*, **247**, 169–174.

Wegner, Z. (1970). Lice (Anoplura) of small mammals caught in Dobrogea (Roumania). *Societăţii de Ştiinţe Biologia din Republica Socialistă România, Zoologia*, **20**, 305–314.

Wenk, P. (1953). Der Kopf von *Ctenocephalus canis* (Curt.) (Aphaniptera). *Zoologische Jahrbücher, Abteilung für Anatomie und Ontogenie der Tiere*, **73**, 103–164.

Wenzel, R. L. & Tipton, V. J. (1966). Some relationships between mammal hosts and their ectoparasites. In *Ectoparasites of Panama*, ed. R. L. Wenzel & V. J. Tipton. Chicago, IL: Field Museum of Natural History, pp. 677–723.

Wesołowski, T. & Stańska, M. (2001). High ectoparasite loads in hole-nesting birds: a nestbox bias? *Journal of Avian Biology*, **32**, 281–285.

Wessels, W. (1998). Gerbillidae from the Miocene and Pliocene of Europe. *Mitteilungen bayerische Staatssammlung für Paläontologie und historische Geologie*, **38**, 187–207.

Whitehead, M. D., Burton, H. R., Bell, P. J., Arnould, J. P. Y. & Rounsevell, D. E. (1991). A further contribution on the biology of the Antarctic flea, *Glaciopsyllus antarcticus* (Siphonaptera: Ceratophyllidae). *Polar Biology*, **11**, 379–383.

Whiteman, N. K. & Parker, P. G. (2004). Body condition and parasite load predict territory ownership in the Galapagos hawk. *Condor*, **106**, 915–921.

Whiteman, N. K. & Parker, P. G. (2005). Using parasites to infer host population history: a new rationale for parasite conservation. *Animal Conservation*, **8**, 175–181.

Whiting, M. F., Carpenter, J. C., Wheeler, Q. D. & Wheeler, W. C. (1997). The Strepsiptera problem: phylogeny of the holometabolous insect orders inferred from 18S and 28S ribosomal DNA sequences and morphology. *Systematic Biology*, **46**, 1–68.

Whiting, M. F. (2002a). Phylogeny of the holometabolous insect orders: molecular evidence. *Zoologica Scripta*, **31**, 3–15.

Whiting, M. F. (2002b). Mecoptera is paraphyletic: multiple genes and phylogeny of Mecoptera and Siphonaptera. *Zoologica Scripta*, **31**, 93–104.

Whiting, M. F., Whiting, A. S. & Hastriter, M. W. (2003). A comprehensive phylogeny of Mecoptera and Siphonaptera. *Entomologische Abhandlungen*, **61**, 169.

Widmann, O. (1922). Extracts from the diary of Otto Widmann. *Transactions of the Academy of Science of St Louis*, **24**, 1–77.

Wikel, S. K. (1984). Immunomodulation of host responses to ectoparasite infestation: an overview. *Veterinary Parasitology*, **14**, 321–339.

Wikel, S. K. (ed.) (1996). *The Immunology of Host–Ectoparasitic Arthropod Relationships*. Wallingford, UK: CAB International.

Wikel, S. K. & Alarcon-Chaidez, F. J. (2001). Progress toward molecular characterization of ectoparasite modulation of host immunity. *Veterinary Parasitology*, **101**, 275–287.

Willadsen, P. (1980). Immunity to ticks. *Advances in Parasitology*, **18**, 293–313.

Willadsen, P. (1987). Immunological approaches to the control of ticks. *International Journal for Parasitology*, **17**, 671–677.

Willadsen, P. (2001). The molecular revolution in the development of vaccines against ectoparasites. *Veterinary Parasitology*, **101**, 353–367.

Willadsen, P. (2006). Vaccination against ectoparasites. *Parasitology*, **133**, S9–S25.

Willadsen, P., Bird, P. E., Cobon, G. & Hungerford, J. (1995). Commercialization of a recombinant vaccine against *Boophilus microplus*. *Parasitology*, **110**, 543–550.

Williams, B. (1991). Adaptations to endoparasitism in the larval integument and respiratory system of the flea *Uropsylla tasmanica* Rothschild (Siphonaptera, Pygiopsyllidae). *Australian Journal of Zoology*, **39**, 77–90.

Williams, B. (1993). Reproductive success of cat fleas, *Ctenocephalides felis*, on calves as unusual hosts. *Medical and Veterinary Entomology*, **7**, 94–98.

Williams, R. T. (1971). Observations on the behaviour of the European rabbit flea, *Spilopsyllus cuniculi* (Dale), on a natural population of wild rabbits, *Oryctolagus cuniculus* (L.), in Australia. *Australian Journal of Zoology*, **19**, 41–51.

Williams, R. T. & Paper, I. (1971). Observations on the dispersal of the European rabbit flea, *Spilopsyllus cuniculi* (Dale), through a natural population of wild rabbits, *Oryctolagus cuniculus* (L.). *Australian Journal of Zoology*, **19**, 129–140.

Willmann, R. (1981a). Das Exoskelett der männlichen Genitalien der Mecoptera (Insecta). I. Morphologie. *Zeitschrift für zoologische Systematik und Evolutionsforschung*, **19**, 96–150.

Willmann, R. (1981b). Das Exoskelett der männlichen Genitalien der Mecoptera (Insecta). II. Die phylogenetischen Beziehungen der Schnabelfliegen-Familien. *Zeitschrift für zoologische Systematik und Evolutionsforschung*, **19**, 153–174.

Wilson, D. E. & Reeder, D. M. (eds.) (2005). *Mammal Species of the World: A Taxonomic and Geographic Reference*, 3rd edn. Baltimore, MD: Johns Hopkins University Press.

Wilson, K., Bjørnstad, O. N., Dobson, A. P., et al. (2001). Heterogeneities in macroparasite infections: patterns and processes. In *The Ecology of Wildlife Diseases*, ed. P. J. Hudson, A. Rizzoli, B. T. Grenfell, H. Heesterbeek & A. P. Dobson. Oxford, UK: Oxford University Press, pp. 6–44.

Wilson, N. A. (1961). The ectoparasites (Ixodides, Anoplura and Siphonaptera) of Indiana Mammals. Unpublished Ph.D. thesis, Purdue University, West Lafayette, IN.

Wilson, N. A. & Durden, L. A. (2003). Ectoparasites of terrestrial vertebrates inhabiting the Georgia Barrier Islands, USA: an inventory and preliminary biogeographical analysis. *Journal of Biogeography*, **30**, 1207–1220.

Wilson, N. A., Telford, S. R. & Forrester, D. J. (1991). Ectoparasites of a population of urban gray squirrels in northern Florida. *Journal of Medical Entomology*, **28**, 461–464.

Windsor, D. A. (1990). Heavenly hosts. *Nature*, **348**, 104.

Windsor, D. A. (1995). Equal rights for parasites. *Conservation Biology*, **9**, 1–2.

Winkel, W. (1975a). Vergleichend-brutbiologische Untersuchungen an fünf Meisenarten (*Parus* spp.) in einem niedersächsischen Aufforstungsgebiet mit japanischer Lärche *Larix leptolepis*. *Die Vogelwelt*, **96**, 41–63.

Winkel, W. (1975b). Vergleichend-brutbiologische Untersuchungen an fünf Meisenarten (*Parus* spp.) in einem niedersächsischen Aufforstungsgebiet mit japanischer Lärche *Larix leptolepis*. *Die Vogelwelt*, **96**, 104–114.

Withers, P. C. (1992). *Comparative Animal Physiology*. Fort Worth, TX: Saunders College Publishing.

Witt, L. H., Linardi, P. M., Meckes, O., et al. (2004). Blood-feeding of *Tunga penetrans* males. *Medical and Veterinary Entomology*, **18**, 439–441.

Worthen, W. B. (1996). Community composition and nested-subsets analyses: basic descriptors for community ecology. *Oikos*, **76**, 417–426.

Worthen, W. B. & Rohde, K. (1996). Nested subsets analyses of colonization-dominated communities: metazoan ectoparasites of marine fishes. *Oikos*, **75**, 471–478.

Wright, D. H., Patterson, B. D., Mikkelson, G. M., Cutler, A. & Atmar, W. (1998). A comparative analysis of nested subset patterns of species composition. *Oecologia*, **113**, 1–20.

Xun, H. & Qi, Y.-M. (2004). Histochemistry of three enzymes in newly emerged and engorged adults of rat fleas *Monopsyllus anisus* (Rothschild) and *Leptopsylla segnis* (Schönherr). *Acta Entomologica Sinica*, **47**, 444–448 (in Chinese).

Xun, H. & Qi, Y.-M. (2005). Histochemistry of fat and nonspecific esterase in newly emerged and sucked adults of rat fleas *Monopsyllus anisus* (Rothschild) and *Leptopsylla segnis* (Schoenherr). *Acta Entomologica Sinica*, **48**, 829–832 (in Chinese).

Yadav, A., Khajuria, J. K. & Devi, J. (2006). Cat flea infestation in goats. *Indian Veterinary Journal*, **83**, 439–440.

Yakunin, B. M. & Kunitskaya, N. T. (1980). On the effect of high relative humidity on reproduction and longevity of fleas *Xenopsylla skrjabini*. In *Problems of Natural Focality of Plague: Proceedings of the 4th Soviet–Mongol Conference of Specialists from the Anti-Plague Establishments*, vol. 1, ed. E. P. Golubinsky. Irkutsk, USSR: Irkutsk State Scientific Anti-Plague Institute of Siberia and Far East, pp. 88–89 (in Russian).

Yakunin, B. M., Zolotova, S. I., Serzhanov, O. S., et al. (1971). On density dynamics and reproduction of *Pulex irritans*. In *Proceedings of the 7th Scientific Conference of the Anti-Plague Establishments of the Middle Asia and Kazakhstan*, ed. M. A. Aikimbaev. Alma-Ata, USSR: The Middle Asian Scientific Anti-Plague Institute, pp. 450–452 (in Russian).

Yakunin, B. M., Tchernova, N. A. & Kunitskaya, N. T. (1979). Annual number of generations of fleas *Xenopsylla skrjabini* in the Mangyshlak Peninsula (Aphaniptera). *Parazitologiya*, **13**, 510–515 (in Russian).

Yamauchi, T. (2005). Human dermatitis caused by the house-martin flea, *Ceratophyllus farreni chaoi* (Siphonaptera: Ceratophyllidae) in Shimane Prefecture, Japan. *Medical Entomology and Zoology*, **56**, 49–52 (in Japanese).

Yensen, E., Baird, C. R. & Sherman, P. W. (1996). Larger ectoparasites of the Idaho ground squirrel (*Spermophilus brunneus*). *Great Basin Naturalist*, **56**, 237–246.

Yeruham, I. & Koren, O. (2003). Severe infestation of a she-ass with the cat flea *Ctenocephalides felis felis* (Bouché, 1835). *Veterinary Parasitology*, **115**, 365–367.

Yeruham, I., Rosen, S. & Hadani, A. (1989). Mortality in calves, lambs and kids caused by severe infestation with the cat flea *Ctenocephalides felis felis* (Bouché, 1835) in Israel. *Veterinary Parasitology*, **30**, 351–356.

Ying, B., Kosoy, M. Y, Maupin, G. O., Tsuchiya, K. R. & Gage, K. L. (2002). Genetic and ecological characteristics of *Bartonella* communities in rodents in southern China. *American Journal of Tropical Medicine and Hygiene*, **66**, 622–627.

Yinon, U., Shulov, A. & Margalit, Y. (1967). The hygroreaction of the larvae of the Oriental rat flea, *Xenopsylla cheopis* Rothsch. (Siphonaptera: Pulicidae). *Parasitology*, **57**, 315–319.

Yudin, B. S., Krivosheev, V. G. & Belyaev, V. G. (1976). *Small Mammals of the Northern Far East*. Novosibirsk, USSR: Nauka, Siberian Branch (in Russian).

Yue, B.-S., Zou, F.-D., Sun, Q.-Z. & Li, J. (2002). Mating behavior of the cat flea, *Ctenocephalides felis* Bouché (Siphonaptera: Pulicidae) and male response to female extract on an artificial feeding system. *Acta Entomologica Sinica*, **9**, 29–34 (in Chinese).

Yurgenson, I. A. & Maksimov, V. N. (1981). Effect of air temperature and relative humidity on the pre-imaginal development of fleas *Ctenophthalmus teres* (Siphonaptera). *Parazitologiya*, **15**, 38–46 (in Russian).

Yushchenko, G. V. (1965). On the problem of the pseudotuberculosis research in the USSR. *Proceedings of the Scientific Institute of Vaccines and Sera and Tomsk Medicine Institute*, **16**, 167–173 (in Russian).

Zagniborodova, E. N. (1960). Fauna and ecology of fleas in western Turmenistan. In *Problems of Natural Foci and Epizootology of Plague in Turkmenistan*, ed. B. K. Fenyuk. Saratov, USSR: Turkmenian Anti-Plague Station and All-Union Scientific Anti-Plague Institute 'Microb', pp. 320–334 (in Russian).

Zagniborodova, E. N. (1965). Epizootological importance of migrating fleas of the great gerbil in Turkmenistan. *Proceedings of the Academy of Sciences of the Turkmenian SSR, Biology*, **5**, 65–70 (in Russian).

Zagniborodova, E. N. (1968). Long-term study of the ecology of fleas of the great gerbil in the southern part of the central Kara-Kum Desert. In *Rodents and their Ectoparasites*, ed. B. K. Fenyuk. Saratov, USSR: Saratov University Press, pp. 78–86 (in Russian).

Zahavi, A. (1977). The cost of honesty (further remarks on the handicap principle). *Journal of Theoretical Biology*, **67**, 603–605.

Zahn, A. & Rupp, D. (2004). Ectoparasite load in European vespertilionid bats. *Journal of Zoology*, **262**, 383–391.

Zakson-Aiken, M., Gregory, L. M. & Shoop, W. L. (1996). Reproductive strategies of the cat flea (Siphonaptera: Pulicidae): parthenogenesis and autogeny? *Journal of Medical Entomology*, **33**, 395–397.

Zatsarinina, G. V. (1972). Salvic alakurt (*Dorcadia dorcadia* Roth., Aphaniptera, Vermipsillidae): the pest of the deer breeding. *Proceedings of the Entomological Sector [Trudy Entomologisheskogo Sektora]*, **3**, 1–108 (in Russian).

Zavala-Velazquez, J. E., Ruiz-Sosa, J. A., Sanchez-Elias, R. A., Becerra-Carmona, G. & Walker, D. H. (2000). *Rickettsia felis* rickettsiosis in Yucatan. *Lancet*, **356**, 1079–1080.

Zeigler, R. & Ibrahim, M. M. (2001). Formation of lipid reserves in fat body and eggs of the yellow fever mosquito, *Aedes aegypti*. *Journal of Insect Physiology*, **47**, 623–627.

Zenuto, R. R., Antinuchi, C. D. & Busch, C. (2002). Bioenergetics of reproduction and pup development in a subterranean rodent (*Ctenomys talarum*). *Physiological and Biochemical Zoology*, **75**, 469–478.

Zhang, T., Yu, X.-M. & Zhang, S.-Y. (2005). Investigation on the community composition of the small mammals and the parastic fleas in Xingning, Guangdong. *Chinese Journal of Vector Biology and Control*, **16**, 446–447 (in Chinese).

Zhang, Y., Jin, S., Quan, G., et al. (1997). *Distribution of Mammalian Species in China*. Beijing: China Forestry Publishing House.

Zhang, Y.-Z., Gong, Z.-D., Feng, X.-G., et al. (2002). Study on the relationship between fleas and hosts in Mt. Baicaoling, Yunnan Province, China. *Endemic Diseases Bulletin*, **17**, 22–23 (in Chinese).

Zhao, L., Jin, H.-L., She, R.-P., et al. (2006). A rodent model for allergic dermatitis induced by flea antigens. *Veterinary Immunology and Immunopathology*, **114**, 285–296.

Zhovty, I. F. (1963). Some contradictory questions of ecology of the rodent fleas in association with their epidemiological importance. *Transactions of the Irkutsk State Scientific Anti-Plague Institute of Siberia and Far East*, **6**, 96–104 (in Russian).

Zhovty, I. F. (1967). Effect of rodents' life history on ecological conditions of their fleas. *Proceedings of the Irkutsk State Scientific Anti-Plague Institute of Siberia and Far East*, **27**, 195–210 (in Russian).

Zhovty, I. F. (1970). Essays on the ecology of fleas parasitic on rodents in Siberia and Far East. II. Fleas of marmots. In *Vectors of Particularly Dangerous Diseases and Their Control*, ed. V. E. Tiflov. Stavropol, USSR: Scientific Anti-Plague Institute of Caucasus and Trans-Caucasus, pp. 253–283 (in Russian).

Zhovty, I. F. & Kopylova, O. A. (1957). Fleas of the Daurian pika in the period of the increase of its density. *Proceedings of the Irkutsk State Scientific Anti-Plague Institute of Siberia and Far East*, **15**, 293–298 (in Russian).

Zhovty, I. F. & Leonov, Y. A. (1958). Abundance of fleas on the Norway rats in the settlements of the southern part of the Primorie Region (Far East) and some patterns of its variation. *Proceedings of the Irkutsk State Scientific Anti-Plague Institute of Siberia and Far East*, **17**, 75–89 (in Russian).

Zhovty, I. F. & Peshkov, B. I. (1958). Observations on the overwintering of fleas parasitic on the grey marmots in the Trans-Baikalia. *Proceedings of the Irkutsk State Scientific Anti-Plague Institute of Siberia and Far East*, **17**, 27–32 (in Russian).

Zhovty, I. F. & Vasiliev, G. I. (1962a). Temperature conditions of rodents' fur as an environment for fleas. *Transactions of the Irkutsk State Scientific Anti-Plague Institute of Siberia and Far East*, **4**, 152–156 (in Russian).

Zhovty, I. F. & Vasiliev, G. I. (1962b). On the self-cleaning from fleas in rodents. *Transactions of the Irkutsk State Scientific Anti-Plague Institute of Siberia and Far East*, **4**, 156–160 (in Russian).

Zhovty, I. F., Netchaeva, L. K., Koshkin, S. M., et al. (1983). Patterns of seasonal density fluctuations in populations of the rat fleas in the Primorie Region (Far East) and a search for the criteria for their prognosis. In *Prophylaxis of Diseases in the Natural Foci*, ed. I. F. Taran. Stavropol, USSR: Scientific Anti-Plague Institute of Caucasus and Trans-Caucasus, pp. 234–235 (in Russian).

Zolotova, S. I. (1968). Comparative ecological descriptions of fleas of the great gerbil, *Xenopsylla gerbilli minax* Jord., 1926 and *Ctenophthalmus dolichus* Ioff, 1953. Unpublished Ph.D. thesis, The Middle Asian Scientific Anti-Plague Institute, Alma-Ata, USSR (in Russian).

Zolotova, S. I. & Afanasieva, O. V. (1969). Materials on the ecology of fleas of the great gerbil. IV. Duration of pre-imaginal development in *Ctenophthalmus dolichus*. In *Proceedings of the 6th Scientific Conference of the Anti-Plague Establishments of the Middle Asia and Kazakhstan*, vol. 2, ed. M. A. Aikimbaev. Alma-Ata, USSR: The Middle Asian Scientific Anti-Plague Institute, pp. 66–68 (in Russian).

Zolotova, S. I. & Iskhanova, Z. A. (1979). Relationships between fleas *Xenopsylla gerbilli minax* and *X. skrjabini* in the area of overlapping of their geographic ranges. In *Proceedings of the 10th Scientific Conference of the Anti-Plague Establishments of the Middle Asia and Kazakhstan*, ed. M. A. Aikimbaev. Alma-Ata, USSR: The Middle Asian Scientific Anti-Plague Institute, pp. 99–101 (in Russian).

Zolotova, S. I. & Varshavskaya, P. N. (1974). Age composition of imago in the population of *Xenopsylla skrjabini* in the northern Cis-Aral Region. In *Proceedings of the 8th Scientific Conference of the Anti-Plague Establishments of the Middle Asia and Kazakhstan*, ed. M. A. Aikimbaev. Alma-Ata, USSR: The Middle Asian Scientific Anti-Plague Institute, pp. 316–318 (in Russian).

Zolotova, S. I., Pavlova, A. E. & Yakunin, B. M. (1971). Longevity of *Pullex irritans* after feeding on a non-specific host. In *Proceedings of the 7th Scientific Conference of the Anti-Plague Establishments of the Middle Asia and Kazakhstan*, ed. M. A. Aikimbaev. Alma-Ata, USSR: The Middle Asian Scientific Anti-Plague Institute, pp. 377–378 (in Russian).

Zolotova, S. I., Bibikova, V. I. & Murzakhmetova, K. (1979). On the fecundity of fleas *Xenopsylla gerbilli minax* parasitic on the great gerbil (Aphaniptera). *Parazitologiya*, **13**, 497–502.

Zuk, M. & McKean, K. A. (1996). Sex differences in parasite infections: patterns and processes. *International Journal for Parasitology*, **26**, 1009–1024.

Index

abdomen 103
abundance–distribution
    relationship 295, 352,
    354, 384
accidental host 116
acid phosphatase 175
Acomys 438, 439
    cahirinus 35, 47, 137, 138,
    139, 164, 165, 166, 167,
    168, 170, 174, 185, 198,
    222, 246, 253, 259, 278,
    307, 336, 338, 354, 438,
    441
    russatus 185, 343, 438
acquired resistance 198, 255,
    258–61, 367
adenosine
    triphosphotase 175
age-intensity pattern 346
age-prevalence pattern 346
aggregation 322, 328–35,
    343, 345, 346, 347, 354,
    365, 367, 380
aggregation model of
    coexistence 396–401
Aix sponsa 341
Akodon montensis 452
Alectoris rufa 386
alkaline phosphatase 175
Allactaga
    bobrinskii 38

bullata 38
elater 38, 168, 189,
    383
sibirica 38, 417
tetradactyla 38
vinogradovi 38
williamsi 38
Amalaraeus
    arvicolae 452
    penicilliger 69, 254, 255,
    291, 362, 371, 372, 377,
    402, 450
Amphalius
    clarus 358
    runatus 149
Amphipsylla 310
    anceps 298
    asiatica 27
    kuznetzovi 27
    marikovskii 27
    primaris 27, 90, 149
    rossica 65, 208, 298, 361,
    450
    schelkovnikovi 372
    sibirica 27
    vinogradovi 358
Amphipsyllini 26
Ancistropsylla 144
Ancistropsyllidae 22, 33
Anomiopsyllus 245, 402
    amphibolus 140

falsicalifornicus 144
nudatus 72
Anoplura 432
anti-nestedness 383, 384
aphagous larvae 50
Aphelocoma coerulescens 225,
    228
apocrine secretion 145
Apodemus 347, 349, 350,
    406
    agrarius 70, 91, 147, 246,
    329, 378, 379, 406, 408,
    414, 447, 450
    flavicollis 141, 378, 414,
    450, 464
    mystacinus 317
    speciosus 451
    sylvaticus 267, 371, 402,
    415, 450, 451
    uralensis 64, 297, 378, 381,
    406, 414, 447, 450
apparent facilitation 395,
    434
Archaeopsylla
    erinacei 65, 72, 145, 155,
    194
    sinensis 144
Arvicanthis niloticus 361
Arvicola amphibius 383, 406,
    408
Arvicolinae 16, 26, 424

Ascaridia galli 265
assembly rules 377
association by colonization 121
association by descent 121
asymmetric competition 391, 405
Atyphloceras multidentatus 328, 392
auxiliary host 117, 118, 119, 120
average daily metabolic rate (ADMR) 219, 412, 413
Aviostivalius klossi 75

Bandicota bengalensis 189
bartonellosis 71-2, 73
basal metabolic rate (BMR) 339, 340, 399, 412, 413, 461
basiconic sensilla 104, 156
begging 231
biogeographical realm
 Afrotropical 19, 32
 Australian 19, 32
 Holarctic 32, 430
 Nearctic 19, 421, 422, 424, 430
 Neotropic 19, 33
 Oriental 19
 Palaearctic 19, 421, 422, 424, 429, 430
Blarina brevicauda 452
blood digestion, duration of 166-7
'body' fleas 147, 148, 149, 151
body mass change 223, 226
Boreidae 31, 32, 108
bottom—up effect 421-2, 424
Brooks parsimony analysis 36, 38
brucellosis 71
Buteo galapagoensis 195

Caenopsylla laptevi ibera 94
Callopsylla
 caspia 94
 dolabris 358
Callosciurus erythraeus 22
Calomyscinae 26
cannibalism 49, 183
carrier host 117
cat leukaemia 73
Catallagia charlottensis 328, 392
cat-flea typhus 72
Cediopsylla simplex 35, 47, 81, 156, 191, 193
cell-mediated immunity 258, 267, 272, 339, 434
Cenozoic 306
Ceratophyllidae 19, 22, 23, 33, 197, 307, 309, 310, 424
Ceratophylloidea 34
Ceratophyllomorpha 33
Ceratophyllus 241
 aviciteli 179
 celsus 231
 ciliatus 22, 121
 columbae 75
 farreni 91, 303, 368, 372
 fringillae 75
 gallinae 5, 45, 46, 70, 73, 74, 75, 113, 118, 130, 131, 197, 198, 202, 209, 224, 225, 229, 230, 231, 232, 235, 236, 240, 241, 243, 252, 265, 267, 268, 278, 280, 314, 328, 368, 372, 462
 garei 70
 hirundinis 45, 75, 91, 140, 303, 368, 372
 idius 54, 58, 245
 lunatus 10, 117
 petrochelidoni 140

 rusticus 75, 91, 368, 372
 styx 34, 105, 131, 140, 358
Cervus elaphus sibiricus 144
Chaetopsylla homoea 144
chaetotaxy, abnormalities of 213
Charadriiformes 16
chequerboard species pairs, numbers of 378, 379
Chimaeropsyllidae 5, 19, 22
Chiroptera 5, 16
chorion 47
Choristopsylla tristis 113
Chriopteropsyllini 32
circadian rhythm 179, 215
circulating immune complexes 258, 267, 271
Citellophilus
 martinoi 452
 simplex 434, 452
 tesquorum 45, 46, 47, 64, 66, 83, 86, 90, 99, 113, 140, 144, 145, 148, 150, 174, 178, 179, 189, 197, 211, 255, 326, 328, 357, 358, 368, 369, 372
 trispinus 71, 214, 346
climate, effect of 43
clutch size 47
coevolution 34, 36-8
colonization—extinction dynamics 384, 434
comb *see* ctenidia
comparative approach 462
competition 26, 176, 188, 189, 203-4, 205, 334, 367-8, 388, 391, 396, 402-5, 409, 424
component community 375
compound community 375
concealed antigens 465
cophylogeny 121

Coptopsylla
 africana 271, 438
 bairamaliensis 136
 lamellifer 83, 174, 211, 214, 358, 402
 olgae 136
Coptopsyllidae 22, 23
core populations 293
core–satellite organization 377
Corrodopsylla curvata 35
cospeciation 38, 40
Craneopsyllidae 19
Cricetinae 16, 26, 424
Cricetulus
 barabensis 358
 migratorius 383, 406, 408, 428
cross-resistance 261–2, 380, 395, 413
'crowns of thorns' 255
C-score 378, 379, 381
ctenidia 30, 35, 81, 82, 129, 254
Ctenocephalides
 canis 21, 35, 71, 72, 73, 74, 84, 189, 456
 felis 21, 46, 51, 72, 73, 74, 81, 100, 104, 113, 131, 141, 144, 145, 189, 191, 209, 210, 215, 244, 246, 247, 257, 258, 261, 372, 456, 462
 felis damarensis 194
Ctenophthalmus 156, 310, 405
 agyrtes 64, 71, 75, 97, 247, 329, 450, 453
 andorrensis 303
 assimilis 70, 75, 254, 255, 361, 450, 453
 bisoctodentatus 147, 450, 452
 breviatus 54, 317

 congeneroides 58, 69, 91, 147, 149
 dolichus 46, 52, 159, 179, 211, 402, 451
 golovi 83, 174, 298
 machadoi 76
 nobilis 49, 371, 402
 orientalis 70, 434
 pollex 148
 pseudagyrtes 140, 452
 secundus 70
 shovi 317
 solutus 329, 447, 450, 453
 strigosus 58
 teres 207, 211
 uncinatus 64, 246, 254, 298
 wagneri 70, 147, 207
 wladimiri 207, 211
Cyanistes caeruleus 130, 202, 225, 228, 230, 231, 232, 241, 243, 252, 253
Cyanoliseus patagonus 143
Cynomys ludovicianus 223, 392, 464

Dasypsyllus gallinulae 372
Dasyuromorphia 16
delayed hypersensitivity 257
Delichon ubica 303, 368
density compensation 388
Dermacentor variabilis 259
developmental rate 92, 100, 185, 209, 210, 211
Didelphimorphia 16, 310
diffuse competition 392
digestible energy intake 219
digestion stages 159
Dinopsyllus
 apistus 451
 ellobius 303
 lypusus 303
 smiti 76
Dipodidae 38

Dipodillus dasyurus 47, 55, 125, 126, 128, 129, 137, 138, 139, 164, 165, 166, 167, 168, 174, 176, 177, 178, 185, 187, 188, 189, 218, 220, 221, 259, 262, 267, 271, 272, 329, 336, 338, 343, 358, 360, 361, 368, 413, 428, 438, 439, 441
Dipodomys
 ingens 121
 ordii 98
 spectabilis 98
Diprotodontia 16
Diptera 31
Dipus sagitta 38, 40, 417
Dipylidium caninum 74
dispersal 140–1, 231, 364
dispersal host 117
disperser 337, 341, 350, 360, 361, 380
distance decay of similarity 406–8
divorce rate 231
Doratopsylla
 blarinae 452
 dasycnema 64, 97, 116, 329, 356, 450
Dorcadia
 dorcadia 144, 174
 ioffi 47, 53, 144, 156, 215
Dryomys nitedula 383
duplication 34, 36, 41

Echidnophaga 405
 gallinacea 45, 46, 72, 73, 74, 143, 155, 173, 174, 191, 209, 213, 215, 225, 228, 347
 iberica 452
 myrmecobii 104, 147, 194

Echidnophaga (cont.)
 oschanini 144, 156, 174, 191, 215, 346
 perilis 147, 194
egg production 81, 194, 199, 200, 202, 203, 204, 206, 208, 211, 215, 275, 278
egg size 81, 199
Eliomys melanurus 438, 439
energetic cost of digestion 160, 165
environmental stimuli 131
Eocene 29, 32, 33, 41
epidemiological model 352, 354, 356, 362, 369, 460
epipharynx 155, 156
Eremodipus lichtensteinii 38
Erinaceomorpha 16
erysepeloid 71
Euchoplopsyllus glacialis 50
evolutionary rate 43
exceptional host 117
exposed antigens 465
extensor tibiae muscles 108

facilitation 379, 391, 393, 395, 396, 397, 409
Fahrenholz's rule 314
Falconiformes 16
Falco tinnunculus 341
fat body 184
favoured species combinations 377
feeding frequency 84, 174
Ficedula
 albicollis 241
 hypoleuca 225, 241
Fisher's theory of sex allocation 100, 101
fitness 115, 117, 123, 124, 125, 128–9, 157, 160, 188, 201, 217, 235–7, 278, 341, 395

fixed-equiprobable (FE) algorithm 379
fixed-fixed (FF) algorithm 379
flea allergy dermatitis 69, 73
fledgling recruitment 235, 236
fluctuating asymmetry 229
food provisioning 230–1
fore coxae 104
Francisella tularensis 70, 71
frontal tubercle 52
Frontopsylla
 aspiniformis 358
 elata 69, 91, 94, 174, 207
 frontalis 10, 179
 luculenta 113, 147, 149, 255, 358, 361
 ornata 298, 310
 semura 174, 208, 209
 wagneri 38, 418
Fulmarus glacialoides 50
functional niche 289
fundamental niche 115

Gallus gallus 265
generalized parsimony method 36
Genetic Algorithm for Rule-Set Production (GARP) 465
geographical range size 22–3, 27, 302, 303, 425
Geomyidae 16
geotaxis 49, 90, 131, 132
Gerbillinae 16, 424
Gerbillus
 andersoni 125, 126, 128, 129, 192, 221, 223, 226, 227, 232, 234, 236, 262, 267, 271, 272, 329, 343, 350, 462
 gerbillus 438

 henleyi 343, 438
 nanus 329, 343
 pyramidum 125, 126, 128, 343
Geusibia 40, 309
'ghost of competition past' 405
glaciation 306
Glaciopsyllus antarcticus 18, 50
Glaucomys volans 369
Glis glis 347
glucose consumption technique 262
Gondwanaland 33
grooming 81, 129, 144, 166, 245–55, 277, 291, 334, 339, 351, 367, 379
Gymnomeropsylla
 margaretamydis 317

habitat selection, theory of 124, 125
haematin 159
haematocrit 227, 278
haemorrhagic fever with renal syndrome 70
Halobaena caerulea 50
Hectopsylla
 narium 104, 143
 psittaci 143
Heligmosomoides polygyrus 464
Hepatozoon erhardovae 74
Herpetosoma 74
Heteromyidae 16, 422
hierarchical Bayesian model 464
Hirundo rustica 229
Hoplopsyllus anomalus 121, 347, 374
host
 body size 306, 339, 399, 411, 412, 413, 461
 density 356–64, 381, 416

food availability 194, 197–8
hormones 48, 173, 191, 193–4, 225
location 90, 130
mortality 217
odour 215
search 113, 137, 138
sexual size dimorphism 342, 345, 415
social status 337
social structure 417
spectrum 115, 119, 295, 309, 314
host-compatibility filter 118, 120, 291
host-encounter filter 120
host-induced parasite mortality 364, 367
host-opportunistic 116, 121, 168, 187, 283, 285, 291, 295, 296, 298, 300, 303, 304, 308, 309, 314, 332, 333, 362, 384
host-specific 116, 121, 122, 163, 168, 185, 283, 285, 291, 295, 297, 298, 300, 303, 304, 306, 308, 309, 314, 315, 332, 333, 384
host specificity, index of 283
host-switching 34, 36, 38, 41, 43, 44, 121, 309
host-to-host transfer 141, 285
humoral immunity 272, 339, 434
Hydrobates pelagicus 225
Hydrochoerus hydrochaeris 195
Hymenolepis diminuta 74
Hystrichopsylla
  dippiei 22

kris 52
orientalis 75, 452
talpae 71, 371, 372, 450
Hystrichopsyllidae 5, 19, 22, 23, 33, 310

ideal free distribution (IFD) 124, 130, 140, 141, 188, 201
immune 'readiness' 270, 273
immune defence 119, 169, 170, 171, 172–3, 177, 194, 204, 221, 225, 228, 230, 239, 255–78, 291, 334, 339, 346, 379, 413, 414, 432, 465
immunocompetence 170, 198, 263–6, 275, 278, 308, 341, 345, 379, 415, 422, 434
immunodepression 379, 380, 434
immunoglobulins 258, 268, 269, 271
'immunohandicap' hypothesis 267
immunosenescence 346
immunosuppression 169, 276, 341, 395
inertia 34, 38
infracommunity 321, 375, 376, 382, 384, 387, 388, 390, 395, 414, 415, 418, 450
infrapopulation 321
insular fauna 22
interaction networks 294–5, 307, 461
interactive community 334, 375, 431
interspecific aggregation 396, 397, 398, 399

intraspecific aggregation 324, 329, 339, 396, 397, 398, 399
introduced host 371
Ischnopsyllidae 5, 19, 22, 52, 65, 105, 147
Ischnopsyllini 32
Ischnopsyllus 156
  hexactenus 82
  octactenus 82
  petropolitanus 82
island biogeography, theory of 425
isodar approach 124–5, 126, 129, 187
isolationist community 375, 432
Ixodes ricinus 267
Ixodidae 351, 432

Jaccard index of similarity 406
Jaculus
  blanfordi 38
  jaculus 38, 438
  orientalis 38
'jack-of-all-trades' 296, 298, 300
Jordanopsylla
  allredi 106
  becki 8
Junco hyemalis 267

labial palps 155
labrum 156
laciniae 155, 156
Lagomorpha 15, 16, 310, 420
Lagurus lagurus 254
latency of feeding 166
latitudinal gradient 302, 430
Lemniscomys striatus 452
Leptopsylla 452
  algira 303

Leptopsylla (cont.)
  segnis 21, 45, 46, 47, 54, 70, 71, 72, 75, 83, 118, 146, 156, 157, 159, 174, 175, 179, 200, 202, 208, 211, 215, 254, 255, 347
  sexdentata 136
  taschenbergi 45, 46, 47, 157, 159, 174, 208, 303
Leptopsyllidae 19, 22, 33, 310
leptospirosis 71
Lepus
  arcticus 50
  capensis 372
  saxatilis 347
leukocyte blast transformation 258, 262, 269
lifespan 94, 411, 413
lifetime fecundity 47, 183
light regime 179
lineage sorting 34, 36, 38
lines of defence 239
listeriosis 70
Lophuromys 452
Lycopsyllidae 19, 22, 23
Lyme disease 71
lymphocytic chiriomenengitis 70

Macropsyllidae 19, 22
major histocompatibility complex 271
Malacopsyllidae 5, 19, 22, 53
Malaraeus telchinus 98, 102, 141, 328
Margaretamys parvus 318
Marmota
  baibacina 99, 145, 347
  bobac 317
  flaviventris 226, 232
  himalayana 358
  sibirica 255
Mastomys 191

mating 46
matrix 'temperature' 383
maxillary plates 104
mean crowding 324, 365
Mecoptera 31
Megabothris
  acerbus 35, 81
  advenarius 364
  asio 364
  calcarifer 91, 149
  quirini 98
  rectangulatus 69, 377
  turbidus 64, 75, 298, 354, 372, 447, 450
  walkeri 71
Meles meles 242, 244
melioidosis 71
Meringis nidi 98
Meriones
  crassus 94, 125, 126, 128, 129, 168, 170, 172, 176, 177, 178, 185, 187, 188, 189, 194, 204, 227, 247, 251, 253, 258, 262, 265, 267, 269, 275, 276, 278, 293, 343, 402, 428, 438, 439, 441
  meridianus 141, 358, 383, 452
  tamariscinus 383, 452
  unguiculatus 151, 159, 169, 173, 358, 372
Mesopsylla
  eucta 159
  hebes 418
Mesostigmata 432
mesothorax 103
metabolic rate 51, 54, 83, 85–7, 108, 109–11, 112, 161, 174, 175, 178, 218–22, 411
metabolizable energy intake 219

metapopulation 321, 322
  dynamics 307
  model 464
  theory 360
metathorax 103
Microtus 347, 350, 406
  agrestis 328, 377, 463
  arvalis 246, 254, 347, 349, 361, 378, 383, 406, 408, 414, 447, 450
  brandti 150
  californicus 99, 102, 151
  canicaudus 328, 392
  gregalis 27, 383, 406, 408
  oeconomus 328, 377, 406, 408
  subterraneus 362, 450, 452
'migration' 136
Milankovitch oscillations 302
Miocene 29, 40
Mioctenopsylla traubi kurilensis 309
Morisita–Horn index of similarity 406
Molossus molossus 143
monophyly 29
Monopsyllus
  anisus 22, 64, 70, 175, 208, 211, 372
  indages 140
  sciurorum 5, 49, 113, 176, 450
multiple matings 46
Muridae 36
Murinae 16, 424
murine typhus 72, 73
Mus musculus 22, 75, 141, 254, 292, 383, 406, 408, 438
Mustela 117
  erminea 118
  eversmanni 144

Myodes 406
   glareolus 70, 116, 141, 246, 254, 267, 328, 329, 342, 347, 349, 350, 356, 362, 371, 378, 381, 402, 406, 414, 450
   rufocanus 328, 383
   rutilus 328, 406, 408
Myodopsylla insignis 342
Myospalacinae 26
Myotis
   lucifugus 341
   myotis 263, 342
   mystacinus 342
   nattereri 342
Myoxopsylla laverani 438
myxoma 73

Nannochoristidae 31, 108
Nearctopsylla genalis 452
negative binomial distribution 323, 328, 329, 353, 357
Neomys fodiens 356, 406
Neopsylla 310
   abagaitui 113, 358
   bidentatiformis 54, 69, 91, 99, 113, 140, 144, 145, 150, 197, 211, 255, 328, 357, 358, 361
   pleskei 255, 298
   setosa 46, 47, 54, 64, 65, 71, 83, 86, 148, 159, 174, 178, 179, 211, 357, 434
   teratura 297
neosomy 50, 53, 69, 79, 113, 144
Neotoma
   albigula 98
   fuscipes 144, 244, 402
   lepida 140
   mexicana 122
   micropus 98

Neotunga euloidea 143
nest aggregation 307
'nest' fleas 147, 148, 151
'nest fumigation' hypothesis 245
nest sanitation 243–4
nestedness 377, 382–6
nest-site selection 242
niche
   conservatism 312
   filtering 396, 401
   segregation 401
Niwratia elongata 30
non-randomness 377, 378, 381, 383
normal host 117
Nosopsyllus 405
   barbarus 303
   consimilis 21, 46, 47, 65, 70, 83, 159, 174, 201
   fasciatus 5, 21, 45, 46, 54, 64, 70, 71, 75, 86, 106, 118, 140, 146, 148, 165, 174, 179, 190, 211, 246, 254, 261, 372, 450
   iranus 47, 55, 66, 81, 97, 112, 125, 126, 129, 271, 360, 361, 438, 441
   laeviceps 47, 65, 83, 92, 141, 148, 159, 169, 173, 174, 179, 200, 206, 211, 326, 346, 358, 372, 402
   mokrzeckyi 45, 47, 141, 159
   oranus 303
   pumilionis 329, 333
   tersus 47, 136, 207, 402
   turkmenicus 136, 402
Notiopsylla kerguelensis 18
null models 378–9
Nyctalus
   leisleri 341
   noctula 342
Nycteribiidae 103
Nycteridopsyllini 33

Ochotona
   daurica 358, 361
   pricei 144
Ochotonidae 40
Ochotonobius hirticrus 144, 149, 156, 358
Oligocene 32
Omsk haemorrhagic fever 70
Onychomys leucogaster 122, 413
Ophthalmopsylla
   kiritschenkovi 418
   kukuschkini 358
   praefecta 418
   volgensis 159
Opisodasys pseudarctomys 113
optimal virulence concept 333
Orchopeas 245, 452
   howardi 35, 81, 85, 98, 116, 369, 452
   leucopus 35
   sexdentatus 98, 144
ordination space 438
origin of parasitism 32
Ornithonyssus bursa 229
Oropsylla
   alaskensis 49, 50, 64
   arctomys 35
   bruneri 35, 361
   hirsuta 223, 392
   montana 341, 347, 352, 374
   rupestris 361
   silantiewi 49, 50, 94, 99, 144, 145, 255, 326, 347, 358
   tuberculata 223, 392
Oryctolagus cuniculus 47, 147, 193, 326, 358, 452

Pachyptila belcheri 50
Palaeopsylla
   baltica 29
   dissimilis 29
   klesbiana 29

Palaeopsylla (cont.)
   minor 147
   similis 450, 452
   soricis 64, 97, 298, 361, 450
panbiogeographical
   analysis 35
Paraceras melis 112, 131, 137, 157, 242
Paradipus ctenodactylus 38
Paradoxopsyllus 310, 405
   repandus 47, 208
   teretifrons 174, 208
Paramelemorphia 16
Parapsyllus
   heardi 18, 50
   nestoris 118
Parapulex chephrenis 35, 47, 81, 83, 85, 91, 131, 132, 137, 138, 164, 165, 167, 168, 170, 174, 185, 198, 222, 246, 253, 259, 262, 307, 354, 438, 441
parasite-induced host mortality 332, 333, 346, 354, 364, 365, 367
parasite-load dependent mortality 346
parasitic larvae 50
Paridae 314
Parotomys brantsii 242, 244
Parus
   ater 232
   major 197, 198, 224, 225, 226, 227, 229, 230, 231, 232, 235, 236, 240, 241, 242, 243, 265, 267, 268, 269, 278, 280, 328, 368
Passeriformes 15, 16
pasteurellosis 71
Pectinoctenus pavlovskii 46
peripheral populations 293
peritrophic membrane 173, 258

Peromyscopsylla
   bidentata 75, 450
   hesperomys 122
   silvatica 372
Peromyscus
   boylii 122
   gossipinus 22
   leucopus 122, 259
   nasutus 122
   truei 99, 122
Petrochelidon pyrrhonota 231
phagocytic activity of leukocytes 258, 276
Phodopus sungorus 247
Pholidota 143
photoperiod 215
phototaxis 49, 90, 131, 132
phytohaemagglutinin (PHA) 198, 262, 264, 265, 267, 269, 272, 273, 277, 278, 308
Piciformes 16
Pipistrellus pipistrellus 342
placental–foetal circulation 268
plague 70, 73, 90, 318–20, 463, 464, 465
   blockage 90, 318, 464
   origin of 50
pleural arch 105, 106
pleural height 106, 108
pleurosternum 104
pneumococcosis 71
Poecile palustris 222, 225, 241
Polygenis
   atopus 415
   bolhsi 262
   tripus 91, 148, 452
population size 150, 188
Porribiini 32
post-invasive immune response 273
'potpourri' hypothesis 244
preferred host 117

pre-invasive immune response 273
primary host 116
Primates 15
principal component analysis 438
principal host 117, 118, 119, 121, 122, 169
proboscis 155
Procellariiformes 16
productivity 427–8
programmed grooming 250, 251
'proportional sampling' 387
Prosoptes cuniculi 262
prothorax 104
Psammomys obesus 438
pseudotuberculosis 71
pterothorax 103
Pulex
   irritans 5, 21, 54, 69, 72, 73, 74, 104, 105, 155, 157, 168, 174, 190, 257, 261, 372
   larimerius 29
   simulans 257, 261
Pulicidae 19, 22, 25, 33, 38, 197, 310
Pulicomorpha 33
Pygeretmus
   platyurus 38
   pumilio 38
   zhitkovi 38
Pygiopsyllidae 19, 22, 23, 310
Pygiopsyllomorpha 33

Quarternary 306

rabbit haemorrhagic disease 73
randomization test 291
Rapoport's rule 302

Rattus
  exulans 170
  norvegicus 64, 75, 137, 169, 328, 347, 358, 372
  rattus 22, 75, 137, 189, 191, 347
rectal sac 51, 214
reinfestation analysis 150
Reithrodontomys megalotis 99
Rensch's rule 79–80
repeatability analysis 286, 326, 333, 338, 450
reproductive diapause 58
resident 337, 360, 361, 369, 380
resilin 105, 108
resource breadth hypothesis 295, 296, 298, 301
respiratory water loss 210
Rhadinopsylla 405, 450
  cedestis 174, 208
  dahurica 358
  dives 358, 372
  insolita 91
  li 326
  masculana 271, 310, 438
  ucrainica 326
Rhombomys opimus 65, 122, 136, 139, 168, 189, 292, 307, 346, 358, 362, 372, 376, 383, 392, 405, 412, 451, 464
Rhopalopsyllidae 19, 22, 25
Rhopalopsyllus lutzi 76
Rhynchopsyllus pulex 143
Rickettsia
  felis 47, 72
  prowazekii 72
  typhi 72
Riparia riparia 34, 140, 357
Rodentia 16, 310, 420
Rostropsylla daca 136
Rousettus aegyptiacus 168

salmonellosis 71
saturated community 388, 391
Saurophthiroides mongolicus 30
Saurophthirus longipes 30
scan-grooming 251
Scapteromys aquaticus 342, 415
Sciuridae 16
Sciurus
  carolinensis 22, 452
  vulgaris 113, 303
scratch-grooming 251
secondary host 117
Sekeetamys calurus 164, 185, 438, 439
sessile 53, 105, 142
sex allocation 100, 101
sexually oriented signals 228
Sigmodontinae 16
skin associated lymphoid tissues (SALT) 173
Sorex 116, 406
  araneus 298, 317, 328, 329, 383, 415, 447
  minutus 415
  satunini 317
Soricidae 424
Soricomorpha 16, 36, 310, 420
specific dynamic effect (SDE) 160
specific eradication threshold 361
Spermophilus
  beecheyi 150, 341, 347, 352, 374
  brunneus 358
  citellus 434, 452
  columbianus 226, 236
  dauricus 49, 99, 113, 150, 247, 358
  fulvus 189
  musicus 66

  pygmaeus 64, 66, 178, 189, 357
  richardsoni 361
  undulatus 49, 369, 372
Sphinctopsylla ares 47
Spilopsyllus cuniculi 22, 46, 47, 73, 106, 131, 141, 146, 155, 159, 170, 191, 193, 199, 211, 212, 259, 303, 328, 452
Spinturnix myoti 263
spleen 264
Staphanocircidae 22
starvation, resistance to 92, 94, 162, 191, 194, 197
Stenoponia 405
  americana 66
  ponera 66
  sidimi 66
  tokudai 451
  tripectinata 66, 81, 86, 98, 106, 108, 111, 112, 118, 125, 126, 129, 271, 326, 438
  vlasovi 451
Stephanocircidae 19, 23, 310
Sternopsylla 113
  distincta 342
Sternopsyllini 32
stick-tight fleas 73, 104, 140, 143
stimulus-driven grooming 251
Stivaliidae 22
Strashila incredibilis 30
Strigiformes 16
Sturnus vulgaris 225, 244
Stylodipus
  andrewsi 417
  telum 38
'supergenome' 459
'superorganism' 459
suprapopulation 321
sustaining host 117

Sylvilagus floridanus 48, 193
Synopsyllus fonquerniei 97
Synosternus
  cleopatrae 81, 112, 125, 126, 128, 129, 192, 220, 223, 226, 227, 232, 234, 236, 262, 269, 271, 273, 329, 350, 462
  longispinus 136
  pallidus 155
Szidat's rule 36

Tachycineta bicolor 225, 231, 241, 245
Tadarida brasiliensis 342
Talpa europaea 147, 452
Tamias townsendii 121
Tamiasciurus
  douglasii 122
  hudsonicus 122, 303
Tarsopsylla
  octodicemdentata 113, 303
Tarwinia australis 30
'tasty chick' hypothesis 278
taxonomic distance 118
taxonomic distinctness 119, 284, 286, 310, 312, 393, 395, 401, 411, 417, 425
Taylor's power law 323–4, 329, 331, 332, 334, 343, 352, 354, 356, 365
T-cells 265, 266, 308
testicular plug 45
testosterone 267, 269
thermal coefficient 178
thorax 3, 29, 103
Thrassis
  bacchii 85
  stanfordi 226
threshold of establishment 369
tick-borne encephalitis 69
Toxoplasma gondii 74
trade-off hypothesis 295, 296

tree reconciliation analysis 36, 40
Trichodectes melis 242
true host 116
Trypanosoma
  lewisi 74
  microti 463
trypanosomosis 74
tularaemia 70, 73
Tunga
  caecata 143
  caecigena 143
  monositus 46, 47, 50, 143, 258
  penetrans 5, 46, 47, 49, 53, 68, 73, 75, 142, 155, 328
tungiasis 68, 73
Tungidae 20, 22, 52, 53

underdispersion 314, 354
unfavoured species combinations 377
unsaturated community 388
Uropsylla tasmanica 49, 50

variance-to-mean ratio 323
Vermipsylla alakurt 49, 73, 75, 144, 155, 156, 215
Vermipsyllidae 22, 52, 53
virulence 239, 333
V-ratio 378, 379, 381
Vulpex macrotis 190

water content 83, 89
water vapour uptake 89
'winter immunoenhancement' hypothesis 276

xenocommunity 375, 376, 382, 384, 387, 388, 390, 395, 397, 399, 406, 418, 436, 450, 461

xenopopulation 322, 328, 334, 354, 368
Xenopsylla 310, 405, 464
  astia 83, 146, 189, 194, 358
  bantorum 59, 65, 361
  blanci 303
  brasiliensis 89, 210, 301
  cheopis 21, 46, 47, 59, 64, 69, 71, 72, 75, 83, 85, 86, 89, 90, 91, 94, 105, 106, 111, 137, 140, 145, 146, 148, 155, 156, 159, 173, 174, 176, 179, 189, 191, 194, 207, 209, 210, 211, 258, 259, 261, 301, 328, 347, 368, 456, 462
  conformis 47, 51, 52, 54, 58, 81, 83, 91, 94, 111, 125, 126, 129, 131, 132, 136, 170, 174, 176, 177, 179, 187, 188, 189, 200, 204, 206, 213, 215, 247, 251, 253, 262, 269, 271, 276, 278, 326, 358, 372, 402, 403, 438, 441, 442, 443
  cunicularis 194, 199, 214, 326, 358
  debilis 59
  difficilis 59
  dipodilli 47, 55, 81, 91, 97, 137, 138, 185, 262, 271, 272, 273, 329, 358, 360, 361, 368, 438
  eridos 242
  gerbilli 47, 58, 65, 71, 92, 139, 140, 159, 179, 189, 200, 211, 358, 372, 392, 402, 451
  gratiosa 225
  hirtipes 47, 58, 65, 122, 136, 358, 372, 392, 402
  nubica 303

nuttalli 58, 86, 87, 179, 189, 200, 207, 211, 212, 372
ramesis 51, 52, 54, 81, 85, 89, 91, 94, 97, 111, 113, 125, 126, 128, 129, 131, 132, 136, 168, 172, 176, 177, 179, 187, 188, 189, 194, 204, 206, 213, 215, 218, 220, 222, 227, 262, 265, 267, 269, 271, 273, 275, 276, 293, 296, 402, 403, 438, 441, 442, 443
skrjabini 47, 58, 65, 86, 87, 145, 159, 168, 174, 179, 189, 200, 207, 208, 211, 214, 292, 358, 362, 451
vexabilis 170, 341
Xiphiopsyllidae 19, 22

Yersinia pestis 50, 90, 318
yersiniosis 71
yolk 268, 269